Electromagnetics and Calculation of Fields

Springer
*New York
Berlin
Heidelberg
Barcelona
Budapest
Hong Kong
London
Milan
Paris
Santa Clara
Singapore
Tokyo*

Nathan Ida   João P.A. Bastos

Electromagnetics and Calculation of Fields

Second Edition

With 500 Illustrations

Springer

Nathan Ida
Department of Electrical Engineering
The University of Akron
Akron, OH 44325-3904, USA

João P.A. Bastos
Department of Electrical Engineering
Federal University of Santa Catarina
Florianopolis, SC 880549, Brazil

Consulting Editor
R. Mittra, Ph.D.
Department of Electrical and Computer Engineering
University of Illinois at Urbana-Champaign
Urbana, IL 61801-2991, USA

Library of Congress Cataloging-in-Publication Data
Ida, Nathan.
 Electromagnetics and calculation of fields / Nathan Ida, João
P.A. Bastos. — 2nd ed.
  p.   cm.
 Includes bibliographical references and index.
 ISBN 0-387-94877-5 (alk. paper)
 1. Electromagnetism. 2. Electromagnetic fields. I. Bastos,
João.
 QC760.I56   1997
 537—dc20   96-42356

Printed on acid-free paper.

© 1997 Springer-Verlag New York, Inc.
All rights reserved. This work may not be translated or copied in whole or in part without the written permission of the publisher (Springer-Verlag New York, Inc., 175 Fifth Avenue, New York, NY 10010, USA), except for brief excerpts in connection with reviews or scholarly analysis. Use in connection with any form of information storage and retrieval, electronic adaptation, computer software, or by similar or dissimilar methodology now known or hereafter developed is forbidden.
The use of general descriptive names, trade names, trademarks, etc., in this publication, even if the former are not especially identified, is not to be taken as a sign that such names, as understood by the Trade Marks and Merchandise Marks Act, may accordingly be used freely by anyone.

Production managed by Lesley Poliner; manufacturing supervised by Jeffrey Taub.
Camera-ready copy prepared from the authors' MS Word files.
Printed and bound by Braun-Brumfield, Inc., Ann Arbor, MI.
Printed in the United States of America.

9 8 7 6 5 4 3 2 1

ISBN 0-387-94877-5 Springer-Verlag New York Berlin Heidelberg   SPIN 10552732

Preface

This second edition of *Electromagnetics and Calculation of Fields* was undertaken with the intention of both updating the first edition and, most importantly, introducing the new subject of edge or vector finite elements. We felt that edge elements have matured enough to be included in a text of this type. In addition, the original text was expanded to include an introduction to the finite element method as a general design tool with in-depth discussion of the method's practical aspects, before its application to specific electromagnetic phenomena is introduced. Many of the features of the previous edition as well as the general appearance and flavor were retained while the coverage was extended and the presentation improved.

The text is intended as an introduction to electromagnetics and computation of electromagnetic fields. While many texts on electromagnetics exist, the subject of computation of electromagnetic fields is normally not treated or is treated in a number of idealized examples, with the main emphasis on development of theoretical relations.

"Why another book on electromagnetics?" This is perhaps the first question the reader may ask when opening this book. It is a valid question, because among the many books on electromagnetics some are excellent. We have two answers to this question, answers that have motivated the writing of this book.

The first concerns the method of presentation of electromagnetism. Generally, in classical books the material is presented in the following sequence: electrostatics, magnetostatics, magnetodynamics, and wave propagation, using integral forms of the field equations. As a primary effect of this presentation, the reader is led to think that the knowledge of this science is synonymous with memorizing dozens of formulas. Additionally, an impression that there is no firm connection between these equations lingers in the reader's mind since at each step new postulates are added, seemingly unrelated to previous material. Our opinion is, and we shall try to convey this to the reader, that the electromagnetic formalism is extremely simple and based on very few equations. They are the four "Maxwell equations" which include practically all the existent relationships between the electromagnetic quantities. The only additional relationships that need be considered is the Lorentz force and the material constitutive relations.

Maxwell's four equations are presented in the point, or local form. This treatment provides a powerful tool, because it is valid for any situation and is independent of geometry. When we wish to apply them to a real device, it is more practical to use the integral form of the equations, because applications lead quite naturally to the notions of line, surface, and volume. In this formalism, the "dozens of formulae" associated with the classical methods of presentation are particular cases of Maxwell's equations as applied to specific geometries. Thus, it seems to us to be more useful to present the Electromagnetic theory based on a few equations, together with the tools needed for their general use. We will show that the electrostatic, magnetostatic, and magnetodynamic applications are particular cases of these four equations. Under certain conditions, Maxwell's equations can be decoupled into two principal groups (electrostatic and magnetic) which can be studied, and used, independently. However, while the electromagnetic formalism is simple, it does not mean that all its phenomena are necessarily easy to solve. There are many complex problems whose solutions are the subject of extensive current inquiry. However, this does not prevent one from trying to present the Electromagnetic theory in a simple, generally applicable way.

Many maintain that this is a science with a definite theoretical character. This is certainly true, but we will try to emphasize its applied aspects associated with a variety of applications. However, it is not an exercise book. There are many good books dedicated exclusively to problem solving. The final sections of most chapters contain a few, carefully chosen examples. These are presented with comments, in order to complement and simplify the understanding of the theoretical material presented in the chapter.

The facts discussed above have motivated the writing of the first part of the book, titled "The Electromagnetic Field and Maxwell's Equations." This part is divided into seven chapters.

In the first chapter we present the mathematical basis necessary to work with the Maxwell's equations in integral and differential forms.

Chapter 2 describes the common basis of electromagnetics, or Maxwell's equations, with special attention to the conditions under which they can be decoupled into two independent groups of equations: one for electrostatic and the other for magnetic phenomena.

In Chapter 3, the electrostatic field equations are studied. The electric field intensity, dielectric strength, and capacitance are presented. In its final section, the finite element method is described, with the goal of proposing a solution of the Laplace equation related to the electric field.

The fourth chapter treats the second group of the decoupled Maxwell equations: the magnetostatic field equations. Here the static aspects of magnetism is presented. The Biot-Savart law, magnetic materials, the notion of inductance, etc. are presented. Permanent magnets are discussed in detail, because of their importance in design of high-performance magnetic devices.

Chapter 5 deals with magnetodynamic phenomena. Field penetration in conductors, eddy currents, eddy current losses, hysteresis losses, and induced currents due to movement are discussed.

In Chapter 6, we discuss the interactions between electromagnetic and mechanical quantities. Among these, we mention here the force on conductors, force on charges (Lorentz force), magnetic field energy, and the Poynting vector.

Special attention is given to Maxwell's force tensor. This aspect of electromagnetic fields is rarely presented in classical books. It is very useful, because it permits, in a simple and elegant way, the calculation of the force exerted on a body by the magnetic field on the surface enclosing the body. We have a particular interest in calculation of forces using Maxwell's force tensor because of its usefulness in numerical calculation of forces.

Chapter 7 deals with what we call high-frequency electromagnetic fields and wave propagation aspects. The distinction between low and high-frequencies is not on frequency but rather on displacement currents. High-frequency applications are those in which displacement currents cannot be neglected, regardless of frequency. Wave propagation, scattering, reflection, transmission, and refraction of plane waves are discussed first, followed by fields in waveguides and cavity resonators. Propagation in dielectrics, lossy dielectrics, and conductors are also presented.

The second answer to the question posed above relates to the second part of the book, entitled "Introduction to the Finite Element Method in Electromagnetics."

The analysis and design of an electrical device requires the knowledge of the electromagnetic phenomena throughout the device. The difficulty of solving Laplace's and Poisson's equations for electrical and magnetic fields was an enormous obstacle in achieving this important goal for a long time. The solution of second order partial differential equations is a special domain in applied mathematics. Methods of solution were limited to a few simple cases until very recently. By 1950, the finite elements method began to emerge as a general method for solution of mechanical engineering problems. Around 1970 this method began to be adapted to electrical engineering applications. The methods employed prior to this, either empirical or approximate, were not satisfactory, especially when the structure had a complex geometry. The use of the finite element method opened a new era in applied electromagnetism and, as a consequence, in electrical engineering, which had important needs in computer-aided design tools. The knowledge of field distribution permits the rational design construction of structures efficiently, economically, and within safety standards.

One of the principal goals of this work is to present the Finite Element method and its application to electromagnetic problems. The didactic approach used here is quite different than in other books on the subject: we present the problems to be solved and then the application of the finite elements to these problems. This means that the principal goal of this book is to solve electromagnetic problems, and the use of finite elements technique is the tool of choice in achieving this end. In this sense, this part of the book can be viewed as a tutorial on problem solving using the finite element method. Although problem solving is the major aspect, the material is presented in detailed fashion to allow the reader to form a firm basis in this important aspects of applied electromagnetics.

The second part of the book is divided into six chapters. The main objective of this part is the didactic presentation of the finite element method which by now has become an indispensable method for analysis and design of electromagnetic devices.

In Chapter 8, we introduce the finite element method as a general engineering design tool in preparation for specific applications in the following chapters. The concepts associated with the finite element method are discussed in depth and are accompanied by simple examples. A small finite element program for second order elements is included to show how the concepts are linked to practical implementation.

In Chapter 9, we present the variational method. This theoretical formulation, associated with finite elements, constitutes a well-adapted method to solve static problems. The axi-symmetric formulation, the Newton-Raphson method (for nonlinear problems) as well as several examples are described in this chapter.

Chapter 10 introduces Galerkin's residual method as it relates to time dependent applications. Two-dimensional (2D) and three-dimensional (3D) finite elements and their application are shown.

Chapter 11 introduces the concept of edge elements. The hexahedral edge element shape functions are presented in a didactic manner and subroutines showing how they are constructed are listed. The application of edge elements to static 3D and eddy current problems is explained and graphical results are provided to show their utility.

In Chapter 12, we describe several computational aspects associated with the implementation of Finite Element programs, with special attention to the matrix system, to memory use, and to the solution methods needed to solve the system of equations obtained. Iterative, direct, and eigenvalue problems are discussed.

In the thirteenth and last chapter, many aspects of the general organization of field calculation software are discussed. Pre- and post-processing methods (such as mesh generation and equipotential lines drawing) are described. The more modern systems (extension of traditional software packages) and trends in this research field are also discussed in the final sections of this chapter.

Finally, we would like to thank R. Carlson, B. Davat, R. Mesquita, G. Meunier, A. Raizer, C. Rioux, and N. Sadowski who have, in different ways, contributed to the realization of this work. Special thanks goes to many others, and in particular students, who pointed out errors, misspellings, and possible ways of improving the presentation.

N. Ida
J. P. A. Bastos
November 1996

Contents

Preface v

Part I. The Electromagnetic Field and Maxwell's Equations

1. Mathematical Preliminaries
  1.1. Introduction 1
  1.2. The Vector Notation 1
  1.3. Vector Derivation 2
     1.3.1. The Nabla (∇) Operator 2
     1.3.2. Definition of the Gradient, Divergence, and Curl 2
  1.4. The Gradient 3
     1.4.1. Example of Gradient 5
  1.5. The Divergence 6
     1.5.1. Definition of Flux 6
     1.5.2. The Divergence Theorem 8
     1.5.3. Conservative Flux 9
     1.5.4. Example of Divergence 11
  1.6. The Curl 12
     1.6.1. Circulation of a Vector 12
     1.6.2. Stokes' Theorem 14
     1.6.3. Example of Curl 17
  1.7. Second Order Operators 18
  1.8. Application of Operators to More than One Function 19
  1.9. Expressions in Cylindrical and Spherical Coordinates 20

2. The Electromagnetic Field and Maxwell's Equations
  2.1. Introduction 22
  2.2. Maxwell's Equations 22
     2.2.1. Fundamental Physical Principles of the Electromagnetic
            Field 23

  2.2.2. Point Form of the Equations 29
  2.2.3. The Equations in Vacuum 32
  2.2.4. The Equations in Media with $\varepsilon=\varepsilon_0$ and $\mu=\mu_0$ 34
  2.2.5. The Equations in General Media 35
  2.2.6. The Integral Form of Maxwell's Equations 37
 2.3. Approximations to Maxwell's Equations 43
 2.4. Units 45

3. Electrostatic Fields

 3.1. Introduction 47
 3.2. The Electrostatic Charge 47
  3.2.1. The Electric Field 48
  3.2.2. Force on an Electric Charge 48
  3.2.3. The Electric Scalar Potential V 49
 3.3. Nonconservative Fields: Electromotive Force 53
 3.4. Refraction of the Electric Field 55
 3.5. Dielectric Strength 59
 3.6. The Capacitor 61
  3.6.1. Definition of Capacitance 61
  3.6.2. Energy Stored in a Capacitor 64
  3.6.3. Energy in a Static, Conservative Field 64
 3.7. Laplace's and Poisson's Equations in Terms of the Electric Field 65
 3.8. Examples 67
  3.8.1. The Infinite Charged Line 67
  3.8.2. The Charged Spherical Half-Shell 70
  3.8.3. The Spherical Capacitor 71
  3.8.4. The Spherical Capacitor with Two Dielectric Layers . . . 72
 3.9. A Brief Introduction to the Finite Element Method: Solution of the
   Two-Dimensional Laplace Equation 74
  3.9.1. The Finite Element Technique for Division of a Domain . 75
  3.9.2. The Variational Method 77
  3.9.3. A Finite Element Program 80
  3.9.4. Example for Use of the Finite Element Program 84
 3.10. Tables of Permittivities, Dielectric Strength, and Conductivities 88

4. Magnetostatic Fields

 4.1. Introduction 90
 4.2. Maxwell's Equations in Magnetostatics 91
  4.2.1. The Equation $\nabla\times\mathbf{H}=\mathbf{J}$ 91
  4.2.2. The Equation $\nabla\cdot\mathbf{B}=0$ 93
  4.2.3. The Equation $\nabla\times\mathbf{E}=0$ 93

- 4.3. The Biot-Savart Law 94
- 4.4. Boundary Conditions for the Magnetic Field 96
- 4.5. Magnetic Materials 98
  - 4.5.1. Diamagnetic Materials 99
  - 4.5.2. Paramagnetic Materials 100
  - 4.5.3. Ferromagnetic Materials 100
  - 4.5.4. Permanent Magnets 104
- 4.6. The Analogy between Magnetic and Electric Circuits 115
- 4.7. Inductance and Mutual Inductance 119
  - 4.7.1. Definition of Inductance 119
  - 4.7.2. Energy in a Linear System 120
  - 4.7.3. The Energy Stored in the Magnetic Field 122
- 4.8. Examples 123
  - 4.8.1. Calculation of Field Intensity and Inductance of a Long Solenoid 123
  - 4.8.2. Calculation of H for a Circular Loop 125
  - 4.8.3. Field of a Rectangular Loop 127
  - 4.8.4. Calculation of Inductance of a Coaxial Cable 128
  - 4.8.5. Calculation of the Field Inside a Cylindrical Conductor .. 129
  - 4.8.6. Calculation of the Magnetic Field Intensity in a Magnetic Circuit 130
  - 4.8.7. Calculation of the Magnetic Field Intensity of a Saturated Magnetic Circuit 133
  - 4.8.8. Magnetic Circuit Incorporating Permanent Magnets 135
- 4.9. Laplace's Equation in Terms of the Magnetic Scalar Potential .. 138
- 4.10. Properties of Soft Magnetic Materials 140

5. Magnetodynamic Fields
- 5.1. Introduction 142
- 5.2. Maxwell's Equations for the Magnetodynamic Field 143
- 5.3. Penetration of Time Dependent Fields in Conducting Materials . 146
  - 5.3.1. The Equation for **H** 146
  - 5.3.2. The Equation for **B** 147
  - 5.3.3. The Equation for **E** 147
  - 5.3.4. The Equation for **J** 148
  - 5.3.5. Solution of the Equations 148
- 5.4. Eddy Current Losses in Plates 153
- 5.5. Hysteresis Losses 156
- 5.6. Examples 160
  - 5.6.1. Induced Currents Due to Change in Induction 160
  - 5.6.2. Induced Currents Due to Changes in Geometry 163
  - 5.6.3. Inductive Heating of a Conducting Block 165

       5.6.4. Effect of Movement of a Magnet Relative to a Flat Conductor 169
       5.6.5. Visualization of Penetration of Fields as a Function of Frequency 171
       5.6.6. The Voltage Transformer 172

6. Interaction between Electromagnetic and Mechanical Forces

  6.1. Introduction 175
  6.2. Force on a Conductor 175
  6.3. Force on Moving Charges: The Lorentz Force 178
  6.4. Energy in the Magnetic Field 180
  6.5. Force as Variation of Energy (Virtual Work) 182
  6.6. The Poynting Vector 184
  6.7. Maxwell's Force Tensor 188
  6.8. Examples 195
       6.8.1. Force between Two Conducting Segments 195
       6.8.2. Torque on a Loop 198
       6.8.3. The Hall Effect 200
       6.8.4. The Linear Motor and Generator 202
       6.8.5. Attraction of a Ferromagnetic Body 205
       6.8.6. Repulsion of a Diamagnetic Body 206
       6.8.7. Magnetic Levitation 207
       6.8.8. The Magnetic Brake 209

7. Wave Propagation and High-Frequency Electromagnetic Fields

  7.1. Introduction 212
  7.2. The Wave Equation and Its Solution 215
       7.2.1. The Time Dependent Equations 215
       7.2.2. The Time Harmonic Wave Equations 220
       7.2.3. Solution of the Wave Equation 222
       7.2.4. Solution for Plane Waves 222
       7.2.5. The One-Dimensional Wave Equation in Free Space and Lossless Dielectrics 223
  7.3. Propagation of Waves in Materials 227
       7.3 1. Propagation of Waves in Lossy Dielectrics 227
       7.3.2. Propagation of Plane Waves in Low-Loss Dielectrics ... 229
       7.3.3. Propagation of Plane Waves in Conductors 230
       7.3.4. Propagation in a Conductor: Definition of the Skin Depth 232
  7.4. Polarization of Plane Waves 233

- 7.5. Reflection, Refraction, and Transmission of Plane Waves 235
  - 7.5.1. Reflection and Transmission at a Lossy Dielectric Interface: Normal Incidence 236
  - 7.5.2. Reflection and Transmission at a Conductor Interface: Normal Incidence 239
  - 7.5.3. Reflection and Transmission at a Finite Conductivity Conductor Interface 240
  - 7.5.4. Reflection and Transmission at an Interface: Oblique Incidence 241
  - 7.5.5. Oblique Incidence on a Conducting Interface: Perpendicular Polarization 242
  - 7.5.6. Oblique Incidence on a Conducting Interface: Parallel Polarization 244
  - 7.5.7. Oblique Incidence on a Dielectric Interface: Perpendicular Polarization 245
  - 7.5.8. Oblique Incidence on a Dielectric Interface: Parallel Polarization 248
- 7.6. Waveguides 249
  - 7.6.1. TEM, TE, and TM Waves 249
  - 7.6.2. TEM Waves 251
  - 7.6.3. TE Waves 251
  - 7.6.4. TM Waves 252
  - 7.6.5. Rectangular Waveguides 253
  - 7.6.6. TM Modes in Waveguides 253
  - 7.6.7. TE Modes in Waveguides 256
- 7.7. Cavity Resonators 258
  - 7.7.1. TM and TE Modes in Cavity Resonators 259
  - 7.7.2. TE Modes in a Cavity 261
  - 7.7.3. Energy in a Cavity 261
  - 7.7.4. Quality Factor of a Cavity Resonator 263
  - 7.7.5. Coupling to Cavities 263

Part II. Introduction to the Finite Element Method in Electromagnetics

8. Introduction to the Finite Element Method
- 8.1. Introduction 265
- 8.2. The Galerkin Method – Basic Concepts 266
- 8.3. The Galerkin Method – Extension to 2D 277
  - 8.3.1. The Boundary Conditions 278
  - 8.3.2. Calculation of the 2D Elemental Matrix 279
- 8.4. The Variational Method – Basic Concepts 281

xiv   Contents

    8.5. The Variational Method – Extension to 2D 284
        8.5.1. The Variational Formulation 284
        8.5.2. Calculation of the 2D Elemental Matrix 289
    8.6. Generalization of the Finite Element Method 291
        8.6.1. High-Order Finite Elements: General 292
        8.6.2. High-Order Finite Elements: Notation 293
        8.6.3. High-Order Finite Elements: Implementation 296
        8.6.4. Continuity of Finite Elements 298
        8.6.5. Polynomial Basis 298
        8.6.6. Transformation of Quantities – the Jacobian 300
        8.6.7. Evaluation of the Integrals 302
    8.7. Numerical Integration 306
        8.7.1. Evaluation of the Integrals 306
        8.7.2. Basic Principles of Numerical Integration 307
        8.7.3. Accuracy and Errors in Numerical Integration 311
    8.8. Some Specific Finite Elements 313
        8.8.1. 1D Elements 314
        8.8.2. 2D Elements 315
        8.8.3. 3D Elements 318
    8.9. Coupling Different Finite Elements; Infinite Elements 323
        8.9.1. Coupling Different Types of Finite Elements 323
        8.9.2. Infinite Elements 325
    8.10. Calculation of Some Terms in Poisson's Equation 327
        8.10.1. The Stiffness Matrix 327
        8.10.2. Evaluation of the Second Term in Eq. (8.130) 329
        8.10.3. Evaluation of the Third Term in Eq. (8.130) 329
        8.10.4. Evaluation of the Source Term 330
    8.11. A Simplified 2D Second-Order Finite Element Program 330
        8.11.1. The Problem to Be Solved 330
        8.11.2. The Discretized Domain 332
        8.11.3. The Finite Element Program 333

9. The Variational Finite Element Method: Some Static Applications

    9.1. Introduction 343
    9.2. Some Static Applications 343
        9.2.1. Electrostatic Fields: Dielectric Materials 343
        9.2.2. Stationary Currents: Conducting Materials 345
        9.2.3. Magnetic Fields: Scalar Potential 346
        9.2.4. The Magnetic Field: Vector Potential 347
        9.2.5. The Electric Vector Potential 352
    9.3. The Variational Method 353

        9.3.1. The Variational Formulation 354
        9.3.2. Functionals Involving Scalar Potentials 355
        9.3.3. The Vector Potential Functionals 359
  9.4. The Finite Element Method 362
  9.5. Application of Finite Elements with the Variational Method . . . 366
        9.5.1. Application to the Electrostatic Field 367
        9.5.2. Application to the Case of Stationary Currents 370
        9.5.3. Application to the Magnetic Field: Scalar Potential 370
        9.5.4. Application to the Magnetic Field: Vector Potential 371
        9.5.5. Application to the Electric Vector Potential 373
  9.6. Assembly of the Matrix System 373
  9.7. Axi-Symmetric Applications 375
  9.8. Nonlinear Applications 383
        9.8.1. Method of Successive Approximation 383
        9.8.2. The Newton-Raphson Method 384
  9.9. The Three-Dimensional Scalar Potential 387
        9.9.1. The First-Order Tetrahedral Element 388
        9.9.2. Application of the Variational Method 389
        9.9.3. Modeling of 3D Permanent Magnets 389
  9.10. Examples 390
        9.10.1. Calculation of Electrostatic Fields 391
        9.10.2. Calculation of Static Currents 392
        9.10.3. Calculation of the Magnetic Field: Scalar Potential . . . 394
        9.10.4. Calculation of the Magnetic Field: Vector Potential . . . 396
        9.10.5. Three-Dimensional Calculation of Fields of Permanent
                Magnets 398

10. Galerkin's Residual Method: Applications to Dynamic Fields

  10.1. Introduction 400
  10.2. Application to Magnetic Fields in Anisotropic Media 401
  10.3. Application to 2D Eddy Current Problems 405
        10.3.1. First-Order Element in Local Coordinates 405
        10.3.2. The Vector Potential Equation Using Time
                Discretization 409
        10.3.3. The Complex Vector Potential Equation 417
        10.3.4. Structures with Moving Parts 420
        10.3.5. The Axi-Symmetric Formulation 422
        10.3.6. A Modified Complex Vector Potential Formulation for
                Wave Propagation 425
        10.3.7. Formulation of Helmholtz's Equation 427
        10.3.8. Advantages and Limitations of 2D Formulations 430
  10.4. Application of the Newton-Raphson Method 432

10.5. Examples 434
 10.5.1. Eddy Currents: Time Discretization 434
 10.5.2. Moving Conducting Piece in Front of an
  Electromagnet 437
 10.5.3. Modes and Fields in a Waveguide 440
 10.5.4. Resonant Frequencies of a Microwave Cavity 442

11. Hexahedral Edge Elements – Some 3D Applications

11.1. Introduction 445
11.2. The Hexahedral Edge Element Shape Functions 448
11.3. Construction of the Shape Functions 456
11.4. Application of Edge Elements to Low-Frequency
 Maxwell's Equations 460
 11.4.1. Static Cases 461
 11.4.2. Listing of the Matrix Construction Code 464
 11.4.3. Modeling of Permanent Magnets 466
 11.4.4. Eddy Currents – the Time-Stepping Procedure 466
 11.4.5. Eddy Currents – The Complex Formulation 468
 11.4.6. The Newton-Raphson Method 469
 11.4.7. The Divergence of **J** and Other Particulars 471
11.5. Modeling of Waveguides and Cavity Resonators 472
11.6. Examples 473
 11.6.1. Static Calculations (TEAM Problem 13) 474
 11.6.2. A Linear Motor with Permanent Magnets 475
 11.6.3. Eddy Current Calculations (TEAM Problem 21) 477
 11.6.4. Calculation of Resonant Frequencies
  (TEAM Problem 19) 480

12. Computational Aspects in Finite Element Software Implementation

12.1. Introduction 484
12.2. Geometric Repetition of Domains 484
 12.2.1. Periodicity 484
 12.2.2. Anti-Periodicity 486
12.3. Storage of the Coefficient Matrix 487
 12.3.1. Symmetry of the Coefficient Matrix 487
 12.3.2. The Banded Matrix and Its Storage 487
 12.3.3. Compact Storage of the Matrix 489
12.4. Insertion of Dirichlet Boundary Conditions 490
12.5. Quadrilateral and Hexahedral Elements 491

12.6. Methods of Solution of the Linear System 493
    12.6.1. Direct Methods 493
    12.6.2. Iterative Methods 497
12.7. Methods of Solution for Eigenvalues and Eigenvectors 500
    12.7.1. The Jacobi Transformation 500
    12.7.2. The Givens Transformation 503
    12.7.3. The QR and QZ Methods 504
12.8. Diagram of a Finite Element Program 506

13. General Organization of Field Computation Software

13.1. Introduction 509
13.2. The Pre-Processor Module 510
    13.2.1. The User/System Dialogue 510
    13.2.2. Domain Discretization 511
13.3. The Processor Module 516
13.4. The Post-Processor Module 517
    13.4.1. Visualization of Results 517
    13.4.2. Calculation of Numerical Results 519
13.5. The Computational Organization of a Software Package 524
    13.5.1. The EFCAD Software 525
13.6. Evolving Software 528
    13.6.1. The Adaptive Mesh Method 528
    13.6.2. A Coupled Thermal/Electrical System 533
    13.6.3. A Software Package for Electrical Machines 537
    13.6.4 A System for Simultaneous Solution of Field Equations and External Circuits 541
    13.6.5. Computational Difficulties and Extensions to Field Computation Packages 547
13.7. Recent Trends 547

Bibliography 549

Subject Index 559

1
Mathematical Preliminaries

1.1. Introduction

In this chapter we review a few ideas from vector algebra and calculus, which are used extensively in future chapters. We assume that operations like integration and differentiation as well as elementary vector calculus operations are known.

This chapter is written in a concise fashion, and therefore, only those subjects directly applicable to this work are included. Readers wishing to expand on material introduced here can do so by consulting specialized books on the subject. It should be emphasized that we favor the geometrical interpretation rather than complete, rigorous mathematical exposition.

We look with particular interest at the ideas of gradient, divergence, and curl as well as at the divergence and Stokes' theorems. These notions are of fundamental importance for the understanding of electromagnetic fields in terms of Maxwell's equations. The latter are presented in local or point form in this work.

1.2. The Vector Notation

Many physical quantities posses an intrinsic vector character. Examples are velocity, acceleration, and force, with which we associate a direction in space. Other quantities, like mass and time, lack this quality. These are scalar quantities. Another important concept is the vector field. A force, applied at a point on a body is a vector; however, the velocity of a gas inside a tube is a vector defined throughout a region (i.e., the cross-section of the tube, or a volume), not only at one point. In the later case, we have a vector field. We use this concept extensively since many of the electromagnetic quantities (electric and magnetic fields, for example) are vector fields.

1.3. Vector Derivation

1.3.1. *The Nabla (∇) Operator*

First, we recall that a scalar function may depend on more than one variable. For example, in the Cartesian system of coordinates the function can be denoted as

$$f(x,y,z)$$

Its partial derivatives, if these exist, are

$$\frac{\partial f}{\partial x}, \quad \frac{\partial f}{\partial y}, \quad \frac{\partial f}{\partial z}$$

The nabla (∇) operator is a vector, which, in Cartesian coordinates, has the following components:

$$\nabla = \left(\frac{\partial}{\partial x}, \frac{\partial}{\partial y}, \frac{\partial}{\partial z}\right)$$

The operator is frequently written as

$$\nabla = \hat{\mathbf{i}}\frac{\partial}{\partial x} + \hat{\mathbf{j}}\frac{\partial}{\partial y} + \hat{\mathbf{k}}\frac{\partial}{\partial z}$$

where $\hat{\mathbf{i}}, \hat{\mathbf{j}}, \hat{\mathbf{k}}$, are the orthogonal unit vectors in the Cartesian system of coordinates.

The nabla is a mathematical operator to which, by itself, we cannot associate any geometrical meaning. It is the interaction of the nabla operator with other quantities that gives it geometric significance.

1.3.2. *Definition of the Gradient, Divergence, and Curl*

We define a scalar function $U(x,y,z)$ with nonzero first-order partial derivatives with respect to the coordinates x, y, and z at a point M

$$\frac{\partial U}{\partial x}, \quad \frac{\partial U}{\partial y}, \quad \frac{\partial U}{\partial z}$$

and a vector **A** with components A_x, A_y, and A_z which depend on x, y, and z; ∇ is a vector which can interact with a vector or a scalar, as shown below:

$$\nabla \text{(Vector)} \begin{cases} \mathbf{A} \text{ (Vector)} \begin{cases} \text{- scalar product: } \nabla \cdot \mathbf{A} \text{ or } div \ \mathbf{A} \text{ (scalar)} \\ \text{- vector product: } \nabla \times \mathbf{A} \text{ or } curl \ \mathbf{A} \text{ or } rot \ \mathbf{A} \text{ (vector)} \end{cases} \\ U \text{ (Scalar)} \text{ - product: } \nabla U \text{ or } grad \ U \text{ (vector)} \end{cases}$$

These three products can now be calculated:

$$div \ \mathbf{A} = \nabla \cdot \mathbf{A} = \left(\hat{\mathbf{i}} \frac{\partial}{\partial x} + \hat{\mathbf{j}} \frac{\partial}{\partial y} + \hat{\mathbf{k}} \frac{\partial}{\partial z} \right) \cdot \left(\hat{\mathbf{i}} A_x + \hat{\mathbf{j}} A_y + \hat{\mathbf{k}} A_z \right)$$

or

$$\nabla \cdot \mathbf{A} = \frac{\partial A_x}{\partial x} + \frac{\partial A_y}{\partial y} + \frac{\partial A_z}{\partial z} \qquad (1.1)$$

$$rot \ \mathbf{A} = curl \ \mathbf{A} = \nabla \times \mathbf{A} = det \begin{bmatrix} \hat{\mathbf{i}} & \hat{\mathbf{j}} & \hat{\mathbf{k}} \\ \frac{\partial}{\partial x} & \frac{\partial}{\partial y} & \frac{\partial}{\partial z} \\ A_x & A_y & A_z \end{bmatrix}$$

or

$$\nabla \times \mathbf{A} = \hat{\mathbf{i}} \left(\frac{\partial A_z}{\partial y} - \frac{\partial A_y}{\partial z} \right) + \hat{\mathbf{j}} \left(\frac{\partial A_x}{\partial z} - \frac{\partial A_z}{\partial x} \right) + \hat{\mathbf{k}} \left(\frac{\partial A_y}{\partial x} - \frac{\partial A_x}{\partial y} \right) \qquad (1.2)$$

$$grad \ U = \nabla U = \hat{\mathbf{i}} \frac{\partial U}{\partial x} + \hat{\mathbf{j}} \frac{\partial U}{\partial y} + \hat{\mathbf{k}} \frac{\partial U}{\partial z} \qquad (1.3)$$

After defining the gradient, curl, and divergence as algebraic entities, we will gain some insight into their geometric meaning in the following sections.

1.4. The Gradient

Given a scalar function $U(x,y,z)$, with partial derivatives $\partial U/\partial x$, $\partial U/\partial y$, $\partial U/\partial z$, and dependent on a point M, with coordinates x,y,z, denoted as $M(x,y,z)$, we can calculate the differential of U as dU. This is done by considering the point $M(x,y,z)$ and another point, infinitely close to M, $M'(x+dx, y+dy, z+dz)$, and using the total differential

$$dU = \frac{\partial U}{\partial x} dx + \frac{\partial U}{\partial y} dy + \frac{\partial U}{\partial z} dz \qquad (1.4)$$

4   1. Mathematical Preliminaries

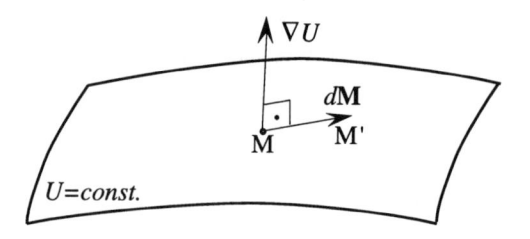

Figure 1.1. The gradient is orthogonal to a constant potential surface.

Defining the vector $d\mathbf{M}=\mathbf{M}'-\mathbf{M}$ which possesses the components

$$d\mathbf{M} = (dx,\ dy,\ dz)$$

dU can be written as

$$dU = \left(\hat{\mathbf{i}}\frac{\partial U}{\partial x} + \hat{\mathbf{j}}\frac{\partial U}{\partial y} + \hat{\mathbf{k}}\frac{\partial U}{\partial z}\right) \cdot (\hat{\mathbf{i}}dx + \hat{\mathbf{j}}dy + \hat{\mathbf{k}}dz)$$

or

$$dU = (\nabla U) \cdot d\mathbf{M} \qquad (1.5)$$

As for the geometrical significance of the gradient, assume that there exists a surface with points $M(x,y,z)$ and that on all these points, $U=constant$ (see Figure 1.1). Hence, for all differential displacements M and M' on this surface, we can write $dU=0$. From Eq. (1.5) we have

$$(\nabla U) \cdot d\mathbf{M} = 0$$

From the definition of the scalar product, it is clear that ∇U and $d\mathbf{M}$ are orthogonal. Assume that the displacement of $\mathbf{M}$ to $\mathbf{M}'$ is in the direction of increasing U, as shown in Figure 1.2. In this case, $dU>0$, or

$$(\nabla U) \cdot d\mathbf{M} > 0$$

Note that the vectors ∇U and $d\mathbf{M}$ form an acute angle.

From the foregoing arguments we conclude that ∇U is a vector, perpendicular to a surface on which U is constant and that it points in the direction of increasing U. We also note that ∇U points in the direction of maximum change in U, since $dU=(\nabla U) \cdot d\mathbf{M}$ is maximum when $d\mathbf{M}$ is in the same direction as ∇U.

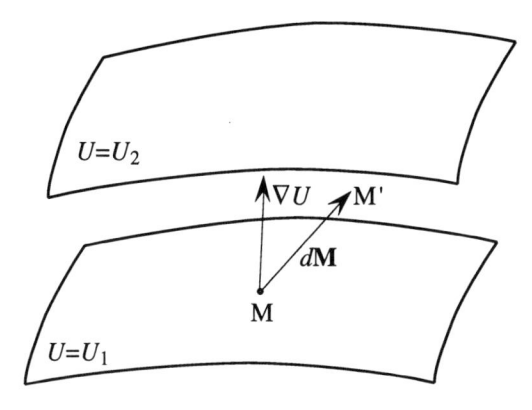

Figure 1.2. Geometrical representation of the gradient.

1.4.1. *Example of Gradient*

Given a function r, as the distance of a point $M(x,y,z)$ from the origin $O(0,0,0)$, determine the gradient of this function.

The surface $r=constant$ is a sphere of radius r with center at $O(0,0,0)$, whose equation is

$$r = \sqrt{x^2 + y^2 + z^2}$$

The components of ∇r are:

$$\frac{\partial r}{\partial x} = \frac{x}{\sqrt{x^2 + y^2 + z^2}} = \frac{x}{r}$$

$$\frac{\partial r}{\partial y} = \frac{y}{r}$$

$$\frac{\partial r}{\partial z} = \frac{z}{r}$$

We obtain

$$\nabla r = \frac{1}{r}(\hat{\mathbf{i}}x + \hat{\mathbf{j}}y + \hat{\mathbf{k}}z)$$

The magnitude of the gradient is

$$|\nabla r| = \sqrt{\frac{1}{r^2}(x^2 + y^2 + z^2)} = 1$$

6   1. Mathematical Preliminaries

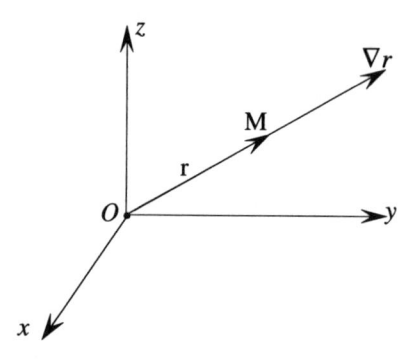

Figure 1.3. Definition of direction of ∇r.

As for the direction of ∇r, we define a vector **OM** = **M** − **O**, as shown in Figure 1.3.

Noting that $\nabla r = $ **OM**/r; where r is the distance (scalar) between M and O, we conclude that ∇r and **OM** are collinear vectors. Therefore, ∇r points in the direction of increasing r, or towards spheres with radii larger than r, as was indicated formally above.

1.5. The Divergence

1.5.1. *Definition of Flux*

Consider a point M in the vector field **A**, as well as a differential surface ds at this point, as in Figure 1.4.

We choose a point N such that the vector **MN** is perpendicular to ds. We call $\hat{\mathbf{n}}$ the normal unit vector, given by the expression

$$\hat{\mathbf{n}} = \frac{\mathbf{MN}}{|MN|}$$

A vector **ds**, with magnitude equal to ds and direction identical to $\hat{\mathbf{n}}$ is defined as

$$d\mathbf{s} = \hat{\mathbf{n}} ds$$

The flux of the vector **A** through the surface ds is now defined by the following scalar product

$$d\Phi = \mathbf{A} \cdot d\mathbf{s} = A \, ds \, \cos\theta \qquad (1.6)$$

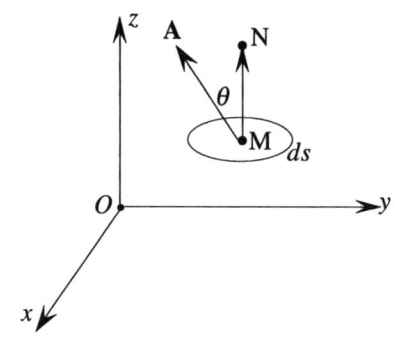

Figure 1.4. Definition of normal unit vector to a surface *ds*.

where θ is the smaller angle between **A** and $\hat{\mathbf{n}}$. The flux is maximum when **A** and *ds* are parallel, or, when **A** is perpendicularly incident on the surface *ds*.

Since *ds* is a vector, it possesses three components which represent the projections of the vector on the three planes of the system *Oxyz*, (see Figure 1.5). Thus, *ds* has the following components

$$ds_x = dydz$$

$$ds_y = dzdx$$

$$ds_z = dxdy$$

With the components of **A**

$$\mathbf{A} = (A_x,\ A_y,\ A_z)$$

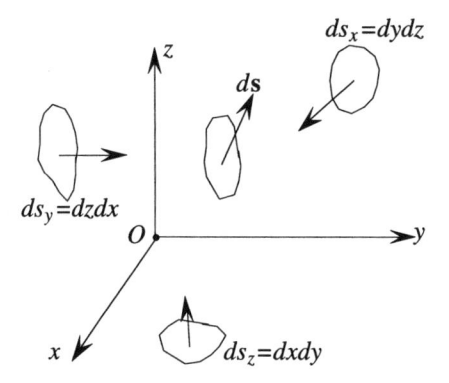

Figure 1.5. Components of the vector *ds*.

8   1. Mathematical Preliminaries

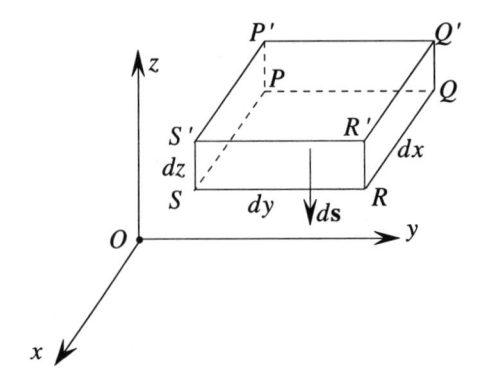

Figure 1.6. A closed surface in Cartesian coordinates: definition of divergence.

we obtain

$$d\Phi = A_x dydz + A_y dxdz + A_z dxdy \qquad (1.7)$$

Note that the flux can be defined only when the direction of $\hat{\mathbf{n}}$ is defined. In the case of a closed surface S, $\hat{\mathbf{n}}$ always points in the outward direction to the volume enclosed by the surface S.

1.5.2 The Divergence Theorem

Consider the surface of a rectangular box whose sides are dx, dy, and dz, and are parallel to the planes of $Oxyz$. The area of the lower face $PQRS$ is $dxdy$ (see Figure 1.6), and its vector $d\mathbf{s}$ is

$$d\mathbf{s} = (0,\ 0,\ -dxdy)$$

From Eq. (1.7), the flux of $\mathbf{A}$ crossing this surface is

$$-A_z dxdy \qquad (1.8)$$

On the upper face, $P'Q'R'S'$, an analogous expression is found. On the upper surface the normal to the surface is positive, and the component A_z of the vector $\mathbf{A}$, is augmented by dA_z. Therefore, A_z on the upper face is

$$A_z + dA_z = A_z + \frac{\partial A_z}{\partial z} dz$$

and the flux is

$$\left(A_z + \frac{\partial A_z}{\partial z} dz\right) dxdy \qquad (1.9)$$

The sum of the two fluxes gives

$$\frac{\partial A_z}{\partial z} dv \qquad (1.10)$$

where dv is the volume of the rectangular box.

Using the same rationale on the other two pairs of parallel surfaces, yields the expression

$$d\Phi = \left[\frac{\partial A_x}{\partial x} + \frac{\partial A_y}{\partial y} + \frac{\partial A_z}{\partial z}\right] dv \qquad (1.11)$$

Observing that

$$\frac{\partial A_x}{\partial x} + \frac{\partial A_y}{\partial y} + \frac{\partial A_z}{\partial z} = \nabla \cdot \mathbf{A}$$

The expression for the flux becomes

$$d\Phi = (\nabla \cdot \mathbf{A}) dv$$

With $d\Phi = \mathbf{A} \cdot d\mathbf{s}$, the integration over the whole surface S gives the total flux

$$\Phi = \oint_S \mathbf{A} \cdot d\mathbf{s} = \int_V (\nabla \cdot \mathbf{A}) dv \qquad (1.12)$$

This equality between the two integrals means that the flux of the vector $\mathbf{A}$ through the closed surface S is equal to the volume integral of the divergence of $\mathbf{A}$ over the volume enclosed by the surface S.

1.5.3. *Conservative Flux*

Consider a tube of flux, such that the field of vectors $\mathbf{A}$ defines a volume in which the vectors are tangent to the lateral walls, as shown in Figure 1.7. S_1 and S_2 are arbitrary surfaces which section the flux tube. S_3 is the lateral surface of the tube section. We denote A_1, A_2, and A_3, the vectors $\mathbf{A}$ on surfaces S_1, S_2, and S_3, as indicated in Figure 1.7. To simplify the discussion we assume the vectors $\mathbf{A}_1$ and $\hat{\mathbf{n}}_1$ are in the same direction and $\mathbf{A}_2$ and $\hat{\mathbf{n}}_2$ are in the same direction. Now the different fluxes can be calculated:

10   1. Mathematical Preliminaries

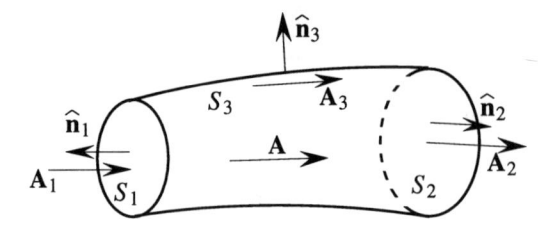

Figure 1.7. A tube of flux.

$$\Phi = \oint_S \mathbf{A} \cdot d\mathbf{s} = \int_V (\nabla \cdot \mathbf{A}) dv$$

$$d\Phi_1 = \mathbf{A}_1 \cdot d\mathbf{s}_1 = -A_1 ds_1$$

$$d\Phi_2 = \mathbf{A}_2 \cdot d\mathbf{s}_2 = A_2 ds_2$$

$$d\Phi_3 = \mathbf{A}_3 \cdot d\mathbf{s}_3 = 0$$

The total flux is therefore

$$\Phi_t = -\int_{S_1} A_1 ds_1 + \int_{S_2} A_2 ds_2$$

Because S_1 and S_2 are arbitrary surfaces, they can be approximated geometrically. It is clear that if S_2 tends to S_1 and at the same time $\mathbf{A}_2$ tends to $\mathbf{A}_1$, the sum above tends to zero. Since the flux entering the tube is equal to the flux leaving it, we conclude that the flux through the closed surface, in this case, is zero. Utilizing the divergence theorem

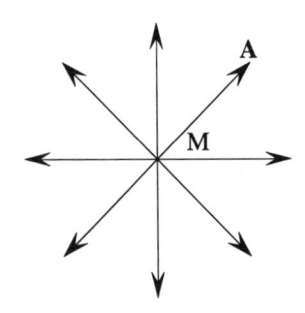

Figure 1.8. A radial vector field.

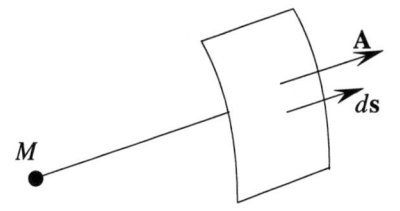

Figure 1.9. Directions of **A** and **ds** for the field in Figure 1.8.

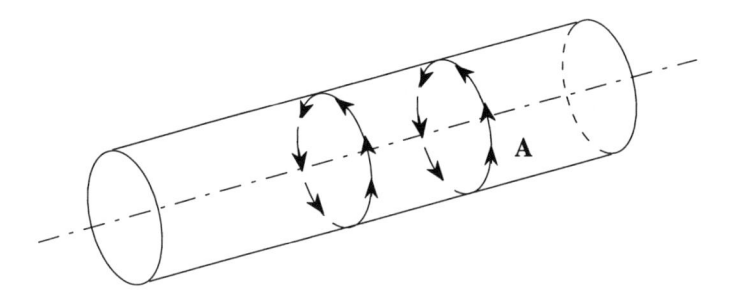

Figure 1.10. A circumferential vector field.

$$\Phi = \oint_S \mathbf{A} \cdot d\mathbf{s} = \int_V (\nabla \cdot \mathbf{A}) dv$$

from which we note that $\nabla \cdot \mathbf{A} = 0$. This leads to the conclusion that the flux is conservative (the flux in the tube is conserved). From the discussion here we conclude that when the flux is conservative, the divergence of the field is zero.

1.5.4. *Example of Divergence*

Consider a radial vector field as shown in Figure 1.8, and assume the magnitude of **A** is constant at all points on a sphere centered at M. To calculate the flux of the vector **A** through a spherical shell of radius R, we note that $d\mathbf{s}$ and **A** are collinear and in the same direction (as in Figure 1.9). We get

$$\Phi = \oint_S \mathbf{A} \cdot d\mathbf{s} = \oint_S A\,ds = A\oint_S ds = 4\pi R^2 A$$

From the divergence theorem, (the flux of **A** is nonzero) we have:

$$\nabla \cdot \mathbf{A} \neq 0$$

We now calculate the flux of a different vector field **A**, through the lateral surface of the cylinder shown in Figure 1.10.

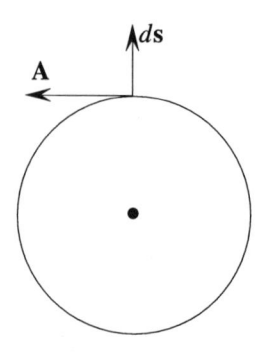

Figure 1.11. Directions of **A** and *ds* for the field in Figure 1.10.

Examining the transverse cross-section (Figure 1.11) we note that the vectors **A** and *ds* are perpendicular. Consequently, from the divergence theorem we conclude that in this case, $\nabla \cdot \mathbf{A} = 0$.

Observing these two examples, we note that in the first case, the vectors of the field are literally divergent at point M and, therefore, the divergence of the field is different than zero. In the second example, the vectors of the field do not depart but rather suggest rotation or a vortex, not divergence. In this case the divergence of the vector is zero.

1.6. The Curl

1.6.1. *Circulation of a Vector*

The circulation of a vector field **A** along a contour L, between points P and Q is given by the line integral

$$C_{PQ} = \int_P^Q \mathbf{A} \cdot d\mathbf{l} \qquad (1.13)$$

where $d\mathbf{l}$ is a differential length vector along the contour.

If **A** and $d\mathbf{l}$ are parallel, as in Figure 1.13a, the circulation is maximum. However, if **A** and $d\mathbf{l}$ are perpendicular to each other, the circulation is zero.

In the first case, the vector **A** circulates along the contour L, while in the second case, where **A** is perpendicular to the contour, it does not circulate along the contour L.

If the vector **A** is the gradient of a scalar function, $\mathbf{A} = \nabla U$, we have

$$C_{PQ} = \int_P^Q \mathbf{A} \cdot d\mathbf{l} = \int_P^Q (\nabla U) \cdot d\mathbf{l}$$

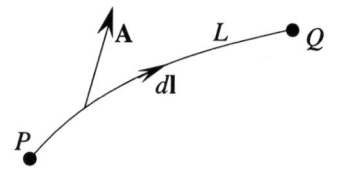

Figure 1.12. Definition of circulation of a vector **A** along contour L.

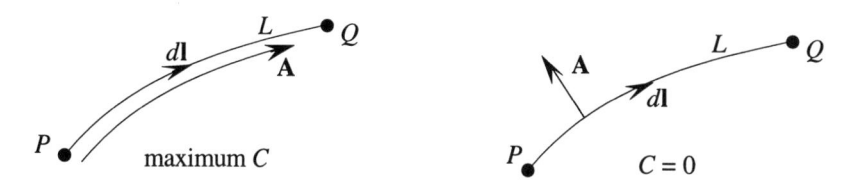

Figure 1.13a. Maximum circulation.

Figure 1.13b. Zero circulation.

Assuming that d**l** is equal to d**M** in Eq. (1.5) we get

$$dU = (\nabla U) \cdot d\mathbf{l}$$

and the circulation is

$$C_{PQ} = \int_P^Q dU = U_Q - U_P$$

Thus, if **A**=∇U, the circulation along the contour depends only on the initial and final points of the contour (P and Q), not on the contour itself. If the contour L is closed, or, if Q and P coincide, the circulation of **A** is zero.

It remains to be verified if all vectors can be identified by a gradient. To do so, suppose that a general vector has components $\partial U/\partial x$, $\partial U/\partial y$, $\partial U/\partial z$. Note that

$$\frac{\partial}{\partial y}\left(\frac{\partial U}{\partial z}\right) = \frac{\partial}{\partial z}\left(\frac{\partial U}{\partial y}\right)$$

$$\frac{\partial^2 U}{\partial y \partial z} = \frac{\partial^2 U}{\partial z \partial y}$$

or that

$$\frac{\partial^2 U}{\partial y \partial z} - \frac{\partial^2 U}{\partial z \partial y} = 0$$

This is valid for any gradient. Writing **A**=∇U, gives

14   1. Mathematical Preliminaries

$$A_x = \frac{\partial U}{\partial x}, \quad A_y = \frac{\partial U}{\partial y}, \quad A_z = \frac{\partial U}{\partial z}$$

if

$$\frac{\partial^2 U}{\partial y \partial z} - \frac{\partial^2 U}{\partial z \partial y} = \frac{\partial}{\partial y}\left(\frac{\partial U}{\partial z}\right) - \frac{\partial}{\partial z}\left(\frac{\partial U}{\partial y}\right) = 0$$

we get the condition

$$\frac{\partial A_z}{\partial y} - \frac{\partial A_y}{\partial z} = 0$$

Analogously for the indices x, y and x, z we obtain

$$\frac{\partial A_x}{\partial z} - \frac{\partial A_z}{\partial x} = 0$$

$$\frac{\partial A_y}{\partial x} - \frac{\partial A_x}{\partial y} = 0$$

These three terms are the components of the curl $\nabla \times \mathbf{A}$, [see equation (1.2)]. Therefore, if $\mathbf{A} = \nabla U$, the curl of $\mathbf{A}$ is identically zero.

1.6.2. Stokes' Theorem

Assuming that the flux of a vector $\mathbf{B}$ is conservative, it is possible to define a vector $\mathbf{A}$ so that $\mathbf{B} = \nabla \times \mathbf{A}$, since $\nabla \cdot \mathbf{B} = 0$. By substitution we get $\nabla \cdot (\nabla \times \mathbf{A}) = 0$. This quantity, $\nabla \cdot (\nabla \times \mathbf{A})$ is identically zero for any vector $\mathbf{A}$, and the definition $\mathbf{B} = \nabla \times \mathbf{A}$ is correct for any vector $\mathbf{B}$ if $\nabla \cdot \mathbf{B} = 0$.

Assume that there is a closed contour defining a surface. This surface is divided into small areas s. We examine one infinitely small rectangle, with sides dx, dy, parallel to the axes of the coordinate system (Ox and Oy) as in Figure 1.14. The circulation of $\mathbf{A} = (A_x, A_y, A_z)$ along the contour L which encloses the area s can now be calculated. This is done separately for each section of the contour.

- Between points P and Q

$$C_1 = \int_P^Q \mathbf{A} \cdot d\mathbf{l} = \int_P^Q (\hat{\mathbf{i}} A_x + \hat{\mathbf{j}} A_y + \hat{\mathbf{k}} A_z) \cdot \hat{\mathbf{i}} \, dx$$

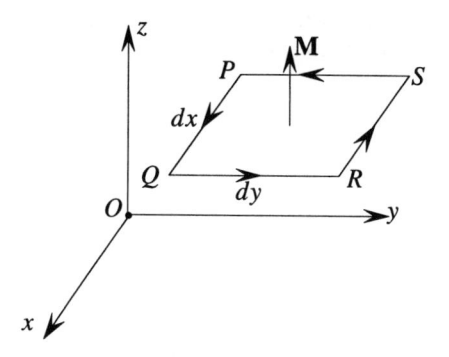

Figure 1.14. A small surface defined by a closed contour.

or

$$C_1 = A_x dx$$

- Between points R and S, the vector $\mathbf{A}$ is

$$\mathbf{A} = (A_x + dA_x,\ A_y + dA_y,\ A_z + dA_z)$$

where

$$A_x + dA_x = A_x + \frac{\partial A_x}{\partial y} dy$$

and, therefore, the circulation is

$$C_2 = -\left(A_x + \frac{\partial A_x}{\partial y} dy\right) dx$$

The sum of C_1 and C_2 is

$$C_{1,2} = -\frac{\partial A_x}{\partial y} dx dy$$

In a completely analogous manner, the sum of the circulations along lines SP and QR is

$$C_{3,4} = \frac{\partial A_y}{\partial x} dx dy$$

16   1. Mathematical Preliminaries

The sum of $C_{1,2}$ and $C_{3,4}$ gives the total circulation along the rectangle *PQRS*. Denoting this circulation as C_z we have

$$C_z = \left(\frac{\partial A_y}{\partial x} - \frac{\partial A_x}{\partial y}\right) dx dy$$

However, since $\mathbf{B} = \nabla \times \mathbf{A}$, the expression in brackets is equal to B_z:

$$C_z = B_z \, dx dy$$

Analogously, C_x and C_y can be calculated on rectangles parallel to Oyz and Ozx, to obtain

$$C_x = B_x \, dy dz$$

$$C_y = B_y \, dx dz$$

The total circulation on a curve composed of these three differential rectangles is [see Eq. (1.7)]

$$C = C_x + C_y + C_z = \int_S \mathbf{B} \cdot d\mathbf{s} = \int_S (\nabla \times \mathbf{A}) \cdot d\mathbf{s}$$

This expression was obtained through the calculation of the circulation of **A**, and, therefore, we can write

$$\oint_L \mathbf{A} \cdot d\mathbf{l} = \int_S \mathbf{B} \cdot d\mathbf{s} = \int_S (\nabla \times \mathbf{A}) \cdot d\mathbf{s} \qquad (1.14)$$

This is

$$\oint_L \mathbf{A} \cdot d\mathbf{l} = \int_S (\nabla \times \mathbf{A}) \cdot d\mathbf{s} \qquad (1.15)$$

Carrying out the calculation above over the whole surface of the global domain validates this result in a global sense. Therefore, we conclude that the flux of a vector **A** through the open surface *S* is equal to the circulation of the vector **A** along the contour *L*, encircling the surface.

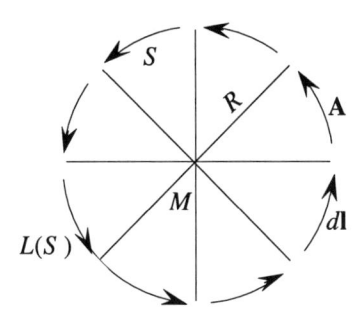

Figure 1.15. A vector field **A** with constant magnitude at distance R from point M.

1.6.3. *Example of Curl*

Consider the field of vectors **A**, at a point M, as indicated in Figure 1.15. Assume that at a constant distance R, the magnitude of **A** is constant. Defining a surface S such that it is enclosed by the circle of radius R, we can calculate the circulation of **A** along the contour L. **A** and $d\mathbf{l}$ are collinear vectors, and we have

$$\oint_L \mathbf{A} \cdot d\mathbf{l} = \oint_L A\, dl = A \oint_L dl = 2\pi R A$$

Since this circulation is nonzero, we conclude from Stokes' theorem that $\nabla \times \mathbf{A}$ is also nonzero.

Because the vector $\nabla \times \mathbf{A}$ is perpendicular to **A** (by definition of the cross product), we observe that the geometrical positions of **A** and $\nabla \times \mathbf{A}$ are as shown in Figure 1.16.

Another example is shown in Figure 1.17. Choosing the same surface S, and noting that **A** and $d\mathbf{l}$ are orthogonal vectors, gives

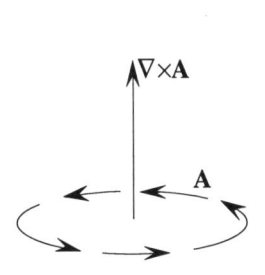

Figure 1.16. Geometric representation of the curl of **A**.

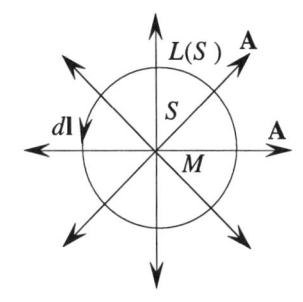

Figure 1.17. A radial vector field.

$$\oint_L \mathbf{A} \cdot d\mathbf{l} = 0$$

and we conclude that $\nabla \times \mathbf{A} = 0$.

In this particular vector field, there is no rotation or vortex. On the other hand, the field is divergent.

1.7. Second-Order Operators

It is possible to combine two first-order operators operating on scalar functions U and vector functions $\mathbf{A}$. The possible combinations are:

$$\nabla \cdot (\nabla U) \tag{1.16}$$

$$\nabla \times (\nabla U) \tag{1.17}$$

$$\nabla (\nabla \cdot \mathbf{A}) \tag{1.18}$$

$$\nabla \cdot (\nabla \times \mathbf{A}) \tag{1.19}$$

$$\nabla \times (\nabla \times \mathbf{A}) \tag{1.20}$$

Note that, for example, $\nabla \times (\nabla \cdot \mathbf{A})$, cannot exist by definition since the curl operates only on vectors, while $\nabla \cdot \mathbf{A}$ is, by definition, a scalar.

We now calculate, as an example, the operator $\nabla \cdot (\nabla U)$. Given:

$$\nabla U = \hat{\mathbf{i}} \frac{\partial U}{\partial x} + \hat{\mathbf{j}} \frac{\partial U}{\partial y} + \hat{\mathbf{k}} \frac{\partial U}{\partial z}$$

we can write

$$\nabla \cdot (\nabla U) = (\hat{\mathbf{i}} \frac{\partial}{\partial x} + \hat{\mathbf{j}} \frac{\partial}{\partial y} + \hat{\mathbf{k}} \frac{\partial}{\partial z}) \cdot (\nabla U)$$

or, after performing the dot product:

$$\nabla \cdot (\nabla U) = \frac{\partial^2 U}{\partial x^2} + \frac{\partial^2 U}{\partial y^2} + \frac{\partial^2 U}{\partial z^2}$$

Defining the Laplace operator (or, in short, the Laplacian) as

$$\nabla^2 = \frac{\partial^2}{\partial x^2} + \frac{\partial^2}{\partial y^2} + \frac{\partial^2}{\partial z^2} \qquad (1.21)$$

we have

$$\nabla \cdot (\nabla U) = \nabla^2 U \qquad (1.22)$$

The calculation of $\nabla \times (\nabla U)$ and $\nabla \cdot (\nabla \times \mathbf{A})$ represent operations which are identically equal to zero, without any particular condition being imposed on U and $\mathbf{A}$.

Finally, Eq. (1.18) and (1.19) are generally used together such that

$$\nabla^2 \mathbf{A} = \nabla(\nabla \cdot \mathbf{A}) - \nabla \times (\nabla \times \mathbf{A}) \qquad (1.23)$$

where $\nabla^2 \mathbf{A}$ is called the "vector Laplacian" of $\mathbf{A}$. This is written as

$$\nabla^2 \mathbf{A} = \hat{\mathbf{i}} \nabla^2 A_x + \hat{\mathbf{j}} \nabla^2 A_y + \hat{\mathbf{k}} \nabla^2 A_z \qquad (1.24)$$

where, for example, the component in the Ox direction is

$$\nabla^2 A_x = \frac{\partial^2 A_x}{\partial x^2} + \frac{\partial^2 A_x}{\partial y^2} + \frac{\partial^2 A_x}{\partial z^2}$$

These operators define second-order partial differential equations, and constitute a very important domain in mathematics and physics. Written in an appropriate form, they describe the phenomena of diffusion of fields, either electromagnetic (electric or magnetic fields), or mechanical (diffusion of heat, flow of fluids, etc.).

1.8. Application of Operators to More than One Function

If U and Q are scalar functions and $\mathbf{A}$ and $\mathbf{B}$ are vector functions, all dependent on x, y, and z, we can show that:

$$\nabla(UQ) = U(\nabla Q) + Q(\nabla U) \qquad (1.25)$$

$$\nabla \cdot (U\mathbf{A}) = U(\nabla \cdot \mathbf{A}) + (\nabla U) \cdot \mathbf{A} \qquad (1.26)$$

$$\nabla \cdot (\mathbf{A} \times \mathbf{B}) = -\mathbf{A} \cdot (\nabla \times \mathbf{B}) + (\nabla \times \mathbf{A}) \cdot \mathbf{B} \qquad (1.27)$$

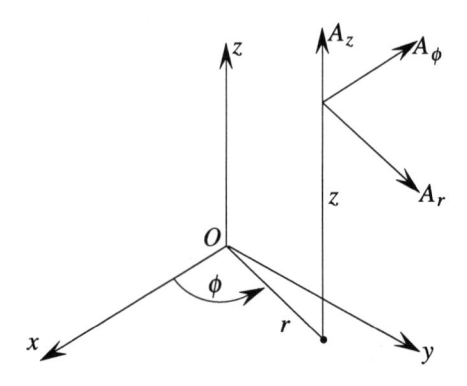

Figure 1.18. Cylindrical coordinate system, its relation to the Cartesian system, and components of a vector **A**.

$$\nabla \times (U\mathbf{A}) = U(\nabla \times \mathbf{A}) + (\nabla U) \times \mathbf{A} \qquad (1.28)$$

1.9. Expressions in Cylindrical and Spherical Coordinates

The expressions below give the principal operators in cylindrical and spherical coordinates, applied to a vector function **A** and a scalar function U.

a. Cylindrical coordinates r, ϕ, z (Figure 1.18)

$$\nabla U = \hat{\mathbf{r}} \frac{\partial U}{\partial r} + \hat{\boldsymbol{\phi}} \frac{1}{r} \frac{\partial U}{\partial \phi} + \hat{\mathbf{z}} \frac{\partial U}{\partial z} \qquad (1.29)$$

$$\nabla \cdot \mathbf{A} = \frac{1}{r} \frac{\partial}{\partial r}(rA_r) + \frac{1}{r} \frac{\partial A_\phi}{\partial \phi} + \frac{\partial A_z}{\partial z} \qquad (1.30)$$

$$\nabla \times \mathbf{A} = \hat{\mathbf{r}}\left(\frac{1}{r}\frac{\partial A_z}{\partial \phi} - \frac{\partial A_\phi}{\partial z}\right) + \hat{\boldsymbol{\phi}}\left(\frac{\partial A_r}{\partial z} - \frac{\partial A_z}{\partial r}\right) + \hat{\mathbf{z}}\left(\frac{1}{r}\frac{\partial (rA_\phi)}{\partial r} - \frac{1}{r}\frac{\partial A_r}{\partial \phi}\right) \qquad (1.31)$$

$$\nabla^2 U = \frac{\partial^2 U}{\partial r^2} + \frac{1}{r}\frac{\partial U}{\partial r} + \frac{1}{r^2}\frac{\partial^2 U}{\partial \phi^2} + \frac{\partial^2 U}{\partial z^2} \qquad (1.32)$$

b. Spherical coordinates R, θ, ϕ (Figure 1.19)

Expressions in Cylindrical and Spherical Coordinates 21

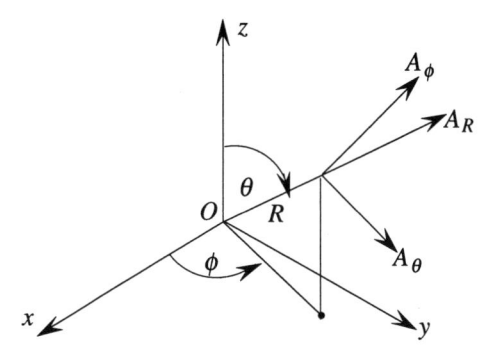

Figure 1.19. Spherical coordinate system, its relation to the Cartesian system, and components of a vector **A**.

$$\nabla U = \hat{\mathbf{R}} \frac{\partial U}{\partial R} + \hat{\theta} \frac{1}{R} \frac{\partial U}{\partial \theta} + \hat{\phi} \frac{1}{R \sin\theta} \frac{\partial U}{\partial \phi} \qquad (1.33)$$

$$\nabla \cdot \mathbf{A} = \frac{1}{R^2} \frac{\partial}{\partial R}(R^2 A_R) + \frac{1}{R \sin\theta} \frac{\partial}{\partial \theta}(A_\theta \sin\theta) + \frac{1}{R \sin\theta} \frac{\partial A_\phi}{\partial \phi} \qquad (1.34)$$

$$\nabla \times \mathbf{A} = \hat{\mathbf{R}} \frac{1}{R\sin\theta} \left(\frac{\partial}{\partial \theta}(A_\phi \sin\theta) - \frac{\partial A_\theta}{\partial \phi} \right) + \hat{\theta} \frac{1}{R} \left(\frac{1}{\sin\theta} \frac{\partial A_R}{\partial \phi} - \frac{\partial}{\partial R}(RA_\phi) \right)$$
$$+ \hat{\phi} \frac{1}{R} \left(\frac{\partial (RA_\theta)}{\partial R} - \frac{\partial A_R}{\partial \theta} \right) \qquad (1.35)$$

$$\nabla^2 U = \frac{1}{R} \frac{\partial^2 (RU)}{\partial R^2} + \frac{1}{R^2 \sin^2\theta} \frac{\partial^2 U}{\partial \phi^2} + \frac{1}{R^2 \sin\theta} \frac{\partial}{\partial \theta}\left(\sin\theta \frac{\partial U}{\partial \theta}\right) \qquad (1.36)$$

2
The Electromagnetic Field and Maxwell's Equations

2.1. Introduction

In this chapter we present the main body of electromagnetism, through the use of Maxwell's equations

James Clerk Maxwell (1831-1879), a Scottish physicist, based his work on the work and experiments of Ampere, Gauss, and Faraday and elaborated on their theories. His great contribution lies in that he unified in four equations the studies of his predecessors. He introduced the idea of "displacement currents," which generalizes "Ampere's law" and makes it valid in all situations. The introduction of displacement currents allowed him to foresee the physical phenomenon of propagation of electromagnetic waves. Nine years after Maxwell's death, Heinrich Hertz discovered electromagnetic waves experimentally, proving the global view of Maxwell's theory. While Maxwell dedicated part of his life to the theory of electromagnetic fields he also contributed to many other areas, among which is his work on the kinetic theory of gases.

The electromagnetic formalism is extremely simple and is based primarily on Maxwell's four equations. This idea is of great importance in this work since in our presentation of the electromagnetic fields we make considerable use of these equations. We hope that the opposite approach, with which this science is often associated, namely the use of many dozens of equations and formulae, is by now completely abandoned. However, in spite of the simplicity intrinsic in Maxwell's equations, some of the more useful results to specific problems, generated through physical situations associated with Maxwell's equations, can be very complex. Their ongoing solutions are still the object of inquiry.

2.2. Maxwell's Equations

We distinguish, at the outset, two domains in electromagnetism, although both are included in Maxwell's equations:

- The high-frequency domain, which includes the study of electromagnetic waves and propagation of energy through matter. While a high-frequency domain is difficult to define, we will normally view any aspect of electromagnetic fields in which the displacement currents cannot be neglected as belonging to the high-frequency domain.
- The low-frequency domain, which includes the major part of the electromagnetic devices like motors, relays, and transformers. These are all applications at power frequencies (frequencies below a few tens of kHz). Strictly speaking, any application in which displacement currents can be neglected is a low-frequency application. In this domain, corresponding to a quasi-static state, we can, in general, study electric fields and magnetic fields as separate quantities. At high frequencies, the electric and magnetic fields are interdependent.

Equally well, we can divide electromagnetism by means of the diagram below, where each block represents a particular aspect of Maxwell's equations.

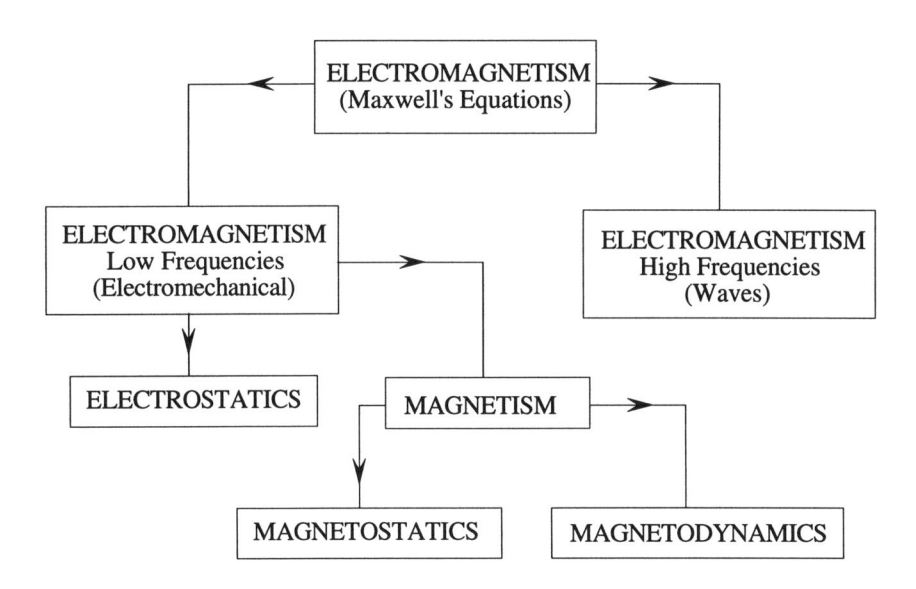

```
        ┌──────────────────────────┐
     ◄──┤   ELECTROMAGNETISM       ├──►
     │  │   (Maxwell's Equations)  │     │
     │  └──────────────────────────┘     │
     ▼                                   ▼
┌────────────────────┐           ┌────────────────────┐
│  ELECTROMAGNETISM  │           │  ELECTROMAGNETISM  │
│   Low Frequencies  ├──────────►│   High Frequencies │
│  (Electromechanical)│          │      (Waves)       │
└────────┬───────────┘           └────────────────────┘
         │                                  │
         ▼                                  │
┌────────────────┐    ┌──────────────┐      │
│ ELECTROSTATICS │    │  MAGNETISM   ├──────┘
└────────────────┘    └──────┬───────┘
                             │
                             ▼
                ┌──────────────────┐  ┌──────────────────┐
                │  MAGNETOSTATICS  │  │  MAGNETODYNAMICS │
                └──────────────────┘  └──────────────────┘
```

2.2.1. *Fundamental Physical Principles of the Electromagnetic Field*

Maxwell's equations are a set of partial differential equations, linear in space and time, applied to electromagnetic quantities. When electromagnetic fields interact with materials, the equations can assume nonlinear forms. This aspect of the field equations will be treated separately in later sections. To Maxwell's equations we attribute properties of "principles" or "postulates" based on the realization that experience has not contradicted them.

24   2. The Electromagnetic Field and Maxwell's Equations

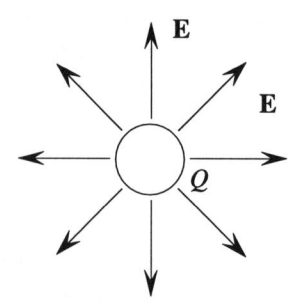

Figure. 2.1. Electric field due to a charge Q, or an equivalent charge distribution.

The electromagnetic quantities involved in Maxwell's equations are:

- The electric field intensity **E**
- The electric flux density or electric induction **D**
- The magnetic field intensity **H**
- The magnetic flux density or magnetic induction **B**
- The (surface) current density **J**
- The (volume) charge density ρ

In addition to these we define:

- The magnetic permeability μ
- The electric permittivity ε
- The electric conductivity σ

The significance of each of these quantities is considered next. To do so we assume here that the notions of electric charge and electric current are known.

2.2.1a. The Electric Field Intensity **E**

An electric charge or an assembly of electric charges Q, stationary in space, has the property of creating an electric quantity in space, called an electric field intensity **E** as shown in Figure 2.1.

The electric field intensity is a vector quantity and obeys the rules of vector fields. The manner in which the electric field intensity **E** is calculated will be shown in subsequent chapters.

2.2.1b. The Magnetic Field Intensity **H**

Suppose that the charge or assembly of charges in Figure 2.1 are not stationary in space but move at a given velocity. In this case, a magnetic field intensity **H** is generated as shown in Figure 2.2.

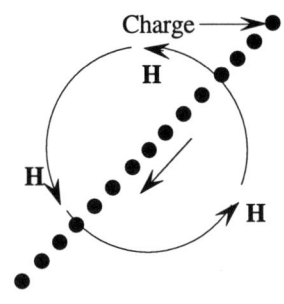

Figure 2.2. A moving charge and the magnetic field intensity it generates.

A moving charge or charges lead to the idea of the electric current. This is the ultimate result of the vector field **H**, whose calculation we will present shortly. If this movement of charges occurs in a conducting wire (as is the case in a majority of real situations), the electric field intensity is practically nonexistent since the electrons move between vacant positions in the atoms of the conducting material and the net sum of charges is essentially zero.

Later on we will see that variations in electric field intensities also generate magnetic field intensities.

2.2.1c. The Magnetic Flux Density **B** and the Magnetic Permeability μ

Since **B** is a vector, we can define a flux Φ crossing a surface S as

$$\Phi = \int_S \mathbf{B} \cdot d\mathbf{s}$$

The flux Φ is called the magnetic flux.

The permeability μ of a material expresses an intrinsic capacity of the material and indicates how much or how little it is susceptible to the passage of the magnetic flux. To show the meaning of μ in a simple manner, we introduce the relation

$$\mathbf{B} = \mu \mathbf{H}$$

Consider two media with identical geometries which have different permeabilities μ_1 and μ_2 such that $\mu_1 > \mu_2$ as in Figure 2.3. Suppose that, by external means, we create a magnetic field intensity **H** in both materials and that **H** is constant throughout the cross-section S. Then,

$$\mathbf{B}_1 = \mu_1 \mathbf{H} \quad \text{and} \quad \mathbf{B}_2 = \mu_2 \mathbf{H}$$

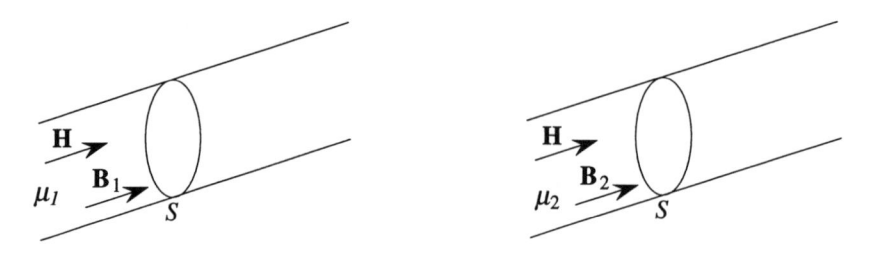

26   2. The Electromagnetic Field and Maxwell's Equations

Figure 2.3. Two materials of different permeabilities maintain different magnetic flux densities for the same field intensity.

The fluxes Φ_1 and Φ_2 are

$$\Phi_1 = B_1 S = \mu_1 H S \quad \text{and} \quad \Phi_2 = B_2 S = \mu_2 H S$$

We obtain

$$\frac{\Phi_1}{\Phi_2} = \frac{B_1}{B_2} = \frac{\mu_1}{\mu_2}$$

Note that the larger the permeability of the medium, the larger its magnetic flux density and the larger the flux that can pass through its cross-section S. In other words, **B** is called "magnetic flux density" or "induction" since this quantity expresses the capacity to induce flux within a medium, or the density of the flux in the medium. As in the example above, a high flux density is associated with a high permeability μ. Using the literal meaning of the terms "induction" and "permeability," we can say that a large flux is "induced" in a medium and that the medium is highly "permeable" to flux. The permeability of air is $\mu_0 = 4\pi \times 10^{-7}$ *Henry/meter*.

2.2.1d. The Electric Flux Density **D** and Electric Permittivity ε

There is a direct parallel between the pairs **D**,ε and **B**,μ, shown above. There are, however, a few salient differences between them. The first difference is the fact that ε varies little between materials, in contrast to the permeability μ. In useful dielectric materials ε varies no more than by a factor of 100 while the variation in μ can often attain factors of 10^4 or more. A second observation is that, in general, when solving problems with electric fields and electric flux densities, we are especially interested in the electric field intensity, while in magnetism the magnetic flux density assumes a predominant role in the analysis of the phenomenon. The permittivity of air is $\varepsilon_0 = 8.854 \times 10^{-12}$ *Farad/meter*.

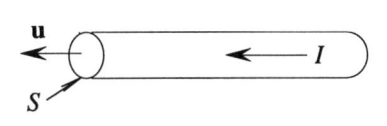

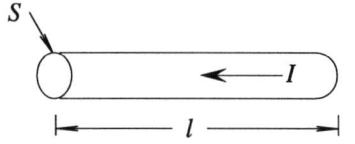

Figure 2.4a.

Figure 2.4b.

2.2.1e. The Surface Current Density J

Consider a straight conductor with a uniform cross-sectional area S and a current I crossing the cross-section in the direction indicated in Figure 2.4a.

We define a unit vector $\hat{\mathbf{u}}$ perpendicular to the surface S. The mean (average) surface current density crossing the area S is given by

$$J = \frac{I}{S}$$

If we assume that the surface S is small, the current density J can be considered to be constant over the surface. We define a vector which has a magnitude equal to J and its direction is given by $\hat{\mathbf{u}}$

$$\mathbf{J} = \hat{\mathbf{u}}\, J$$

The calculation of the flux of $\mathbf{J}$ crossing the surface S defines the current I, since

$$I = \int_S \mathbf{J} \cdot d\mathbf{s}$$

where ds is the differential area. In general, $\mathbf{J}$ varies throughout the cross-section.

2.2.1f. Volume Charge Density ρ

Assuming that a number of charges Q occupy a volume Vol, a uniform volume charge density is defined as

$$\rho = \frac{Q}{Vol}$$

Nonuniform charge distributions can similarly be taken into account observing that

$$Q = \int_v \rho\, dv$$

where dv is the volume differential.

2.2.1g. The Electric Conductivity σ

In general, when analyzing electric field problems we distinguish between two types of materials: dielectric or insulating materials and conducting materials. Insulating materials are characterized by their permittivity ε and their dielectric strength (which will be discussed later). Conducting materials are characterized by their conductivity σ. The later expresses the material's capacity to conduct electric current. We define the relation

$$\mathbf{J} = \sigma \mathbf{E}$$

which is Ohm's law in point form. In the case of a linear conductor of length l and cross-sectional area S, as in Figure 2.4b, this assumes the familiar form shown below.

We saw earlier that

$$J = \frac{I}{S}$$

and we state that the electric field in this case is

$$E = \frac{V}{l}$$

where V is the electric potential difference on this section of the conductor. Substituting these two relations in the equation $\mathbf{J} = \sigma \mathbf{E}$ we have

$$\frac{I}{S} = \sigma \frac{V}{l}$$

or

$$V = \frac{lI}{\sigma S}$$

where $l/\sigma S$ is the resistance R of the conductor. We now have $V = RI$ which is Ohm's law in a more common form. The difference between these two forms of expressing Ohm's law, is that the first, $\mathbf{J} = \sigma \mathbf{E}$, is a "point form" expression. It defines a quantity at any point in space. On the other hand, in the form $V = RI$, it is necessary to introduce the dimensions of the conductor (S and l) making this an "integral" form of Ohm's law.

The relation, $\mathbf{J} = \sigma \mathbf{E}$, as well as $\mathbf{B} = \mu \mathbf{H}$ and $\mathbf{D} = \varepsilon \mathbf{E}$ are called constitutive equations or constitutive relations and are used in addition to Maxwell's equations. They describe the relations between field quantities based on the electric and magnetic properties of materials (ε, μ, and σ).

2.2.2. Point Form of the Equations

Maxwell's four equations are as follows:

$$\nabla \times \mathbf{H} = \mathbf{J} + \frac{\partial \mathbf{D}}{\partial t} \tag{2.1}$$

$$\nabla \cdot \mathbf{B} = 0 \tag{2.2}$$

$$\nabla \times \mathbf{E} = -\frac{\partial \mathbf{B}}{\partial t} \tag{2.3}$$

$$\nabla \cdot \mathbf{D} = \rho \tag{2.4}$$

from these equations, we can define a fifth relation. Applying the divergence on both sides of Eq. (2.1) gives

$$\nabla \cdot (\nabla \times \mathbf{H}) = \nabla \cdot \mathbf{J} + \nabla \cdot \frac{\partial \mathbf{D}}{\partial t}$$

or, using the fact that $\nabla \cdot (\nabla \times \mathbf{H}) = 0$, we have

$$0 = \nabla \cdot \mathbf{J} + \frac{\partial}{\partial t}(\nabla \cdot \mathbf{D})$$

Utilizing Eq. (2.4) gives

$$\nabla \cdot \mathbf{J} = -\frac{\partial \rho}{\partial t} \tag{2.5}$$

This equation is called the electrical continuity equation. We observe that in general, $\partial \rho / \partial t$ is zero and therefore we normally obtain $\nabla \cdot \mathbf{J} = 0$. This is significant in that the flux of the vector or, similarly, the conduction current is conservative. In other words, the current entering a given volume is equal to the current leaving the volume. In fact, in practically all electromagnetic devices, the current injected into the device is equal to the current leaving it. When this does not happen, there is an accumulation of charges in the device, or a certain amount of charge is extracted from the device. This is shown schematically in Figure 2.5.

Assuming that $I_2 < I_1$, we have an accumulation of charges in the volume V. As a consequence, $\partial \rho / \partial t \neq 0$ in this volume. The negative sign of the expression

$$\nabla \cdot \mathbf{J} = -\frac{\partial \rho}{\partial t}$$

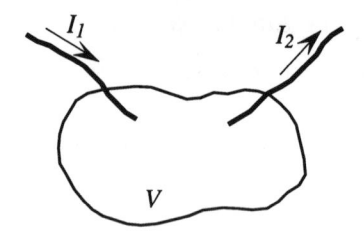

Figure 2.5. Accumulation of charges in a volume due to nonzero divergence of the current density.

signifies that the sum of the flux of **J** is negative ($I_2<I_1$) and it augments the volume charge density ρ with time. In fact, applying integration over the volume and utilizing the divergence theorem gives:

$$\int_V \nabla \cdot \mathbf{J} \, dv = \oint_{S(V)} \mathbf{J} \cdot d\mathbf{s} = -\int_V \frac{\partial \rho}{\partial t} \, dv$$

or

$$-I_1 + I_2 = -\frac{dQ}{dt}$$

The first term is therefore negative, resulting in $dQ/dt>0$; thus, an augmentation of the charge in the volume occurs with time.

2.2.2a. Equation (2.1)

The equation

$$\nabla \times \mathbf{H} = \mathbf{J} + \frac{\partial \mathbf{D}}{\partial t}$$

expresses the manner by which a magnetic field can create a split into conduction current (associated with **J**) and a time variation of the electric flux density (associated with $\partial \mathbf{D}/\partial t$). We assume first the situation in Figure 2.6, where there is no electric flux density, or, alternatively, the electric flux density is constant in time.

Now the equation is $\nabla \times \mathbf{H} = \mathbf{J}$. As we have seen in the previous chapter, **H** and **J** are connected by a rotation or curl relationship. The geometric relation between these quantities is demonstrated in Figure 2.6. The flux of the vector **J** is the conduction current. It is in general the dominant term in the relation while the term $\partial \mathbf{D}/\partial t$, which will be discussed in more detail in subsequent paragraphs, is small.

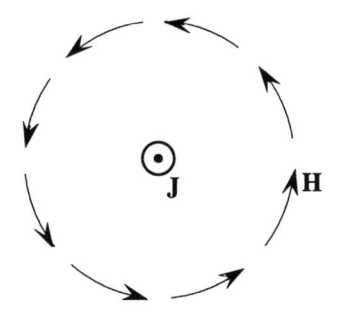

Figure 2.6. Relation between conduction current density and magnetic field intensity.

2.2.2b. Equation (2.2)

The equation

$$\nabla \cdot \mathbf{B} = 0$$

signifies, as shown in the previous chapter, that the magnetic flux is conservative. To understand this we can say that the magnetic flux entering a volume is equal to the magnetic flux leaving the volume. This relation corresponds to a condition which allows understanding of the field behavior and serves, in various cases, a means of determining the magnetic field intensity. However, Eq. (2.1) also established a relation between the magnetic field intensity **H** and **J**, and the same relation permits the determination of **H** as a function of **J** in a large number of practical cases.

2.2.2c. Equation (2.3)

The equation

$$\nabla \times \mathbf{E} = -\frac{\partial \mathbf{B}}{\partial t}$$

is analogous to Eq. (2.1), showing that the time derivative of the magnetic flux density is capable of generating an electric field intensity **E**.

The geometrical situation connecting these quantities is shown in Figure 2.7. Assuming that **B** increases as it comes out of the plane of Figure 2.7, the electric field intensity **E** is in the direction shown in Figure 2.7.

32   2. The Electromagnetic Field and Maxwell's Equations

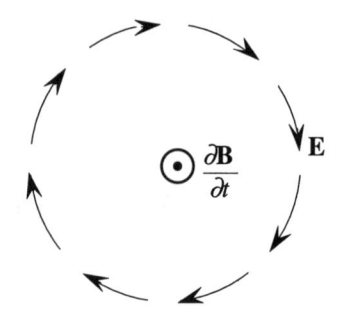

Figure 2.7. Relation between the time derivative of the magnetic flux density and the electric field intensity.

2.2.2d Equation (2.4)

The observation that the divergence of **D**

$$\nabla \cdot \mathbf{D} = \rho$$

is not zero, demonstrates that the flux of the vector **D** is not conservative. We can easily imagine a volume in which there is a difference between the electric fluxes entering and leaving the volume. This situation is shown in Figure 2.8 where an electric charge is located at the center of a sphere. The flux traversing the volume is outward oriented. **D** and ρ are related through the divergence, according to the relations shown in Chapter 1. The geometrical relation between the two quantities is shown in Figure 2.8. The flux of the vector **D** traversing the surface that encloses the volume V of the sphere is nonzero.

2.2.3. *The Equations in Vacuum*

In vacuum, there are no molecules and there can be no physical support for charges and conduction currents. We can assume that **J**=0 and ρ=0. Similarly, $\varepsilon=\varepsilon_0$ and $\mu=\mu_0$. Maxwell's equations assume the following form:

$$\nabla \times \mathbf{H} = \frac{\partial \mathbf{D}}{\partial t}$$

$$\nabla \cdot \mathbf{B} = 0$$

$$\nabla \times \mathbf{E} = -\frac{\partial \mathbf{B}}{\partial t}$$

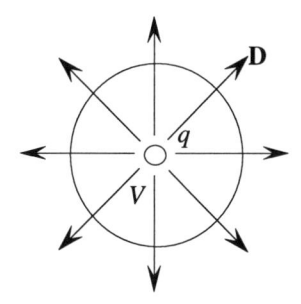

Figure 2.8. The nonconservative nature of the electric flux.

$$\nabla \cdot \mathbf{D} = \rho$$

In addition, the permittivity and permeability can be considered to be constants whereby the equations become

$$\nabla \times \mathbf{H} = \varepsilon_0 \frac{\partial \mathbf{E}}{\partial t} \tag{2.6}$$

$$\nabla \cdot \mathbf{H} = 0 \tag{2.7}$$

$$\nabla \times \mathbf{E} = -\mu_0 \frac{\partial \mathbf{H}}{\partial t} \tag{2.8}$$

$$\nabla \cdot \mathbf{E} = 0 \tag{2.9}$$

Note that the flux of **H** and that of **E** are conservative. In other words, in the high-frequency domain, and in particular for electromagnetic waves, the time derivative of the electric field intensity is responsible for the generation of **H** since, under the conditions outlined above, there is no current. Conversely, the time derivative of **H** generates **E**. The term $\partial \mathbf{D}/\partial t$, which was proposed by Maxwell, is the ultimate proof of the existence of electromagnetic waves which were later realized by Hertz. This term has a second connotation, that of "displacement currents," as will be seen below.

If we consider only static phenomena, the equations assume the following form:

$$\nabla \times \mathbf{H} = 0$$

$$\nabla \cdot \mathbf{H} = 0$$

$$\nabla \times \mathbf{E} = 0$$

$$\nabla \cdot \mathbf{E} = 0$$

Applying the curl to the first equation gives

$$\nabla \times (\nabla \times \mathbf{H}) = 0$$

Using the identity in Eq. (1.23)

$$\nabla \times (\nabla \times \mathbf{H}) = -\nabla^2 \mathbf{H} + \nabla(\nabla \cdot \mathbf{H}) = 0$$

with $\nabla \cdot \mathbf{H} = 0$, we have

$$\nabla^2 \mathbf{H} = 0$$

or

$$\nabla^2 H_x = 0, \quad \nabla^2 H_y = 0, \quad \nabla^2 H_z = 0$$

These equations express the diffusion of the magnetic field in space. It is important to note that within the domain under consideration there are no currents or charges and the generation of the magnetic field can be seen as originating from the conditions imposed on the boundaries of the domain. The equation $\nabla^2 \mathbf{E} = 0$ is obtained in an analogous fashion and the observations above are valid for this equation as well.

2.2.4. The Equations in Media with $\varepsilon = \varepsilon_0$ and $\mu = \mu_0$

In this case, $\mathbf{J}$ and ρ can be nonzero, but $\varepsilon = \varepsilon_0$ and $\mu = \mu_0$. The equations are:

$$\nabla \times \mathbf{H} = \mathbf{J} + \varepsilon_0 \frac{\partial \mathbf{E}}{\partial t} \qquad (2.10)$$

$$\nabla \cdot \mathbf{H} = 0 \qquad (2.11)$$

$$\nabla \times \mathbf{E} = -\mu_0 \frac{\partial \mathbf{H}}{\partial t} \qquad (2.12)$$

$$\nabla \cdot \mathbf{E} = \frac{\rho}{\varepsilon_0} \qquad (2.13)$$

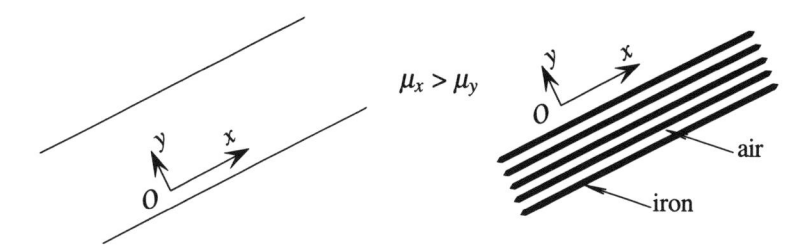

Figure 2.9. Two types of anisotropic materials. The material on the left has grain-oriented structure while that on the right is made of thin insulated sheets.

2.2.5. The Equations in General Media

It is possible to apply Maxwell's equations in various situations and in combinations of different materials. However, instead of discussing all possible applications, we prefer to present the equations through a general situation. For this purpose it is necessary to introduce the concept of magnetic anisotropy. Consider a material whose magnetic permeability is dominant in a certain direction. One such material is a sheet of iron with grain-oriented structure or thin plates made of sheet metal which form, for example, the core of a transformer, as in Figure 2.9.

It is reasonable to assume that in both cases, the magnetic flux flows with more ease in the direction Ox. In the first case, this is due to the orientation of the grains and in the second due to the presence of small gaps between the layers of sheet metal. Assuming a field intensity **H** whose components H_x and H_y are equal to H and if μ_x and μ_y are the permeabilities in the directions Ox and Oy respectively, we have

$$B_x = \mu_x H$$

and

$$B_y = \mu_y H$$

We note that B_x is larger than B_y. In this case, there is an angle between **H** and **B**. If $H_x = H_y$, **H** forms a 45° angle with Ox. At the same time, **B** forms an angle different than 45° since B_x and B_y are different. We conclude that the relation

$$\mathbf{B} = \mu \mathbf{H}$$

where μ is a scalar, is not general since it does not satisfy the cases above. Because of this, we introduce the concept of the "permeability tensor" denoted by $\|\mu\|$. In matrix algebra, a vector, for example **B**, is expressed as

2. The Electromagnetic Field and Maxwell's Equations

$$\mathbf{B} = \begin{bmatrix} B_x \\ B_y \\ B_z \end{bmatrix}$$

The tensor $\|\mu\|$ is a 3×3 matrix

$$\|\mu\| = \begin{bmatrix} \mu_x & 0 & 0 \\ 0 & \mu_y & 0 \\ 0 & 0 & \mu_z \end{bmatrix}$$

where we have assumed, for the moment, that the off-diagonal terms are zero or, that we have a diagonal tensor. The general expression $\mathbf{B}=\|\mu\|\mathbf{H}$ is, in matrix form,

$$\begin{bmatrix} B_x \\ B_y \\ B_z \end{bmatrix} = \begin{bmatrix} \mu_x & 0 & 0 \\ 0 & \mu_y & 0 \\ 0 & 0 & \mu_z \end{bmatrix} \begin{bmatrix} H_x \\ H_y \\ H_z \end{bmatrix}$$

By appropriate matrix operations, we can write

$$B_x = \mu_x H_x, \qquad B_y = \mu_y H_y, \qquad B_z = \mu_z H_z$$

We observe that when the material is isotropic, or if $(\mu_x=\mu_y=\mu_z=\mu)$ the equation $\mathbf{B}=\|\mu\|\mathbf{H}$ assumes the scalar form $\mathbf{B}=\mu\mathbf{H}$. We also observe that when the nondiagonal terms of $\|\mu\|$ are nonzero, there is an interdependency between variables. While in the example above, B_x depended only on H_x, now it may depend on all three components of $\mathbf{H}$. In general, if the tensor $\|\mu\|$ is not a diagonal tensor, we can write a more complex relation as

$$B_x = \mu_x H_x + \mu_{xy} H_y + \mu_{xz} H_z$$

Besides the concept of anisotropy, which complicates the study of magnetic materials, we introduce another phenomenon, frequently encountered in electromagnetic devices. In these devices, the magnetic permeability is not constant but depends on the particular value of $\mathbf{H}$ in the magnetic material in question. This phenomenon is called "nonlinearity." The general relation between $\mathbf{B}$ and $\mathbf{H}$ is now

$$\mathbf{B} = \|\mu(H)\|\mathbf{H}$$

The anisotropy concept can be extended in a similar manner to the electric permittivity ε, as

$$\mathbf{D} = \|\varepsilon\|\mathbf{E}$$

and to conductivity σ, for which the relation is

$$\mathbf{J} = \|\sigma\|\mathbf{E}$$

In both cases, the nonlinear behavior is normally negligible.

Maxwell's equations with the constitutive relations for a general material are:

$$\nabla \times \mathbf{H} = \mathbf{J} + \frac{\partial \mathbf{D}}{\partial t} \qquad (2.14)$$

$$\nabla \cdot \mathbf{B} = 0 \qquad (2.15)$$

$$\nabla \times \mathbf{E} = -\frac{\partial \mathbf{B}}{\partial t} \qquad (2.16)$$

$$\nabla \cdot \mathbf{D} = \rho \qquad (2.17)$$

$$\mathbf{B} = \|\mu\|\mathbf{H} \qquad (2.18)$$

$$\mathbf{D} = \|\varepsilon\|\mathbf{E} \qquad (2.19)$$

$$\mathbf{J} = \|\sigma\|\mathbf{E} \qquad (2.20)$$

2.2.6. *The Integral Form of Maxwell's Equations*

Maxwell's equations in differential form are extremely convenient for mathematical manipulation since they are written in compact form, in terms of the gradient, the divergence, and the curl, whose properties are simple and well understood. With little difficulty, by direct substitution we can obtain second-order partial differential equations in terms of any desired electromagnetic quantity. However, for the analytic solution of problems where, for example, we wish to understand the generation of $\mathbf{H}$ by $\mathbf{J}$, it is more convenient to utilize Maxwell's equations in integral form. In this case it is necessary to associate the equations with volumes or surfaces over which the integration can be performed. The use of the divergence and Stokes' theorems, as fundamental relations allow exact manipulation of the equations in terms of the divergence and curl. Using these relations, we now present Maxwell's four equations in integral form.

2.2.6a. Equation (2.1)

Using the same notation as in section 2.2.2., we write the first equation

$$\nabla \times \mathbf{H} = \mathbf{J} + \frac{\partial \mathbf{D}}{\partial t}$$

We define a surface S where we wish to study the currents and magnetic fields. We call C a contour surrounding this surface. Applying a surface integral on both sides of the equation above gives

$$\int_S (\nabla \times \mathbf{H}) \cdot d\mathbf{s} = \int_S \mathbf{J} \cdot d\mathbf{s} + \int_S \frac{\partial \mathbf{D}}{\partial t} \cdot d\mathbf{s}$$

Using Stokes' theorem gives

$$\oint_C \mathbf{H} \cdot d\mathbf{l} = \int_S \mathbf{J} \cdot d\mathbf{s} + \int_S \frac{\partial \mathbf{D}}{\partial t} \cdot d\mathbf{s}$$

The left side of this equation is the circulation of the field intensity $\mathbf{H}$ along C. The right side includes two terms. The first is called the "conduction current" I.

$$I = \int_S \mathbf{J} \cdot d\mathbf{s}$$

which represents the currents traversing the surface S. These currents, in general, flow in conductors. The second term is called the "displacement current" I_d.

$$I_d = \int_S \frac{\partial \mathbf{D}}{\partial t} \cdot d\mathbf{s}$$

This term can have two connotations. The first relates to electromagnetic waves and represents the time variation of the electric field ($\varepsilon_0 \partial \mathbf{E} / \partial t$) which generates the magnetic field. The second relates to particular devices called capacitors, that will be discussed shortly.

As an example for the use of this equation, consider an infinitely long wire carrying a current I. Assume that we wish to calculate the magnetic field intensity generated by the wire at a distance r from the wire, as shown in Figure 2.10a.

In Figure 2.10b we present a second view of the wire in which the current comes out of the plane of the figure. For the surface S we choose a circle of radius r. The field intensity $\mathbf{H}$ can now be obtained at any point on the circumference. In this case only a conduction current I exists. The integral equation takes the form

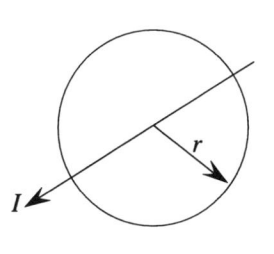

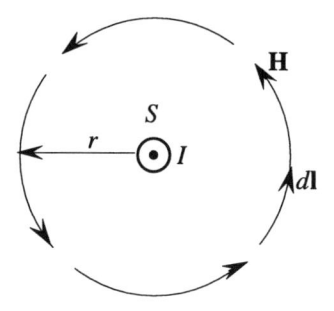

Figure 2.10a

Figure 2.10b

$$\oint_C \mathbf{H} \cdot d\mathbf{l} = \int_S \mathbf{J} \cdot d\mathbf{s}$$

Since **H** and **J** are related through a curl, **H** is tangential and, for homogeneous material properties, we can assume that its magnitude for a constant r is constant. Because **H** and $d\mathbf{l}$ are collinear and in the same direction (Figure 2.10b), the equation above can be written as

$$\oint_C \mathbf{H} \cdot d\mathbf{l} = \oint_C H dl = H \oint_C dl = H 2\pi r$$

For the second part of the equation, we can divide S in a surface S_f of the wire where $J \neq 0$ and the remainder of the surface $S - S_f$ where $J = 0$. It is evident that the integration is nonzero in the section of the wire that has a current I. We obtain therefore

$$\oint_C \mathbf{H} \cdot d\mathbf{l} = I$$

This equation is called "Ampere's law." In this particular case, evaluating the left side of the equation gives

$$H = \frac{I}{2\pi r}$$

We look now at an example involving a device made of two parallel conducting plates separated by an insulating material (see Figure 2.11a). This type of device, known as a "capacitor" will be discussed in detail in the following chapter.

Assuming that at time $t=0$, the switch is closed, a current I_d is generated. This current cannot cross the insulator, and its existence causes accumulation of charges on one of the plates. As a consequence, a corresponding opposite charge on the other plate is generated.

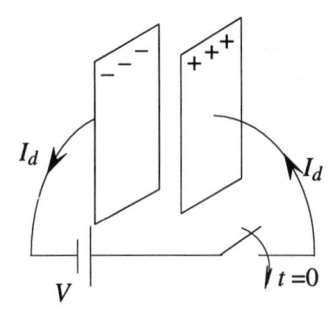

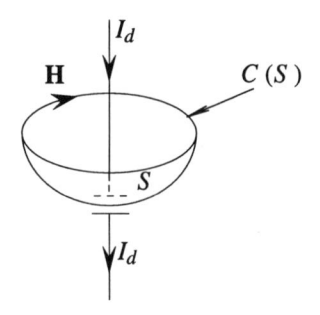

Figure 2.11a. A parallel plate capacitor under transient excitation.

Figure 2.11b. "Proof" for the existence of displacement currents.

This situation is represented schematically in Figure 2.11b. Having a current in the wire, generates a magnetic field intensity H outside the wire. At the same time there is no reason we cannot pass a surface S, bounded by a contour C through the insulating material where $J=0$ and the term $\partial \mathbf{D}/\partial t$ (whose integration furnishes the displacement current) is responsible for the generation of $\mathbf{H}$. This current is called displacement current because it is calculated as a time derivative and this, in turn, is normally associated with the notion of displacement.

2.2.6b. Equation (2.2)

The equation $\nabla \cdot \mathbf{B}=0$ is associated with a volume in which $\mathbf{B}$ is defined. $S(V)$ is the surface that encloses V. Utilizing the divergence theorem we get

$$\int_V \nabla \cdot \mathbf{B}\, dv = \oint_S \mathbf{B} \cdot d\mathbf{s} = 0$$

signifying that the total flux of $\mathbf{B}$ passing through a closed surface is zero, or, alternatively, that the magnetic flux is conservative.

2.2.6c. Equation (2.3)

The equation $\nabla \times \mathbf{E}=-\partial \mathbf{B}/\partial t$ is certainly the more complex of Maxwell's equations. A brief example to its use is shown here. This example will be discussed in more detail in Chapter 5.

Consider a surface S on which $\mathbf{E}$ and $\mathbf{B}$ are defined, and apply the integration

$$\int_S (\nabla \times \mathbf{E}) \cdot d\mathbf{s} = \int_S -\frac{\partial \mathbf{B}}{\partial t} \cdot d\mathbf{s}$$

Through use of Stokes' theorem we get

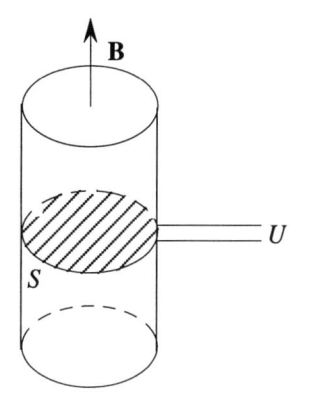

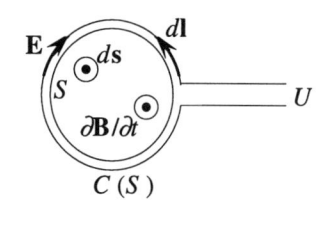

Figure 2.12a. Induced *emf* in a loop due to time variation of the magnetic field.

Figure 2.12b. The relation between the magnetic and electric fields.

$$\oint_C \mathbf{E} \cdot d\mathbf{l} = \int_S -\frac{\partial \mathbf{B}}{\partial t} \cdot d\mathbf{s}$$

where C is the contour that limits the surface S.

We look now at the application of the expression above in an example as defined in Figures 2.12a and 2.12b. The geometry in Figure 2.12a consists of a core of a magnetic material where we assume that **B** is constant on the surface S. Furthermore, **B** is assumed to be time dependent. For the surface S we use an appropriate section of the core. A loop of conducting material encloses the core (electrically isolated from the core) such that it forms a contour $C(S)$. Making an analogy between the equations $\nabla \times \mathbf{H} = \mathbf{J}$ and $\nabla \times \mathbf{E} = -\partial \mathbf{B}/\partial t$ we note that the variation of **B** generates an electric field intensity **E** in the wire loop, in the same fashion that **J** generates **H**. The electric field intensity has a rotation (curl) equal to $-\partial \mathbf{B}/\partial t$. First note that assuming the direction of $d\mathbf{s}$ as in Figure 2.12b, the direction of $d\mathbf{l}$ must be as shown. With the direction of $-\partial \mathbf{B}/\partial t$ as shown in Figure 2.12b, we have

$$-\frac{\partial \mathbf{B}}{\partial t} \cdot d\mathbf{s} < 0 \quad \text{and} \quad \mathbf{E} \cdot d\mathbf{l} < 0$$

which makes **E** opposite in direction to $d\mathbf{l}$.

The circulation of **E** along C leads to an electromotive force (*emf*), detectable by a voltmeter as a potential U (this concept will be discussed later in detail). The electromotive force is

$$U = \oint_C \mathbf{E} \cdot d\mathbf{l}$$

Considering the other term,

$$-\int_S \frac{\partial \mathbf{B}}{\partial t} \cdot d\mathbf{s}$$

because **B** is only dependent on time and not on its position on the surface S, we conclude that $\partial \mathbf{B}/\partial t = d\mathbf{B}/dt$. Since the integration over the surface S and the time derivative are independent operations, we have

$$-\int_S \frac{\partial \mathbf{B}}{\partial t} \cdot d\mathbf{s} = -\frac{d}{dt}\int_S \mathbf{B} \cdot d\mathbf{s} = -\frac{d\Phi}{dt}$$

Equating the two terms we obtain the expression

$$U = -\frac{d\Phi}{dt}$$

which is known as "Faraday's law" since Faraday was the first to show evidence of this phenomenon (in 1831). According to this law, "the time variation of the flux generates an electromotive force in the loop that bounds the surface through which the flux varies." We note here that Faraday's law corresponds to one of Maxwell's equations.

2.2.6d. Equation (2.4)

To treat equation (2.4) ($\nabla \cdot \mathbf{D} = \rho$) we define a volume, enclosed by a surface $S(V)$ and assume that a vector flux density **D** is defined in V and that there are charges inside this volume. Integrating both sides of the equation over the volume V gives

$$\int_V \nabla \cdot \mathbf{D} \, dv = \int_V \rho \, dv$$

Using the divergence theorem we obtain

$$\oint_S \mathbf{D} \cdot d\mathbf{s} = \int_V \rho \, dv$$

The total charge in the volume is Q:

$$Q = \int_V \rho \, dv$$

Therefore,

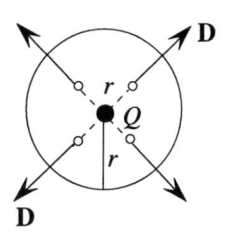

Figure 2.13a.

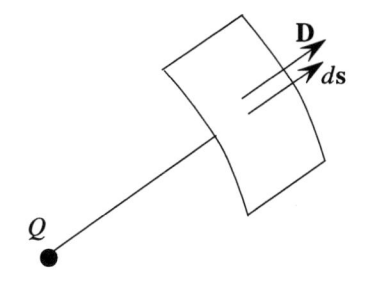

Figure 2.13b.

$$\oint_S \mathbf{D} \cdot d\mathbf{s} = Q$$

This expression shows that the total flux of **D** crossing a surface that encloses a volume is equal to the charge contained in this volume. The equation above is called "Gauss' theorem."

As an example to the use of this equation consider the case of a point charge in space, as in Figure 2.13a. The volume V is a sphere of radius r. Because **D** and r are related through the divergence, **D** is a radial vector in relation to the charge Q and **ds** is also a radial vector as shown in Figure 2.13b,

$$\oint_S \mathbf{D} \cdot d\mathbf{s} = \oint_S D ds = D \oint_S ds = 4\pi r^2 D$$

The magnitude of **D** is identical at any point on $S(V)$. It is therefore independent of $d\mathbf{s}$. Using the relations for vacuum ($\mathbf{D}=\varepsilon_0 \mathbf{E}$), and equating the two results, we obtain

$$E = \frac{Q}{4\pi\varepsilon_0 r^2}$$

2.3. Approximations to Maxwell's Equations

The complete set of Maxwell's equations is:

$$\nabla \times \mathbf{H} = \mathbf{J} + \frac{\partial \mathbf{D}}{\partial t}$$

$$\nabla \cdot \mathbf{B} = 0$$

44   2. The Electromagnetic Field and Maxwell's Equations

$$\nabla \times \mathbf{E} = -\frac{\partial \mathbf{B}}{\partial t}$$

$$\nabla \cdot \mathbf{D} = \rho$$

This is a closed system, in which there is an interdependency between all variables in the equations. They have, therefore, a unique solution.

For practical purposes, it is often useful to simplify these equations based on the conditions under which we anticipate operating. One of the most important simplifications we can make is to neglect the displacement current density term $\partial \mathbf{D}/\partial t$.

For static cases ($\partial \mathbf{D}/\partial t = 0$) this current density disappears. At frequencies of the order of 10^5 *Hz*, and with an electric field intensity of the order of 10^5 *V/m*, the displacement current density is of the order of 10^{-4} *A/mm²*. This is much smaller than conduction current densities which are of the order of 1 *A/mm²*. For this reason, we can neglect parasitic capacitances in wires. We cannot neglect this term if, in the domain under consideration, we have large capacitors and large electric fields. In these cases, the displacement current can be large. However, this constitutes a sufficiently particular case to be treated separately.

In general, at low frequencies, the equations can be written in the following form

$$\nabla \times \mathbf{H} = \mathbf{J}$$

$$\nabla \cdot \mathbf{B} = 0$$

$$\nabla \times \mathbf{E} = -\frac{\partial \mathbf{B}}{\partial t}$$

$$\nabla \cdot \mathbf{D} = \rho$$

The neglection of the term $\partial \mathbf{D}/\partial t$ in the first equation has important significance beyond the mere neglection of a small term: this operation allows decoupling of the system of equations into two parts from which we obtain two systems of equations. The two systems can then be studied separately.

The first system, consists of the equations below:

$$\nabla \times \mathbf{H} = \mathbf{J} \tag{2.21}$$

$$\nabla \cdot \mathbf{B} = 0 \tag{2.22}$$

$$\nabla \times \mathbf{E} = -\frac{\partial \mathbf{B}}{\partial t} \tag{2.23}$$

$$\mathbf{B} = \|\mu\|\mathbf{H} \qquad (2.24)$$

$$\mathbf{J} = \|\sigma\|\mathbf{E} \qquad (2.25)$$

which have the properties needed to treat magnetic problems. If there is no time variation of quantities $(\partial \mathbf{B}/\partial t=0)$ we are in the magnetostatic domain. Otherwise the fields are magnetodynamic.

The second system, with the remaining equations is

$$\nabla \cdot \mathbf{D} = \rho \qquad (2.26)$$

$$\mathbf{D} = \|\varepsilon\|\mathbf{E} \qquad (2.27)$$

and constitutes the electrostatic system.

It is important to point out that the electric field intensities that appear in Eq. (2.23) and in Eq. (2.27) have a totally different nature. The first is generated by a magnetic effect ($\partial \mathbf{B}/\partial t$, the time variation of $\mathbf{B}$) while the second is a result of the presence of static charges.

The approximate forms of Maxwell's equations shown above permit study of electrostatic and magnetic phenomena separately. It is, however, very important to understand the conditions under which this separation is valid and its limitations.

2.4. Units

The MKS system of units is used throughout this work. Although many still use the cgs system for electromagnetic quantities, the international system of units (MKS) is more or less standard in electromagnetics and is expected to become more so in the future.

The MKS units of principal electromagnetic quantities are given in Table 2.1. The table also gives the abbreviations for the various units and the defining dimensional equation where applicable.

2. The Electromagnetic Field and Maxwell's Equations

Table 2.1. Principal electromagnetic units.

Quantity	Dimensional Equation	Unit	Abbreviation
Magnetic field intensity **H**	$H=I/L$	Ampere/meter	A/m
Magnetic flux density **B**		Tesla	T
Electric field intensity **E**	$E=V/L$	Volt/meter	V/m
Electric flux density **D**		Coulomb/m^2	C/m^2
Magnetic flux Φ	$\Phi=BS$	Wb=Tesla.meter^2	Wb
Magnetic permeability μ		Henry/meter	H/m
Electric permittivity ε		Farad/meter	F/m
Electric conductivity σ		1/(Ohms.meter) or Siemens/meter	$(\Omega m)^{-1}$ or S/m
Inductance L *	$L=\Phi/I$	Webber/Ampere = Henry	H
Capacitance C *	$C=Q/V$	Coulomb/Volt = Farad	F

(*) These concepts will be discussed in later sections

3
Electrostatic Fields

3.1. Introduction

With the approximation to Maxwell's equations presented in the previous chapter, it is possible to decouple the equations in two groups. One group describes electrostatic phenomena, the other magnetic phenomena. In this chapter we look at the first group, and at the electrostatic fields it describes.

The fundamental equations in electrostatics are:

$$\nabla \cdot \mathbf{D} = \rho \qquad (3.1)$$

$$\mathbf{D} = \|\varepsilon\| \mathbf{E} \qquad (3.2)$$

We make here the assumption that in dielectric materials we can use the relation $\mathbf{D}=\varepsilon\mathbf{E}$, where ε is a scalar. The concept of relative permittivity ε_r is defined as:

$$\varepsilon_r = \frac{\varepsilon}{\varepsilon_0}$$

where ε is the permittivity of a material and ε_0 is the permittivity of free space ($\varepsilon_0 = 8.854 \times 10^{-12} F/m$). It is normally sufficient to specify the permittivity of a material through ε_r; when it becomes necessary to use the permittivity ε of the materials, we use the relation $\varepsilon = \varepsilon_0 \varepsilon_r$.

3.2. The Electrostatic Charge

Electrostatic charge is of considerable interest in Electrical Engineering. However, we will not dwell here on its applications but rather on its use in defining some concepts of major importance and utilization relevant beyond the electrostatic field problem.

3.2.1. *The Electric Field*

Consider a charge or an assembly of static charges q. It was shown in the previous chapter that q can generate an electric field in the space around it. We have also seen that the equation $\nabla \cdot \mathbf{D} = \rho$, becomes, in integral form, the expression for "Gauss' theorem"

$$\oint_S \mathbf{D} \cdot d\mathbf{s} = q \quad (3.3)$$

where S is the surface enclosing the volume v, which contains the charge q. In this case, we choose for the volume v a sphere with radius r. Assuming free space ($\varepsilon = \varepsilon_0$), and applying Gauss' theorem, we obtain (see section 2.2.6d.)

$$E = \frac{q}{4\pi\varepsilon_0 r^2} \quad (3.4)$$

To understand the meaning of the vector $\mathbf{E}$, we define the vector $\mathbf{r}$ as $\mathbf{r} = \mathbf{P} - \mathbf{O}$ as shown in Figure 3.1.

The unit vector $\hat{\mathbf{u}}$ in the direction of $\mathbf{r}$ can be obtained from $\hat{\mathbf{u}} = \mathbf{r}/r$. With this notation, $\mathbf{E}$, which is associated with the charge q through the divergence, can be written in the following form:

$$\mathbf{E} = \hat{\mathbf{u}} E$$

$$\mathbf{E} = \frac{q\mathbf{r}}{4\pi\varepsilon_0 r^3} \quad (3.5)$$

3.2.2. *Force on an Electric Charge*

It has been observed experimentally that the force $\mathbf{F}$ exerted on a charge q' under the action of an electric field intensity $\mathbf{E}$ is given by the expression

$$\mathbf{F} = q'\mathbf{E} \quad (3.6)$$

Assuming that the electric field intensity $\mathbf{E}$ is generated by a charge q, as in Figure 3.2, we have

$$\mathbf{F} = \frac{qq'}{4\pi\varepsilon_0 r^3}\mathbf{r} \quad (3.7)$$

The magnitude of $\mathbf{F}$ is

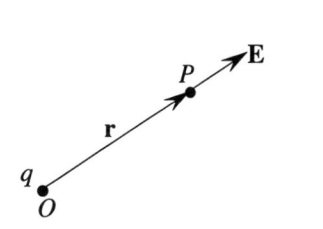

Figure 3.1. 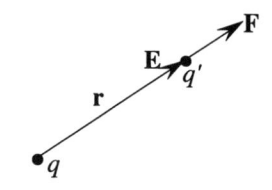   Figure 3.2.

$$F = \frac{qq'}{4\pi\varepsilon_0 r^2}$$

This expression is known as "Coulomb's law," according to which the force is directly proportional to the product of the charges, and inversely proportional to the square of the distance between them.

3.2.3. *The Electric Scalar Potential V*

From the force defined in the previous paragraph it is possible to calculate the work dW performed by this force by the expression

$$dW = \mathbf{F} \cdot d\mathbf{l}$$

where $d\mathbf{l}$ is the displacement of the point on which the charge q applies the force $\mathbf{F}$. We are particularly interested in the work per unit of the charge q'. Viewing the charge in this manner, allows the definition of an expression, related to work, but with the idea of representing the capacity of the charge to perform work. We have

$$\frac{dW}{q'} = \frac{\mathbf{F}}{q'} \cdot d\mathbf{l}$$

Defining the work per unit charge as V and using $\mathbf{F} = q'\mathbf{E}$ gives

$$dV = -\mathbf{E} \cdot d\mathbf{l}$$

The reason for the negative sign will become evident shortly. Assuming that the electric field intensity varies along a trajectory, the energy per unit charge in moving the charge from l_1 to l_2 can be expressed as

$$V_2 - V_1 = -\int_{l_1}^{l_2} \mathbf{E} \cdot d\mathbf{l}$$

50   3. Electrostatic Fields

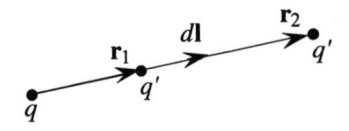

Figure 3.3.

Consider now an example consisting of a point charge, as shown in Figure 3.3. We wish to move the test charge q' from a position $\mathbf{r_1}$ to a position $\mathbf{r_2}$. Moving q' towards q and denoting V_1 the potential at $\mathbf{r_1}$ and V_2 the potential at $\mathbf{r_2}$ gives

$$V_2 - V_1 = -\int_{r_1}^{r_2} \frac{q}{4\pi\varepsilon_0 r^3} \mathbf{r}\cdot d\mathbf{l}$$

Choosing the path $dl=dr$ gives: $\mathbf{r}\cdot d\mathbf{l}=rdl=rdr$. Upon evaluation of this integral we have

$$V_2 - V_1 = \frac{q}{4\pi\varepsilon_0}\left(\frac{1}{r_2} - \frac{1}{r_1}\right) \qquad (3.8)$$

We note that the negative sign introduced in the expression $dV=-\mathbf{E}\cdot d\mathbf{l}$, comes from the fact that the right-hand side of this expression is negative. This implies that V_1 is larger than V_2. Effectively, the adopted convention is that the potential close to the source is larger than the potential at a larger distance. The electric field points from the higher potential to the lower. The electric field intensity is, therefore, divergent in relation to the charge q.

On the other hand, the value of the integral depends only on the initial and final points on the trajectory. In the case of a closed contour we have

$$\oint_C \mathbf{E}\cdot d\mathbf{l} = 0 \qquad (3.9)$$

This can be easily verified using Figure 3.4a. Note that where $\mathbf{E}$ is perpendicular to $d\mathbf{l}$ we have $\mathbf{E}\cdot d\mathbf{l}=0$. On the other two parts of the path, the direction is radial, where $d\mathbf{l}$ is in the direction of $\mathbf{E}$. On one part of the path $\mathbf{E}\cdot d\mathbf{l}>0$ while on the other the magnitude is the same but negative in sign. This occurs on any arbitrary contour, as shown in Figure 3.4b, where any segment $d\mathbf{l}$ can be decomposed into a radial and a tangential component. This decomposition allows simple evaluation of the scalar product $\mathbf{E}\cdot d\mathbf{l}$, and produces a zero sum whenever the contour is closed.

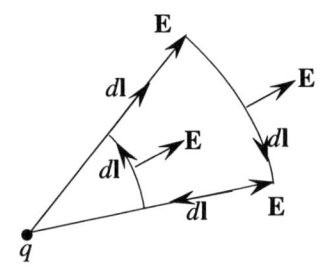

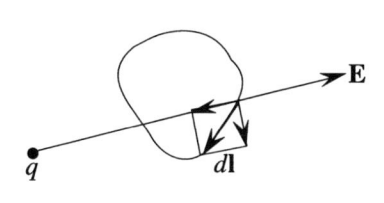

Figure 3.4a. A closed contour in an electric field. The closed contour integral of **E** is zero.

Figure 3.4b. Decomposition of an arbitrary contour into radial and tangential components.

The integral

$$\int \mathbf{E} \cdot d\mathbf{l}$$

represents the work per unit charge. In the example above, when a charge is moved in such a way that it returns to the starting point, the total work performed is zero. In this case, we can say that the field is conservative. This indicates that, if the charge returns to the starting point, there is no energy lost or gained in the process. In other words, the electric field intensity **E** can be derived from a scalar potential V as

$$\mathbf{E} = -\nabla V$$

By calculating the circulation of both sides of this relation we obtain an expression for the absolute potential as

$$\int \mathbf{E} \cdot d\mathbf{l} = -\int \nabla V \cdot d\mathbf{l}$$

Taking into account the definitions in Chapter 1 [Eq. (1.5)] gives

$$V = -\int \mathbf{E} \cdot d\mathbf{l} + K \qquad (3.10)$$

where K is a constant.

In the case of a point charge q, and setting $d\mathbf{l}=d\mathbf{r}$, we obtain

3. Electrostatic Fields

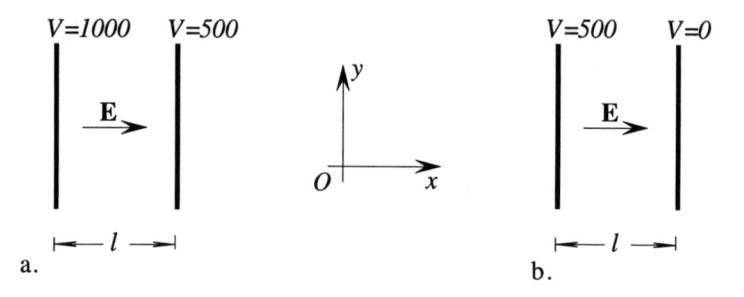

Figure 3.5. Two examples of different potential references that produce the same electric field.

$$V = -\int \frac{q}{4\pi\varepsilon_0 r^3} \mathbf{r} \cdot d\mathbf{r} + K$$

or

$$V = \frac{q}{4\pi\varepsilon_0 r} + K$$

By setting $r=\infty$, the potential is $V=0$, and therefore the constant K is zero. This gives

$$V = \frac{q}{4\pi\varepsilon_0 r} \qquad (3.11)$$

It is always necessary to define a value for the constant K, since different values of V can generate the same field intensity **E**. This is reflected from the examples in Figure 3.5a and 3.5b. Assuming **E** is constant in the direction Ox and that it does not cross the borders of the device, we get

$$\mathbf{E} = -\nabla V = -\hat{\mathbf{i}}\frac{\partial V}{\partial x}$$

therefore $E_x = \Delta V/l$.

In both cases, $E_x = 500/l$; however, the values of V are different. We conclude that in order to define V uniquely in the whole domain it is necessary to fix V at some point in this domain. This value of the potential is a reference potential.

3.3. Nonconservative Fields: Electromotive Force

In addition to the conservative electrostatic field there is another type of field, called "nonconservative." Suppose there is a closed current loop due to a conductor with contour C and enclosing a surface S. From Ohm's law:

$$E = \frac{J}{\sigma} = J\frac{RS}{L} \qquad (3.12)$$

where $R = L/\sigma S$ represents the resistance R of the conductor.

The circulation of **E** along the circuit is

$$\oint_C \mathbf{E} \cdot d\mathbf{l} = I\frac{R}{L}\oint_C dl = IR$$

where the relation $I = JS$ was used. As explained above, the circulation of **E** along the closed contour is zero because it is a conservative electrostatic field. Therefore,

$$0 = IR$$

This indicates that the current due to a conservative field is zero. To see this, we use the example in Figure 3.6. A block of material is located between the two charged plates. Because of the electric field between the plates, there is an instantaneous movement of charges in the block as shown; however, this electrostatic field is not capable of maintaining a current. The charges simply move to a new location where they are again stationary.

Sustaining a current requires a source of energy since a current is a continuous movement of electrons in a circuit. This movement is impeded by the resistance of the circuit and is characterized by dissipation of energy as heat (Joule's effect). The energy must originate in a nonconservative electric field. As an example, the chemical reaction in a battery can produce a nonconservative force by which the electrons can circulate in a loop, until the energy in the battery is exhausted.

Thus, the electric field intensity is the sum of a conservative electrostatic field intensity $\mathbf{E}_l$ and another field intensity $\mathbf{E}_f$.

$$\mathbf{E}_t = \mathbf{E}_l + \mathbf{E}_f$$

Again using Ohm's law in Eq. (3.12) we get

$$\mathbf{E} = \mathbf{J}\frac{RS}{L}$$

54   3. Electrostatic Fields

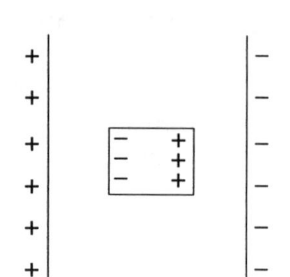

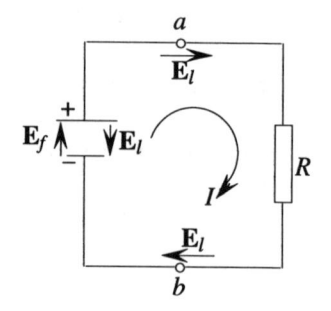

Figure 3.6. Movement of charges due to an electrostatic field. A current cannot be sustained under these conditions.

Figure 3.7. Conservative and non-conservative electric fields in a circuit. Only the nonconservative field causes flow of current.

These electric fields are shown in the circuit of Figure 3.7, where R represents the total resistance of the circuit.

Integrating the expression above along the circuit gives

$$\oint_C \mathbf{E}_t \cdot d\mathbf{l} = \oint_C \mathbf{E}_l \cdot d\mathbf{l} + \oint_C \mathbf{E}_f \cdot d\mathbf{l} = I\frac{R}{L}\oint_C dl = IR$$

Since $\mathbf{E}_l = -\nabla V$, the first integral is zero and we are left with

$$\oint_C \mathbf{E}_f \cdot d\mathbf{l} = IR$$

We define

$$U = \oint_C \mathbf{E}_f \cdot d\mathbf{l} \qquad (3.13)$$

as the electromotive force of the battery (*emf*). U is present between the terminals of the battery and the vector $\mathbf{E}_f$ is indicated in Figure 3.7. The quantities U and $\mathbf{E}_f$ are related through the internal chemical reaction of the battery

$$U = \oint_C \mathbf{E}_f \cdot d\mathbf{l} = \int_a^b \mathbf{E}_f \cdot d\mathbf{l} \qquad (3.14)$$

Therefore, we can view the equation $U=RI$ as evidence to the existence of a nonconservative electric field $\mathbf{E}_f$.

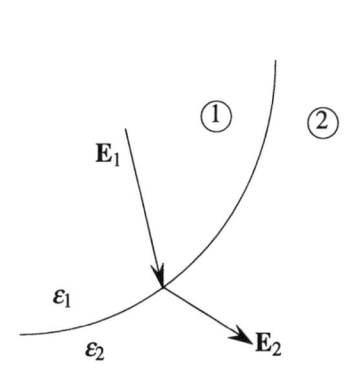

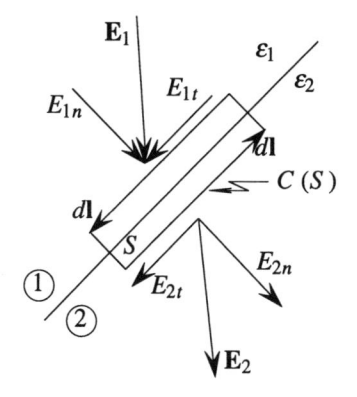

Figure 3.8a. Refraction of the electric field at an interface between two different materials.

Figure 3.8b. A contour enclosing the boundary used to evaluate the tangential components of the electric field intensity.

3.4. Refraction of the Electric Field

The electric field as it passes from one material to another undergoes a change in direction at the interface between the two materials. This effect is called "refraction," and is similar to refraction of light rays passing between two materials with different indices of refraction.

Figure 3.8a shows two materials with different permittivities: ε_1 in material 1 and ε_2 in material 2. We assume that a uniformly distributed static charge density exists on the boundary between the two materials. Since this charge is distributed on the interface it is called a surface charge density ρ_s.

Because the fields are static in time, $\nabla \times \mathbf{E} = 0$. To evaluate this we define a small surface S shown in Figure 3.8b, infinitely close to the interface. For this reason $\mathbf{E}_1$ and $\mathbf{E}_2$ are considered constant on S.

Using Stokes' theorem, we have

$$\int_S (\nabla \times \mathbf{E}) \cdot d\mathbf{s} = \oint_C \mathbf{E} \cdot d\mathbf{l} = 0$$

Neglecting the contribution to the circulation of the smaller sides (perpendicular to the interface) we obtain

$$\oint_C \mathbf{E} \cdot d\mathbf{l} = \int_{C_1} \mathbf{E}_1 \cdot d\mathbf{l} + \int_{C_2} \mathbf{E}_2 \cdot d\mathbf{l} = 0$$

where C_1 and C_2 are the parts of the contour in materials 1 and 2 respectively. We note that

3. Electrostatic Fields

$$\mathbf{E}_1 \cdot d\mathbf{l} = (\mathbf{E}_{1t} + \mathbf{E}_{1n}) \cdot d\mathbf{l} = E_{1t} dl$$

and analogously

$$\mathbf{E}_2 \cdot d\mathbf{l} = -E_{2t} dl$$

If C_1 and C_2 are equal, we obtain

$$E_{1t} \int_{C_1} dl - E_{2t} \int_{C_2} dl = 0$$

or

$$E_{1t} = E_{2t} \qquad (3.15)$$

indicating that the tangential components of the electric field intensity are conserved.

We turn now to the equation $\nabla \cdot \mathbf{D} = \rho$, with which we associate an infinitesimal volume at the interface as shown in Figure 3.9.

Integrating this equation over the volume and using the divergence theorem gives

$$\oint_S \mathbf{D} \cdot d\mathbf{s} = q$$

where q is the total charge enclosed in the volume. This charge is located on the part of the surface S_f within the cylinder:

$$q = \rho_s S_f \qquad (3.16)$$

In other words, the flux of the vector $\mathbf{D}$ is divided in two parts; because the flux on the lateral surface of the cylinder is zero, we have, for an infinitesimally long cylinder

$$\oint_S \mathbf{D} \cdot d\mathbf{s} = \int_{S_1} \mathbf{D}_1 \cdot d\mathbf{s} + \int_{S_2} \mathbf{D}_2 \cdot d\mathbf{s}$$

where S_1 and S_2 are the surfaces of the bases of the cylinder. Since

$$\mathbf{D}_1 \cdot d\mathbf{s} = (\mathbf{D}_{1n} + \mathbf{D}_{1t}) \cdot d\mathbf{s} = -D_{1n} ds$$

and analogously

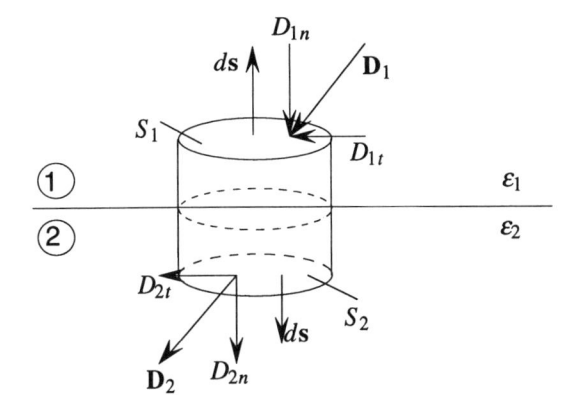

Figure 3.9. A contour enclosing the boundary used to evaluate the normal components of the electric flux density.

$$\mathbf{D_2} \cdot \mathbf{ds} = D_{2n} ds$$

we have, using Eq. (3.16)

$$-D_{1n} \int_{S_1} ds + D_{2n} \int_{S_2} ds = \rho_s S_f$$

Since in this case $S_1 = S_2 = S_f$, we obtain

$$D_{2n} - D_{1n} = \rho_s \qquad (3.17)$$

or

$$\varepsilon_2 E_{2n} - \varepsilon_1 E_{1n} = \rho_s \qquad (3.18)$$

Thus, the change in the normal component of the electric flux density passing through the surface is equal to the surface charge at the interface between the two materials. In the particularly common case, when there are no static charges on the interface ($\rho_s = 0$), we get

$$E_{1t} = E_{2t} \quad \text{and} \quad \varepsilon_1 E_{1n} = \varepsilon_2 E_{2n}$$

Using Figure 3.10 and the relations above we can write the following expressions

$$tan\theta_1 = \frac{E_{1t}}{E_{1n}}$$

3. Electrostatic Fields

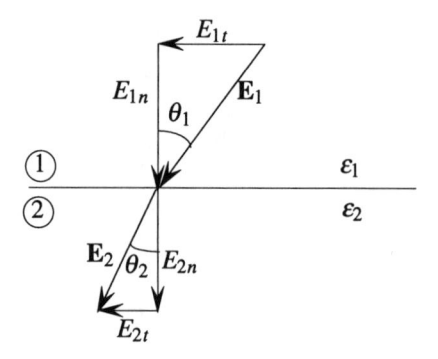

Figure 3.10. Relations between the components of the electric field intensity at the boundary between two different dielectrics.

and

$$tan\theta_2 = \frac{E_{2t}}{E_{2n}}$$

$$\frac{tan\theta_1}{tan\theta_2} = \frac{E_{2n}}{E_{1n}} = \frac{D_{2n}/\varepsilon_2}{D_{1n}/\varepsilon_1}$$

From this, we obtain

$$\frac{tan\theta_1}{tan\theta_2} = \frac{\varepsilon_1}{\varepsilon_2} \qquad (3.19)$$

The larger the change in material properties, the larger the angular change between $\mathbf{E}_1$ and $\mathbf{E}_2$. However, we must point out that the variation in ε between dielectric materials is rather small.

As an example, the maximum ratio between permittivities of air and insulating mineral oil (these two materials are frequently used in transformers) is no higher than 4.

Figure 3.11 shows the angular change in the electric field intensity in a structure with two dielectric materials. In material 1, $\varepsilon_r=1$ and in material 2, $\varepsilon_r=4$. This plot was obtained with the EFCAD computer program. (EFCAD is a finite element computer software designed for numerical solution of electromagnetic field problems. A simple approach to numerical solution of electrostatic problems is discussed at the end of this chapter. More general solutions using EFCAD are discussed in Chapters 8, 9, and 10).

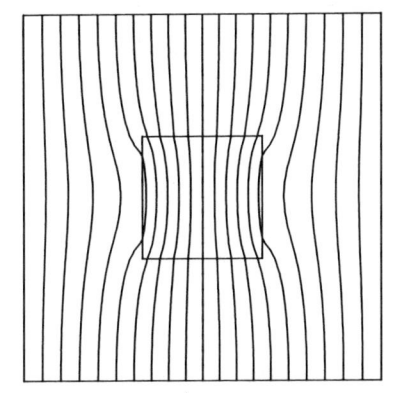

Figure 3.11. The electric field in a geometry with two dielectric materials. The material at center has a permittivity four times higher that the surrounding material.

3.5. Dielectric Strength

In many devices subjected to large variations in potential, in particular in high-voltage equipment, the expression for the electrostatic field $\mathbf{E}=-\nabla V$ assumes a very important role. We look at this role by using Figure 3.12.

Suppose that one part of the equipment is at ground potential ($V=0$). Another part of the device is at a high voltage $V=V_a$. As a good approximation we can calculate the electric field intensity as

$$\mathbf{E} = -\nabla V$$

and therefore,

$$E_1 = \frac{V_a - 0}{l_1} \quad \text{and} \quad E_2 = \frac{V_a - 0}{l_2}$$

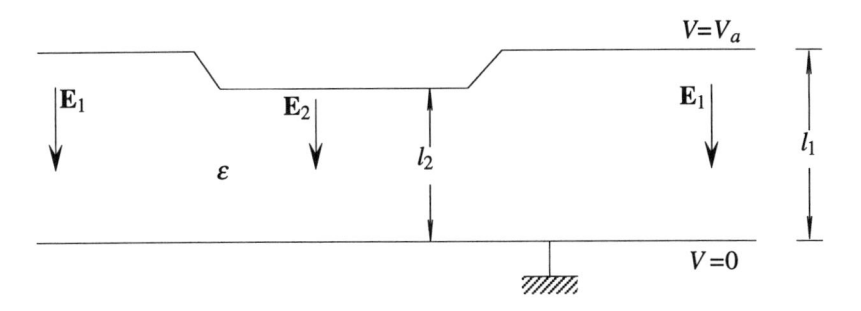

Figure 3.12. A schematic representation of a high-voltage device with low- and high-intensity electric fields.

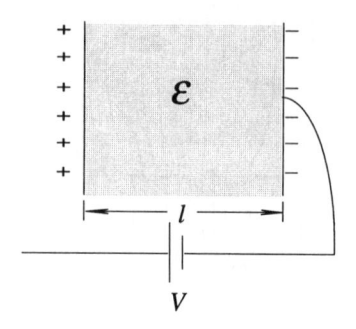

Figure 3.13. Definition of dielectric strength. The electric field at which the insulator breaks down is defined as the dielectric strength of the insulator.

With $l_1 > l_2$, it is evident that $E_2 > E_1$. Large field intensities (or gradients of potential) may exist in certain parts of the equipment. If these fields exceed allowable limits, they could cause harmful effects or damage to the equipment.

We define now the dielectric strength K of an insulator. Consider an insulating material between two metallic plates separated by a distance l and subjected to a potential difference V, as shown in Figure 3.13.

Due to the application of the potential V there is an accumulation of positive and negative charges on the two plates as shown. If we increase the potential V, a critical potential V_c is eventually reached at which the accumulated charge between the plates creates a current (or an electric arc) between the plates, penetrating or "breaking" the insulator. When this happens, the insulating properties of the material are lost. The dielectric strength is therefore defined as

$$K = \frac{V_c}{l} \quad (V/m) \qquad (3.20)$$

K represents the maximum electric field intensity (and therefore the maximum potential difference) an insulator can support without breaking down. Note that the units of K are the same as the units of the electric field intensity. Hence, returning to Figure 3.12, it is important that the highest field intensity in the equipment (in this case E_2) does not exceed the dielectric strength of the material in which this field is encountered. In this sense, we observe that it is very important to know the electric fields in the equipment, in particular the high-intensity fields. A good, detailed knowledge of the field distribution allows the design of the device and optimization of its various dimensions such that the design is safe, is compact, and is done at a reasonable cost.

Finally, we point out that an excessive electric field intensity not only damages equipment but can also be dangerous to personnel and to livestock that happen to be in the area of high field intensities.

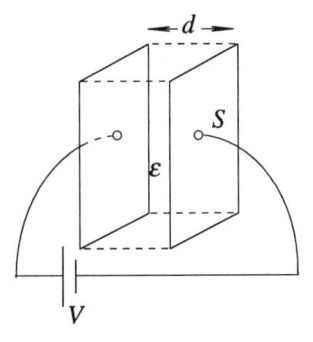

Figure 3.14. The parallel plate capacitor.

3.6. The Capacitor

3.6.1. *Definition of Capacitance*

Consider the two conducting plates shown in Figure 3.14. The plates have a surface area S and are separated a distance d apart. The space between the plates is filled with a dielectric with permittivity ε. The plates are connected to an external source V as shown.

A charge $+Q$ appears on one plate and a charge $-Q$ on the second. This phenomenon occurs because of the structure and the potential V. If we double the potential, the charge doubles. The ratio between charge and potential remains constant, provided the critical potential beyond which breakdown of the insulator occurs is not reached.

$$\frac{Q}{V} = \frac{Q_1}{V_1} = \frac{Q_2}{V_2} = constant$$

This constant, or coefficient of proportionality is called the "capacitance" of the device and is denoted by C. We have, therefore

$$C = \frac{Q}{V} \qquad (3.21)$$

As an example we calculate the capacitance of the device in Figure 3.14 known as a "parallel plate capacitor." We will assume that the electric field is not distorted at the edges of the capacitor (no "edge effects" at the geometrical edges of the capacitor). This approximation allows the consideration of each plate as an infinite, charged plane. To calculate the electric field intensity **E** due to the charged plate we define the surface charge density on the plane (see Figure 3.15).

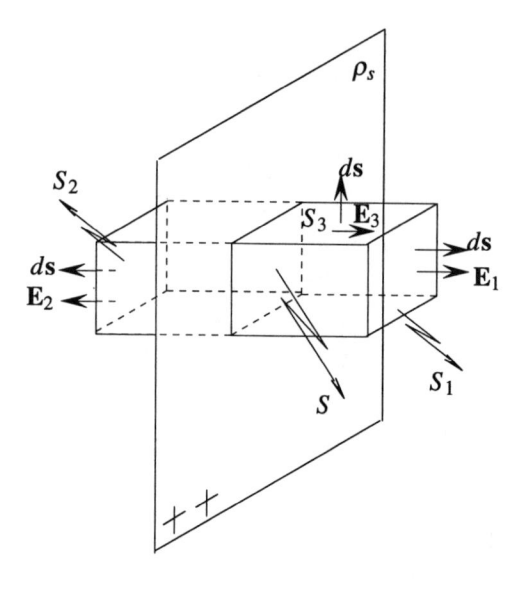

Figure 3.15. The Gauss surface used to calculate the electric field intensity due to one plate of the capacitor. The plate is assumed to be an infinite charged plane.

The divergence $\nabla \cdot \varepsilon \mathbf{E} = \rho$ which in integral form is Gauss' theorem, gives

$$\oint_s \varepsilon \mathbf{E} \cdot d\mathbf{s} = Q$$

Choosing the volume to be a parallelepiped with its furthest surface parallel to the plane, the flux of $\mathbf{E}$ is divided between three surfaces, corresponding to surfaces S_1, S_2, and S_3. S_3 represents all lateral surfaces. This gives

$$\int_{S_1} \varepsilon \mathbf{E}_1 \cdot d\mathbf{s} + \int_{S_2} \varepsilon \mathbf{E}_2 \cdot d\mathbf{s} + \int_{S_3} \varepsilon \mathbf{E}_3 \cdot d\mathbf{s} = Q$$

Note that $\mathbf{E}_3$ and $d\mathbf{s}$ are perpendicular vectors. In other words, considering $\mathbf{E}_1$ and $\mathbf{E}_2$ to be constants, we obtain

$$\varepsilon E_1 S_1 + \varepsilon E_2 S_2 = Q$$

where S_1 and S_2 are equidistant to the plane and are equal to S. Assuming $E_1 = E_2 = E$, allows us to write

$$\varepsilon E S + \varepsilon E S = Q$$

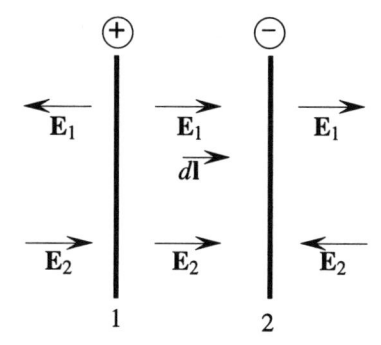

Figure 3.16. Superposition of the fields of the two parallel plates. The electric field intensity outside the plates is zero.

To conclude, we note that the charge contained in this volume is the same as that contained in the intersection of the volume and the plane, or that charge distributed on the surface S, as indicated in Figure 3.15. This gives

$$2\varepsilon E S = \rho_s S \quad \text{or} \quad E = \frac{\rho_s}{2\varepsilon}$$

The field intensity generated by the second plate, which is negatively charged, has the same magnitude but it converges towards the plate (opposite in direction to that of the first plate), as shown in Figure 3.16.

By direct superposition of the fields of the two plates we get a zero field outside the plates and a value $E = \rho_s/\varepsilon$ between the plates. Obtaining the field intensity is important because it is necessary in order to calculate the potential difference between the two plates

$$V = -\int_1^2 \mathbf{E} \cdot d\mathbf{l}$$

which is the circulation of $\mathbf{E}$ from one plate to another. $d\mathbf{l}$ is as shown in Figure 3.16. If $\mathbf{E}$ is constant between the plates we obtain

$$|V| = \frac{\rho_s}{\varepsilon} d$$

The total charge contained within the device is $Q = \rho_s S$ where S is the surface of one plate. Therefore,

$$C = \frac{Q}{V} = \frac{\varepsilon S}{d} \qquad (3.22)$$

One important observation is the fact that while the capacitance is defined by Q and V, the final expression for capacitance depends only on the physical properties and dimensions of the device. In this case the capacitance depends on the permittivity of the dielectric between the plates and on the dimensions S and d.

3.6.2. Energy Stored in a Capacitor

It is obvious that in charging a capacitor with a charge Q a certain amount of energy must be dispensed. Suppose that the charging operation takes some time t, say, starting at time zero until time T. Also, assume that while the charge increases from zero to Q, the potential v changes from zero to V. The energy required to do so is

$$W = \int_0^T ivdt$$

Noting that $i=dq/dt$ and substituting this in the equation above, we obtain

$$W = \int_0^Q vdq$$

From $C=q/v=Q/V$, we have

$$W = \int_0^V vd(Cv)$$

Since C is a constant (dependent only on physical properties and dimensions) this can be written as

$$W = \frac{1}{2}CV^2 \qquad (3.23)$$

3.6.3. Energy in a Static, Conservative Field

The energy calculated above can be viewed as the energy required to generate the field inside the capacitor. The volumetric energy density in the case of the parallel plate capacitor can be calculated as

$$\frac{W}{vol} = \frac{W}{Sd} = \frac{1}{2}\frac{CV^2}{Sd}$$

Noting that $C=\varepsilon S/d$ and $V=Ed$, we obtain

$$\frac{W}{vol} = \frac{1}{2}\varepsilon E^2 \qquad (3.24)$$

Although this result was obtained for a particular example, it can be shown that it applies in general to all electric fields in a dielectric with permittivity ε.

3.7. Laplace's and Poisson's Equations in Terms of the Electric Field

Assuming that in the domain under study there are no time-dependent quantities we can define an electric scalar potential V from which a conservative electric field intensity $\mathbf{E}=-\nabla V$ can be derived. This relation is valid for the electrostatic field because $\nabla \times \mathbf{E}=0$, $(\nabla \times (-\nabla V)=0)$. Since $\nabla \times (-\nabla V)$ is always zero, this definition of the electric field intensity is correct. However, if $\partial \mathbf{B}/\partial t$ is not zero, we cannot use this definition. In the static case we have

$$\nabla \cdot \mathbf{D} = \rho$$

$$\nabla \cdot \varepsilon \mathbf{E} = \rho$$

$$\nabla \cdot \varepsilon (\nabla V) = -\rho$$

which, in explicit form is

$$\frac{\partial}{\partial x}\varepsilon\frac{\partial V}{\partial x} + \frac{\partial}{\partial y}\varepsilon\frac{\partial V}{\partial y} + \frac{\partial}{\partial z}\varepsilon\frac{\partial V}{\partial z} = -\rho \qquad (3.25)$$

In two dimensions this equation becomes

$$\frac{\partial}{\partial x}\varepsilon\frac{\partial V}{\partial x} + \frac{\partial}{\partial y}\varepsilon\frac{\partial V}{\partial y} = -\rho \qquad (3.26)$$

This is Poisson's equation and it defines the electric potential distribution in the dielectric domain where an electrostatic field exists. To solve this equation we must first impose the boundary conditions, or in other words, specify the potentials on the boundaries of the solution domain. In addition we must specify the geometry, the dielectric materials, as well as any static charge densities in the domain.

66   3. Electrostatic Fields

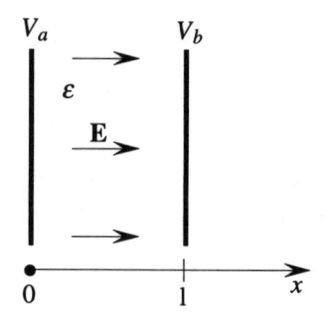

Figure 3.17. Electric field in a parallel plate capacitor due to a voltage difference on the plates.

If there are no static charges ($\rho=0$) and a single dielectric material exists in the domain, the equation becomes

$$\frac{\partial^2 V}{\partial x^2} + \frac{\partial^2 V}{\partial y^2} = 0 \qquad (3.27)$$

This is Laplace's equation. In this case the source of the electric field in the domain of study is the boundary conditions through which potential differences are imposed.

It must be pointed out that the analytic solution to this equation is extremely difficult for the majority of even the simplest realistic problems, and, in the case of complex geometries, practically impossible. For the time being we present the solution to this equation for a very simple problem.

We use again the example of the parallel plate capacitor, where we wish to find the field intensity between the plates. Edge effects are neglected. The problem geometry is shown in Figure 3.17.

The conditions on the boundaries of the geometry are $V=V_a$ at $x=0$, and $V=V_b$ at $x=l$. With the assumption that there are no edge effects, the problem is one-dimensional with variation in the Ox direction. Laplace's equation is therefore

$$\frac{\partial^2 V}{\partial x^2} = 0$$

The solution to this problem can be written as $V(x)=ax+b$ by direct integration. With the known boundary conditions we get

$$V_a = a.0 + b$$

and

$$V_b = a.l + b$$

which allows calculation of the constants a and b. Substituting these in the solution we obtain

$$V(x) = \frac{V_b - V_a}{l} x + V_a$$

With $\mathbf{E} = -\nabla V$, we have in this case

$$\mathbf{E} = -\hat{\mathbf{i}} \frac{\partial V}{\partial x} \quad \text{or} \quad E_x = -\frac{\partial V}{\partial x}$$

and, therefore,

$$E_x = \frac{V_a - V_b}{l}$$

If $V_a > V_b$, E_x is directed in the positive x direction, or, $\mathbf{E}$ is directed in the direction of decreasing potential, as required.

3.8. Examples

3.8.1. *The Infinite Charged Line*

An infinitely long line of charges is defined with a linear charge density ρ_l as shown in Figure 3.18. We wish to calculate the electric field intensity $\mathbf{E}$ at a distance R from the line.

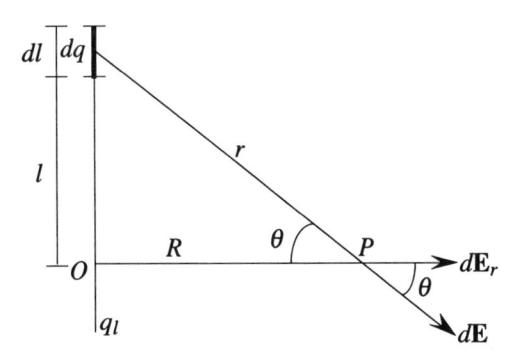

Figure 3.18. Calculation of the electric field intensity of an infinite charged line.

This problem can be solved in two different ways. In the first method, we view the line as being made of infinitesimal point charges dq. Each charge dq produces a field intensity $d\mathbf{E}$ at point P in Figure 3.18. At the beginning of this chapter we saw that a point charge produces a field intensity $\mathbf{E}$ at a distance r, given by

$$E = \frac{q}{4\pi\varepsilon_0 r^2}$$

Analogously, we can write in this case

$$dE = \frac{dq}{4\pi\varepsilon_0 r^2}$$

The charge on the line is defined by the linear charge density ρ_l. The differential point charge is therefore

$$dq = \rho_l dl$$

We note that the vertical component of $d\mathbf{E}$ cancels because for any point charge dq at a distance l along the line there is a symmetric point charge at a point $-l$. The field component of interest is therefore

$$dE_r = \frac{\rho_l dl}{4\pi\varepsilon_0 r^2} \cos\theta \qquad (3.28)$$

At this point, we note that l, r, and q are interdependent variables: we must explicitly rewrite them as a function of a single variable before the calculation can proceed. We note that

$$\tan\theta = \frac{l}{R} \quad \text{and} \quad \cos\theta = \frac{R}{r}$$

and therefore,

$$dl = R \sec^2\theta \, d\theta \quad \text{and} \quad r = \frac{R}{\cos\theta}$$

Substituting dl and r in Eq. (3.28) we obtain

$$dE_r = \frac{\rho_l}{4\pi\varepsilon_0 R} \cos\theta \, d\theta$$

Examples 69

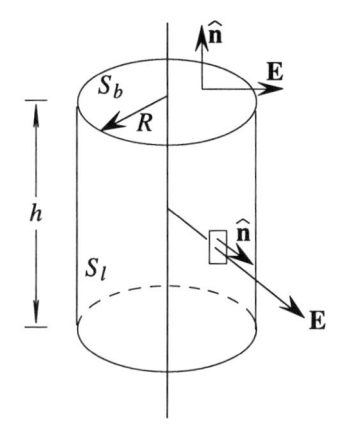

Figure 3.19. A cylindrical Gauss surface is used to calculate the electric field intensity due to an infinite charged line.

The electric field intensity is obtained by integration:

$$E_r = \int_{-\frac{\pi}{2}}^{\frac{\pi}{2}} \frac{\rho_l}{4\pi\varepsilon_0 R} \cos\theta d\theta = \frac{\rho_l}{2\pi\varepsilon_0 R}$$

The angles $-\pi/2$ and $\pi/2$ define the infinite dimensions of the conductor. If the line is finite, these angles will change. They define the length of the conductor as well as the position of the point P in relation to the conductor.

The second method that can be used to solve this problem is by making use of Gauss' theorem. First we note that there are an infinite number of points at a distance R from the line. These points form a cylindrical surface of radius R as shown in Figure 3.19. The equation $\nabla \cdot \mathbf{D} = \rho$, in integral form is

$$\oint_S \mathbf{D} \cdot d\mathbf{s} = q$$

Calculating $\mathbf{D} \cdot d\mathbf{s}$ over the surface of the cylinder, we note that this product is zero on the bases of the cylinder. On the lateral surface, where $\mathbf{E}$ and $d\mathbf{s}$ are parallel vectors we obtain

$$\int_{S_l} \varepsilon_0 \mathbf{E} \cdot d\mathbf{s} = \varepsilon_0 E 2\pi R h$$

70   3. Electrostatic Fields

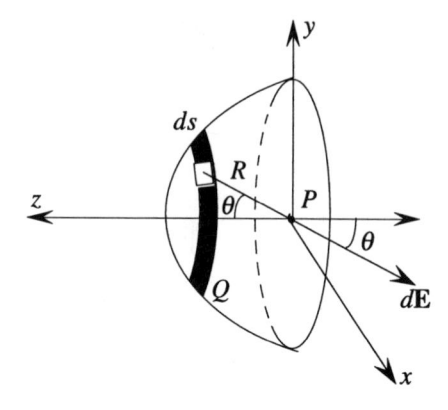

Figure 3.20. Calculation of the electric field intensity due to a charged spherical half-shell.

The charge contained within the volume of the cylinder is $\rho_l h$. Therefore,

$$E = \frac{\rho_l}{2\pi\varepsilon_0 R}$$

The application of Gauss' theorem is much simpler, but, the question is: when can it be applied? There is no fixed rule for the application of Gauss' law that will allow a simple answer to this question. Intuition and the practical considerations of the problem lead to the general direction of solution. However, whenever there is a surface along which the field is constant (in direction and magnitude) Gauss' theorem can be applied. In the case above, if the line were finite in length, the field intensity **E** would vary axially and Gauss' theorem could not be applied.

3.8.2. The Charged Spherical Half-Shell

We wish to calculate the electric field intensity at point P, created by a spherical conducting half-shell, charged with a charge Q. The geometry is shown in Figure 3.20. This is a case to which Gauss' theorem cannot be applied and we need to use direct integration over the charge. In spite of this, Gauss' theorem is always correct regardless of our ability to use it because it is merely an integral representation of one of Maxwell's equations.

To solve this problem we define a surface charge density $\rho_s = Q/2\pi R^2$. An elementary charge dq, corresponding to a differential surface ds, generates a field intensity $d\mathbf{E}$ at point P; the only nonzero component of this field is

$$dE_r = \frac{\rho_s ds}{4\pi\varepsilon_0 R^2} \cos\theta$$

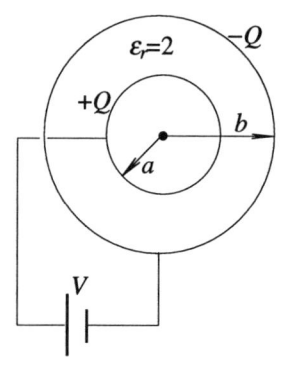

Figure 3.21. The spherical capacitor.

Because of the spherical geometry of the problem, it is convenient to solve it in spherical coordinates. ds in this system of coordinates is

$$ds = R^2 \sin\theta \, d\phi d\theta$$

and

$$E_r = \frac{\rho_s}{4\pi\varepsilon_0} \int_0^{\frac{\pi}{2}} \cos\theta \sin\theta \, d\theta \int_0^{2\pi} d\phi$$

noting that $\cos\theta \sin\theta = (\sin 2\theta)/2$ yields

$$E_r = \frac{\rho_s}{4\pi\varepsilon_0} \frac{1}{2} 2\pi$$

and substituting $Q/2\pi R^2$ for ρ_s, the value of E_r is

$$E_r = \frac{Q}{8\pi\varepsilon_0 R^2}$$

3.8.3. The Spherical Capacitor

Two concentric, conducting spheres with radii a and b, and separated by a dielectric with $\varepsilon_r = 2$ as shown in Figure 3.21 are given. We wish to calculate the capacitance of this device.

To calculate the electric field intensity between the two conducting spheres we use Gauss' theorem

72   3. Electrostatic Fields

$$\oint_S \varepsilon \mathbf{E} \cdot d\mathbf{s} = Q$$

where the surface S is the spherical shell of radius r such that $a<r<b$. We have

$$4\pi\varepsilon E r^2 = Q \quad \text{and} \quad E = \frac{Q}{4\pi\varepsilon r^2}$$

It should be pointed out that for $r<a$ and $r>b$, the electric field intensity is zero since in both cases the total charge contained in the volume enclosed by the surface S is zero.

To calculate V, it is necessary to find $\mathbf{E}$ which is obtained through the definition of the unit vector $\hat{\mathbf{u}}=\mathbf{r}/r$, (section 2.1). We have therefore [from Eq. 3.5].

$$\mathbf{E} = \frac{Q}{4\pi\varepsilon} \frac{\mathbf{r}}{r^3}$$

V is now calculated as

$$|V| = \int_a^b \frac{Q}{4\varepsilon\pi} \frac{\mathbf{r} \cdot d\mathbf{r}}{r^3} = \frac{Q}{4\varepsilon\pi}\left(\frac{1}{a}-\frac{1}{b}\right)$$

where

$$C = \frac{Q}{V} \quad \text{or} \quad C = \frac{4\pi\varepsilon a b}{b-a} \qquad (3.29)$$

with $\varepsilon = 2\varepsilon_0$ we obtain

$$C = \frac{8\pi\varepsilon_0 a b}{b-a}$$

3.8.4. *The Spherical Capacitor with Two Dielectric Layers*

Suppose it is required to double the capacitance of the spherical capacitor of the previous example by introducing a layer of dielectric material with relative permittivity $\varepsilon_r = 5$, as shown in Figure 3.22.

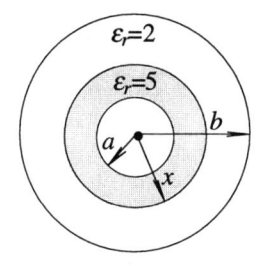

Figure 3.22. A spherical capacitor with two dielectric layers.

The thickness of this layer, $(x-a)$, needs to be calculated. We can calculate x as a function of a, assuming, as an example that $b=2a$. Denoting C_1 the capacitance calculated in the previous example

$$C_1 = \frac{8\pi\varepsilon_0 ab}{b-a} = 16\pi\varepsilon_0 a$$

Introduction of a dielectric layer in the capacitor can be viewed as constituting two capacitors in series. In fact there is no reason we cannot view the interfaces of the dielectric as being very thin conducting plates with charges $+Q$ and $-Q$ as shown in Figure 3.23. The capacitance of the device, denoted by C_2 is the total capacitance of two capacitors C_2' and C_2'' connected in series. These capacitances are [from Eq. (3.29)]

$$C_2' = \frac{20\pi\varepsilon a x}{x-a} \quad \text{and} \quad C_2'' = \frac{16\pi\varepsilon_0 a x}{2a-x}$$

From the formula for capacitors in series

$$\frac{1}{C_2} = \frac{1}{C_2'} + \frac{1}{C_2''} = \frac{x-a}{20\pi\varepsilon_0 a x} + \frac{2a-x}{16\pi\varepsilon_0 a x}$$

With $C_2 = 2C_1$, we have

$$\frac{1}{C_2} = \frac{1}{2C_1}$$

Substitution of the values of C_1 and C_2, gives

$$x = \frac{12a}{7}$$

74   3. Electrostatic Fields

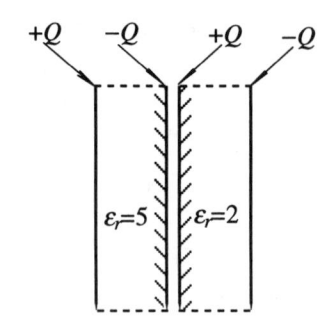

Figure 3.23. A capacitor with two dielectric layers viewed as two capacitors in series.

3.9. A Brief Introduction to the Finite Element Method: Solution of the Two-Dimensional Laplace's Equation

We pointed out above how important it is to be able to understand the electric field intensity distribution in a device. In section 3.7, instead of calculating **E** directly, we approached the same problem by calculating the potential V. This procedure is simple and effective since V is a scalar, and there is only one unknown value at any point in the solution domain.

Practically any classical book on electromagnetic fields contains a chapter on the solution of Laplace's equation through the calculation of the potential V. Some of these are analytic methods, other approximate, still others empirical. Considerable talent and effort went into these solutions in the pre-computer period. We note here a few of these methods:

- Analytic methods using separation of variables
- Point by point iterative methods
- Method of images
- Methods based on equivalency with resistive calculations
- Lehmann's method
- Relaxation methods

In general, these methods provide a solution to problems with relatively simple geometries. Existence of complex geometries and presence of different dielectric materials make the application of these methods practically impossible. The advent of modern numerical methods and computers made these methods of essentially historical value.

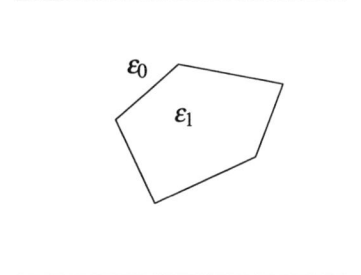

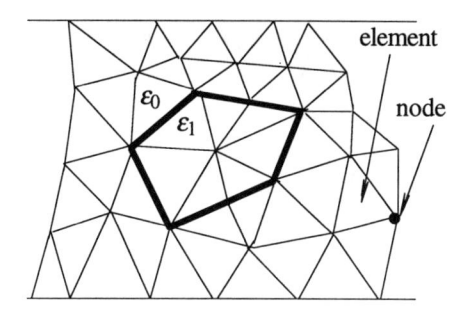

Figure 3.24a. A simple geometry with two materials to be discretized into finite elements.

Figure 3.24b. A finite element discretization of the geometry in Figure 3.24a using triangular finite elements.

The two main numerical methods used for solution of Laplace's (and Poisson's) equation are the finite element method and the finite difference method.

We will present here the first because it is more flexible with respect to description of the geometry of the problem. However, the presentation of the finite element method in this chapter is only a brief introduction. A more detailed presentation is outlined in Chapters 8 through 13.

3.9.1. *The Finite Element Technique for Division of a Domain*

The principle of the finite element method (FEM) consists of the division of the solution domain into small subdomains, called "finite elements." These elements can be of different sizes such that, in regions where we anticipate larger variations in fields, the number of elements and their sizes can be changed to obtain higher element densities. In regions of lower field gradients, the number of elements can be lower. Each element can contain a different material, therefore, the interface between two different materials must coincide with element boundaries, as shown in Figure 3.24a and 3.24b.

The intersection of two edges is called a "node." The surface contained within a triangle is called an "element" (see Figure 3.24b). The collection of elements forms a "mesh" which must be "compatible" (see Figures 3.25a and 3.25b).

Using this type of finite elements (triangular), the potential V varies linearly within the triangle, through an expression of the following type:

$$V(x,y) = a + bx + cy \qquad (3.30)$$

Note that for the nodes i, j, and k of a generic element (Figure 3.26), the potentials satisfy Eq. (3.30). Writing this equation for the three nodes yields the following system:

$$V_i = a + bx_i + cy_i$$

3. Electrostatic Fields

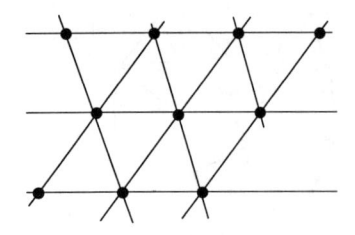

Figure 3.25a. A compatible mesh. Only triangle vertices form nodes.

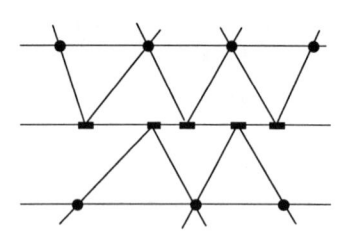

Figure 3.25b. A noncompatible mesh. Some of the nodes are located along edges.

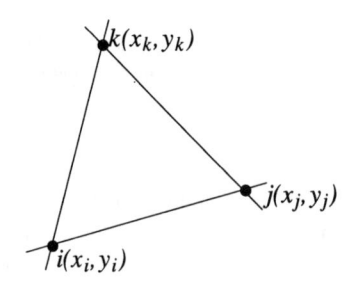

Figure 3.26. The three nodes of a general triangular element.

$$V_j = a + bx_j + cy_j$$

$$V_k = a + bx_k + cy_k$$

Solution to this system of equations provides a, b, and c, as a function of x_i, x_j, x_k, y_i, y_j, y_k, V_i, V_j, and V_k. This can be done by simple use of Cramer's rule as:

$$a = \frac{1}{D} \det \begin{bmatrix} V_i & x_i & y_i \\ V_j & x_j & y_j \\ V_k & x_k & y_k \end{bmatrix} \quad b = \frac{1}{D} \det \begin{bmatrix} 1 & V_i & x_i \\ 1 & V_j & x_j \\ 1 & V_k & x_k \end{bmatrix} \quad c = \frac{1}{D} \det \begin{bmatrix} 1 & x_i & V_i \\ 1 & x_j & V_j \\ 1 & x_k & V_k \end{bmatrix}$$

where

$$D = \det \begin{bmatrix} 1 & x_i & y_i \\ 1 & x_j & y_j \\ 1 & x_k & y_k \end{bmatrix} \tag{3.31}$$

D is equal to twice the area of the triangle. Substituting the values of a, b, and c in Eq. (3.30) gives, after some simple algebra:

$$V(x,y) = \sum_{l=i,j,k} \frac{1}{D}(p_l + q_l x + r_l y) V_l \tag{3.32}$$

where the coefficients of the linear equation are

$$p_i = x_j y_k - x_k y_j$$

$$q_i = y_j - y_k$$

$$r_i = x_k - x_j$$

and the terms p_j, q_j, r_j, p_k, q_k, and r_k are obtained by cyclic permutations of the indices above.

For the relation $\mathbf{E} = -\nabla V$, we have

$$\mathbf{E} = \hat{\mathbf{i}} E_x + \hat{\mathbf{j}} E_y = -\hat{\mathbf{i}} \frac{\partial V}{\partial x} - \hat{\mathbf{j}} \frac{\partial V}{\partial y} \qquad (3.33)$$

where $\hat{\mathbf{i}}$ and $\hat{\mathbf{j}}$ are the unit vectors in the orthogonal system of coordinates. Using Eq. (3.33) we get

$$E_x = -\sum_{l=i,j,k} \frac{1}{D} q_l V_l \quad \text{and} \quad E_y = -\sum_{l=i,j,k} \frac{1}{D} r_l V_l \qquad (3.34)$$

These components are independent of x and y; they are constant in an element.

3.9.2. The Variational Method

The principles of the variational method are explained in detail in Chapter 8. For now, we note that the basis of the method consists of minimization of an energy functional instead of solving directly the physical equation (in this case $\nabla^2 V = 0$). This functional is the energy associated with the electric field intensity in the solution domain

$$F = \int_s \frac{1}{2} \varepsilon E^2 ds \qquad (3.35)$$

where S is the surface of the solution domain. Minimization of the functional is done with respect to the N unknown potentials at the nodes. The energy is minimized when the N equations satisfy the condition

$$\frac{\partial F}{\partial V_n} = 0 \quad (n = 1, 2, \dots, N)$$

3. Electrostatic Fields

The discretization of the domain into M elements allows us to write the energy in the system as the sum of the energy of the M elements. Minimization of the functional becomes

$$\frac{\partial F}{\partial V_n} = \sum_{m=1}^{M} \frac{\partial F_m}{\partial V_n} \qquad (3.36)$$

F_m is the energy associated with an element "m." It is different than zero in the element in question but is zero outside the element. Performing these operations we obtain $\partial F_m/\partial V_n$. Assuming that ε is constant within the whole element and from the fact that $\mathbf{E}$ is also constant, we can write

$$\frac{\partial F}{\partial V_n} = \frac{\partial}{\partial V_n} \int_s \frac{1}{2} \varepsilon E^2 ds = \frac{\partial}{\partial V_n}\left[\frac{1}{2}\varepsilon E^2 \int_s ds\right] = \frac{1}{2}\varepsilon \frac{D}{2}\frac{\partial E^2}{\partial V_n} \qquad (3.37)$$

We calculate $\partial E^2/\partial V_n$, taking into account that V_n is one of the three potentials V_i, V_j, or V_k of element m.

The electric field intensity $\mathbf{E}$ can be written as

$$\mathbf{E} = -\frac{1}{D}\sum_{l=i,j,k}(\hat{\mathbf{i}}q_l + \hat{\mathbf{j}}r_l)V_l$$

$$\mathbf{E} = -\frac{1}{D}\left[\hat{\mathbf{i}}(q_iV_i + q_jV_j + q_kV_k) + \hat{\mathbf{j}}(r_iV_i + r_jV_j + r_kV_k)\right]$$

Since $E^2 = \mathbf{E}\cdot\mathbf{E}$, we get

$$E^2 = \frac{1}{D^2}\left[(q_iV_i + q_jV_j + q_kV_k)^2 + (r_iV_i + r_jV_j + r_kV_k)^2\right]$$

and

$$\frac{\partial E^2}{\partial V_n} = \frac{2}{D^2}\left[(q_iV_i + q_jV_j + q_kV_k)q_n + (r_iV_i + r_jV_j + r_kV_k)r_n\right]$$

which can be assembled in the matrix form:

$$\frac{\partial E^2}{\partial V_n} = \frac{2}{D^2}\left[(q_iq_n + r_ir_n) \quad (q_jq_n + r_jr_n) \quad (q_kq_n + r_kr_n)\right]\begin{bmatrix}V_i\\V_j\\V_k\end{bmatrix} \qquad (3.38)$$

Performing this operation for the three nodes of the triangle and substituting the result in Eq. (3.37) the following is obtained:

$$\begin{bmatrix} \dfrac{\partial F_m}{\partial V_i} \\ \dfrac{\partial F_m}{\partial V_j} \\ \dfrac{\partial F_m}{\partial V_k} \end{bmatrix} = \dfrac{\varepsilon}{2D} \begin{bmatrix} q_i q_i + r_i r_i & q_i q_j + r_i r_j & q_i q_k + r_i r_k \\ & q_j q_j + r_j r_j & q_j q_k + r_j r_k \\ \text{symmetric} & & q_k q_k + r_k r_k \end{bmatrix} \begin{bmatrix} V_i \\ V_j \\ V_k \end{bmatrix} \qquad (3.39)$$

With the objective of calculating

$$\sum_M \dfrac{\partial F_m}{\partial V_n} = 0$$

the matrices $S(3,3)$ above are assembled into a global matrix $SS(N,N)$. This matrix is multiplied by the column vector $V(N)$ of the unknown potentials. This operation transforms the system

$$\sum_{m=1}^{M} \dfrac{\partial F_m}{\partial V_n} = 0$$

into the matrix system

$$[SS]\{V\} = \{0\}$$

The elemental matrices $S(3,3)$ are assembled into the global matrix in the following manner: If the global node numbers i, j, k of a particular element in the mesh are, for example, 20, 25, and 28, the terms of the elemental matrix $S(3,3)$ are added to the line and column numbers corresponding to the node numbers as demonstrated below:

$S(1,1)$ _____ line 20, column 20
$S(1,2)$ _____ line 20, column 25

.
.
.

$S(3,3)$ _____ line 28, column 28

In the problem to be solved there are boundary conditions that must be imposed. To include these boundary conditions, we set the diagonal of the row in the matrix corresponding to the node at which the boundary condition is known to 1. All the rest of the coefficients of the row are set to zero. In the right-hand side vector we set the value corresponding to the same node number to the value of the boundary condition at the node.

80   3. Electrostatic Fields

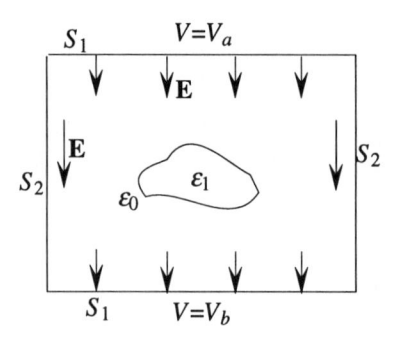

Figure 3.27. Dirichlet and Neumann boundary conditions. Dirichlet boundary conditions are prescribed on S_1 and Neumann boundary conditions on S_2.

Doing this forces the product of the unknown at the node to be equal to the given value on the right-hand side. The known potential on the boundary is called the "Dirichlet boundary condition." On boundaries where nothing is specified, we have a "Neumann boundary condition." The latter indicates that there are no normal components of the field at the boundary. These two types of boundary conditions are shown in Figure 3.27.

On boundaries S_1 Dirichlet boundary condition are prescribed as $V=V_a$ and $V=V_b$. With these two equipotential boundaries, the electric field intensity **E** is perpendicular to the two boundaries S_1. On lines S_2, where nothing has been specified, a Neumann boundary condition exists. On this line the electric field intensity is parallel to the boundary. Again, these aspects will be discussed in more detail in the second part of this work.

3.9.3. A Finite Element Program

In this section we present a finite element program in the most elementary form possible to demonstrate the essential aspects of the method. More sophisticated computer implementations of the method use properties of the matrix [SS] as well as other extensions that will be discussed later.

First, we define the variables involved:

- NNO   number of nodes
- NEL   number of elements
- NCON   number of boundary lines on which potentials are known
- NMAT   number of dielectric materials
- KTRI(NEL,3)   array that indicates the node numbers of the three nodes of each element
- MAT(NEL)   indicates the material number of each element

Introduction to the Finite Element Method   81

- PERM(NMAT)   permittivities of the NMAT materials
- X(NNO) x coordinates of the NNO node numbers
- Y(NNO) y coordinates of the NNO node numbers
- VI(NCON)   imposed potentials on the NCON boundary lines
- NOCC(NCON,20)   node numbers at which the potential VI is imposed
                       (maximum 20 nodes per equipotential line)
- SS(NNO×NNO)   global matrix of coefficients of the system of equations
- VV(NNO)   vector of unknown potentials
- VDR(NNO)   vector of the right-hand side of the matrix

A listing of the program, written in FORTRAN 77 is reproduced below:

```
C---------------MAIN PROGRAM
        COMMON/DATA/KTRI(200,3),MAT(200),X(150),Y(150),PERM(10)
       *VI(10),NOCC(10,20)
        COMMON /MATRIX/SS(150,150),VV(150),VDR(150)
C---------------CALL ZERO TO NULL THE VARIOUS ARRAYS
        CALL ZERO
C---------------CALL INPUT TO READ DATA
        CALL INPUT(NNO,NEL,NCON)
C---------------CALL FORM TO FORM THE MATRIX SS()
        CALL FORM(NEL)
C---------------CALL CONDI TO INSERT BOUNDARY CONDITIONS
        CALL CONDI(NCON,NNO)
C---------------CALL ELIM TO SOLVE THE MATRIX SYSTEM
        CALL ELIM(NNO)
C---------------CALL OUTPUT TO PRINT THE RESULTS
        CALL OUTPUT(NNO,NEL)
        STOP
        END
C
C
        SUBROUTINE ZERO
        COMMON/DATA/KTRI(200,3),MAT(200),X(150),Y(150),PERM(10)
       *VI(10),NOCC(10,20)
        COMMON /MATRIX/SS(150,150),VV(150),VDR(150)
        MAX1=150
        MAX2=10
        MAX3=20
        DO 1 I=1,MAX1
        VV(I)=0.
        VDR(I)=0.
        DO 1 J=1,MAX1
   1    SS(I,J)=0.
        DO 2 I=1,MAX2
        DO 2 J=1,MAX3
   2    NOCC(I,J)=0
        RETURN
        END
C
C
```

82   3. Electrostatic Fields

```
          SUBROUTINE INPUT(NNO,NEL,NCON)
          COMMON/DATA/KTRI(200,3),MAT(200),X(150),Y(150),PERM(10)
         *VI(10),NOCC(10,20)
          READ(0,*)NNO,NEL,NCON,NMAT
C---------------READ THE MESH STRUCTURE
          DO 1 I=1,NEL
  1       READ(0,*)KTRI(I,1),KTRI(I,2),KTRI(I,3), MAT(I)
C---------------READ NODE COORDINATES
          DO 2 I=1,NNO
  2       READ(0,*)X(I),Y(I)
C---------------READ BOUNDARY CONDITIONS
          DO 3 I=1,NCON
          READ(0,*)VI(I)
          READ(0,*)(NOCC(I,J),J=1,20)
  3       CONTINUE
C---------------READ PERMITTIVITIES OF MATERIALS
          DO 4 I=1,NMAT
  4       READ(0,*)PERM(I)
          RETURN
          END
C
C
          SUBROUTINE FORM(NEL)
          COMMON/DATA/KTRI(200,3),MAT(200),X(150),Y(150),PERM(10)
         *VI(10),NOCC(10,20)
          COMMON /MATRIX/SS(150,150),VV(150),VDR(150)
          DIMENSION NAUX(3), S(3,3)
C---------------DO FOR NEL ELEMENTS
          DO 1 I=1,NEL
          N1=KTRI(I,1)
          2=KTRI(I,2)
          N3=KTRI(I,3)
          NM=MAT(I)
C---------------CALCULATE Qi, Qj, Qk, Ri, Rj, Rk
          Q1=Y(N2)-Y(N3)
          Q2=Y(N3)-Y(N1)
          Q3=Y(N1)-Y(N2)
          R1=X(N3)-X(N2)
          R2=X(N1)-X(N3)
          R3=X(N2)-X(N1)
          XPERM=PERM(NM)
C---------------CALCULATE DETERMINANT, TWICE THE AREA OF TRIANGLE
          DET=X(N2)*Y(N3)+X(N1)*Y(N2)+X(N3)*Y(N1)
         *-X(N1)*Y(N3)-X(N3)*Y(N2)-X(N2)*Y(N1)
          COEFF=XPERM/DET/2.
C---------------CALCULATE THE TERMS S(3,3)
          S(1,1)=COEFF*(Q1*Q1+R1*R1)
          S(1,2)=COEFF*(Q1*Q2+R1*R2)
          S(1,3)=COEFF*(Q1*Q3+R1*R3)
          S(2,1)=S(1,2)
          S(2,2)=COEFF*(Q2*Q2+R2*R2)
          S(2,3)=COEFF*(Q2*Q3+R2*R3)
          S(3,1)=S(1,3)
          S(3,2)=S(2,3)
          S(3,3)=COEFF*(Q3*Q3+R3*R3)
```

```
C---------------ASSEMBLE THE S(3,3) INTO THE MATRIX SS(NNO,NNO)
      NAUX(1)=N1
      NAUX(2)=N2
      NAUX(3)=N3
      DO 2 K=1,3
      KK=NAUX(K)
      DO 2 J=1,3
      JJ=NAUX(J)
    2 SS(KK,JJ)=SS(KK,JJ)+S(K,J)
    1 CONTINUE
      RETURN
      END
C
C
      SUBROUTINE CONDI(NCON,NNO)
      COMMON/DATA/KTRI(200,3),MAT(200),X(150),Y(150),PERM(10)
     *VI(10),NOCC(10,20)
      COMMON /MATRIX/SS(150,150),VV(150),VDR(150)
      DO 1 I=1,NCON
      DO 2 J=1,20
      NOX=NOCC(I,J)
      IF(NOX.EQ.0)GOTO 1
C---------------ZERO THE COEFFICIENTS IN LINE OF MATRIX SS
      DO 3 L=1,NNO
    3 SS(NOX,L)=0.
C---------------SET THE DIAGONAL TO 1.
      SS(NOX,NOX)=1.
C---------------PLACE IMPOSED POTENTIALS IN THE RIGHT HAND SIDE
      VDR(NOX)=VI(I)
    2 CONTINUE
    1 CONTINUE
      RETURN
      END
C
C
      SUBROUTINE ELIM(NNO)
      COMMON /MATRIX/SS(36,36),VV(36),VDR(36)
C---------------GAUSSIAN ELIMINATION
      NN=NNO-1
      DO 1 I=1,NN
      DO 1 M=I+1,NNO
      FACT=SS(M,I)/SS(I,I)
      VDR(M)=VDR(M)-VDR(I)*FACT
      DO 5 J=I+1,NNO
    5 SS(M,J)=SS(M,J)-SS(I,J)*FACT
    1 CONTINUE
      VV(NNO)=VDR(NNO)/SS(NNO,NNO)
      DO 7 I=NN,1,-1
      SUM=0.
      DO 8 J=I+1,NNO
    8 SUM=SUM+SS(I,J)*VV(J)
      VV(I)=(VDR(I)-SUM)/SS(I,I)
    7 CONTINUE
      RETURN
      END
```

```
C
C
        SUBROUTINE OUTPUT(NNO,NEL)
        COMMON/DATA/KTRI(200,3),MAT(200),X(150),Y(150),PERM(10)
       *VI(10),NOCC(10,20)
        COMMON /MATRIX/SS(150,150),VV(150),VDR(150)
C---------------PRINT THE POTENTIALS NOS NOS
        DO 1 I=1,NNO
   1    WRITE(*,100)I,VV(I)
   100  FORMAT(' NODE-',I3,'   POTENTIAL=',E10.4)
C---------------PRINT THE FIELDS IN THE ELEMENTS
        DO 2 I=1,NEL
C---------------CALL GRAD TO CALCULATE THE FIELDS OR GRADIENTS
        N1=KTRI(I,1)
        N2=KTRI(I,2)
        N3=KTRI(I,3)
        CALL GRAD(N1,N2,N3,EX,EY)
        EMOD=SQRT(EX*EX+EY*EY)
   2    WRITE(*,101)I,EX,EY,EMOD
   101  FORMAT(' ELEMENT-',I3,'  EX=',E10.4,'  EY=',E10.4,'
       *EM=',E10.4)
        RETURN
        END
C
C
        SUBROUTINE GRAD(N1,N2,N3,EX,EY)
        COMMON/DATA/KTRI(200,3),MAT(200),X(150),Y(150),PERM(10)
       *VI(10),NOCC(10,20)
        COMMON /MATRIX/SS(150,150),VV(150),VDR(150)
        Q1=Y(N2)-Y(N3)
        Q2=Y(N3)-Y(N1)
        Q3=Y(N1)-Y(N2)
        R1=X(N3)-X(N2)
        R2=X(N1)-X(N3)
        R3=X(N2)-X(N1)
C---------------CALCULATE DETERMINANT, TWICE THE AREA OF TRIANGLE
        DET=X(N2)*Y(N3)+X(N1)*Y(N2)+X(N3)*Y(N1)
       *-X(N1)*Y(N3)-X(N3)*Y(N2)-X(N2)*Y(N1)
        EX=-(Q1*VV(N1)+Q2*VV(N2)+Q3*VV(N3))/DET
        EY=-(R1*VV(N1)+R2*VV(N2)+R3*VV(N3))/DET
        RETURN
        END
C
```

3.9.4. *Example for Use of the Finite Element Program*

Suppose that we wish to find the electric field intensity and the potential distributions within the geometry shown in Figure 3.28 where $\varepsilon_1 = 5\varepsilon_0$. A finite element mesh is shown in Figure 3.29 with element and node numbers shown.

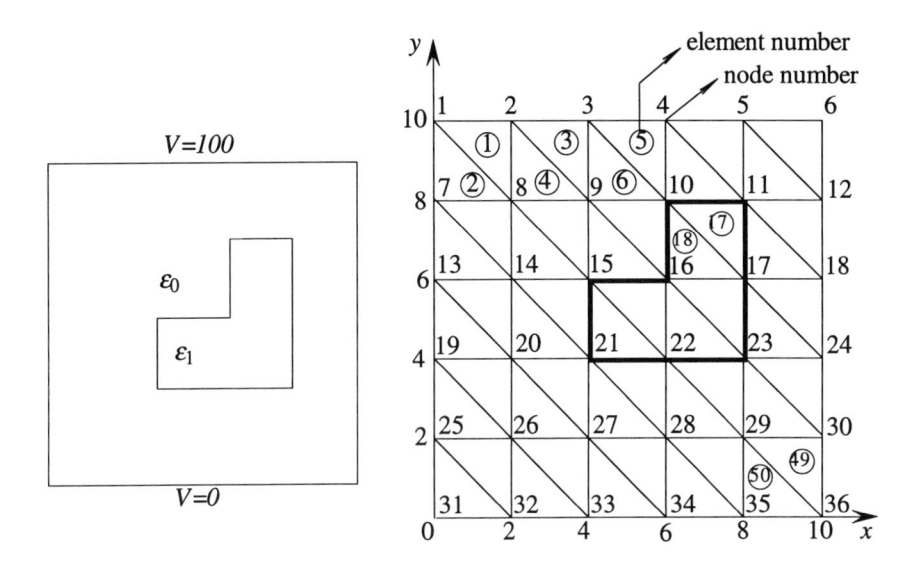

Figure 3.28. A simple geometry used to demonstrate the use of the finite element program.

Figure 3.29. Finite element discretization of the geometry in Figure 3.28. Circled number are element numbers, others are node numbers. Numbers on the axes are dimensions.

The various variables for this mesh are:

NNO = 36   (number of nodes)
NEL = 50   (number of elements)
NCON = 2   (number of equipotential boundary lines)
NMAT = 2   (number of materials)

The arrays KTRI and MAT corresponding to the elements are:

1 8 2 1   (1st triangle)
1 7 8 1
2 9 3 1
2 8 9 1
.
.
.
10 17 11 2   (material 2)
.
.
29 35 36 1   (triangle 50)

3. Electrostatic Fields

The array X and Y are given as:

0. , 10.   (Node 1)
2. , 10.   (Node 2)
.
.
.
10. , 0.   (Node 36)

The boundary conditions are:

100.   (potential on upper boundary)
1 2 3 4 5 6 0 0....   (nodes with potential 100 V)
0   (potential on lower boundary)
31 32 33 34 35 36 0 0....   (nodes with potential 0 V)

Permittivities are given as:

1.   (ε_r of material 1)
5.   (ε_r of material 2)

The results obtained from this program are listed below:

NODE- 1   POTENTIAL= .1000E+03
NODE- 2   POTENTIAL= .1000E+03
NODE- 3   POTENTIAL= .1000E+03
NODE- 4   POTENTIAL= .1000E+03
NODE- 5   POTENTIAL= .1000E+03
NODE- 6   POTENTIAL= .1000E+03
NODE- 7   POTENTIAL= .8010E+02
NODE- 8   POTENTIAL= .7960E+02
NODE- 9   POTENTIAL= .7740E+02
NODE- 10 POTENTIAL= .7145E+02
NODE- 11 POTENTIAL= .7162E+02
NODE- 12 POTENTIAL= .7612E+02
NODE- 13 POTENTIAL= .6120E+02
NODE- 14 POTENTIAL= .6089E+02
NODE- 15 POTENTIAL= .5856E+02
.
.
.
NODE- 26 POTENTIAL= .2274E+02

Introduction to the Finite Element Method 87

NODE- 27 POTENTIAL= .2466E+02
NODE- 28 POTENTIAL= .2565E+02
NODE- 29 POTENTIAL= .2544E+02
NODE- 30 POTENTIAL= .2453E+02
NODE- 31 POTENTIAL= .0000E+00
NODE- 32 POTENTIAL= .0000E+00
NODE- 33 POTENTIAL= .0000E+00
NODE- 34 POTENTIAL= .0000E+00
NODE- 35 POTENTIAL= .0000E+00
NODE- 36 POTENTIAL= .0000E+00
ELEMENT- 1 EX= .0000E+00 EY=-.1020E+02 EM= .1020E+02
ELEMENT- 2 EX= .2508E+00 EY=-.9950E+01 EM= .9953E+01
ELEMENT- 3 EX= .0000E+00 EY=-.1130E+02 EM= .1130E+02
ELEMENT- 4 EX= .1098E+01 EY=-.1020E+02 EM= .1026E+02
ELEMENT- 5 EX= .0000E+00 EY=-.1427E+02 EM= .1427E+02

.
.
.
.

ELEMENT-37 EX= .4643E+00 EY=-.1307E+02 EM= .1308E+02
ELEMENT-38 EX= .1063E+00 EY=-.1343E+02 EM= .1343E+02
ELEMENT-39 EX= .2174E+01 EY=-.1135E+02 EM= .1156E+02
ELEMENT-40 EX= .4560E+00 EY=-.1307E+02 EM= .1308E+02
ELEMENT-41 EX=-.3197E+00 EY=-.1137E+02 EM= .1138E+02
ELEMENT-42 EX= .0000E+00 EY=-.1105E+02 EM= .1105E+02
ELEMENT-43 EX=-.9604E+00 EY=-.1233E+02 EM= .1237E+02
ELEMENT-44 EX= .0000E+00 EY=-.1137E+02 EM= .1137E+02
ELEMENT-45 EX=-.4951E+00 EY=-.1283E+02 EM= .1284E+02
ELEMENT-46 EX= .0000E+00 EY=-.1233E+02 EM= .1233E+02
ELEMENT-47 EX= .1063E+00 EY=-.1272E+02 EM= .1272E+02
ELEMENT-48 EX= .0000E+00 EY=-.1283E+02 EM= .1283E+02
ELEMENT-49 EX= .4560E+00 EY=-.1226E+02 EM= .1227E+02
ELEMENT-50 EX= .0000E+00 EY=-.1272E+02 EM= .1272E+02

Equipotential lines obtained for a similar case with the EFCAD finite element program (a much more sophisticated software package), using a larger number of nodes are shown in Figure 3.30. The electric field intensity lines are shown in Figure 3.31.

As a final note we observe that the program presented above contains approximately 150 lines of code. This program can calculate most realistic problems and is much more flexible than all the analytic methods mentioned at the beginning of this section.

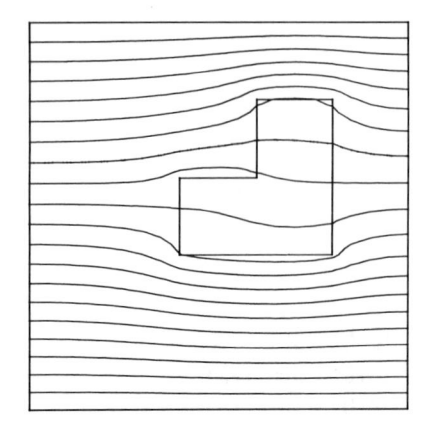

Figure 3.30. Equipotential lines for the geometry in Figure 3.28.

Figure 3.31. Field intensity lines for the geometry in Figure 3.28.

3.10. Tables of Permittivities, Dielectric Strength, and Conductivities

Relative permittivity, dielectric strength, and conductivities of some materials are given in the following two tables.

Table 3.1. Relative permittivities and dielectric strength of some materials.

Material	Relative Permittivity ε_r	Dielectric Strength K (V/m)
Air	1	3×10^6
Mineral oil	2.3	1.5×10^7
Insulating paper	3	2×10^7
Porcelain	7	2×10^7
Glass	6	3×10^7
Paraffin	2	3×10^7
Quartz	4	4×10^7
Polystyrene	2.6	5×10^7
Mica	6	2×10^8

Table 3.2. Conductivities of some common materials.

Material	Conductivity $\sigma\,(\Omega.m)^{-1}$
Silver	6.1×10^7
Copper	5.7×10^7
Gold	4.1×10^7
Aluminum	3.5×10^7
Brass	1.1×10^7
Iron	1×10^7
Mercury	1×10^6
Graphite	1×10^5
Carbon	3.0×10^4

4
Magnetostatic Fields

4.1. Introduction

We have shown in Chapter 2 that Maxwell's equations can be divided into two groups based on the separation of electric and magnetic properties. A special subset of Maxwell's equations consists of the equations

$$\nabla \times \mathbf{H} = \mathbf{J}$$

$$\nabla \cdot \mathbf{B} = 0$$

$$\nabla \times \mathbf{E} = -\frac{\partial \mathbf{B}}{\partial t}$$

If the equations are independent of time, we have

$$\nabla \times \mathbf{H} = \mathbf{J} \qquad (4.1)$$

$$\nabla \cdot \mathbf{B} = 0 \qquad (4.2)$$

$$\nabla \times \mathbf{E} = 0 \qquad (4.3)$$

These equations are accompanied by the constitutive relations

$$\mathbf{B} = \mu \mathbf{H} \qquad (4.4)$$

$$\mathbf{J} = \sigma \mathbf{E} \qquad (4.5)$$

This subset of equations describes the phenomena associated with magnetostatics.

In addition, we introduce in this chapter the different types of magnetic materials and their properties as well as define the concept of inductance. Although in some instances it will become necessary to use the notion of time, the results we obtain are static in nature.

4.2. Maxwell's Equations in Magnetostatics

4.2.1. *The Equation $\nabla \times \mathbf{H} = \mathbf{J}$*

This equation defines qualitatively and quantitatively the generation of $\mathbf{H}$ in terms of $\mathbf{J}$. We recall that the same relation in integral form is

$$\int_S (\nabla \times \mathbf{H}) \cdot d\mathbf{s} = \int_S \mathbf{J} \cdot d\mathbf{s} \qquad (4.6)$$

where S is a surface on which $\mathbf{H}$ and $\mathbf{J}$ are defined. Using Stokes' theorem, the left-hand side of the expression can be written as

$$\int_S (\nabla \times \mathbf{H}) \cdot d\mathbf{s} = \oint_C \mathbf{H} \cdot d\mathbf{l} \qquad (4.7)$$

where C is a contour, enclosing the surface S. The right-hand side of Eq. (4.6) represents the flux of the vector $\mathbf{J}$ crossing the surface S. This flux is the conduction current crossing S. That is

$$\oint_C \mathbf{H} \cdot d\mathbf{l} = I \qquad (4.8)$$

which indicates that the circulation of $\mathbf{H}$ along a contour C encircling a surface S is equal to the current crossing this surface. Maxwell's equation $\nabla \times \mathbf{H} = \mathbf{J}$ written in the form above is referred to as "Ampere's law."

We look now at the application of this equation to the case of an infinite wire carrying a current I as shown in Figure 4.1.

By choosing the surface S_1 as a circle of radius R, the application of Ampere's law is simplified:

$$\oint_{C_1} \mathbf{H} \cdot d\mathbf{l} = I$$

Since $\mathbf{H}$ and $d\mathbf{l}$ are collinear vectors in the same direction, the scalar product $\mathbf{H} \cdot d\mathbf{l}$ is equal to the product of the magnitudes of H and dl. Because of homogeneity of material properties, H is identical at all points along C_1 and is not dependent on C_1. Therefore, the integration becomes

$$H \oint_{C_1} dl = I \quad \text{or} \quad H 2\pi R = I$$

and

92   4. Magnetostatic Fields

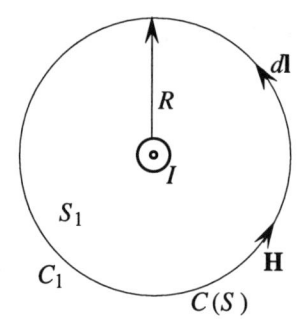

Figure 4.1. The use of Ampere's law to calculate the magnetic field intensity of an infinite wire.

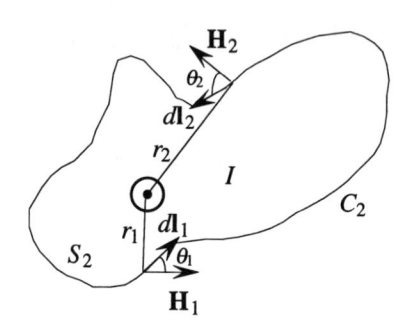

Figure 4.2. The use of Ampere's law with an irregular contour.

$$H = \frac{I}{2\pi R} \quad (4.9)$$

We note here that the fact that we chose S such that $d\mathbf{l}$ coincided with $\mathbf{H}$ facilitated the solution.

Assume now that we choose the surface S shown in Figure 4.2. Since S_2 is an irregular surface, we divide C_2 into n segments and obtain

$$\oint_{C_2} \mathbf{H} \cdot d\mathbf{l} = \int_{l_1} \mathbf{H}_1 \cdot d\mathbf{l}_1 + \int_{l_2} \mathbf{H}_2 \cdot d\mathbf{l}_2 + \ldots \int_{l_n} \mathbf{H}_n \cdot d\mathbf{l}_n$$

This equation has n unknowns. It is important to point out that while Ampere's law is still valid, its application in this particular problem is practically impossible. An additional difficulty is the fact that, since H is always tangential to a circle whose center is at the center of the conductor, the scalar product $\mathbf{H}_n \cdot d\mathbf{l}_n$ changes to $H_n dl_n \cos(\theta_n)$ where the angle θ_n between $\mathbf{H}_n$ and $d\mathbf{l}_n$ varies from point to point.

We can choose still another type of surface, such as S_3 shown in Figure 4.3 which does not contain the conductor. On this surface, the quantity

$$H \oint_{C_3} dl = 0$$

Because the current crossing the surface S_3 is zero, does this also imply that $\mathbf{H}=0$? In reality the field intensity H generated by the current does not depend on the surface we choose for the application of Ampere's law, hence, since $I \neq 0$ in the conductor, $\mathbf{H}$ is also not zero.

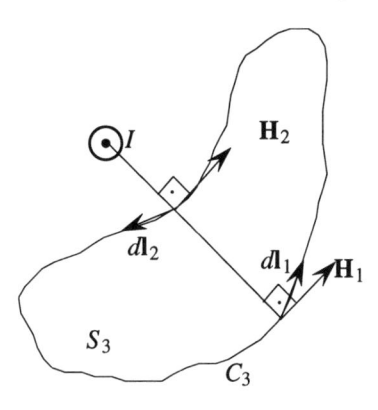

Figure 4.3. A contour that does not include the current.

In fact, we observe in Figure 4.3 that $\mathbf{H}_1 \cdot d\mathbf{l}_1 > 0$ and that $\mathbf{H}_2 \cdot d\mathbf{l}_2 < 0$. Only the sum of all sections yields zero, accounting for the zero net value.

We conclude, pointing out that Ampere's law, as one of Maxwell's equations, is always valid, but its application is not always a simple matter.

An important aspect of the equation $\nabla \times \mathbf{H} = \mathbf{J}$ is apparent when we apply the divergence to both sides of the equation. By doing so, we obtain, as was shown in Chapter 2, the equation

$$\nabla \cdot \mathbf{J} = 0 \qquad (4.10)$$

which is the equation of electric continuity. This indicates that the conduction current, which is the flux of the vector $\mathbf{J}$, is conservative; that is, a current entering a volume is the same as the current leaving the volume.

4.2.2. The Equation $\nabla \cdot \mathbf{B} = 0$

This equation is in a sense analogous to the equation $\nabla \cdot \mathbf{J} = 0$ above. In this case, it is the magnetic flux that is conservative. We note that this equation does not indicate how $\mathbf{B}$ is generated; it only defines the conservative flux condition. We will see in subsequent paragraphs that the application of this condition provides a convenient relation for solution of certain problems.

4.2.3. The Equation $\nabla \times \mathbf{E} = 0$

This equation is a particular case of $\nabla \times \mathbf{E} = -\partial \mathbf{B}/\partial t$ and indicates how the electric field is generated due to the time variation of $\mathbf{B}$. The fact that the curl of the electric field intensity is zero does not mean that in the magnetostatic case the electric field intensity is zero.

94   4. Magnetostatic Fields

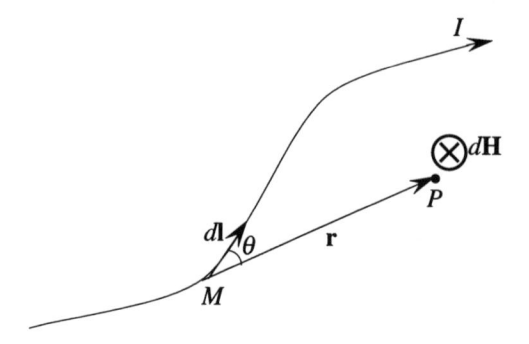

Figure 4.4.

There is no reason why an electric field external to the domain cannot be applied, which we can consider as constant. However, in the domain under study, we cannot have an electric field generated by devices contained within the domain.

4.3. The Biot-Savart Law

The three equations above constitute the main relations of magnetostatics. We can attribute to Ampere's law, derived from $\nabla \times \mathbf{H} = \mathbf{J}$, a certain prominence in relation to the other two equations since it relates the magnetic field intensity **H** to its generating source **J**. Although this law is valid in general situations, its application is limited to a few simple cases, unless we invoke approximations. One of these few applications with exact solutions is the example of an infinite wire, considered above, although, considering an infinite wire, is in itself an approximation.

The Biot-Savart law is an auxiliary expression for the calculation of H as a function of the current that generates it. It is necessary to note that Biot-Savart's law, conceptually, adds absolutely nothing to Maxwell's equations. We can view it as an algebraic variation to Ampere's law. This law was proposed by Biot and Savart as an experimental law. Biot and Savart's law was introduced relatively late in the development of field theory. This derivation is rather complex and involves electromagnetic quantities which we have not yet defined. We do not treat the particulars of their work here as being irrelevant to the purpose of this work. Instead we simply use it here as a given relation.

To introduce Biot-Savart's law, we use Figure 4.4 where we wish to calculate the magnetic field intensity **H** at point *P*. This field intensity is generated by the current *I*, passing through a conductor of arbitrary shape. The Biot-Savart law is written in differential form as

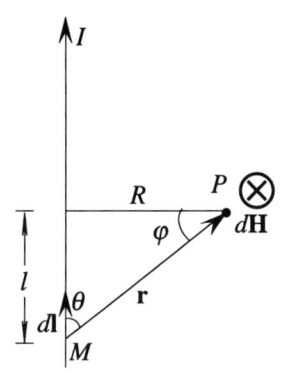

Figure 4.5. The use of Biot-Savart's law for the calculation of the magnetic field intensity of an infinitely long wire carrying a current I.

$$d\mathbf{H} = I \frac{d\mathbf{l} \times \mathbf{r}}{4\pi r^3} \quad (4.11)$$

The wire is divided into small segments with which we can associate a vector $d\mathbf{l}$, whose direction is the same as the current I. We now have to define a vector $\mathbf{r}$ such that $\mathbf{r} = \mathbf{P} - \mathbf{M}$. The summation of the vectors $d\mathbf{H}$ provides the field $\mathbf{H}$ generated by the current I, at point P. The direction of $d\mathbf{H}$ is as indicated in Figure 4.4. One method of obtaining the direction of $d\mathbf{H}$ is to use the cross product $d\mathbf{l} \times \mathbf{r}$. The magnitude of $d\mathbf{H}$ is given by

$$dH = \frac{I \, dl}{4\pi r^2} \sin\theta \quad (4.12)$$

where θ is the angle between $d\mathbf{l}$ and $\mathbf{r}$.

Biot-Savart's law permits the calculation of $\mathbf{H}$ due to a conductor of irregular form. In this case we divide the conductor into a finite number of segments and sum the resulting values of $d\mathbf{H}$ vectorially. It is not difficult to write a computer program that performs these operations automatically.

However, we can apply Biot-Savart's law in an analytic fashion only to a limited number of structures. As an example we look at the calculation of $\mathbf{H}$ generated by an infinite wire carrying current I. $\mathbf{H}$ is calculated at a point P at a distance R from the wire (see Figure 4.5). We note that $d\mathbf{H}$ is perpendicular to the plane of Figure 4.5. Its magnitude is

$$dH = \frac{I \, dl}{4\pi r^2} \sin\theta$$

or

96   4. Magnetostatic Fields

$$dH = \frac{I\,dl}{4\pi r^2} \cos\phi \qquad (4.13)$$

Noting that

$$\tan\phi = \frac{1}{R} \quad \text{and} \quad \cos\phi = \frac{R}{r}$$

we have

$$dl = R\sec^2\phi\,d\phi \quad \text{and} \quad r = \frac{R}{\cos\phi}$$

Substituting these expressions in Eq. (4.13), yields upon simplification

$$H = \frac{I}{4\pi R}\int_{-\frac{\pi}{2}}^{+\frac{\pi}{2}} \cos\phi\,d\phi$$

The limits of integration, $-\pi/2$ and $+\pi/2$, are the angles ϕ corresponding to the limits $-\infty$ and $+\infty$ of the wire. As a result, we obtain

$$H = I/2\pi R$$

This value is identical to that calculated by Ampere's law for the same situation.

The example given above has a didactic benefit. The utilization of Ampere's law is much simpler, and we should employ it whenever possible. However, there are many cases in which only Biot-Savart's law can be employed, as we will see later.

It is important to point out that beyond these two laws there are no other analytical methods for computation of the field **H** as a function of **J** for general applications. Numerical methods alone can determine **H** in most realistic geometries.

4.4. Boundary Conditions for the Magnetic Field

In a manner similar to the electric field intensity **E**, the magnetic field intensity **H** also undergoes an angular change in passing from one material to another, if the two materials have different permeabilities. Consider two materials which possess permeabilities μ_1 and μ_2 respectively, as in Figure 4.6. The relation between θ_1 and θ_2, is obtained by using the equations

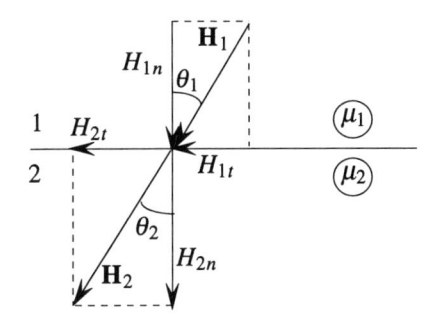

Figure 4.6. Boundary conditions for the magnetic field intensity at the boundary between two materials with different permeabilities.

$$\nabla \times \mathbf{H} = 0 \quad \text{and} \quad \nabla \cdot \mathbf{B} = 0$$

assuming that there are no currents on the boundary between the two materials. Utilizing these equations in a manner analogous to that done for the equations $\nabla \times \mathbf{E}=0$ and $\nabla \cdot \mathbf{D}=0$ in the previous chapter, we obtain equivalent results

- continuity of the tangential component of **H**

$$H_{1t} = H_{2t}$$

- continuity of the normal component of **B**

$$B_{1n} = B_{2n}$$

Similarly we observe that

$$tan\theta_1 = \frac{H_{1t}}{H_{1n}} \quad \text{and} \quad tan\theta_2 = \frac{H_{2t}}{H_{2n}}$$

and

$$\frac{tan\theta_1}{tan\theta_2} = \frac{H_{2n}}{H_{1n}}$$

Recalling that $\mathbf{H}=\mathbf{B}/\mu$, we obtain

$$\frac{tan\theta_1}{tan\theta_2} = \frac{\mu_1}{\mu_2} \qquad (4.14)$$

98   4. Magnetostatic Fields

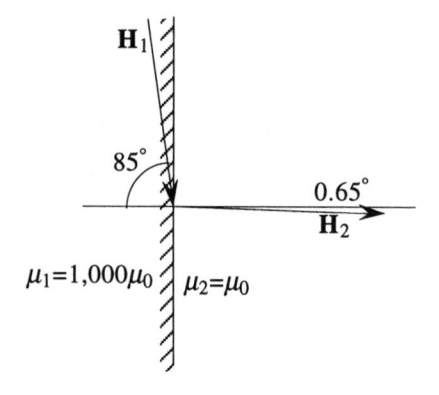

Figure 4.7. The magnetic field intensity relations at the boundary of a high-permeability material and free space.

As we will see in following sections, there are materials with very large permeabilities and materials with low permeabilities. We can obtain important relations based on these permeabilities. Suppose for example, that $\mu_2 = \mu_0$, $\mu_1 = 1000\mu_0$ and that θ_1 is 85°. Then

$$\tan\theta_2 = \frac{\tan 85°}{1000}$$

or

$$\theta_2 = 0.65°$$

This effect is demonstrated in Figure 4.7. Note that, in contrast to electric fields, the angular change is much larger. As an example, in passing the boundary between iron ($\mu = 1000\mu_0$) and air ($\mu = \mu_0$), the magnetic field **H** undergoes an angular change such that, in air, it is practically perpendicular to the iron.

Figures 4.8a and 4.8b show the angular change of electric fields ($\varepsilon_2/\varepsilon_1 = 5$) and magnetic fields ($\mu_2/\mu_1 = 1000$). These examples were obtained through the use of the EFCAD program.

4.5. Magnetic Materials

We first define the relative permeability of a material as

$$\mu_r = \frac{\mu}{\mu_0}$$

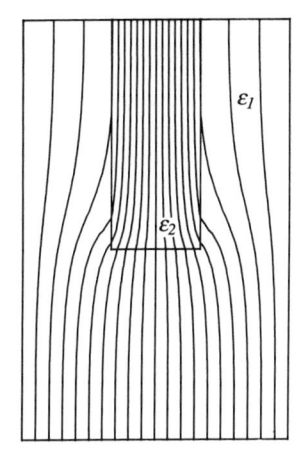

Figure 4.8a. Change in the electric field intensity due to a dielectric material with relative permittivity of 5.

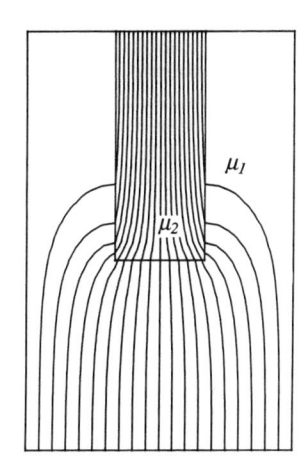

Figure 4.8b. Change in the magnetic flux density due to a magnetic material with relative permeability of 1000.

where μ is the actual permeability of the material and $\mu_0 = 4\pi \times 10^{-7} H/m$ is the permeability of air (actually free space). Therefore μ_r of air is equal to 1.

There are basically two types of magnetic materials. These are:

- Soft magnetic materials: these include diamagnetic, paramagnetic, and ferromagnetic materials.
- Hard magnetic materials: permanent magnets.

It is not our objective here to analyze the microscopic structure of these materials. We will only look at their behavior from a macroscopic point of view.

4.5.1. Diamagnetic Materials

Diamagnetic materials have a relative permeability slightly lower than 1. Notable materials in this group are: mercury, gold, silver, and copper. Copper, for example, has a relative permeability $\mu_r = 0.999991$; the other materials posses μ_r of the same order of magnitude. In practice we can consider these materials as having μ_r equal to 1. One effect of diamagnetism is illustrated in Figure 4.9. A diamagnetic material is placed under the influence of a uniform field. Since $\mu_r < 1$, more of the flux passes through air than through the material, since air is a more permeable material. This causes (as we shall see in Chapter 6) a force that tends to repel the diamagnetic body from the source generating the field. However, because the permeability is very close to 1, this effect is very small and difficult to measure.

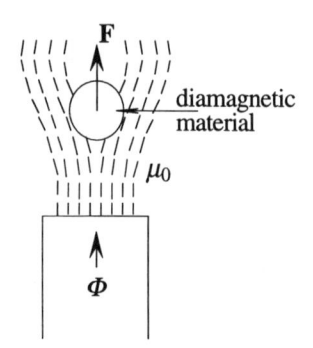

Figure 4.9. A diamagnetic material in a magnetic field. Flux lines tend to pass through free space because it has lower reluctance than the diamagnetic material. The diamagnetic material is repelled by the field.

4.5.2. Paramagnetic Materials

Paramagnetic materials possess a relative permeability μ_r slightly larger than 1. One example to this type of materials is aluminum, for which $\mu_r = 1.00000036$. Therefore, as with the diamagnetic materials, we can consider the materials as having $\mu_r = 1$ for most practical purposes. In general, the effect due to paramagnetism is negligible.

4.5.3. Ferromagnetic Materials

4.5.3a. General

Ferromagnetic materials possess relative permeabilities much larger than 1. As we shall see shortly, these are materials of extreme importance in electromagnetic devices due to their large relative permeabilities μ_r. As an example, iron with 0.2% impurities has a relative permeability of about 6000. A few iron alloys reach a relative permeability of 10^6. It is interesting to note that when a ferromagnetic material is placed in a hot environment, and when the temperature passes over a critical value, called "Curie temperature," the ferromagnetic material changes its magnetic behavior to that of a paramagnetic material. Each material has its specific Curie temperature. For iron this value is approximately 770 ^oC. Another characteristic of this type of materials is that the relative permeability μ_r depends on the magnitude of the magnetic field intensity $|\mathbf{H}|$ in the material. This phenomenon, called "nonlinearity" can be explained with the aid of Figure 4.10. The figure shows a magnetic circuit made of a material with high permeability. A ferromagnetic sample is inserted in this circuit. We wish to obtain the magnetic characteristics of the material.

We will see later that $H = NI/l$ and noting that $B = \Phi/S$, we can measure the current I applied to a coil and the flux Φ passing through the material.

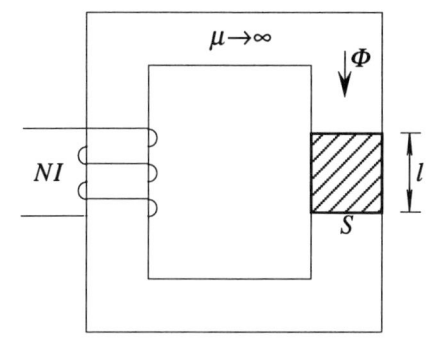

Figure 4.10. Magnetic circuit used to obtain the magnetic characteristics of a ferromagnetic sample.

For low values of I, Φ increases in the same proportion as the magnitude of I. Beyond a certain value of I the increase in Φ is in different proportion than the increase in I. With H and B directly proportional to I and Φ we obtain the curve shown in Figure 4.11. B_s, indicated in the figure, is called the "saturation flux density" or "saturation induction." We point out that the characteristic curve of a magnetic material is always the $B(H)$ curve since the $\Phi(I)$ curve contains, implicitly, the dimensions of surface, and length, and the number of turns. These refer to a particular structure rather than to the material alone.

We can now obtain a permeability curve for the material as a function of H. Using the relation $B=\mu H$ and Figure 4.11, we can write $tan\alpha = B/H = \mu$. Note that up to point 1, the tangent is constant. Beyond this point, α begins to diminish ($\alpha_2 < \alpha_1$); hence the permeability diminishes.

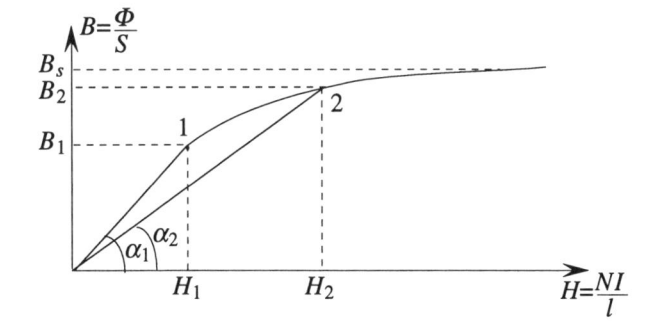

Figure 4.11. The $B(H)$ curve obtained for the sample in Figure 4.10. Only the curve in the first quadrant is shown.

102   4. Magnetostatic Fields

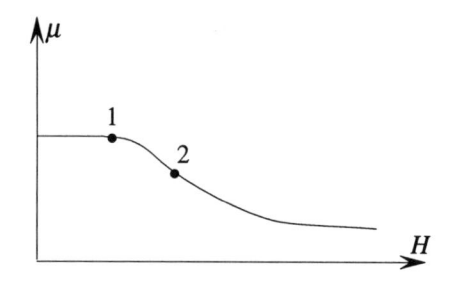

Figure 4.12. Permeability curve corresponding to Figure 4.11.

Figure 4.12 shows the permeability curve. In practice, at the beginning of the $B(H)$ curve, α varies (not shown). However, many properly designed electromagnetic devices work at high flux densities, close to saturation, (therefore far from the beginning of the curve), and we can neglect this perturbation in μ.

If we subject a ferromagnetic material to the influence of a magnetic field, we obtain the situation shown in Figure 4.13. With $\mu_r >> 1$, the magnetic flux is strongly attracted by the ferromagnetic material since it is a highly permeable material. In this case, the ferromagnetic material is physically attracted. The reason for this will be seen later, in Chapter 6.

4.5.3b. The Influence of Iron on Magnetic Circuits

An infinitely long wire, carrying a current I, creates in the surrounding space a field, which, as we have seen, has a magnitude $H=I/2\pi r$ (r is the distance of a point from the wire). I, the current in the wire can be called a magneto-motive force (*mmf*) since it is capable of generating a magnetic field. Consider a second situation as shown in Figure 4.14, where the same wire is surrounded by a ferromagnetic material with high relative permeability μ_r. Assume now that this material has a physical gap. To calculate the magnetic field we use Ampere's law

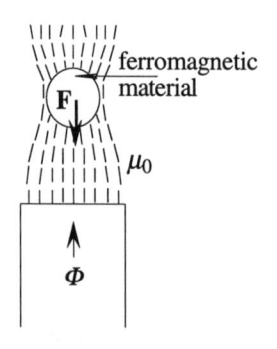

Figure 4.13. Ferromagnetic material in a magnetic field. Flux lines pass through the material and the material is attracted to the source of the field.

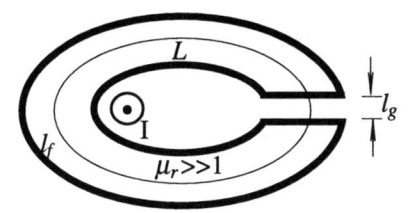

Figure 4.14. A current surrounded by a high-permeability magnetic circuit.

$$\oint_C \mathbf{H} \cdot d\mathbf{l} = I$$

Choosing a contour L that coincides with the field $\mathbf{H}$, and dividing the contour into l_f in iron and l_g in the gap, the expression is

$$\int_{l_f} \mathbf{H}_f \cdot d\mathbf{l} + \int_{l_g} \mathbf{H}_g \cdot d\mathbf{l} = I$$

Assuming that the fields are constant in their respective regions, we obtain

$$H_f l_f + H_g l_g = I \qquad (4.15)$$

There are two unknowns in this equation, requiring the establishment of a second relation. This relation is obtained from the secondary consideration that the flux Φ_f in iron is identical to the flux Φ_g in the gap

$$\Phi_f = \Phi_g \qquad \therefore \mu_f H_f S_f = \mu_g H_g S_g$$

where S_f and S_g are the cross-sectional areas of the iron and the gap (perpendicular to the plane of the figure). Assuming that these two have approximately the same magnitude, we have

$$H_f = \frac{\mu_0}{\mu_f} H_g \qquad (4.16)$$

using (4.16) in (4.15) gives

$$H_g = \frac{I}{\frac{\mu_0}{\mu_f} l_f + l_g} \qquad (4.17)$$

With $\mu_f >> \mu_0$, we get $H_g = I/l_g$. If the circuit is in deep saturation, this approximation is not valid.

As a numerical example, assume that $I=10$ A and $l_g=1$ mm (a typical value in electromechanical structures). We obtain $H_g=10,000$ A/m. Returning to the case of the wire in space, mentioned at the beginning of this section, calculating the field with a 10 A current, we get at 1 mm distance (very close to the wire) from the expression $H=I/2\pi r$, a field intensity $H=1,590$ A/m.

In conclusion, we observe the following phenomena due to the presence of ferromagnetic materials in magnetic circuits:

- Modification of the shape of the magnetic field and conduction of flux to regions where it is required.
- Generation of high magnetic field intensities in gaps. We shall see in Chapter 6 that magnetic forces depend on the field intensity squared. This is one reason for the interest in high-permeability materials.
- The circulation of the magnetic field in the ferromagnetic material is negligible if $\mu_f >> \mu_0$, since $H_f << H_g$. We conclude that, in practice, all the *emf* is applied in the gap, where

$$\int_{l_g} \mathbf{H}_g \cdot d\mathbf{l} \approx nI \qquad (4.18)$$

and we assumed that instead of a single wire there are n wires.

4.5.4. Permanent Magnets

4.5.4a. General Properties of Hard Magnetic Materials

We have already classified magnetic materials as follows:

- Soft magnetic materials: materials which, when the applied magnetic field is canceled, do not retain a significant remnant flux density (or remnant induction). These materials are "passive" to the presence of the magnetic field. If the external field varies in magnitude or direction, the same change takes place in the material with essentially no retardation effect. All materials studied in the previous sections were soft magnetic materials.
- Hard magnetic materials: materials that retain a significant remnant flux density when the external field passing through the material is removed. These hard magnetic materials are called permanent magnets. The magnetic properties of a permanent magnet are described with the help of Figure 4.15a and 4.15b.

The magnetic circuit in Figure 4.15a is made of a ferromagnetic material with high permeability. A hard magnetic material is placed in the gap.

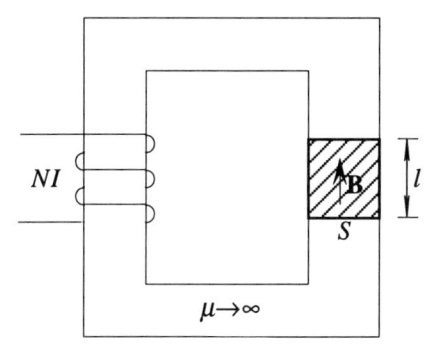

Figure 4.15a. A hard magnetic material sample in the gap of a high-permeability magnetic circuit.

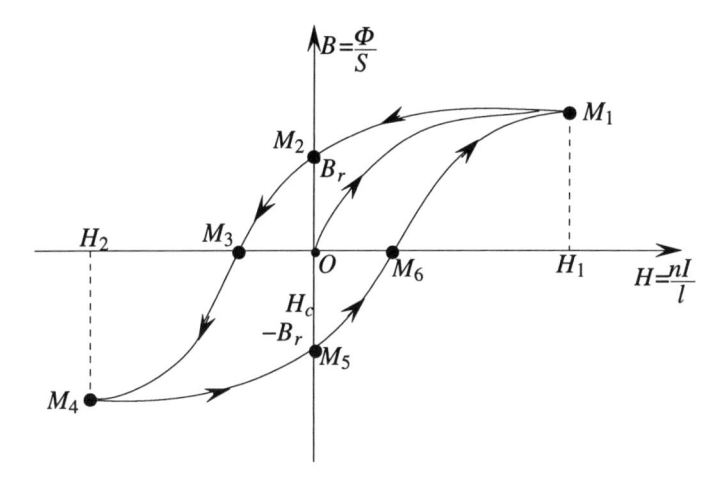

Figure 4.15b. The $B(H)$ curve (hysteresis loop) for the sample in Figure 4.15a.

We assume that this material has not yet been subjected to the effect of any magnetic field. The field generated inside the iron, and therefore in the magnet, is proportional to I, since $H = nI/l$. The flux density B in the magnet is proportional to the measured flux Φ, $(B = \Phi/S)$.

The different stages in the operational cycle of the magnet are as follows:

a) Initially, increasing the current I causes the magnetic field intensity to change from zero to H_1. The curve OM_1 is called the "initial magnetization curve."
b) Decreasing H from H_1 to zero (M_1 to M_2), or decreasing I to 0. At this stage the intrinsic characteristic of hard magnetic materials appears, since the magnet retains a considerable remnant flux density, indicated as B_r in Figure 4.15b. We

note that, at this point, where $I=0$, we have a flux in the magnetic circuit due to the remnant flux density B of the magnet.

c) Increasing the current in the opposite direction, B passes through zero at H_c (interval $M_2 - M_3$). At H_c, the flux due to the permanent magnet is identical in magnitude but opposite in direction to the flux of the coil, canceling the flux in the magnetic circuit, and therefore $B=0$. The field intensity H_c is called the "coercive field intensity" or "coercive force."

d) Continuing the cycle (M_3, M_4, M_5), we again create a magnetic flux, but the remnant flux density B_r at point M_3 is opposite in direction to B_r at point M_2.

This cycle is called the hysteresis loop, whose internal area is significant for hard magnetic materials. Soft ferromagnetic materials also posses hysteresis curves, but the area described by the loop is relatively small.

We look now at the origin of magnetism in hard magnetic materials from a microscopic point of view.

Magnetic materials posses small domains (macroscopically they have dimensions of the order of 10^{-3} to $10^{-6} m$) called "magnetic domains" or "Weiss domains," composed of various molecules of the material in question. Initially, these domains posses magnetic fields directed in arbitrary directions (see Figure 4.16), however, when a large external field H is applied, the magnetic fields of the domains tend to align with the external field. When the external field is removed, the domains maintain their alignment. The cumulative effect of the magnetic domains form the remnant flux density of the permanent magnet.

In soft magnetic materials, the situation is approximately the same, however, the application of a weak external magnetic field in the direction opposite to the internal field is sufficient to eliminate the remnant flux density, as opposed to the situation in permanent magnets.

A magnet is defined by its $B(H)$ curve in the second quadrant (interval M_2 to M_3 in Figure 4.15b), which is reproduced in Figure 4.17. This curve, in general, is known or is supplied by the manufacturer, and indicates the remnant flux density B_r, the coercive field intensity H_c, as well as the manner in which B and H vary between these two points. Its purpose is to define how the magnet's flux density varies as a function of the internal field intensity of the magnet.

We can now explain why the magnet is represented in the second quadrant: in Figure 4.18 a field line of the magnet in air is emphasized.

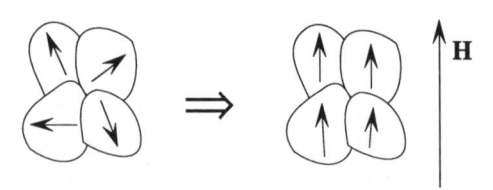

Figure 4.16. Magnetic domains before and after alignment with the external magnetic field.

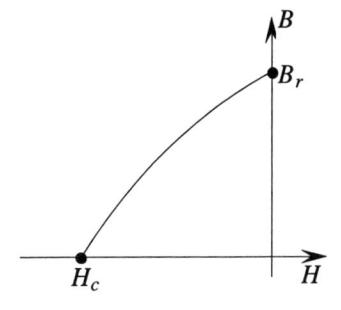

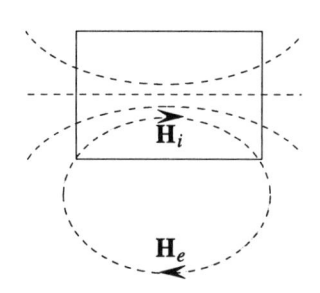

Figure 4.17. The $B(H)$ curve of a permanent magnet.

Figure 4.18. The field of the magnet. A field line in air is shown.

Applying Ampere's law, we calculate the field intensity H along the contour l, where l_i and l_e are the internal and external parts of the contour respectively

$$\oint_l \mathbf{H} \cdot d\mathbf{l} = 0 \quad \text{(no currents)}$$

$$\int_{l_i} \mathbf{H}_i \cdot d\mathbf{l} + \int_{l_e} \mathbf{H}_e \cdot d\mathbf{l} = 0$$

This analysis assumes that the fields are uniform on the two parts of the contour (l_i and l_e) respectively. We get

$$H_i = -H_e \frac{l_e}{l_i}$$

Note that H_i must have an opposite direction to H_e along any contour. Therefore, considering the conservation of flux, the magnetic flux densities B_i and B_e must be in the same directions in relation to the contour of circulation. In other words, since in air, **B**, and **H** have the same direction ($\mathbf{B} = \mu_0 \mathbf{H}$), the directions of **B** and **H** in the magnet must be opposite to each other. The magnetic field intensity **H** and the magnetic flux density **B** in the magnet and the region that it encloses are indicated in Figure 4.19, showing that in the interior of the magnet **H** and **B** are opposite in directions. **B** and **H** must be the same throughout the tube of flux because of the conservation of flux.

In addition, it is evident that the apparent permeability of the magnet is negative, since $H<0$ and $B>0$ (Figure 4.17, in the second quadrant).

108   4. Magnetostatic Fields

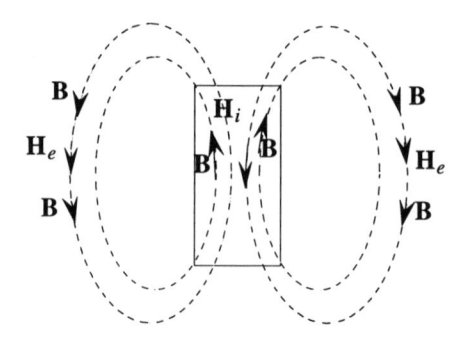

Figure 4.19. The magnetic field intensity and magnetic flux density inside and outside a magnet. The magnetic field intensity inside the magnet is opposite, in direction to the magnetic flux density.

4.5.4b. The Energy Associated with a Magnet

It is obvious that if a magnetostatic field exists, there must also be an energy associated with this field. Provided that the field is defined in a volume V, this energy is given by the expression

$$W = \int_v \frac{1}{2} HB\, dv \qquad (4.19)$$

which will be formally derived in Chapter 6. This expression is also valid for a magnet. We consider the volume v to be a tube of flux with an internal volume v_i and an external volume v_e as shown in Figure 4.20. The flux in this tube is $d\Phi$. If ds is the cross-section of the tube and dl an elemental distance along the tube's length, we can write $dv = ds\,dl$. The energy is now

$$W = \frac{1}{2}\int_{v_i} Hdl\, Bds + \frac{1}{2}\int_{v_e} Hdl\, Bds$$

Substituting $d\Phi$ for its value in the integral, we get

$$W = \frac{1}{2} d\Phi \int_{l_i} Hdl + \frac{1}{2} d\Phi \int_{l_e} Hdl$$

where l_i and l_e are the internal and external contours of circulation.

$$W = \frac{1}{2} d\Phi \left[\int_{l_i} \mathbf{H} \cdot d\mathbf{l} + \int_{l_e} \mathbf{H} \cdot d\mathbf{l} \right] = \frac{1}{2} d\Phi \oint_l \mathbf{H} \cdot d\mathbf{l}$$

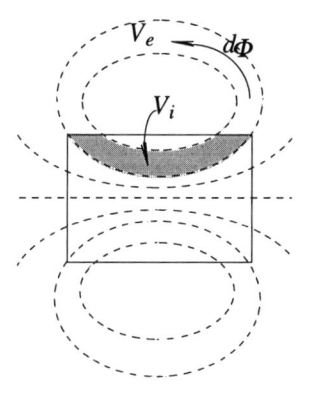

Figure 4.20. A flux tube inside and outside the magnet.

Since there are no currents in the system, we find from Ampere's law that $W=0$, which indicates that the sum of the internal energy (in volume V_i) and external energy (in volume V_e) is zero. Extending this result over the whole magnet, using an assembly of flux tubes, we have

$$W_e = -W_i$$

From this, the internal energy in the magnet is given by

$$W_i = \frac{1}{2}\int_{V_i} BH dv$$

and the free energy (external) is, therefore,

$$W_e = -\frac{1}{2}\int_{V_i} BH dv \qquad (4.20)$$

This energy depends on the product BH at the operating point of the magnet and on the volume of the magnet. In general it is of interest to concentrate this energy in a specific area. Two examples of energy due to magnets are shown in Figure 4.21a and 4.21b.

In Figure 4.21a the energy due to the magnet is distributed in space and, therefore, in the volume surrounding the magnet the volumetric energy density is low. In Figure 4.21b, we will assume that the magnet is identical to that in Figure 4.21a even though the magnetic field of the magnet depends on the surrounding materials (the operating point of the magnet is influenced by the surrounding materials). Now the energy is concentrated in volume V_g, which is much smaller.

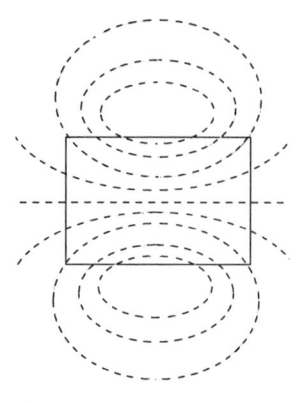

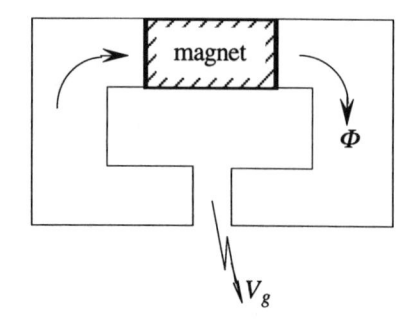

Figure 4.21a. A magnet in an open magnetic circuit. The energy is distributed throughout space.

Figure 4.21b. A magnet in a magnetic circuit. Energy is concentrated in the volume of the gap.

The energy density in this volume is high; the energy of the magnet does not disperse as in the first example.

For a given volume of the magnet, the energy depends only on the product BH. There is a point on the characteristic curve $B(H)$ of the magnet at which this product is maximum, representing the operating point of the magnet at which it can produce maximum energy. At points $B=B_r$ (and, therefore, $H=0$) or $B=0$ (and, therefore, $H=H_c$), the BH product is zero and the magnet does not produce any external energy. The first case $B=B_r$ is represented in Figure 4.22 where a magnet is embedded in a magnetic circuit without a gap. The permeability μ_f of the circuit is considered to be infinite. Considering the fields H_f and H_i to be constant in the iron and magnet, we get from Ampere's law that $H_f l_f + H_i l_i = 0$.

Therefore, provided that $\mu_f >> \mu_0$, we get $H_f = 0$, and $H_i = 0$, which correspond to $B = B_r$ on the magnetization curve. However, this is a purely hypothetical situation, since μ_f has a finite value and, therefore, there is some field dispersion through the sides of the magnet. The field, therefore, is not constant throughout the magnet. This, however, does not change the fact that the flux density B of the magnet is very close to B_r.

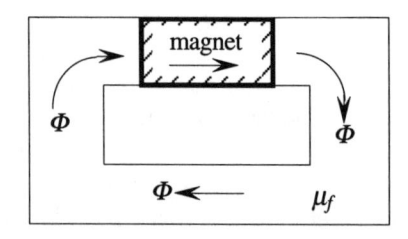

Figure 4.22. A permanent magnet in a closed magnetic circuit.

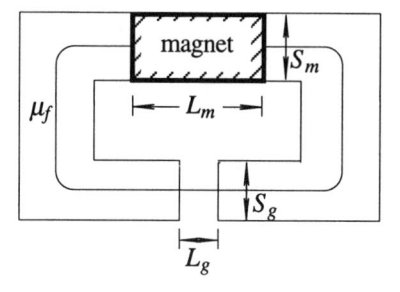

Figure 4.23. A permanent magnet in a magnetic circuit including a gap.

The second case, where $H=H_c$ (and $B=0$) represents a situation where the effect of the magnet is canceled, for example, by the magnetic field of a coil in the magnetic circuit in which the magnet is inserted. This case is shown in Figure 4.15a. The operating point of the magnet is point M_3 in Figure 4.15b.

We now examine an example of the calculation of the magnetic field due to a permanent magnet, as in Figure 4.23. A magnet is inserted in a magnetic circuit with $\mu_r \gg 1$. The magnetic circuit has a gap as shown. Writing Ampere's law, and assuming that the fields are constant in their respective domains, we have for the circulation of H along the contour L

$$H_m L_m + H_f L_f + H_g L_g = 0$$

With $H_f \approx 0$, we have

$$H_m L_m = - H_g L_g \qquad (4.21)$$

In other words, the conservation of flux gives

$$\Phi_m = \Phi_g$$

or

$$B_m S_m = \mu_0 H_g S_g \qquad (4.22)$$

Dividing Eq. (4.22) by Eq. (4.21) gives

$$\frac{B_m}{H_m} = - \mu_0 \frac{S_g}{S_m} \frac{L_m}{L_g} \qquad (4.23)$$

Eq. (4.23) describes B_m/H_m as a function of the dimensional factors of the structure.

4. Magnetostatic Fields

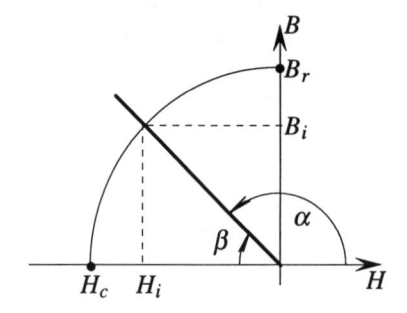

Figure 4.24. Load line of a permanent magnet. The intersection of the load line with the $B(H)$ curve is the operating point of the magnet.

The value is negative as it should be since the magnet operates in the second quadrant. In fact, the value B_m/H_m represents, in the B-H plane, a straight line. This line is called a "load line" (or "operating line"). The intersection of this line with the characteristic curve of the magnet provides the points B_i and H_i at which the magnet operates, as a function of the dimensions of the magnetic circuit (see Figure 4.24).

The angle α in Figure 4.24 is evaluated from the relation

$$\tan\alpha = \frac{B_m}{H_m} = -\mu_0 \frac{S_g}{S_m} \frac{L_m}{L_g}$$

From the properties of the tangent function, we can use the angle β directly

$$\beta = \arctan\left[\mu_0 \frac{S_g}{S_m} \frac{L_m}{L_g}\right]$$

From B_i and H_i, we can determine H_g by multiplying Eq. (4.21) by Eq. (4.22)

$$B_i H_i S_m L_m = -\mu_0 H_g^2 S_g L_g$$

Denoting V_m the volume of the magnet, such that $V_m = S_m L_m$ and V_g the volume of the gap such that $V_g = S_g L_g$, we obtain

$$H_g = \left[-\frac{B_i H_i}{\mu_0} \frac{V_m}{V_g}\right]^{\frac{1}{2}} \quad (4.24)$$

We note here that H_g is larger as the $B_m H_m$ product of the magnet is larger [for this reason we are interested in operating at $(BH)_{max}$ of the magnet] and as

V_m/V_g is larger (therefore we are interested in increasing the volume of the magnet and using small gaps).

4.5.4c. Principal Types of Permanent Magnets

A good permanent magnet is required to have a high coercive field intensity H_c as well as high remnant flux density B_r. A high coercive field intensity is important because it does not allow the magnet to be demagnetized and a high B_r is normally associated with the capacity of the magnet to produce high magnetic fields in magnetic circuits in which it is inserted.

Until 1930, usable magnetic materials were the Chrome-Tungsten and Chrome-Cobalt alloys. Their major problem was a low coercive field intensity H_c ($H_c<20,000$ A/m). In 1940, the Alnico alloys (Fe+Al+Ni+Co) appeared, whose B_r is approximately 1 T and who have a coercive field intensity of $H_c>50,000$ A/m. This type of magnet is still used extensively especially in applications where operation at high temperatures is required. In 1947, with the appearance of ceramic ferrite magnets (SrFe$_{12}$O$_{19}$ or BaFe$_{12}$O$_{19}$), the use of magnets became widespread since these magnets are inexpensive and possess high H_c ($H_c=100,000$ A/m). Despite the fact that B_r is low ($B_r \approx 0.4$ T), their properties prevail and today these magnets are utilized extensively. Another important property of ceramic magnets is the fact that they are nonconducting, and therefore are the preferred magnetic element for application at high frequencies (since there are no induced currents circulating in the magnet).

In 1974, permanent magnets made of rare earth elements were introduced. The Samarium-Cobalt magnets (*SmCo*$_5$ with $B_r \approx 0.8$ T and $H_c \approx 600,000$ A/m and Sm$_2$Co$_{17}$ with $B_r \approx 1$ T and $H_c \approx 600,000$ A/m) represent a revolution in this domain since, in addition to possessing high coercive fields, they have a high value of B_r. However, due to very complex fabrication processes and difficulties in obtaining the raw materials for the magnets, their prices continues to be very high. This does not diminish the considerable interest in these materials. Many of the best servo-motors and other high-performance electromagnetic devices (high-power density devices), utilize these elements. More recently, the Neodymium-Iron-Boron magnets (Nd$_2$Fe$_{14}$B with $B_r \approx 1.2$ T and $H_c \approx 800,000$ A/m) were tauted as high-performance magnets that could be as inexpensive as ceramic magnets and as commonly available. However, these promises did not materialize and these magnets form a small part of the commercially available magnets. Neodymium magnets tend to lose their characteristics at relatively low temperatures (around 140 °C) and, therefore, the interest in these magnets is limited to low-temperature applications.

Developments in high-energy permanent magnets continue with the objective of improving their characteristics as well as increasing their energy density.

Figure 4.25 shows the $B(H)$ curves of the principal permanent magnets available. The magnetic characteristics of the magnets described above are given in Table 4.1.

114   4. Magnetostatic Fields

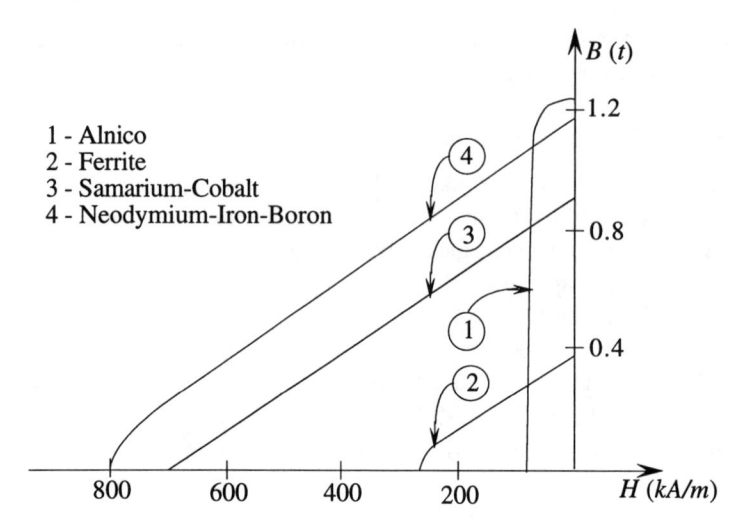

Figure 4.25. $B(H)$ curves for some permanent magnets.

4.5.4.d. Dynamic Operation of Permanent Magnets

Utilization of a permanent magnet requires some precautions to ensure that the magnet is not demagnetized during normal operation. As an example we assume the curve in Figure 4.26 to be the curve for the permanent magnet. The most commonly used magnets (ferrite, Sm-Co, Alnico) have a differential permeability ($tan\alpha = B_a/H_a$), very close to that of air ($\mu_r = 1$). If we operate at point P_1, the magnet preserves its remnant flux density B_r. However, if we operate at point P_2, it loses its remnant flux density B_r and, instead, the magnet has a new remnant flux density B_{r2}. This may cause the performance to fall below an acceptable limit. If the operating point is changed to point P_3, the magnet loses its remnant flux density completely. Therefore, it is important to avoid operating points below P_1 in Figure 4.26.

Table 4.1. Properties of permanent magnets.

Material at (20°C)	B_r [T]	H_c [KA/m]	$(BH)_{max}$ [KJ/m³]	μ_r
Alnico	1.25	60	50	3–5
Ferrite	0.38	240	25	1.1
Sm-Co	0.9	700	150	1.05
Ne-Fe-B	1.15	800	230	1.05

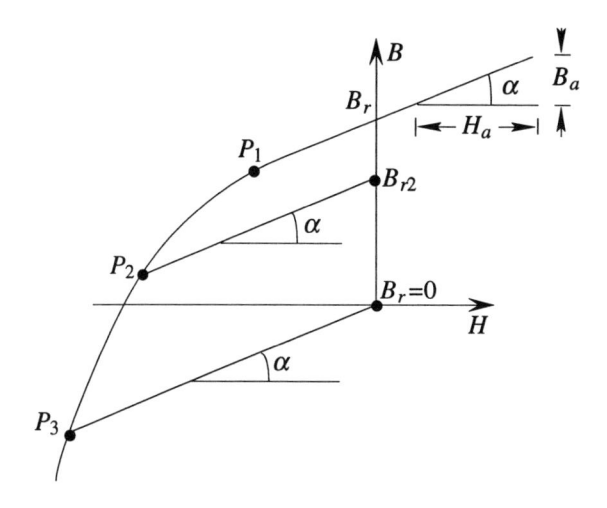

Figure 4.26. Dynamic operation of a permanent magnet.

4.6. The Analogy between Magnetic and Electric Circuits

Because of similarities in form of the electric and magnetic quantities, it is useful to look at the analogy between electric and magnetic quantities. To do so we use the two groups of equations given below:

$$\nabla \times \mathbf{E} = -\frac{\partial \mathbf{B}_e}{\partial t} \qquad \nabla \times \mathbf{H} = \mathbf{J}_m$$

$$\mathbf{E} = \frac{1}{\sigma}\mathbf{J} \qquad \mathbf{H} = \frac{1}{\mu}\mathbf{B}$$

$$\nabla \cdot \mathbf{J} = 0 \qquad \nabla \cdot \mathbf{B} = 0$$

Since the mathematical expressions for the two fields are equivalent, we can define equivalencies between the physical properties of the electric and magnetic fields. These equivalent quantities are:

$$(\mathbf{E} \text{ and } \mathbf{H}); \quad (-\partial \mathbf{B}_e/\partial t \text{ and } \mathbf{J}_m); \quad (\mathbf{J} \text{ and } \mathbf{B}); \quad (\sigma \text{ and } \mu)$$

Within the two groups there are variables with the same name ($\mathbf{J}_m, \mathbf{J}$) and ($\mathbf{B}_e, \mathbf{B}$) but these have different meaning. The variables on the left relate to electric circuits, those on the right to magnetic circuits. As an example, $\mathbf{J}$ is associated with the current that circulates in the conductors of the circuit, while $\mathbf{J}_m$ is the source of the magnetic field. Similarly, the flux density **B** is related to the flux of the magnetic circuit, while $\mathbf{B}_e$ is time dependent and is the source of an electric field.

The equivalency of the magnetic and electric quantities is particularly useful in defining and analyzing magnetic circuits. To see this we show the equivalency between electric and magnetic circuits.

In circuit theory we assume that $-\partial \mathbf{B}_e/\partial t = 0$ and $\mathbf{J}_m = 0$. Under these conditions, we get:

Electric circuits	Magnetic circuits
$\nabla \times \mathbf{E} = 0$	$\nabla \times \mathbf{H} = 0$
An electric scalar potential V is defined, from which **E** is calculated	A magnetic scalar potential V_m is defined, from which **H** is calculated
$\mathbf{E} = -\nabla V$	$\mathbf{H} = -\nabla V_m$
since	since
$\nabla \times (\nabla V) = 0$	$\nabla \times (\nabla V_m) = 0$
The potential V can be obtained, for example, through the use of cells or batteries. The relation	The potential V_m can be obtained, for example, through the use of the magneto-motive force (*mmf*) of an electromagnet. The relation
$\nabla \cdot \mathbf{J} = 0$	$\nabla \cdot \mathbf{B} = 0$
indicates that the flux of **J** is conservative, or that the current entering a volume is equal the current exiting the volume. This is normally referred to in circuit theory as "Kirchoff's nodal law." A current I is defined as	indicates that the flux of **B** is conservative, or that the flux entering a volume is equal to the flux exiting the volume. This equation is equivalent to Kirchoff's law. A magnetic flux Φ is defined as
$I = \int_s \mathbf{J} \cdot d\mathbf{s}$	$\Phi = \int_s \mathbf{B} \cdot d\mathbf{s}$
which is nonzero for an open surface.	which is nonzero for an open surface.

Analogy between Magnetic and Electric Circuits

The electric resistance is defined as	The magnetic reluctance is defined as
$$R = \frac{V}{I} = \frac{1}{\sigma}\frac{l}{S}$$	$$R = \frac{V_m}{\Phi} = \frac{1}{\mu_0}\frac{l}{S}$$
where l and S are the length and cross-sectional area of the resistive material.	where l and S are the length and cross-sectional area of a gap.
For resistances in series and in parallel, we obtain the equivalent resistance R_e.	For calculation of equivalent serial and parallel reluctances, the operations are analogous to those for resistances.
$$R_e = R_1 + R_2 + R_3 + \ldots + R_n$$ $$\frac{1}{R_e} = \frac{1}{R_1} + \frac{1}{R_2} + \frac{1}{R_3} + \ldots + \frac{1}{R_n}$$	$$R_e = R_1 + R_2 + R_3 + \ldots + R_n$$ $$\frac{1}{R_e} = \frac{1}{R_1} + \frac{1}{R_2} + \frac{1}{R_3} + \ldots + \frac{1}{R_n}$$
Example:	Example:
$$R = \frac{1}{\sigma}\frac{l}{S}$$	$$R = \frac{1}{\mu_0}\frac{l}{S}$$
$$I = \frac{V}{R} = \frac{V\sigma S}{l}$$	$$\Phi = \frac{NI}{R} = \frac{NI\mu_0 S}{l}$$
Assuming the conductivity σ of the conductors (i.e., wires) in the circuit to be infinite, the only resistance to current is R.	Assuming the permeability μ of the iron in the circuit to be infinite, the only reluctance to flux is in the gap.

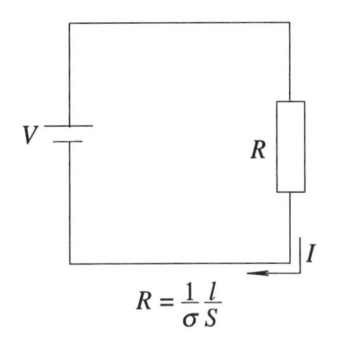

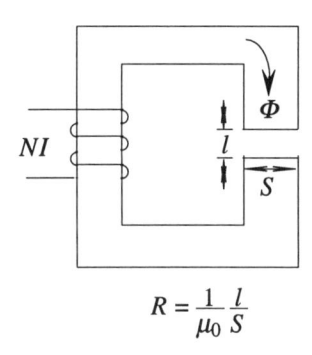

The parallelism between the electric and magnetic circuits indicates that it is possible to calculate the magnetic quantities of a magnetic circuit through the equivalent electric circuit. In this manner, for example, the magnetic circuit of Figure 4.27a can be calculated using the electric circuit of Figure 4.27b.

4. Magnetostatic Fields

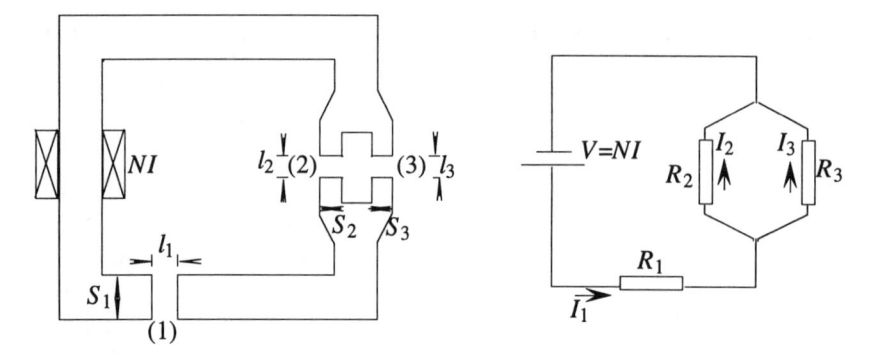

Figure 4.27a. A magnetic circuit with three gaps.

Figure 4.27b. The equivalent electric circuit for the magnetic circuit in Figure 4.27a.

Defining

$$R_1 = \frac{1}{\mu_0}\frac{l_1}{S_1} \qquad R_2 = \frac{1}{\mu_0}\frac{l_2}{S_2} \qquad R_3 = \frac{1}{\mu_0}\frac{l_3}{S_3}$$

we can calculate the currents I_1, I_2, and I_3 proceeding exactly as for an electric circuit. These values, obtained in terms of NI, R_1, R_2, and R_3, are the values of Φ_1, Φ_2, and Φ_3, which are the fluxes passing through gaps (*1*), (*2*), and (*3*). Recalling that, for example, $\Phi_1 = \mu_0 H_1 S_1$, we can obtain H_1, H_2, and H_3.

However, this procedure for the calculation of magnetic circuits is inconvenient and has some limitations:

- If the magnetic circuit is saturated, the establishment of the equivalent electric circuit is much more complicated. The reluctance of iron is nonzero causing additional branches in the circuit and complicating the analysis.
- The algebra involved in solution of the equations is laborious.
- In magnetic circuits, there is dispersion and leakage of the flux passing through the circuit. This does not occur in electric circuits.

This equivalency between electric and magnetic circuits should be considered a valid method for qualitative analysis of magnetic circuits because, in general, the analysis of an electric circuit is simpler. In the example of Figure 4.27a and 4.27b, the fact that $I_1 = I_2 + I_3$, establishes, quite naturally, the idea of conservation of flux $\Phi_1 = \Phi_2 + \Phi_3$, as the equivalent magnetic relation.

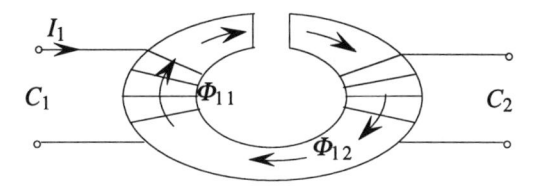

Figure 4.28. Two electric circuits in a magnetic path. The flux linking the two circuits is Φ_{12} and the flux linking only C_1 is Φ_{11}.

4.7. Inductance and Mutual Inductance

4.7.1. *Definition of Inductance*

Consider the situation shown in Figure 4.28. Two electric circuits C_1 and C_2 are located in the presence of ferromagnetic materials. We do not consider any losses in the ferromagnetic material. A time-dependent current I_1 flows in circuit $C1$. Because the frequency is sufficiently low, losses due to time dependency of the current I_1 can be neglected (these losses will be examined later).

The current I_1 in C_1 generates a magnetic flux. Part of this flux links with circuit C_2. This flux, denoted Φ_{12}, is the flux generated by C_1 in C_2. We assume that the ferromagnetic circuit is linear so that, for example, if the current I_1 is doubled Φ_{12} is also doubled. Since circuit C_2 has N_2 turns, the "flux linkage" is defined as the number of times circuit C_2 links the flux Φ_{12}, or $N_2\Phi_{12}$. For a linear circuit, a proportionality factor K between the flux linkage and the current I_1 can be defined as $K = N_2\Phi_{12}/I_1$. This factor is called the mutual inductance between C_1 and C_2 and is denoted as M_{12}:

$$M_{12} = N_2 \frac{\Phi_{12}}{I_1} \qquad (4.25)$$

In addition, circuit C_1, contains N_1 turns and produces a flux, a portion of which only links with itself. This flux is denoted Φ_{11}. We can therefore define a "self inductance" or simply "inductance" of C_1 as

$$L_1 = N_1 \frac{\Phi_{11}}{I_1} \qquad (4.26)$$

If the behavior of the magnetic materials is nonlinear, the proportionality factor between Φ_{12} and I_1 or Φ_{11} and I_1 is not constant. In this case inductance is defined as the ratio of infinitesimal change in flux linkage to infinitesimal change in current producing the flux change. With this loss of proportionality, M_{12} and L_1 vary with the value of I_1 and, therefore, are functions of I_1.

4.7.2. Energy in a Linear System

In the system discussed in the previous section, neglecting ohmic losses, the potential U_1, corresponding to the *emf* at the terminals of circuit C_1 is equal to

$$U_1 = \frac{d\Phi_{11}}{dt} = \frac{d}{dt}(L_1 I_1)$$

The electric power associated with C_1 is

$$P_1 = U_1 I_1 = \frac{d\Phi_{11}}{dt} I_1$$

and the magnitude of the energy (meaning the energy needed in circuit C_1 in order to generate the magnetic flux Φ_{11} in the inductor) is

$$W = \int_0^t P_1 dt = \int_0^t I_1 \frac{d\Phi_{11}}{dt} dt = \int_0^{I_1} I_1 d(L_1 I_1)$$

Assuming that the system is linear, or that there is no saturation in the magnetic circuit, L_1 is constant and we get

$$W = \frac{1}{2} L_1 I_1^2 \tag{4.27}$$

This expression represents the magnetic energy necessary for the coil in C_1 to generate the magnetic field responsible for the flux Φ_{11}.

Assuming now that circuits C_1 and C_2 carry currents I_1 and I_2, respectively, we can write

$$U_1 = \frac{d\Phi_{11}}{dt} + \frac{d\Phi_{21}}{dt}$$

$$U_2 = \frac{d\Phi_{22}}{dt} + \frac{d\Phi_{12}}{dt}$$

In a manner analogous to the calculation above, we can write the total energy in the two circuits as

$$W = \int_0^t (U_1 I_1 + U_2 I_2) dt$$

$$W = \int_0^{\Phi_{11}, \Phi_{21}} (d\Phi_{11} + d\Phi_{21})I_1 + \int_0^{\Phi_{22}, \Phi_{12}} (d\Phi_{22} + d\Phi_{12})I_2$$

Noting that $\Phi_{12}=M_{12}I_1$ and $\Phi_{21}=M_{21}I_2$, as well as $\Phi_{11}=L_1I_1$ and $\Phi_{22}=L_2I_2$, we obtain

$$W = \int_0^{I_1, I_2} (I_1 L_1 dI_1 + M_{21}I_1 dI_2) + \int_0^{I_1, I_2} (I_2 L_2 dI_2 + M_{12}I_2 dI_1)$$

or

$$W = \frac{1}{2} L_1 I_1^2 + \frac{1}{2} L_2 I_2^2 + \int_0^{I_1, I_2} (M_{21}I_1 dI_2 + M_{12}I_2 dI_1) \qquad (4.28)$$

It is easily verified that

$$\int_0^{I_1, I_2} (M_{21}I_1 dI_2 + M_{12}I_2 dI_1) =$$
$$\frac{1}{2}(M_{21} + M_{12})I_1 I_2 + \frac{1}{2}\int_0^{I_1, I_2} (M_{21} - M_{12})(I_1 dI_2 - I_2 dI_1) \qquad (4.29)$$

The fact that we called one circuit C_1 and the other C_2 is purely arbitrary. The magnetic energy is not altered if the notation is reversed. This implies that, in Eq. (4.29), M_{12} must be equal to M_{21}, since otherwise the energy will be altered. Thus:

$$M_{21} - M_{12} = 0 \quad \text{and} \quad M_{21} = M_{12}$$

This result is universally applicable. Now, the integral on the right-hand side of Eq. (4.29) is zero, and we obtain

$$W = \frac{1}{2} L_1 I_1^2 + \frac{1}{2} L_2 I_2^2 + M_{12} I_1 I_2 \qquad (4.30)$$

This expression can be generalized for any number of circuits as

$$W = \sum_{i=1}^{J} \frac{1}{2} L_i I_i^2 + \sum_{i=1}^{J} \sum_{l=i+1}^{J} M_{il} I_i I_l \qquad (4.31)$$

122 4. Magnetostatic Fields

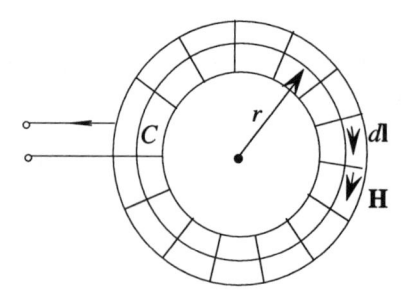

Figure 4.29. The magnetic field intensity in a torus.

4.7.3. The Energy Stored in the Magnetic Field

Before calculating the stored energy in the magnetic field, we calculate first the magnetic field intensity generated in a torus by a coil with N turns, carrying a current I, as shown in Figure 4.29. The cross-sectional area of the torus is S, its mean radius r, and the torus is made of a material with permeability μ. Applying Ampere's law, on the contour C, shown in Figure 4.29, gives

$$\oint_C \mathbf{H} \cdot d\mathbf{l} = NI \quad \text{and} \quad H = \frac{NI}{2\pi r} \qquad (4.32)$$

With $L=N\Phi/I$, we get

$$L = N\mu \frac{NI}{2\pi r} S \frac{1}{I} = \frac{\mu N^2 S}{2\pi r} \qquad (4.33)$$

The energy of the magnetic field $W=LI^2/2$, allows the calculation of the volumetric energy density as

$$W_v = \frac{W}{2S\pi r} = \frac{\frac{1}{2}LI^2}{2S\pi r}$$

Substituting the value of I from Eq. (4.32) and L from Eq. (4.33) gives

$$W_v = \frac{1}{2}\mu H^2$$

This is the volumetric energy density associated with a static magnetic field intensity H. This expression can be rewritten in any of the following forms

$$W_v = \frac{1}{2}\mu H^2 = \frac{1}{2}BH = \frac{1}{2}\frac{B^2}{\mu} \qquad (4.34)$$

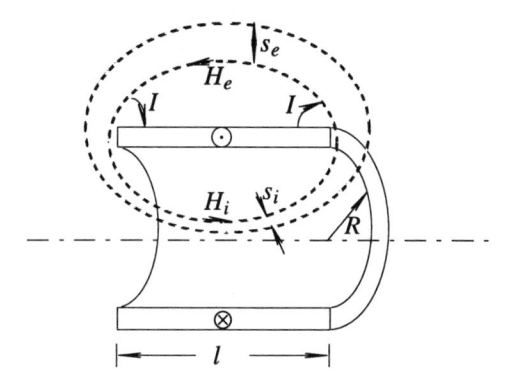

Figure 4.30. The magnetic field intensity due to a finite-length solenoid.

4.8. Examples

4.8.1. Calculation of Field Intensity and Inductance of a Long Solenoid

Consider the solenoid shown in Figure 4.30, where $l>>R$. This characterizes a "long" solenoid. Numerical calculations performed with EFCAD, show that for $l>3R$, the solenoid can be considered as "long" and we obtain the correct value of H in the interior and in the material of the solenoid. Assuming that H_i in the interior of the solenoid is constant and H_e, the exterior field, is also constant, we can apply Ampere's law

$$\oint_c \mathbf{H}\cdot d\mathbf{l} = NI \quad (N=\text{number of turns})$$

Choosing a contour C that coincides with H and dividing it into C_i (interior to the solenoid) and C_e (exterior to the solenoid) gives the total circulation of H around the contour as

$$H_i C_i + H_e C_e = NI \quad (C_i \equiv l)$$

We will show below that $H_e<<H_i$; but for now we assume that this condition is correct, and

$$H_i = \frac{NI}{l}$$

To show that $H_e \ll H_i$, we look at the flux lines, two of which form a tube of flux as shown in Figure 4.30. In a tube the flux is conserved. The cross-sectional area of the tube inside the solenoid is S_i and outside the solenoid it is S_e. It is evident that due to the dispersion of the field outside the solenoid, $S_e \gg S_i$. Since the flux is conserved, we get

$$\mu_0 H_e S_e = \mu_0 H_i S_i$$

and that

$$H_e = H_i \frac{S_i}{S_e}$$

and, therefore $H_e \ll H_i$, as was indicated above.

The self-inductance is given by

$$L = N \frac{\Phi}{I}$$

therefore,

$$\Phi = \mu_0 H_i S = \mu_0 \frac{NI}{l} \pi R^2$$

This gives

$$L = \mu_0 \frac{N^2 \pi R^2}{l}$$

We should point out that, although in the initial expression for the inductance the quantities Φ and I appear, the final result for L is always a function of the geometric dimensions of the device, independent of the current or flux.

We also observe that, although we used Ampere's law, we had to consider a second approximation, namely that the field is constant inside the solenoid. Should the solenoid have been short, the analytical calculation were still possible but the difficulties in doing so were considerable in this case.

Figure 4.31 shows a distribution of the magnetic flux density of a long solenoid, obtained numerically by program EFCAD. This field distribution is quite close to the flux of a long solenoid.

For the dimensions used in this case, calculations with EFCAD, show a field in the interior of the solenoid (point P) that differs from the analytic calculation above by 16%; however, the field in the interior, closer to the edge of the solenoid (point Q) differs by about 44%.

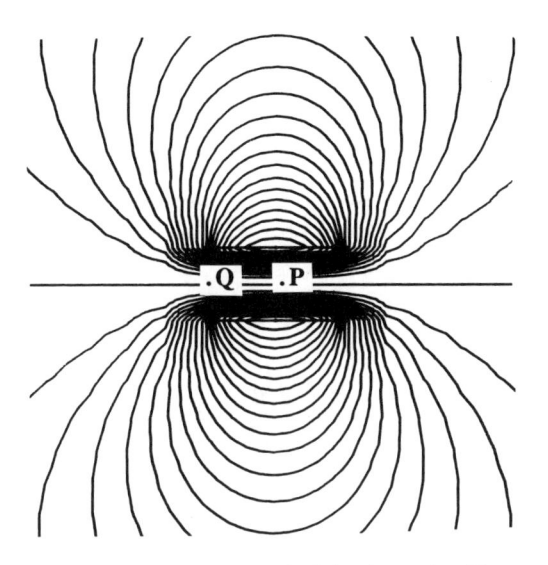

Figure 4.31. Magnetic flux density distribution of a finite, long solenoid.

4.8.2. Calculation of H for a Circular Loop

We wish to calculate the field intensity **H** at point P in Figure 4.32. This field is generated by the current I in the loop. The projection of the point P on the plane of the loop coincides with the center of the loop (point O).

First, we note that the field **H** must be obtained through the use of Biot-Savart's law, since there is no contour C on which the field H is constant in magnitude (this is obvious from symmetry considerations). The application of Ampere's law, while always valid, is more difficult in this case.

To facilitate visualization, we consider $d\mathbf{l}$ shown at the top of Figure 4.32; $d\mathbf{l}$ and **r** are perpendicular to each other; the direction of $d\mathbf{H}$ is orthogonal to the plane formed by $d\mathbf{l}$ and **r**, as shown in the figure. The normal component of $d\mathbf{H}$ is clearly zero if we consider $d\mathbf{H}$ due to a segment $d\mathbf{l}$ diametrically opposite to the segment $d\mathbf{l}$ shown. We get

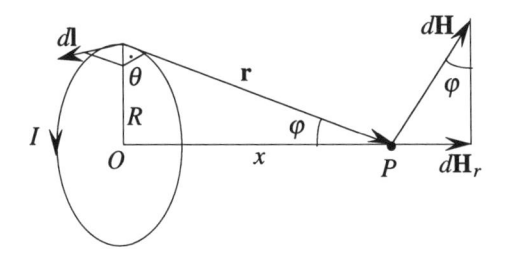

4. Magnetostatic Fields

Figure 4.32. Calculation of the magnetic field intensity for a circular loop.

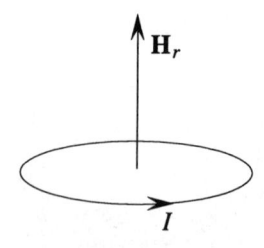

Figure 4.33. The magnetic field intensity of a circular loop.

$$dH = \frac{I\,dl}{4\pi r^2}\sin\theta$$

where $\theta = 90°$.

The component dH_r is

$$dH_r = \frac{I\,dl}{4\pi r^2}\sin\phi$$

Since ϕ is constant, we can write

$$\sin\phi = \frac{R}{\sqrt{R^2+x^2}}$$

$$dH_r = \frac{I\,dl}{4\pi(R^2+x^2)}\frac{R}{\sqrt{R^2+x^2}}$$

$$H_r = \frac{IR}{4\pi(R^2+x^2)^{3/2}}\int_c dl$$

Integration over the contour of the loop is equal to $2\pi R$ and, therefore,

$$H_r = \frac{IR^2}{2(R^2+x^2)^{3/2}}$$

In the particular case where $x=0$, the field intensity H at the center of the loop as shown in Figure 4.33. We obtain

$$H_r = \frac{I}{2R}$$

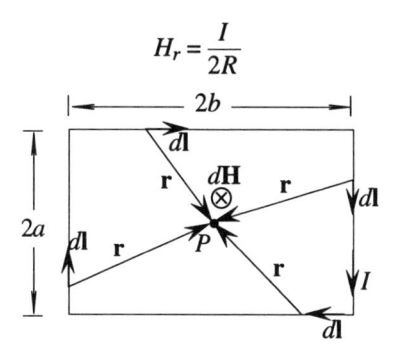

Figure 4.34. Calculation of the magnetic field intensity of a rectangular loop as the sum of the fields of the four segments of the loop.

4.8.3. *Field of a Rectangular Loop*

We wish to calculate the magnetic field at point *P*, located at the intersection of the diagonals of the square loop shown in Figure 4.34. The loop carries a current *I*. For the same reason given in the previous example, we have to use the Biot-Savart law. Note that the vector products *d*l×**r** for the fields *d***H** due to the four segments of the wire have the same direction at point *P*. All four components penetrate into the plane of the figure. The field intensity *H* due to the four segments is calculated separately, as shown in Figure 4.35.

$$dH_1 = \frac{Idl}{4\pi r^2} \sin\theta = \frac{Idl}{4\pi r^2} \cos\phi$$

Changing all variables to be a function of ϕ and observing that $tan\phi = l/b$ gives $dl = b\sec^2\phi d\phi$, and that $cos\phi = b/r$ gives $r = b/cos\phi$. Substituting *dl* and *r* in the expression for *dH*, we obtain

$$H_1 = \frac{I}{4\pi b} \int_{-\phi_1}^{\phi_1} \cos\phi d\phi$$

The limiting angles corresponding to points *Q* and *R* in Figure 4.35 are given by

$$\phi_1 = \pm \arcsin \frac{a}{\sqrt{a^2 + b^2}}$$

we obtain for H_1

$$H_1 = \frac{Ia}{2\pi b \sqrt{a^2 + b^2}}$$

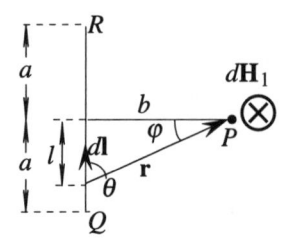

Figure 4.35. Calculation of the magnetic field intensity of a short segment, carrying current I.

Figure 4.36. The magnetic field intensity of a horizontal segment of the loop in Figure 4.34.

H_2 is now the field due to the other segment of the wire, shown in Figure 4.36. By analogy, the field H_2 is

$$H_2 = \frac{Ib}{2\pi a\sqrt{a^2+b^2}}$$

The total field is $H_t = 2H_2 + 2H_1$, or

$$H_t = \frac{I}{\pi\sqrt{a^2+b^2}}\left(\frac{a}{b}+\frac{b}{a}\right) = \frac{I}{\pi ab}\sqrt{a^2+b^2}$$

4.8.4. Calculation of Inductance of a Coaxial Cable

Here we wish to calculate the inductance (per unit length) of the co-axial cable shown in Figure 4.37a and 4.37b. We consider only the flux between the two conductors ($a < r < b$). Considering a straight cable, the field between the two conductors is calculated as for infinite conductors. This gives

$$H = \frac{I}{2\pi r}$$

The flux is obtained through

$$\Phi = \int_s \mathbf{B} \cdot d\mathbf{s}$$

$\mathbf{B}$ and $d\mathbf{s}$ (Figure 4.37b) are collinear vectors in the same direction. Thus

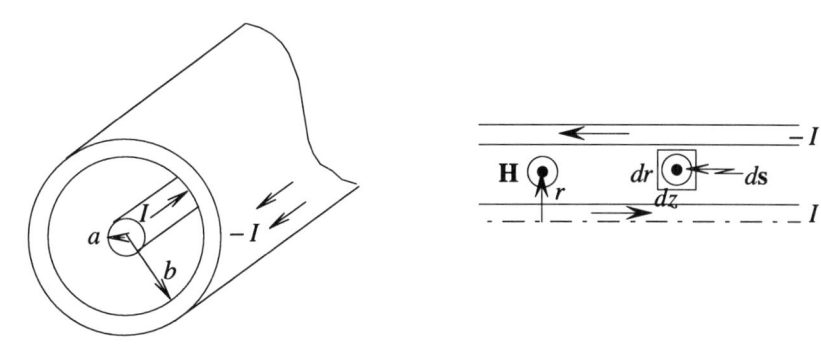

Figure 4.37a. Geometry and dimensions of a coaxial cable.

Figure 4.37b. Cross-section of the cable showing the currents and the magnetic flux density.

$$\Phi = \int_s \mu_0 H ds = \mu_0 \frac{I}{2\pi} \int_a^b \frac{dr}{r} \int_0^z dz$$

or

$$\Phi = \mu_0 \frac{I}{2\pi} \ln \frac{b}{a} z$$

The inductance is, therefore,

$$\frac{L}{z} = \frac{\mu_0}{2\pi} \ln \frac{b}{a}$$

The flux outside the cable is zero since the total current enclosed by a contour C, enclosing both conductors, is zero.

4.8.5. Calculation of the Field inside a Cylindrical Conductor

Assume that the conductor is straight and carries a current I, uniformly distributed throughout the cross sectional area of the conductor as shown in Figure 4.38. To obtain the field at a radius $r<R$ (inside the conductor), we must calculate the current density J

$$J = \frac{I}{\pi R^2}$$

We choose the contour C shown in Figure 4.38 (the circumference of a circle of radius r). Applying Ampere's law,

130   4. Magnetostatic Fields

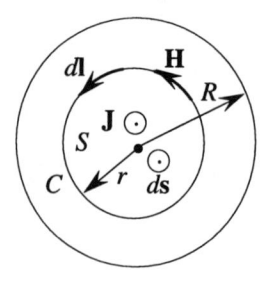

Figure 4.38. Use of Ampere's law to calculate the magnetic field intensity of a cylindrical conductor carrying a current I.

$$\oint_C \mathbf{H} \cdot d\mathbf{l} = \int_S \mathbf{J} \cdot d\mathbf{s}$$

The integral on the right-hand side of this expression gives the current enclosed by C, or, the current I' traversing S

$$I' = J\pi r^2 = I\frac{r^2}{R^2}$$

The circulation of H on C is $2\pi rH$. We obtain

$$H = \frac{Ir}{2\pi R^2}$$

At $r=R$, we get $H=I/2\pi R$.

For $r>R$, the magnitude of H is $H=I/2\pi r$, the familiar expression for the field generated by an infinite wire. Finally, we note that the calculation of the field (and consequently, the flux) in the interior of the conductor given above, can be utilized in the previous example to obtain a more accurate value for the inductance of the coaxial cable of example 4.8.4.

4.8.6. *Calculation of the Magnetic Field Intensity in a Magnetic Circuit*

Given the magnetic circuit of Figure 4.39a, we wish to calculate the magnetic field intensity in the gaps. We use the notation of Figures 4.39b. In this notation S indicates the cross-sectional area of the gap and I the length of the gap. The number in brackets indicates the gap number. As an example, in Figure 4.39a, in gap 1 (cross-sectional area S_1 and length l_1), the magnetic field intensity in the gap is denoted by h_1.

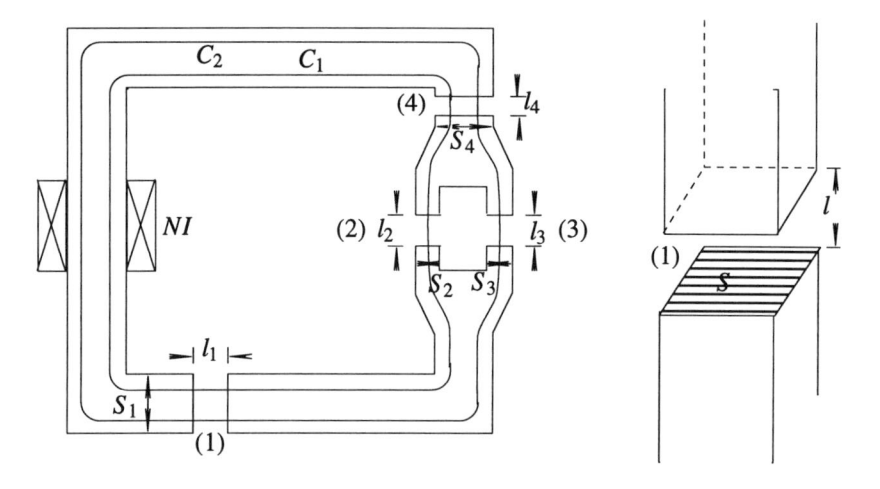

Figure 4.39a. A magnetic circuit with three gaps.

Figure 4.39b. Dimensions of gaps.

Before approaching this problem, we make the following assumptions:

- We assume $\mu_f \gg \mu_0$, which means that h_f (the field in the iron) is negligible compared to the fields in the gaps; the circulation of h_f is, therefore, also negligible.
- The fields in the gaps are considered to be constant, without fringing at the edges of the gaps. Note that if the faces of the iron on two sides of the gap are displaced in relation to each other, this condition cannot be satisfied and the method used here is not applicable.

In the example given in Figure 4.39, we have:

- on contour C_1

$$h_1 l_1 + h_2 l_2 + h_4 l_4 = NI \qquad (4.35)$$

- on contour C_2

$$h_1 l_1 + h_3 l_3 + h_4 l_4 = NI \qquad (4.36)$$

These provide the first two equations needed to evaluate h_1, h_2, h_3, and h_4. The remaining two equations are obtained from the conservation of flux

$$\Phi_1 = \Phi_2 + \Phi_3$$

4. Magnetostatic Fields

or, in terms of the field intensities

$$\mu_0 h_1 S_1 = \mu_0 h_2 S_2 + \mu_0 h_3 S_3 \qquad (4.37)$$

and

$$\Phi_1 = \Phi_4$$

or

$$\mu_0 h_1 S_1 = \mu_0 h_4 S_4 \qquad (4.38)$$

The condition $\Phi_4 = \Phi_2 + \Phi_3$ can also be written but this condition is obviously redundant.

The equations of circulation of the field intensity (including the coil) and the equations of conservation of flux (without redundancies) form a solvable system of equations. The following assumptions are made:

$$S_1 = S_4 = S \quad \text{and} \quad S_2 = S_3 = S/3$$

and, also, $l_1 = l_2 = l_3 = l_4 = l$. The system of equations is, therefore,

$$h_1 + h_2 + h_4 = NI/l$$

$$h_1 + h_3 + h_4 = NI/l$$

$$h_1 = h_2/3 + h_3/3$$

$$h_1 = h_4$$

From the first two equations, we get $h_2 = h_3$. By solving the system, we obtain

$$h_1 = h_4 = \frac{2NI}{7l}$$

$$h_2 = h_3 = \frac{3NI}{7l}$$

The flux generated by the coil is equal to Φ_1:

$$\Phi_1 = \mu_0 \frac{2NI}{7l} S_1$$

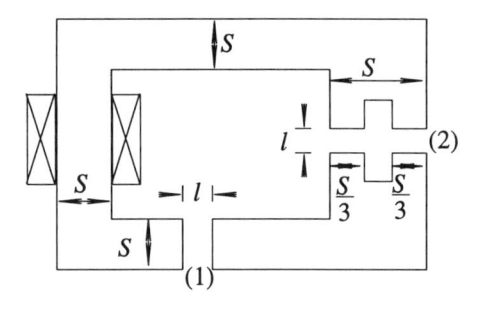

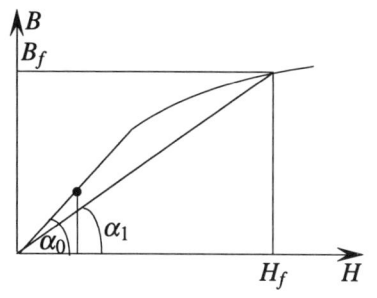

Figure 4.40. A saturated magnetic circuit.

Figure 4.41. The $B(H)$ curve for the magnetic circuit in Figure 4.40.

4.8.7. Calculation of the Magnetic Field Intensity of a Saturated Magnetic Circuit

In the previous example we assumed the permeability of iron is infinite, so that the effect of the field h_f in the iron can be neglected. However, if the magnetomotive force NI of the coil is large, the magnetic circuit attains saturation due to the large flux. In this case, the field h_f is not negligible and the permeability of iron cannot be considered to be infinite. To solve this problem, it is necessary to know the $B(H)$ curve of the material used for the magnetic circuit. The solution is obtained through an iterative process because, although the $B(H)$ curve is known, the operating point on the curve is not.

We consider the circuit of Figure 4.40 and assume saturation of the iron. The $B(H)$ curve for the material is given in Figure 4.41. The equations of circulation of the field and conservation of flux can now be written. For this purpose, we use l_f as the average length of the magnetic circuit (excluding gaps). The circulation is

$$h_1 l + h_2 l + h_f l_f = NI$$

From the conservation of flux:

$$\Phi_f = \Phi_1 \quad \therefore \ b_f S = \mu_0 h_1 S$$

$$\Phi_1 = \Phi_2 \quad \therefore \ \mu_0 h_1 S = 2\mu_0 h_2 \frac{S}{3}$$

The system can be written in the form:

$$h_1 l + h_2 l + h_f l_f = NI \qquad (4.39)$$

$$\mu_f h_f = \mu_0 h_1 \qquad (4.40)$$

134   4. Magnetostatic Fields

$$h_1 = 2\frac{h_2}{3} \tag{4.41}$$

In this system, we must use an iterative process to obtain the solution to the problem. One method of setting up an iterative process is to substitute h_2 from Eq. (4.41) into Eq. (4.39). This results in a reduced system with variables h_f and h_1:

$$\frac{5}{2}h_1 l + h_f l_f = NI \tag{4.42}$$

$$h_f = \frac{\mu_0}{\mu_f} h_1 \tag{4.43}$$

The process can be initiated in the linear domain where $h_{f0}=0$. At this value of h_{f0} we obtain μ_{f0} of the $B(H)$ curve by calculating the tangent of the angle α_0 as indicated in Figure 4.41; the iterative process is:

- from Eq. (4.42), $h_{11}=2NI/5l$
- from Eq. (4.43), $h_{f1}=\mu_0 h_{11}/\mu_{f0}$
- from the $B(H)$ curve with the value h_{f1}, we obtain b_{f1} and calculate

$$\mu_{f1} = \frac{b_{f1}}{h_{f1}}$$

- with the value h_{f1}, we can return to Eq. (4.42) to obtain a new value h_{12} through the relation

$$h_{12} = (NI - h_{f1} l_f)\frac{2}{5l}$$

- in Eq. (4.43), the new value of h_f is now h_{f2}
- we return to the $B(H)$ curve and iterate successively until the solution converges. Convergence is reached when the change in the field is smaller than a specified error. This process can be written in general form as

- $h_{1i} = (NI - h_{f(i-1)} l_f)\frac{2}{5l}$
- $h_{fi} = \frac{\mu_0}{\mu_{f(i-1)}} h_{1i}$

- from the curve, we obtain μ_{fi}

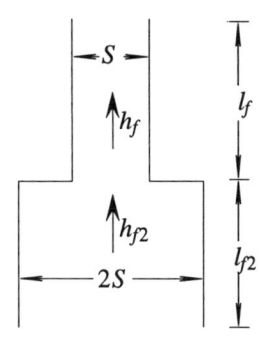

Figure 4.42. A magnetic circuit with varying cross-section.

Once convergence is obtained, and in general this occurs, we can obtain a value for h_2 through use of Eq. (4.41).

In this example, the field in iron is assumed to be the same as in the other parts of the magnetic circuit. This is the case when the cross-section of the iron is constant. It does not occur when the iron is shaped such that different field intensities exist in parts of the iron path as indicated in Figure 4.42. The equation of the circulation of H now contains more than one variable h_{f2}

$$... + h_f l_f + h_{f2} l_{f2} + ... = NI$$

However, we can establish more than one equation for the conservation of flux from which a solution can still be obtained. The additional equation for the geometry in Figure 4.42 is

$$\mu_{f2} h_{f2} 2S = \mu_f h_f S$$

Finally, we remark here that in a well-designed and properly dimensioned magnetic circuit this situation should not occur.

4.8.8. *Magnetic Circuit Incorporating Permanent Magnets*

Figure 4.43 shows a magnetic circuit in which the electromotive force is supplied by a magnet with known characteristics. We wish to calculate the fields in the gaps of the circuit. The two equations describing the circulation of H for the two contours passing through the magnet are

$$H_i L + h_1 l + h_2 l + h_3 l = 0 \quad (4.44)$$

$$H_i L + h_1 l + h_2 l + h_4 2l = 0 \quad (4.45)$$

136   4. Magnetostatic Fields

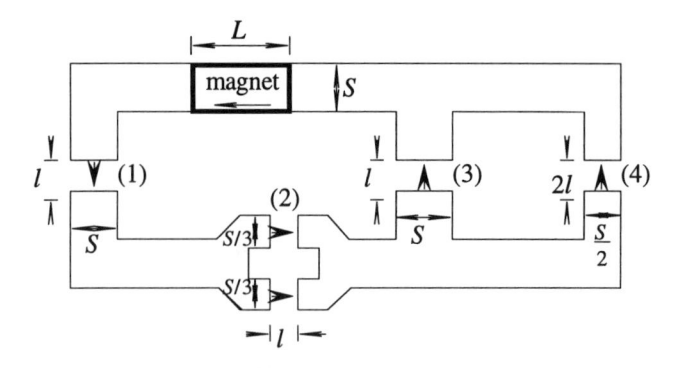

Figure 4.43. A magnetic circuit incorporating a permanent magnet.

We could have used the contour that passes through gaps (3) and (4); however, some precautions must be taken: first, it is necessary to define the direction of the magnetic flux in these gaps due to the flux in the magnet, as indicated in Figure 4.43 (both fluxes point upward); second, in the equation of circulation of the field we must take into account the vector $d\mathbf{l}$, as indicated in Figure 4.44. The equation

$$\oint_c \mathbf{H} \cdot d\mathbf{l} = I = 0$$

gives for the contour in Figure 4.44

$$-h_3 l + h_4 2l = 0 \qquad (4.46)$$

Subtracting Eq. (4.44) from Eq. (4.45), we obtain $h_4 2l - h_3 l = 0$. This means that Eq. (4.46) is redundant, reinforcing our decision to utilize only those circulation contours that pass through the coil (or magnet, in this case). The equations we obtain from these circulations are necessary for the solution of the problem (together with the equations of conservation of flux)

The equations of conservation of flux are

$$\Phi_i = \Phi_1 \qquad \Phi_1 = \Phi_2 \qquad \Phi_2 = \Phi_3 + \Phi_4$$

These can be written as

$$B_i S = \mu_0 h_1 S \qquad (4.47)$$

$$\mu_0 h_1 S = 2\mu_0 h_2 \frac{S}{3} \qquad (4.48)$$

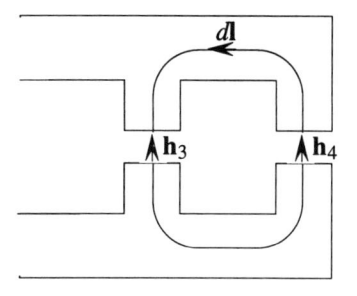

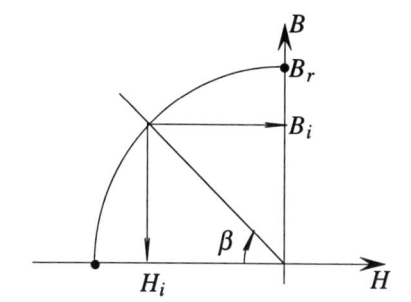

Figure 4.44. A contour through gaps (3) and (4) yields a redundant relation because the contour does not pass through the magnet.

Figure 4.45. The $B(H)$ curve of the magnet used in Figure 4.43.

$$2\mu_0 h_2 \frac{S}{3} = \mu_0 h_3 S + \mu_0 h_4 \frac{S}{2} \qquad (4.49)$$

We obtain the following system of equations

$$H_i L + h_1 l + h_2 l + h_3 l = 0$$

$$H_i L + h_1 l + h_2 l + h_4 2l = 0$$

$$B_i = \mu_0 h_1$$

$$h_1 = \frac{2h_2}{3}$$

$$\frac{2h_2}{3} = h_3 + \frac{h_4}{2}$$

By simple substitution, we obtain

$$\frac{B_i}{H_i} = -\mu_0 \frac{L}{l} \frac{10}{33}$$

With this relation, we can return to the characteristic curve of the magnet and obtain the operating values at the operating point of the magnet in Figure 4.45.

$$\beta = \arctan\left[\mu_0 \frac{L}{l} \frac{10}{33}\right]$$

4. Magnetostatic Fields

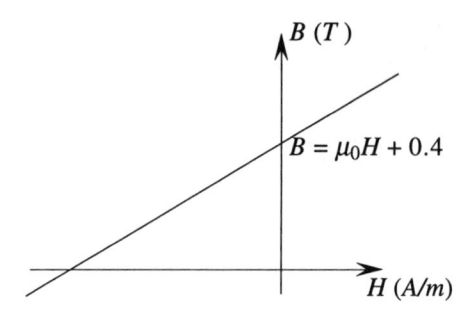

Figure 4.46.

Once the values of B_i and H_i are obtained, these are substituted in the system of equations above to determine h_1, h_2, h_3, and h_4.

As a numerical example, the characteristic curve of the magnet shown in Figure 4.46 is used. This corresponds to a ferrite magnet. The following are given

$$L=4\ cm \quad S=4\ cm^2 \quad l=2\ mm$$

Using the equation

$$B_i = -\mu_0 \frac{L}{l} \frac{10}{33} H_i$$

and

$$B_i = \mu_0 H_i + 0.4$$

we obtain the values

$$B_i = 0.343\ T \text{ and } H_i = -45,082\ At/m$$

From the relation $b_1 = \mu_0 h_1$, we get $h_1 = 272,950$ At/m. With $h_2 = 3h_1/2$, $h_2 = 409,426$ At/m. Using the same method in the other regions, $h_3 = 218,360$ At/m and $h_4 = 109,180$ At/m.

4.9. Laplace's Equation in Terms of the Magnetic Scalar Potential

Assuming that in a given two-dimensional domain there are no currents ($\mathbf{J}=0$), Maxwell's equations are:

Laplace's Equation in Terms of the Magnetic Scalar Potential

$$\nabla \times \mathbf{H} = 0$$

$$\nabla \cdot \mathbf{B} = 0$$

In this case, a scalar potential ψ can be defined, from which the field $\mathbf{H}$ is derived, through the expression

$$\mathbf{H} = -\nabla \psi$$

This definition is valid for any curl-free field. The equation $\nabla \cdot \mathbf{B} = 0$ becomes

$$\nabla \cdot \mathbf{B} = \nabla \cdot \mu \mathbf{H} = \nabla \cdot \mu(-\nabla \psi) = 0$$

which, in explicit form, is

$$\frac{\partial}{\partial x} \mu \frac{\partial \psi}{\partial x} + \frac{\partial}{\partial y} \mu \frac{\partial \psi}{\partial y} = 0$$

This is Laplace's equation for the diffusion of the magnetic field in terms of the magnetic scalar potential ψ. In fact, the equations below are identical mathematically:

$$\nabla \times \mathbf{H} = 0 \qquad \nabla \times \mathbf{E} = 0$$

$$\nabla \cdot \mathbf{B} = 0 \qquad \nabla \cdot \mathbf{D} = 0$$

$$\mathbf{H} = -\nabla \psi \qquad \mathbf{E} = -\nabla V$$

$$\mathbf{B} = \mu \mathbf{H} \qquad \mathbf{D} = \varepsilon \mathbf{E}$$

We can, therefore, in a completely analogous manner, extend the use of the finite element technique presented in the previous chapter. The programs remain unchanged, and the only difference is in the physical interpretation of quantities. In one case, the fields are electric, in the other magnetic. In one the material characteristic is the electric permittivity, in the other the magnetic permeability.

The functional to be minimized in this case is

$$F = \int_S \frac{\mu H^2}{2} ds$$

where S is the surface of the solution domain.

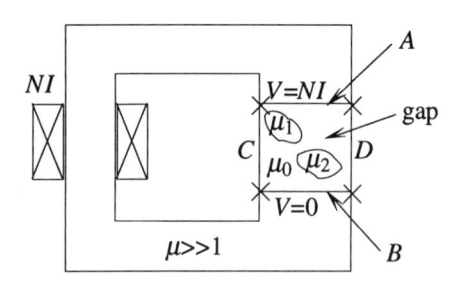

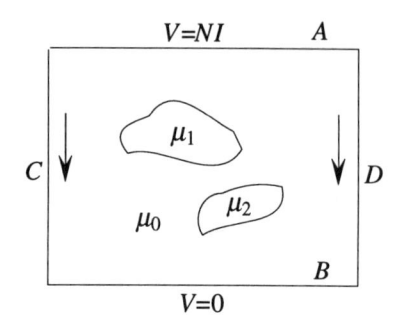

Figure 4.47a. The use of the magnetic scalar potential to calculate the field in the gap of a magnetic circuit in the presence of magnetic materials.

Figure 4.47b. The solution domain, including the applied boundary conditions, on the surfaces of the gap.

We should note here that the generating sources of the field, without currents in the solution domain, are the potential differences imposed as boundary conditions. Examining Figure 4.47a and 4.47b, we can attribute a physical meaning to the magnetic scalar potential. Assuming that with the electromagnet in Figure 4.47a, the magnetic circuit does not reach saturation, the magnetomotive force NI can, in practice, be imposed on the lines A and B which coincide with the boundaries of the gap. By doing so, the source generating the flux is the potential difference NI, between the lines A and B, corresponding to the effect of the coil. In this case, it is possible to study the field distribution in the gap (Figure 4.47b), using the program presented in the previous chapter, by observing that in the gap, $\mathbf{J}=0$. We must also assume that on lines C and D, the magnetic field is tangent to the lines.

There are other forms of Laplace's and Poisson's equations for the diffusion of the magnetic field. These equations, as well as their numerical solution using the finite element method, are discussed in the second part of this book.

4.10. Properties of Soft Magnetic Materials

Some of the properties of soft magnetic materials are summarized in Table 4.2. The most important of the soft magnetic materials is iron and its alloys because of its high relative permeability and its extensive use in electromagnetic devices. The relative permeability of iron varies widely depending on composition. The table also includes some of the more common paramagnetic materials.

Table 4.2 Properties of soft magnetic materials.

Material	Relative permeability (Maximum)	Coercive field [A/m]	Remnant induction [T]	Saturation induction [T]	Conductivity [$\Omega.m$]$^{-1}$
Iron (0.2% impure)	9,000	80	0.77	2.15	10^7
Pure iron (0.05% impure)	2.10^5	4	-	2.15	10^7
Silicon iron (3%Si)	55,000	8	0.95	2.0	2×10^6
Permalloy	10^6	4	0.6	1.08	6×10^6
Supermalloy	10^7	0.16	0.5	0.79	1.6×10^6
Permendur	5,000	160	1.4	2.45	1.4×10^6
Nickel*	-	-	-	-	11.5×10^6
Gold*	-	-	-	-	41.2×10^6
Platinum*	-	-	-	-	94.3×10^6
Lead*	-	-	-	-	45.5×10^6
Tungsten*	-	-	-	-	17.8×10^6
Zinc*	-	-	-	-	17.2×10^6
Silver*	-	-	-	-	61.3×10^6
Aluminum*	-	-	-	-	35.5×10^6
Copper*	-	-	-	-	58.1×10^6
Tin*	-	-	-	-	8.7×10^6
Mercury*	-	-	-	-	1.1×10^6

* Diamagnetic or paramagnetic materials

5
Magnetodynamic Fields

5.1. Introduction

After discussing electrostatic and magnetostatic fields, we now turn to time-dependent phenomena. From the decoupling of Maxwell's equations presented in Chapter 2, the equations for magnetodynamic fields are:

$$\nabla \times \mathbf{H} = \mathbf{J} \quad (5.1)$$

$$\nabla \cdot \mathbf{B} = 0 \quad (5.2)$$

$$\nabla \times \mathbf{E} = -\frac{\partial \mathbf{B}}{\partial t} \quad (5.3)$$

together with the constitutive relations:

$$\mathbf{B} = \mu \mathbf{H} \quad (5.4)$$

$$\mathbf{J} = \sigma \mathbf{E} \quad (5.5)$$

The difference between these equations and the equations for electrostatics is in that $\partial \mathbf{B}/\partial t$ is not zero.

The electromagnetic phenomena associated with the magnetodynamic field can be very complex. On the one hand, a new variable – time – has been introduced. On the other hand, the problems are, in general, three-dimensional in nature. This is because, as we will see shortly, the magnetic flux density **B** and the electric field intensity **E** generated by the flux density variation are in different planes. Due to these facts, magnetodynamic problems are, for the most part, difficult to solve. In most cases we must use approximations in order to solve the problem and, in general, need to adapt the solution methods to the characteristics of the problem. We will treat here many of the more important magnetodynamic effects that occur in devices under time-dependent excitation or devices that contain moving conducting parts in a static magnetic field. The issue of solution of the complex problems of magnetodynamics will be discussed again in Chapter 10.

5.2. Maxwell's Equations for the Magnetodynamic Field

The equations $\nabla \times \mathbf{H} = \mathbf{J}$ and $\nabla \cdot \mathbf{B} = 0$ were introduced in the previous chapter (section 4.2.) and we made considerable use of both. We recall that the first equation, in integral form, is usually referred to as "Ampere's law" and that the second indicates the conservative nature of the flux. Both were part of the set of equations needed for the static magnetic field.

The equation that is unique to the magnetodynamic domain is

$$\nabla \times \mathbf{E} = -\frac{\partial \mathbf{B}}{\partial t}$$

which indicates that the time variation of the flux density $\mathbf{B}$ generates an electric field intensity $\mathbf{E}$. After we gain a better understanding of this equation, we will examine in more detail the example presented in section 2.2.6c.

First, we represent this equation in integral form. To do so, we consider a surface S where $\mathbf{E}$ and $\mathbf{B}$ are defined and apply the integration:

$$\int_S (\nabla \times \mathbf{E}) \cdot d\mathbf{s} = -\int_S \frac{\partial \mathbf{B}}{\partial t} \cdot d\mathbf{s}$$

Using Stokes' theorem gives

$$\oint_C \mathbf{E} \cdot d\mathbf{l} = -\int_S \frac{\partial \mathbf{B}}{\partial t} \cdot d\mathbf{s} \qquad (5.6)$$

An example to this relation is shown in Figure 5.1a and 5.1b. Consider a ferromagnetic cylinder such that the flux density $\mathbf{B}$ is constant throughout its cross-section. The magnetic flux density $\mathbf{B}$ is time dependent and we use the cross-section of the cylinder for the surface S in Eq. (5.6). In addition, a conducting loop encircles the cylinder but the two are electrically isolated. This loop forms the contour C in Eq. (5.6). From the expression

$$\nabla \times \mathbf{E} = -\frac{\partial \mathbf{B}}{\partial t}$$

we note that the variation in $\mathbf{B}$ generates an electric field intensity $\mathbf{E}$ in the loop. This electric field intensity is proportional to $-\partial \mathbf{B}/\partial t$, as shown in Figure 5.1b.

144 5. Magnetodynamic Fields

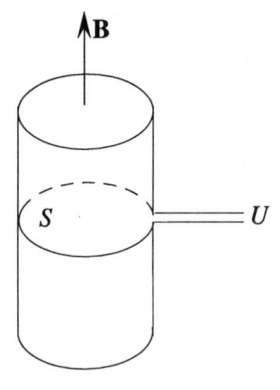

Figure 5.1a. A ferromagnetic cylinder with a constant, time-dependent, magnetic flux density throughout its cross-section. A conducting loop surrounds the cylinder.

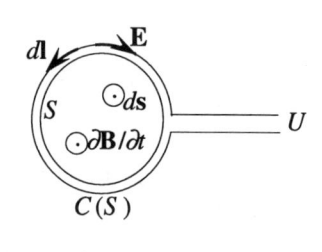

Figure 5.1b. Relations between the magnetic flux density, the electric field intensity and the *emf* in the loop.

The circulation of **E** along contour C leads to an electromotive force, which can be measured by a voltmeter as a voltage U. We have, therefore,

$$U = \oint_C \mathbf{E} \cdot d\mathbf{l}$$

We now turn our attention to the other term in Eq. (5.6), namely,

$$-\int_S \frac{\partial \mathbf{B}}{\partial t} \cdot d\mathbf{s}$$

First, we note that in this case, **B** depends only on time, since it is assumed that it is constant throughout the cross-section S. Therefore, we can write

$$\frac{\partial \mathbf{B}}{\partial t} = \frac{d\mathbf{B}}{dt}$$

Integration over the surface S and differentiation with respect to time are independent operations. Interchanging between the two operations gives

$$-\int_S \frac{\partial \mathbf{B}}{\partial t} \cdot d\mathbf{s} = -\frac{d}{dt}\int_S \mathbf{B} \cdot d\mathbf{s} = -\frac{d\Phi}{dt}$$

Equating the two terms of Eq. (5.6) gives

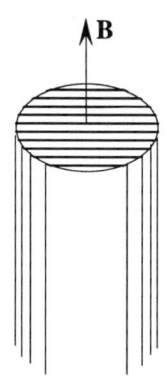

Figure 5.2. The use of laminated cores to reduce losses. The laminations are placed in planes perpendicular to the flow of induced currents.

$$U = -\frac{d\Phi}{dt} \qquad (5.7)$$

This is known as "Faraday's law" in recognition of Faraday who was the first to recognize this phenomenon.

The negative sign in this expression is a simple consequence of the original Maxwell's equation. To demonstrate this, we use Figure 5.1b and note that if the loop is viewed as a short circuit, a current circulates in the loop (since $\mathbf{J} = \sigma\mathbf{E}$). This current, in turn, generates a magnetic field intensity inside the material it encircles. This magnetic field intensity is opposite in direction to the variation $\partial\mathbf{B}/\partial t$. Hence, the generated field in Figure 5.1b penetrates into the plane of the figure, reducing the increase in **B**, whose direction is out of the plane.

The direction of the electric field intensity has been established by Lenz and the rule that defines it is known as "Lenz's law." This is defined as: "the induced current is such that the flux generated by the induced current tends to reduce the flux that generates it." We observe here that Faraday's and Lenz's laws are contained in Maxwell's equation $\nabla \times \mathbf{E} = -\partial\mathbf{B}/\partial t$.

The example above can be extended further; because the loop is short circuited, the current in the loop is called an "induced current" (as opposed to a source current) since it is induced by the time variation of **B**. In reality, the loop is not required to establish this current. The material of the cylinder is conducting, therefore, there are currents generated in the material itself. It is evident that we can associate losses with these currents through Joule's effect. This causes most of the energy used to generate **B** to be converted into heat. In other words, if the initial field in the material is zero, the induced currents in the material are such that they impede the increase in **B**. The net effect is that the total flux density in the material (the sum of the external flux density **B** and the flux density generated by the induced currents) is very low.

Is this effect desirable? The answer depends on the application. If we wish to heat up the material, the answer is positive. In an induction motor, the induced currents are necessary to generate torque. In these two examples, the generation of induced currents is necessary for the functioning of the device. An example in which induced currents are not desired is the transformer. The purpose of the transformer is to transfer energy from its primary coil to its secondary coil (with or without change in potentials). Any energy lost as heat in the ferromagnetic material reduces the efficiency of the transformer. If this loss is excessive, the transformer cannot function adequately. To overcome this difficulty and reduce the induced currents, it is common to build the transformer with laminated cores. This reduces the effective conductivity of the material by breaking the paths through which currents can flow. The laminations are insulated from each other and are placed such that they are perpendicular to the planes in which induced currents tend to flow, as shown in Figure 5.2.

A subtle distinction between induced currents that are useful and those that represent losses exists. Whenever the induced currents degrade the performance of a device, they tend to be termed "eddy currents" and are associated with "eddy current losses" which heat the structure through Joule's effect. Induced currents that are beneficial or necessary for the function of the device are simply called induced currents. It is important, however, to note that the two currents are physically the same phenomenon and are described by the same Maxwell's equation: $\nabla \times \mathbf{E} = -\partial \mathbf{B}/\partial t$.

5.3. Penetration of Time-Dependent Fields in Conducting Materials

The penetration of fields in arbitrarily shaped materials (either ferromagnetic or non-ferromagnetic) is a very complex problem. To solve this problem we assume that the materials are linear and isotropic. Based on this we establish a second order, time-dependent partial differential equation using Maxwell's equations. The penetration of fields in conductors is then found as a solution to this equation.

5.3.1. *The Equation for* **H**

Using the equation $\nabla \times \mathbf{H} = \mathbf{J}$, and applying the curl to both its sides, gives

$$\nabla \times (\nabla \times \mathbf{H}) = \nabla \times \mathbf{J} \qquad (5.8)$$

Using Eq. (1.23), the left side of Eq. (5.8), is

$$\nabla \times (\nabla \times \mathbf{H}) = \nabla(\nabla \cdot \mathbf{H}) - \nabla^2 \mathbf{H} \qquad (5.9)$$

Assuming that the material is linear, we have

$$\nabla \cdot \mathbf{B} = 0 \quad \therefore \quad \mu \nabla \cdot \mathbf{H} = 0 \quad \therefore \quad \nabla \cdot \mathbf{H} = 0$$

which leads to

$$\nabla \times (\nabla \times \mathbf{H}) = -\nabla^2 \mathbf{H} \qquad (5.10)$$

As for the right-hand side of Eq. (5.8), $\nabla \times \mathbf{J}$, we have

$$\nabla \times \mathbf{J} = \nabla \times \sigma \mathbf{E} = \sigma \nabla \times \mathbf{E} = -\sigma \frac{\partial \mathbf{B}}{\partial t} = -\sigma \mu \frac{\partial \mathbf{H}}{\partial t} \qquad (5.11)$$

Substituting Eqs. (5.10) and (5.11) into Eq. (5.8) gives

$$\nabla^2 \mathbf{H} = \sigma \mu \frac{\partial \mathbf{H}}{\partial t} \qquad (5.12)$$

5.3.2. *The Equation for* **B**

Substituting $\mathbf{H} = \mathbf{B}/\mu$ on both sides of Eq. (5.12), we obtain

$$\nabla^2 \mathbf{B} = \sigma \mu \frac{\partial \mathbf{B}}{\partial t} \qquad (5.13)$$

5.3.3. *The Equation for* **E**

Starting with the expression $\nabla \times \mathbf{E} = -\partial \mathbf{B}/\partial t$ and applying the curl on both sides we get

$$\nabla \times (\nabla \times \mathbf{E}) = -\nabla \times \frac{\partial \mathbf{B}}{\partial t} \qquad (5.14)$$

Assuming that there are no static charges in the domain and that ε is constant, the equation $\nabla \cdot \mathbf{D} = \rho$ becomes

$$\nabla \cdot \mathbf{E} = 0 \qquad (5.15)$$

Using Eq. (1.23) for the left-hand side gives

$$\nabla \times (\nabla \times \mathbf{E}) = \nabla (\nabla \cdot \mathbf{E}) - \nabla^2 \mathbf{E}$$

148   5. Magnetodynamic Fields

which, with Eq. (5.15) is

$$\nabla \times (\nabla \times \mathbf{E}) = -\nabla^2 \mathbf{E} \tag{5.16}$$

For the right-hand side of Eq. (5.14), we have

$$-\nabla \times \frac{\partial \mathbf{B}}{\partial t} = -\frac{\partial}{\partial t}\nabla \times \mathbf{B} = -\frac{\partial}{\partial t}\mu\nabla \times \mathbf{H} = -\mu\frac{\partial}{\partial t}\mathbf{J} = -\mu\sigma\frac{\partial \mathbf{E}}{\partial t} \tag{5.17}$$

Substituting Eqs. (5.16) and (5.17) into Eq. (5.14) gives

$$\nabla^2 \mathbf{E} = \sigma\mu\frac{\partial \mathbf{E}}{\partial t} \tag{5.18}$$

5.3.4. *The Equation for* **J**

The equation for **J** is obtained from $\nabla \times \mathbf{E} = -\partial \mathbf{B}/\partial t$ by substituting $\mathbf{J}/\sigma$ for **E**

$$\frac{1}{\sigma}\nabla \times \mathbf{J} = -\mu\frac{\partial \mathbf{H}}{\partial t}$$

Taking the curl on both sides,

$$\nabla \times (\nabla \times \mathbf{J}) = -\mu\sigma\frac{\partial}{\partial t}\nabla \times \mathbf{H} \tag{5.19}$$

The left-hand side, using the relation $\nabla \cdot \mathbf{J} = 0$ is

$$\nabla \times (\nabla \times \mathbf{J}) = \nabla(\nabla \cdot \mathbf{J}) - \nabla^2 \mathbf{J} = -\nabla^2 \mathbf{J} \tag{5.20}$$

and, therefore,

$$\nabla^2 \mathbf{J} = \sigma\mu\frac{\partial \mathbf{J}}{\partial t} \tag{5.21}$$

This equation can also be obtained by simply substituting $\mathbf{J}/\sigma$ for **E** in Eq. (5.18).

5.3.5. *Solution of the Equations*

We note that Eqs. (5.12), (5.13), (5.18), and (5.21) have the general form

$$\nabla^2 \mathbf{P} = \sigma\mu\frac{\partial \mathbf{P}}{\partial t} \tag{5.22}$$

where the vector **P** represents **H**, **B**, **E**, or **J** in the respective equations. We now seek a general solution, applicable to all four vectors. Eq. (5.22), in explicit form for the x component of the vector **P** is

$$\frac{\partial^2 P_x}{\partial x^2} + \frac{\partial^2 P_x}{\partial y^2} + \frac{\partial^2 P_x}{\partial z^2} = \mu\sigma\frac{\partial P_x}{\partial t} \qquad (5.23)$$

The solution of this equation is difficult to obtain. To understand the complexity of the phenomenon involved we remember that, on the one hand, this equation was obtained by assuming linearity and isotropy of materials and, on the other hand, that there are two more equations, identical to (5.23), for the y and z components of **P**. For this reason, and to be able to analyze this equation, we limit ourselves to a simpler case. First, we assume that all fields are sinusoidal in nature. Using the current density **J** as a variable, in complex notation, we have

$$\mathbf{J} = \mathbf{J}_0 e^{j\omega t} \qquad (5.24)$$

where $j=\sqrt{-1}$, $\omega=2\pi f$ is the angular frequency, f the frequency, and $\mathbf{J}_0$ the amplitude of the sinusoidal current density **J**. We can also write

$$\frac{\partial \mathbf{J}}{\partial t} = j\omega \mathbf{J}_0 e^{j\omega t} = j\omega \mathbf{J} \qquad (5.25)$$

Substituting this in Eq. (5.21) yields

$$\nabla^2 \mathbf{J} - j\sigma\mu\omega\mathbf{J} = 0 \qquad (5.26)$$

Using a variable δ such that

$$\delta = \sqrt{\frac{2}{\sigma\mu\omega}} \qquad (5.27)$$

Eq. (5.26) becomes

$$\nabla^2 \mathbf{J} - \frac{2j}{\delta^2}\mathbf{J} = 0 \qquad (5.28)$$

This equation is still difficult to solve and we need to further simplify the problem; we assume that the conductor is a semi-infinite block with the surface on the xy plane as shown in Figure 5.3.

Consider now a sinusoidal electric field intensity $\mathbf{E}_0$, coincident with the Ox direction throughout the surface of the conducting block.

150   5. Magnetodynamic Fields

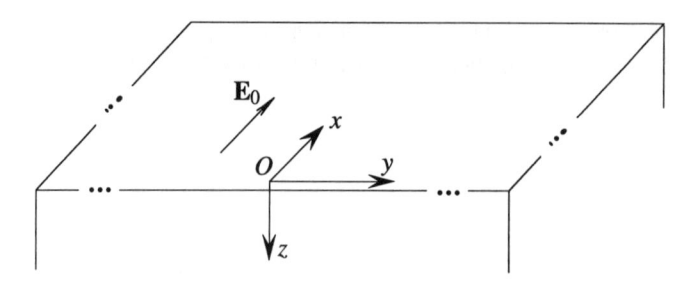

Figure 5.3. An electric field intensity, parallel to the surface of a semi-infinite conducting block.

Since the tangential component of the electric field intensity is continuous, the same electric field intensity exists immediately below the surface, in the conductor. This implies the existence of a current density $\mathbf{J}_0 = \sigma \mathbf{E}_0$.

With these assumptions, **J** has only a component in the *x* direction and varies only in the *z* direction. Eq. (5.28) assumes the following simplified form:

$$\frac{\partial^2 J_x(z)}{\partial z^2} - \frac{2j}{\delta^2} J_x(z) = 0 \qquad (5.29)$$

The solution to this equation is

$$J_x(z,t) = J_0 e^{-z/\delta} \cos(\omega t - z/\delta) \qquad (5.30)$$

In this equation the following are immediately evident:

- The amplitude of the current density is $J_0 e^{-z/\delta}$.
- The phase of the current density is $-z/\delta$.

As *z* increases, or as fields penetrate deeper into the conductor, the amplitude of *J* decreases and the phase changes. The amplitude decreases exponentially. The term δ is called "depth of penetration" or "skin depth." When $z=\delta$, or when the field has penetrated to a depth δ, we get

$$J_0 e^{-\delta/\delta} = J_0/e = 0.37 J_0$$

At this point ($z=\delta$), the value of *J* is 37% of J_0.

For practical purposes, *J* can be neglected for $z=3\delta$ or larger.

From the expression

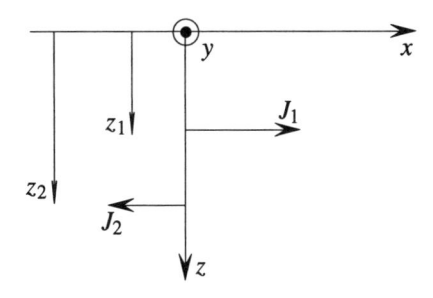

Penetration of Time-Dependent Fields in Conductors 151

Figure 5.4. Relations between the amplitudes of the current density at two locations along the z axis.

$$\delta = \sqrt{\frac{2}{\sigma\mu\omega}} = \frac{1}{\sqrt{\pi f \mu \sigma}}$$

it is obvious that the higher the frequency the smaller the depth of penetration of the field. Similarly, the larger the permeability (in ferromagnetic materials) or the conductivity, the smaller the depth of penetration δ.

As for the phase $-z/\delta$, this varies with z. If we define a reference phase as zero at the surface of the conductor ($z=0$) we can write: $cos(\omega t - 0)$. At another point below the surface, ($z \neq 0$), we have $cos(\omega t - \alpha)$. By way of example, suppose that:

at $z = z_1$, $\omega t - z_1/\delta = 0$ and $J_1 = J_0 e^{-z_1/\delta}$

at $z = z_2$, $\omega t - z_2/\delta = 180°$ and $J_2 = J_0 e^{-z_2/\delta}(-1)$

Figure 5.4 shows the values of $\mathbf{J}$ at $z=z_1$ and $z=z_2$. The vectors $\mathbf{J}$ are in opposite directions ($|J_2|$ is obviously smaller than $|J_1|$ due to the exponential attenuation in amplitude). For these geometrical and temporal considerations the following expression for $\mathbf{J}$:

$$J(x,t) = J_0 e^{-z/\delta} cos(\omega t - z/\delta)$$

describes damped harmonic motion, as for example, in the case of a string, fixed at one end and subjected to a varying motion at the other end. (see Figure 5.5). (The effects of reflection of waves from the fixed point cannot be considered in this comparison). The vectors $\mathbf{J}_1$ and $\mathbf{J}_2$ mentioned above and their relationship are also shown in Figure 5.5.

The considerations above are valid for $\mathbf{H}$, $\mathbf{B}$, and $\mathbf{E}$ as well since Eqs (5.12), (5.13), (5.18), and (5.21) have the same coefficients.

152   5. Magnetodynamic Fields

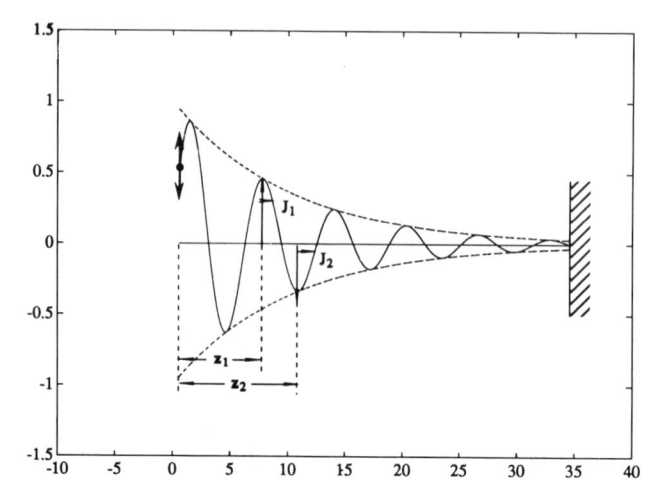

Figure 5.5. Wave motion of a string with one end fixed. The other end is moved up and down.

The depth of penetration δ is the same for the fields above: as one decays due to penetration in the conductor, the others decay at the same rate. The only point that must be remembered is that **H** and **B** are perpendicular to **J** and **E** as defined by the curl relation $\nabla \times \mathbf{H} = \mathbf{J}$, (see Figure 5.6).

These results were obtained by applying a series of assumptions and simplifications. The question is, how good are these results in real applications? The answer to this question is given from two points of view:

- In the semi-infinite block assumed in Figure 5.3, calculation of the depth of penetration δ for copper, at 60 Hz ($\sigma = 5.7 \times 10^7$ S/m; $\mu = \mu_0$), gives $\delta = 8.6$ mm. This dimension is small compared to most low-frequency electromagnetic devices (motors, transformers, power transmission cables, etc.). This justifies the assumption that the dimensions of the device are large compared to the depth of penetration and the error in using the definitions above is small. If we were to establish the correct (and much more complex) differential equations for a given structure, taking into account its geometry, we would almost certainly get different values for δ. However, intuitively, we can sense that these results would not be much different than those obtained from the simplified solution. If the dimensions of the structure are small compared to δ, the calculation above must be re-evaluated. We note, however, that the calculation above is for copper, for which $\mu = \mu_0$, and at a low frequency of 60 Hz. For this reason, depth of penetration δ is relatively large. In a ferromagnetic material, where $\mu_r > 1000$, the depth of penetration is much smaller. In this case, the considerations above are almost always valid.

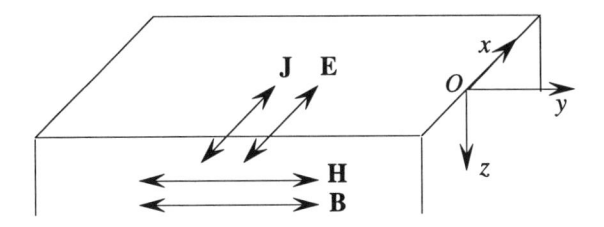

Figure 5.6. Relations between the electric and magnetic fields in a semi-infinite conducting block. The electric field intensity and the current density are perpendicular to the magnetic field intensity.

- As for the assumption that the fields are sinusoidal in nature, it is important to note that many electromagnetic devices operate under sinusoidal excitation. If the excitation is nonsinusoidal, we can decompose the excitation into sinusoidal components through the use of Fourier series. Each component will have different frequencies and therefore, different depths of penetration. However, the fundamental frequency (the same frequency as the original excitation) possesses the largest depth of penetration since it has the lowest frequency. The rest of the harmonics, which normally have lower amplitudes, have lower depths of penetration because of their higher frequencies. Because of this, we normally use the same formula for nonsinusoidal fields as for sinusoidal fields. This is a valid assumption in a majority of practical cases; in other particular cases, where this is not valid, the situation must be considered on a case-by-case basis.

5.4. Eddy Current Losses in Plates

We saw that in a ferromagnetic material subjected to a time-varying field, there are loops of induced currents, or eddy currents. Suppose now that the material is laminated, for the purpose of impeding the flow of the eddy currents. We will see shortly, that in spite of this, there are losses due to Joule's effect. Figure 5.7 shows a thin plate or lamination made of a ferromagnetic material. The plate is subjected to a flux density **B**, parallel to the plate in the Ox direction. The dimensions l_x and l_y are much larger than e. This means that the plate is thin and low-amplitude induced currents do not affect the externally generated flux density **B**. Under these conditions, we can assume that the induced current density **J** is not dependent on x or y. This is shown in Figure 5.8a and 5.8b in which the principal component of **J** is in the direction Oy. From $\mathbf{J}=\sigma\mathbf{E}$, the same is true for **E**. The equation $\nabla\times\mathbf{E}=-\partial\mathbf{B}/\partial t$ applied to this case assumes the form

154   5. Magnetodynamic Fields

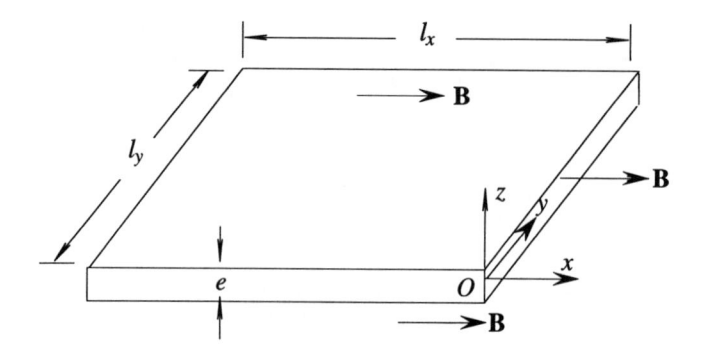

Figure 5.7. A thin lamination in a magnetic field.

$$\det \begin{bmatrix} \hat{i} & \hat{j} & \hat{k} \\ \dfrac{\partial}{\partial x} & \dfrac{\partial}{\partial y} & \dfrac{\partial}{\partial z} \\ 0 & E_y & 0 \end{bmatrix} = \begin{bmatrix} -\dfrac{\partial B_x}{\partial t} \\ 0 \\ 0 \end{bmatrix}$$

and, therefore,

$$\frac{\partial E_y}{\partial z} = \frac{\partial B_x}{\partial t}$$

The solution to this equation is

$$E_y(z) = \frac{\partial B_x}{\partial t} z + k$$

Since there is no discontinuity in **J** or **E** in Figure 5.8b, $E_y=0$ at $z=0$. This makes the constant k zero and the solution is

$$E_y(z) = \frac{\partial B_x}{\partial t} z \qquad (5.31)$$

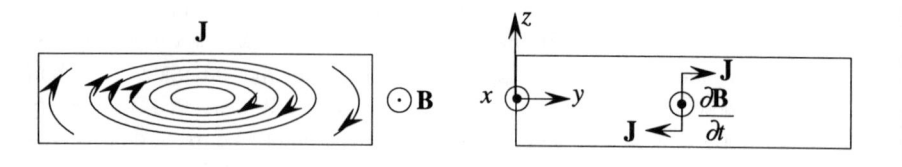

Figure 5.8a. Current density in the thin plate of Figure 5.7.

Figure 5.8b. Relations between the magnetic flux density, the electric field intensity and the *emf* in the loop.

Eddy Current Losses in Plates

The power dissipated in the plate by the Joule effect is (see observation at the end of this paragraph)

$$P_F = \int_V \sigma E_y^2 dv \qquad (5.32)$$

where V is the volume of the plate ($V=l_x l_y e$). Applying Eq. (5.31) in (5.32) gives

$$P_F = \sigma \left(\frac{\partial B_x}{\partial t}\right)^2 \int_0^{l_x} \int_0^{l_y} \int_{-e/2}^{+e/2} z^2 dx dy dz$$

$$P_F = \frac{\sigma}{12}\left(\frac{\partial B_x}{\partial t}\right)^2 l_x l_y e^3 = \frac{\sigma}{12}\left(\frac{\partial B_x}{\partial t}\right)^2 e^2 V \qquad (5.33)$$

Suppose that B_x varies sinusoidally with time as $B_x = B_m \cos\omega t$. This gives $\partial B_x/\partial t = -\omega B_m \sin\omega t$. Eq. (5.33) is now

$$P_F = \frac{\sigma}{12}\omega^2 B_m^2 e^2 V \sin^2\omega t$$

Normally, the instantaneous value of losses is of less interest than the average losses. To calculate the average losses we use the average value of 1/2 for $\sin^2\omega t$. The power loss per unit volume, $p_f = P_F/V$, is therefore

$$p_f = \frac{1}{24}\sigma\omega^2 e^2 B_m^2 \qquad (5.34)$$

From this expression we note the following:

- p_f depends on e^2, (depends quadratically on the thickness of the plate).
- p_f depends on ω^2, (is quadratic in frequency).
- Materials with low conductivities σ have small losses.

Since, in general, the frequency is imposed by the operating conditions of the device, it is possible to adapt the thickness of the lamination, e to the frequency of the field. The higher the operating frequency, the thinner the lamination should be. For dc fields ($\omega=0$) or in nonconducting materials ($\sigma=0$), there are no power losses due to eddy currents.

<u>Observation</u>: The expression for power

156   5. Magnetodynamic Fields

$$P = \int_v \sigma E_y^2 dv$$

for a conducting block, with cross-section S and length l, takes a simpler form. Assuming that the current through the cross-section is uniform, we can write $J=I/S$. With $E=J/\sigma$, the expression becomes

$$P = \int_v \sigma \frac{J^2}{\sigma^2} dsdl = \frac{J^2}{\sigma} S\, l$$

$$P = \frac{I^2}{S^2} \frac{1}{\sigma} S\, l = \frac{I^2}{\sigma} \frac{l}{S}$$

Since the resistance of the block is $R=l/\sigma S$, the dissipated power through Joule's effect is

$$P = RI^2$$

5.5. Hysteresis Losses

Suppose that a ferromagnetic material is under the influence of a periodic, time-dependent magnetic field intensity H, with frequency f and period $T=1/f$. All ferromagnetic materials possess a characteristic curve $B(H)$. This type of curve is always a hysteresis curve, more or less pronounced, depending on the type of material (see Figure 5.9).

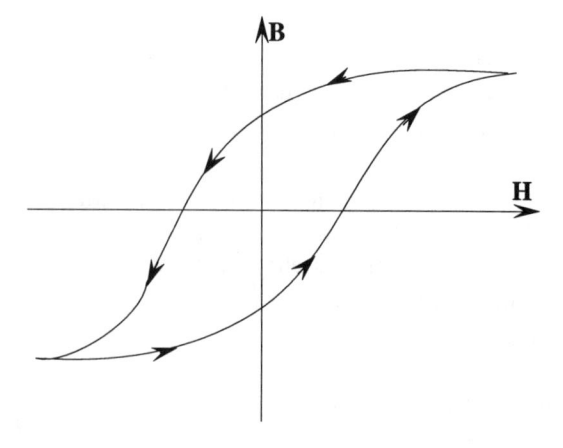

Figure 5.9. Hysteresis curve of a ferromagnetic material.

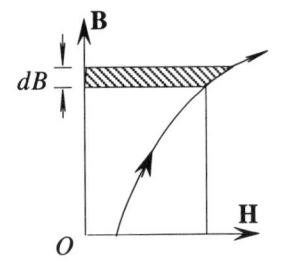

Figure 5.10. 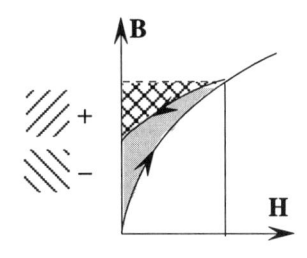   Figure 5.11

As H varies, the material passes through the loop of the hysteresis curve. This means that part of the energy in the device is consumed in the cycling through the hysteresis curve. This is generally detrimental to the operation of the device since it causes heating of the material. Denoting P_h the power associated with the hysteresis curve, the energy W_h consumed in a cycle is

$$W_h = P_h T \quad \text{or} \quad W_h = P_h/f$$

and

$$P_h = W_h f$$

The volumetric energy density w_h, necessary to generate a flux density B in a material starting at $B=0$ is given by the following relation

$$w_h = \int_0^B H db \qquad (5.35)$$

The total energy in a block of material of volume V is therefore $W_h = w_h V$.

We now look at the graphical representation of Eq. (5.35). Because it is an integral, its representation on the $B(H)$ curve is the surface shown in Figure 5.10. If the curve is traced in the positive (counterclockwise) direction, this area is positive. If it is described in the opposite (clockwise) direction, this surface becomes negative, as shown in Figure 5.11.

The difference between the positive and negative areas corresponds to the internal area of the hysteresis loop. The same consideration applies to the whole of the hysteresis loop (Figure 5.12). We conclude that the area represents the total volumetric energy consumed in one loop.

The power due to hysteresis is, therefore,

$$P_h = AVf \qquad (5.36a)$$

where A is the internal area of the hysteresis loop.

158   5. Magnetodynamic Fields

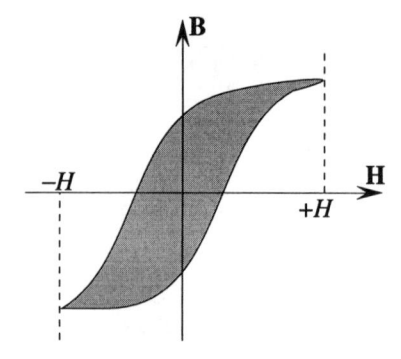

Figure 5.12. Energy consumed in a cycle of the hysteresis curve is the area of the curve.

The dissipated energy per cycle is

$$W = AV \qquad (5.36b)$$

The losses depend on the magnitude of **H** in the material, as in Figure 5.13.

In large electromagnetic devices, the operating point is often such that H is high and the material is in saturation. Using experimental results for values of B between $0.2T$ and $1.5T$, Steinmetz came up with an empirical expression which provides the area of the hysteresis curve. This is known as "Steinmetz's equation," given in W/m^3:

$$P_h = \eta B^{1.6} \qquad (5.37)$$

where η is a material dependent constant.

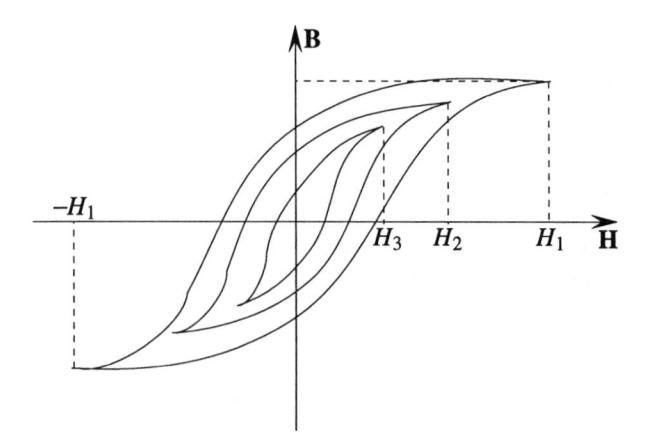

Figure 5.13. The hysteresis curve and losses due to hysteresis depend on the magnitude of **H**.

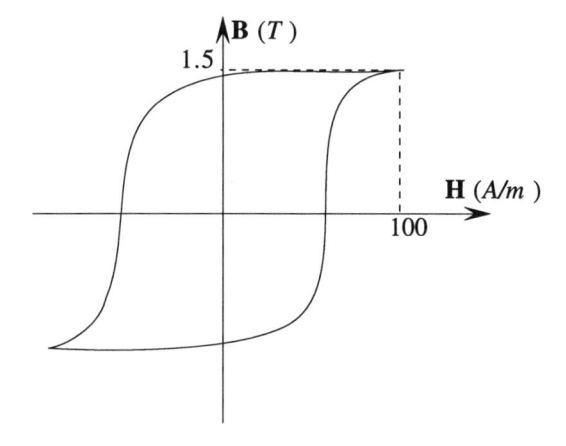

Figure 5.14. B(H) curve for a transformer core.

Some typical values of η are:

$\eta=30$ for silicon iron alloys
$\eta=250–500$ for iron alloys
$\eta=3.8–8.5.\times 10^3$ for hard magnetic materials

This equation represents a good approximation for the conditions specified above.

As an example to the calculation of losses, suppose we have a transformer core, made of an iron alloy with a $B(H)$ curve as shown in Figure 5.14. The volume of the core is $100 cm^3$. Other parameters are given as:

$$f=60Hz; \quad \sigma=10^7 \ S/m; \quad e=1 \ mm; \quad B_{max}=1.5 \ T$$

The losses due to heating by eddy currents are given in Eq. (5.34)

$$p_f = \frac{1}{24} \sigma \omega^2 e^2 B_{max}^2$$

Using the numerical values above (in MKS) gives

$$p_f = 133,240 \ W/m^3$$

To obtain the total power in the volume of the core, the power density is multiplied by the volume

$$P_f = 13.3 \ W$$

To calculate the losses due to hysteresis, we must evaluate the area of the loop in Figure 5.14. This is approximately

$$A = 1.5 \times 100 \times 4 = 600 \ (J/m^3)$$

The power given by Eq. (5.36a) is

$$P_h = AVf = 600 \times 10^{-4} \times 60 = 3.6 \ W$$

The total losses are

$$P_t = P_h + P_f = 16.9 \ W$$

This means that 16.9 W of the power injected into the transformer is dissipated by losses in the iron core.

The area of the hysteresis loop can also be calculated by use of Steinmetz's equation. In this case there is some uncertainty about the value of η. For proper use of this equation, η must be obtained from the manufacturer of the laminations or must be found in laboratory experiments. The process of evaluating the area of the hysteresis loop directly from the cycle is preferred.

5.6. Examples

5.6.1. Induced Currents Due to Change in Induction

Consider a conducting ring as shown in Figure 5.15a. The flux density, $\mathbf{B}=\mathbf{B}_0$ is constant and uniform throughout the area defined by the ring. At a given time $t=0$, the field $\mathbf{B}$ starts to increase as given by the following relation

$$\mathbf{B}(t) = \mathbf{B}_0(1+kt)$$

where k is a constant.

The following numerical data is used in this example:

S = cross-sectional area of ring = 2 mm^2
R = mean radius of ring = 1 cm
$\sigma = 1.0 \times 10^7 \ (\Omega m)^{-1}$
$B_0 = 1 \ T$
$k = 60 \ T/s$

We wish to calculate the induced current in the ring.

Assuming that the external magnetic flux density increases in the direction shown in Figure 5.15b, according to the equation

Examples 161

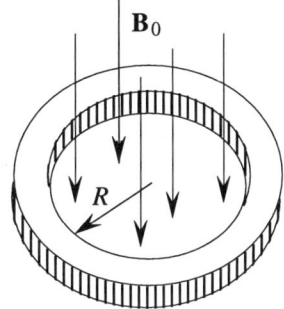

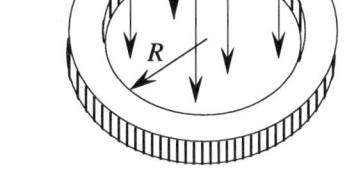

Figure 5.15a. A conducting ring in a time-dependent magnetic field.

Figure 5.15b. The relation between the magnetic field, induced electric field intensity, and induced magnetic flux density in the ring.

$$\nabla \times \mathbf{E} = -\frac{\partial \mathbf{B}}{\partial t} \qquad (5.38)$$

Note that an electric field intensity $\mathbf{E}_i$ and a magnetic flux density $\mathbf{B}_i$ are induced as shown in the figure. To solve this problem, the following assumptions are made:

- The cross-sectional area of the turn is small compared to the radius R. For this reason, $\mathbf{E}$ and, therefore, $\mathbf{J} = \sigma \mathbf{E}$ are considered constant throughout the cross-section of the ring.
- $\mathbf{B}_i$ is much smaller than the internal flux density. This means that flux density $\mathbf{B}$ in Eq. (5.38) is equal to the external flux density $\mathbf{B}(t) = \mathbf{B}_0(1+kt)$.

To evaluate $\mathbf{E}$, $\mathbf{J}$, and the current I in the ring, we apply Eq. (5.38) in integral form

$$\oint_{C(S)} \mathbf{E} \cdot d\mathbf{l} = \int_S -\frac{\partial \mathbf{B}}{\partial t} \cdot d\mathbf{s} \qquad (5.39)$$

where S is the surface defined by the circular ring and $C(S)$ is the circumference of the ring. Since $\mathbf{B}$ depends only on time and vectors $\mathbf{E}$, $\partial \mathbf{B}/\partial t$, and $d\mathbf{s}$ are collinear we get

$$E 2\pi R = -\frac{\partial B}{\partial t} \pi R^2$$

or

5. Magnetodynamic Fields

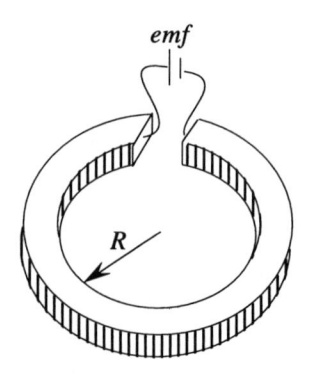

Figure 5.16. The electromotive force generated in the ring can be viewed as a source for the purpose of calculation of current.

$$E = \frac{R}{2}\frac{\partial}{\partial t}(B_0 + B_0 k t) = \frac{R}{2} B_0 k$$

Since $\mathbf{J} = \sigma \mathbf{E}$, we have

$$J = \frac{\sigma k B_0 R}{2}$$

and

$$I = \frac{\sigma k B_0 R S}{2} \qquad (5.40)$$

It is possible to obtain I in a different manner. This is done by evaluating the electromotive force generated in the ring due to the time variation of $\mathbf{B}$. To do so, we return to Eq. (5.39) and write

$$emf = \oint_C \mathbf{E} \cdot d\mathbf{l}$$

which gives

$$emf = -\int_S \frac{\partial \mathbf{B}}{\partial t} \cdot d\mathbf{s} = \pi R^2 B_0 k \qquad (5.41)$$

The electromotive force can be viewed as the voltage source placed in the ring, as shown in Figure 5.16. The *emf* in the ring is

$$emf = Res \times I$$

where *Res* is the resistance of the loop and *I* the current in the loop. Since the resistance depends on conductivity

$$Res = \frac{1}{\sigma}\frac{l}{S} = \frac{1}{\sigma}\frac{2\pi R}{S} \qquad (5.42)$$

Using Eqs. (5.41) and (5.42) gives

$$\pi R^2 B_0 k = \frac{1}{\sigma}\frac{2\pi R}{S} I$$

and

$$I = \frac{\sigma k B_0 S R}{2}$$

which is identical to the result obtained in Eq. (5.40). With the numerical values given above the current is

$$I = 6 \, Amperes$$

Now we can calculate the flux density $\mathbf{B}_i$. Using the result of section 4.8.2, the field intensity at the center of an annular ring carrying a current I is

$$H = \frac{I}{2R} = 300 \, A/m$$

and

$$B_i = \mu_0 H = 0.377 \times 10^{-3} \, T$$

Although $\mathbf{B}_i$ varies within the ring, increasing towards the ring itself, its magnitude is much lower than $\mathbf{B}_0$, justifying our assumptions.

5.6.2. Induced Currents Due to Changes in Geometry

Consider the structure in Figure 5.17a. It is made of four conducting bars, with one of the bars free to move. Assume that this does not create a resistance at the points of contact between the moving and stationary bars. A magnetic flux density **B** exists within the rectangular loop created by the bars.

164   5. Magnetodynamic Fields

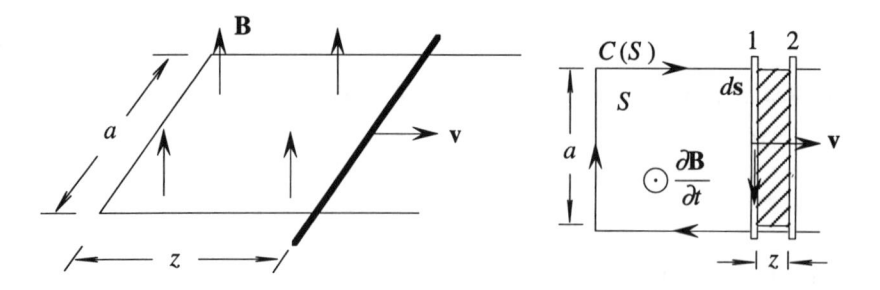

Figure 5.17a. A moving bar on rails. As the bar moves, the geometry changes.

Figure 5.17b. The change in flux due to movement of the bar.

The moving bar moves at a velocity **v** (directions of **B** and **v** are indicated in Figure 5.17a).

An electromotive force is generated in the loop because the flux in the loop varies with time. The current in the loop is such that it opposes the change in flux, as shown in Figure 5.17b. The electromotive force is

$$emf = \oint_C \mathbf{E} \cdot d\mathbf{l} = -\int_S \frac{\partial \mathbf{B}}{\partial t} \cdot d\mathbf{s}$$

where S is the varying surface of the loop and C is the contour limiting this surface. It should be noted that change in flux in this case corresponds to the flux crossing the part of the surface indicated in Figure 5.17b. This surface is created by the moving bar as it moves from point 1 to point 2 at a velocity $v=dz/dt$. Since **B** is not time dependent and since $d\mathbf{s}$ and **B** are collinear, we have (magnitude only)

$$emf = \frac{\partial}{\partial t} \int_S B ds = \frac{\partial}{\partial t}(Baz)$$

with $S=za$. Since B is constant, we get

$$emf = Ba \frac{\partial z}{\partial t} = Bav \qquad (5.43)$$

The magnitude of the current I in the loop is obtained from the expression $emf=RI$. The resistance of the loop depends on conductivity σ, its cross-section and the length l. We note from Eq. (5.43), that while v is constant, the *emf* is also constant. However, as the bar moves, the resistance of the loop increases since l increases. For this reason, the current in the loop $I=emf/R$ decreases with time.

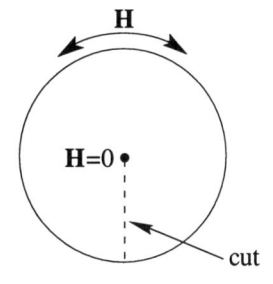

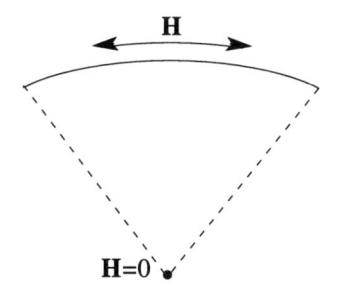

Figure 5.18a. A cylindrical current-carrying conductor.

Figure 5.18b. The conductor after cutting and "opening."

5.6.3. *Some Further Examples of Skin Depth*

5.6.3.a. Perpendicular Field on a Conducting Block

In section 5.3 and especially for the solution of Eq. (5.29) it was shown that for tangential components of fields (as, for example in Figure 5.3), the fields **H**, **B**, **E**, and **J**, decrease exponentially in magnitude as they penetrate into conducting materials, such that for $z=(2/\mu\sigma\omega)^{1/2}$ the amplitude of the fields is reduced to 1/e (0.37%) of the value at the surface of the conductor.

Here we discuss the case of a field perpendicular to the surface instead of the case of section 5.3. in which we assumed the field to be tangential or parallel to the external surface. The remaining assumptions are the same; the block is semi-infinite and excitation is sinusoidal.

Under these assumptions there is only one component of the flux density $B_z=B_0$ (perpendicular to the xy plane) with dependence on the z coordinate only. Taking into account the continuation of the normal component of the magnetic flux density, the flux density immediately under the surface, inside the material is also equal to $B_0=B_z$. The equivalent equation to Eq. (5.29) is

$$\frac{\partial^2 B_z(z)}{\partial z^2} - \frac{2j}{\delta^2} B_z(z) = 0 \qquad (5.44)$$

The solution to this equation is analogous to that of Eq. (5.30)

$$B_z(z,t) = B_0 e^{-z/\delta} \cos(\omega t - z/\delta) \qquad (5.45)$$

This field possesses the same physical characteristics as the tangential field, including an identical depth of penetration.

5.6.3.b. Cylindrical Cable

A conducting cable carrying an *ac* current through its cross-section is shown in Figure 5.18a. We will show, using a qualitative analysis, why the current flows in the external part of the cable.

First, it is obvious from Ampere's law that the magnetic field intensity must be zero at the center of the cable. Similarly, using Ampere's law, we find that the field intensity outside the cable is nonzero, as shown in Figure 5.18a.

Now imagine that we cut the cable of Figure 5.18a and "open" it, arriving at the hypothetical situation of Figure 5.18b. In this figure the magnetic field intensity decreases to zero from the external surface to the center.

Using the results of section 5.3., where we found that **H**, **B**, **E**, and **J** all behave the same, based on the general equation (Eq. 5.22)

$$\nabla^2 \mathbf{P} = \sigma\mu \frac{\partial \mathbf{P}}{\partial t} \tag{5.46}$$

we conclude that the current flowing in the cable must behave similarly to the magnetic field intensity **H** and therefore decreases from the outer surface to the center.

More accurately, the equation describing this case, in cylindrical coordinates and written in terms of **J**, is

$$\frac{\partial^2 \mathbf{J}}{\partial r^2} + \frac{1}{r}\frac{\partial \mathbf{J}}{\partial r} - \sigma\mu \frac{\partial \mathbf{J}}{\partial t} = 0$$

where **J** is directed in the z direction and depends only on the radius r. If **J** is assumed to be sinusoidal, and using the complex notation as for the solution in rectangular coordinates, this equation is a form of Bessel's equation. Its solution is similar to that obtained for the semi-infinite block. This means that the current density is attenuated as the distance from the surface of the conductor increases. As a numerical example, we mention that for cables with radii between 2δ to 4δ, the magnitude of the current density at a distance $z=d$ (from the surface of the conductor) is about $0.42J$ (where J is the magnitude of the current density at the surface). The magnitude for the rectangular block is $0.37J$. As the ratio of radius to skin depth increases, the solution for the cylindrical conductor approaches that for the rectangular block.

5.6.3.c. Shielding

Suppose that a particular region is required to have no external magnetic field which may disturb the condition inside that domain. If external fields exist, it becomes necessary to protect this region from the external fields by means of shielding. From the equation for skin depth,

$$\delta = \sqrt{\frac{2}{\sigma\mu\omega}} \qquad (5.47)$$

it is obvious that for *dc* fields, the depth of penetration is infinite and shielding based on skin depth cannot be accomplished. In such cases, shielding can only be achieved by use of high-permeability materials which channel most of the magnetic flux through the ferromagnetic material.

Shielding of *ac* fields is based on the skin depth and implemented using high conductivity materials. Suppose we chose copper for this purpose. The skin depth in copper is given as

$$\delta = \frac{0.0666}{\sqrt{f}} \qquad (5.48)$$

where *f* is the frequency of the field.

The thickness of the shielding box must be calculated as a function of frequency. To see the behavior of fields, consider the plots in Figure 5.19. The region that must be shielded is placed in the middle of the airgap in Figure 5.19a which shows a homogeneous field generated by the two coils and the magnetic path.

Figure 5.19b shows a *dc* field and a shielding box made of a high-permeability material. Since the ferromagnetic box creates a low reluctance path, the flux passes through the material of the box, rather than through the space in the box creating an effective shield.

Figures 5.19c, d, and e, show the effect of *ac* shielding. In Figure 5.19c, the frequency is zero showing no shielding effect as discussed above. The penetration of the magnetic field through the box is unimpeded as expected from Eq. (5.48).

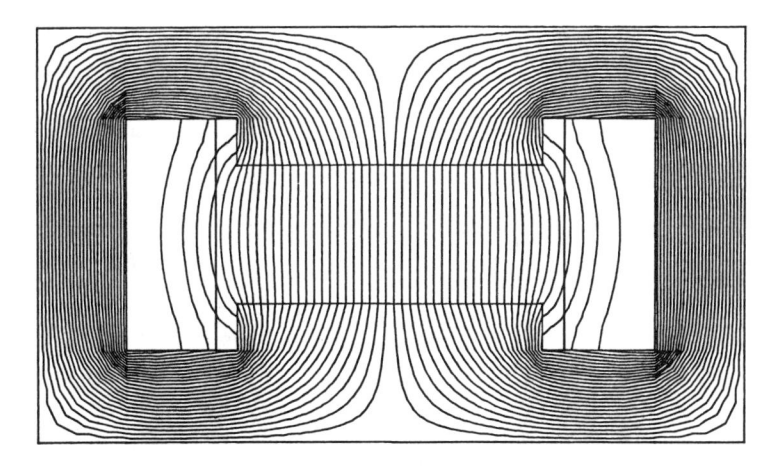

Figure 5.19a. A uniform field in an airgap.

168   5. Magnetodynamic Fields

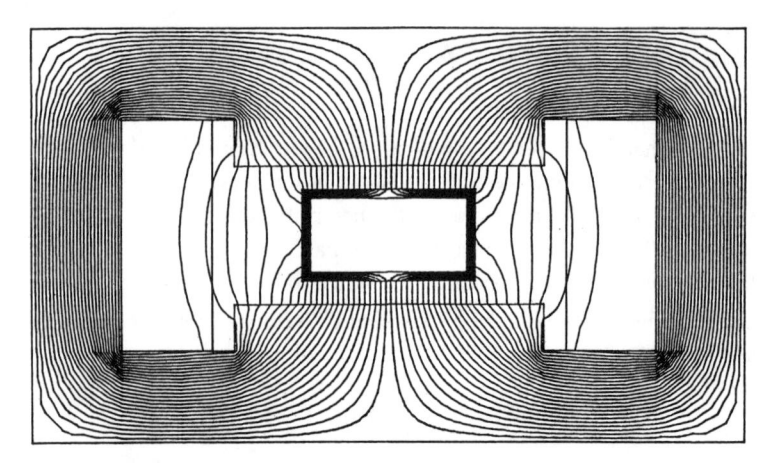

Figure 5.19b. *Dc* shielding using high-permeability ferromagnetic materials.

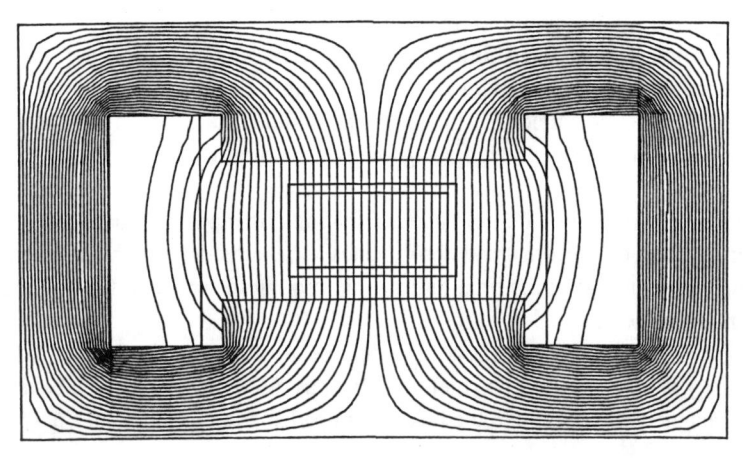

Figure 5.19c. *Ac* shielding at 0 *Hz*. Field penetrates completely through the box.

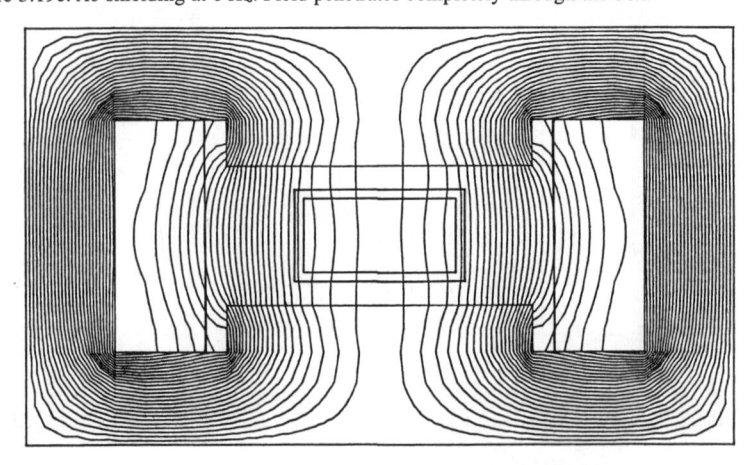

Figure 5.19d. *Ac* shielding at 250 *Hz*. Partial penetration of field is visible.

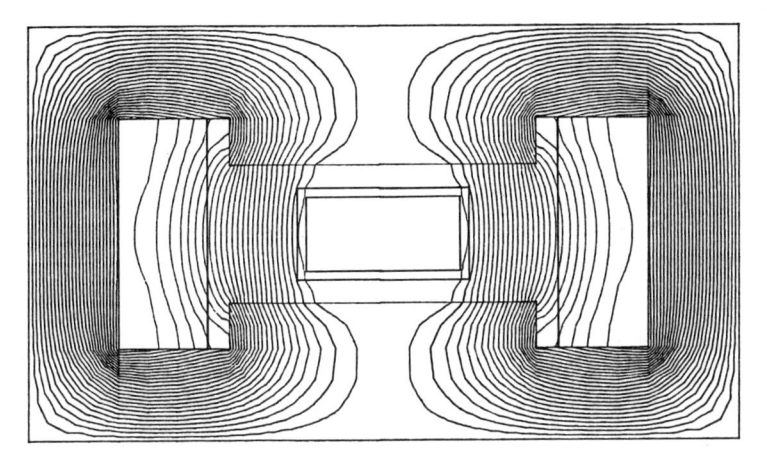

Figure 5.19e. Ac shielding at 1,500 Hz. Field does not penetrate into the box.

In Figure 5.19d, the frequency is 250 *Hz*. The skin depth at this frequency is $\delta=4.22$ *mm*, but since the wall thickness is only 2 *mm*, the field penetrates into the box, although only partially. In Figure 5.19e, the frequency is 1,500 *Hz* (for which $\delta=1.72$ *mm*) and virtually all the external flux is excluded from the interior of the box.

Finally, we note the fundamental difference between the two methods of shielding. In Figure 5.19b, shielding is obtained by attracting the flux using a high-permeability material. In Figure 5.19e, the field is repelled from the interior of the box.

5.6.4. *Effect of Movement of a Magnet Relative to a Flat Conductor*

We analyze now, qualitatively, the particular situation of a permanent magnet moving near the surface of a conducting plate. The first case, shown in Figure 5.20, is that of a permanent magnet, approaching the surface of the plate, while its axis remains perpendicular to the plate. The magnetic flux in the plate increases as the magnet approaches. Current loops are generated in the plate, such that the flux due to these currents opposes the increasing flux due the approaching magnet. Therefore, the flux, which is initially negligible (the magnet is far from the plate), is the sum of the flux of the approaching magnet and the opposing flux due to the induced currents and remains essentially unchanged.

In many cases we wish to prevent the penetration of a magnetic field into a certain volume. In order for this to happen, the required volume is enclosed by a conducting material.

For a time-dependent field, the induced currents in the conducting material create a "magnetic shielding" effect, resulting in a lower field in the volume of interest.

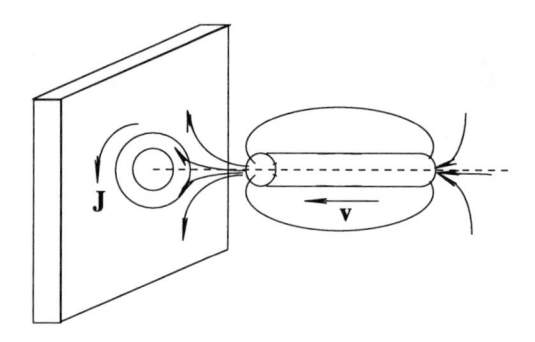

Figure 5.20. A magnet moving at a velocity v towards a conducting plate.

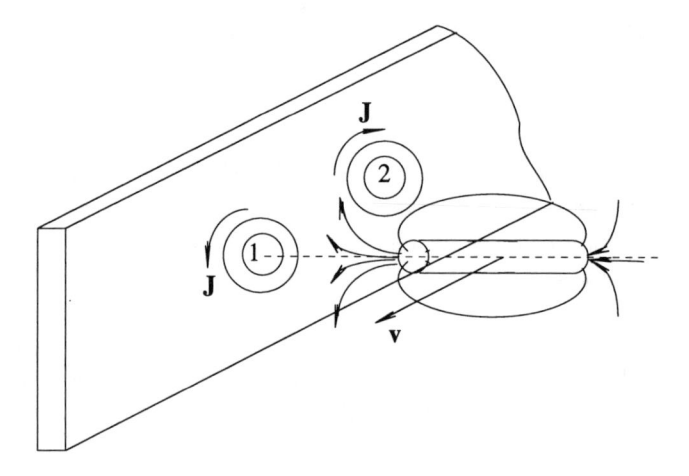

Figure 5.21. A permanent magnet moving at a velocity v along a conducting plate.

It is obvious that for a *dc* field (zero frequency field), shielding does not exist since the depth of penetration is infinite. We note, additionally, that the thickness of the conducting plate should be calculated based on the material used (σ and μ) and on the frequency ω of the external field. Through calculation of the depth of penetration (Eq. 5.27), the thickness necessary for appropriate magnetic shielding can be decided upon.

A second situation that we wish to examine is shown in Figure 5.21, where a permanent magnet is moved parallel to the surface of the plate. We examine the induced currents at location 1. These currents are established in the direction shown since they tend to oppose the change in flux due to the movement of the magnet. But, at location 2, where the flux of the magnet at a previous moment was, the currents generated are such that the flux, which is already established, tends to be maintained.

These two examples show qualitatively the effects of Faraday's and Lenz's laws or, simply, the equation $\nabla \times \mathbf{E} = -\partial \mathbf{B}/\partial t$.

5.6.5. Visualization of Penetration of Fields as a Function of Frequency

Figure 5.22 shows a magnetic structure, consisting of a laminated magnetic circuit (hence, $\sigma=0$), an inducing coil, and a conducting, ferromagnetic part.

In Figure 5.23a, the magnetic flux distribution under *dc* excitation is shown, as obtained by the finite element program EFCAD. The depth of penetration is only limited by the geometry of the domain of study.

Figures 5.23b, 5.23c, 5.23d, and 5.23e show the flux distribution for 50, 100, 250, and 500 Hz. The penetration is reduced as the frequency increases.

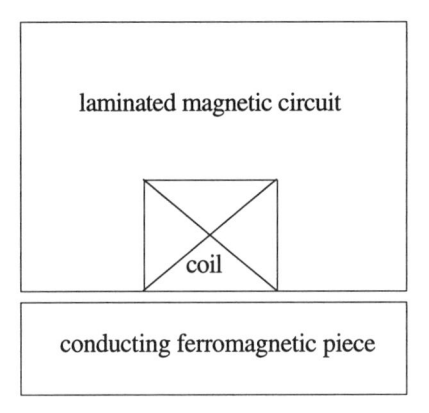

Figure 5.22. A laminated magnetic path used to demonstrate the penetration of *ac* fields in conducting, ferromagnetic materials.

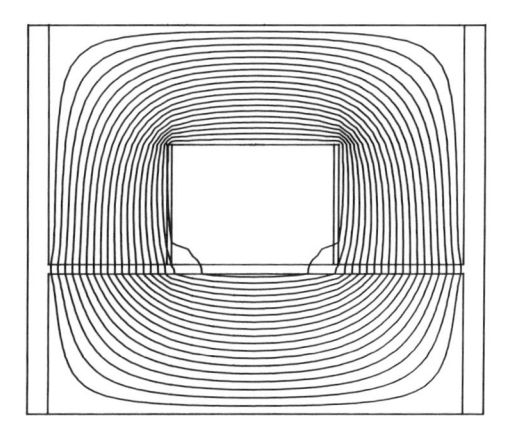

Figure 5.23a Field distribution at 0 Hz. Full penetration of field occurs.

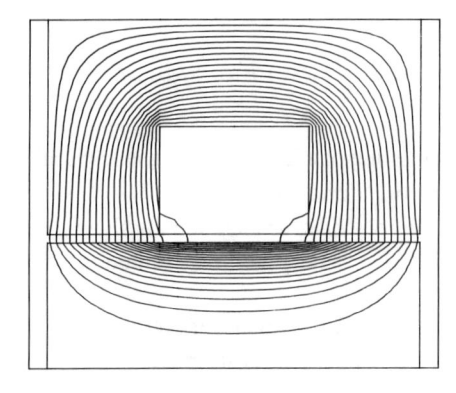

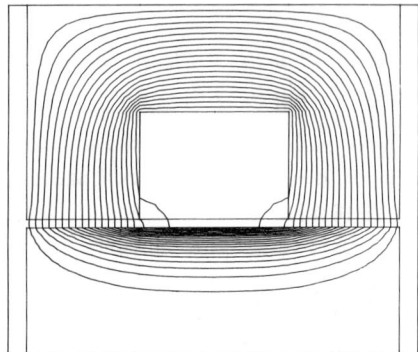

Figure 5.23b. Field distribution at 50 Hz.

Figure 5.23c. Field distribution at 100 Hz.

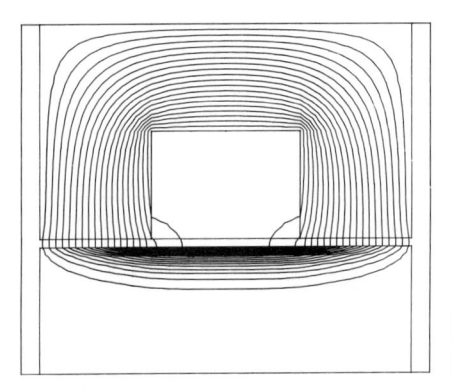

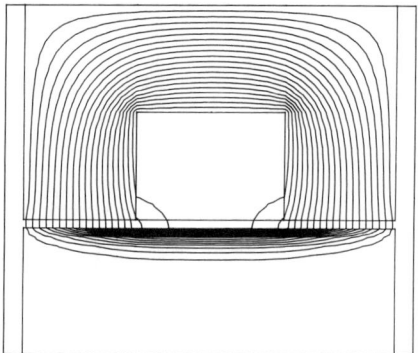

Figure 5.23d. Field distribution at 250 Hz.

Figure 5.23e. Field distribution at 500 Hz.

From the results obtained from the finite element solution, it is also obvious that the total flux generated by the coil is reduced with increasing frequency. This is as expected since with diminishing penetration, the reluctance of the magnetic path increases.

5.6.6. *The Voltage Transformer*

In the magnetic circuit shown in Figure 5.24, the electric circuit C_1 has N_1 turns and is connected to a sinusoidal source.

The flux generated by C_1 can be calculated assuming that the permeability μ of the magnetic circuit is much larger than that of air, and that there is no current in circuit C_2. We start with Ampere's law:

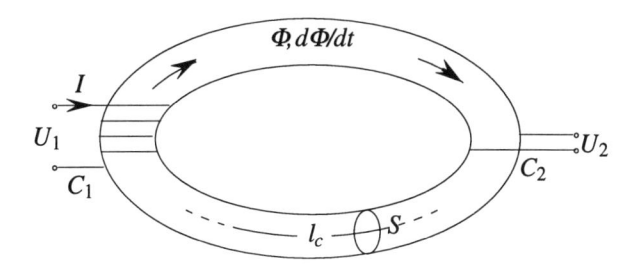

Figure 5.24. Principle of the voltage transformer.

$$\oint_{l_C} \mathbf{H} \cdot d\mathbf{l} = NI$$

where l_c is the average contour of the magnetic circuit. The total current enclosing the contour l_c is equal to the number of turns N_1 multiplied by the current I. Assuming H to be constant in the cross-section of the magnetic path gives

$$H = \frac{N_1 I}{l_C}$$

And, using S for the cross-sectional area of the magnetic circuit, the flux is given as

$$\Phi = \mu \frac{N_1 I}{l_C} S \qquad (5.49)$$

Since I is the only time-dependent quantity,

$$\frac{d\Phi}{dt} = \frac{\mu N_1 S}{l_C} \frac{dI}{dt}$$

and

$$\frac{dI}{dt} = \frac{l_C}{\mu N_1 S} \frac{d\Phi}{dt} \qquad (5.50)$$

Neglecting resistive effects, the potential U_1 is given by

$$U_1 = L\frac{dI}{dt} = L\frac{l_C}{\mu N_1 S} \frac{d\Phi}{dt} \qquad (5.51)$$

5. Magnetodynamic Fields

L, the self induction of C_1, is given in Eq. (4.33)

$$L = \mu \frac{N_1^2 S}{l_C} \tag{5.52}$$

Substituting Eq. (5.52) into Eq. (5.51) gives

$$U_1 = N_1 \frac{d\Phi}{dt} \tag{5.53}$$

Now we can calculate the potential U_2, as detected by a voltmeter connected to circuit C_2. For a single loop in C_2, we use the equation $\nabla \times \mathbf{E} = -\partial \mathbf{B}/\partial t$ in integral form:

$$\oint_{l_m} \mathbf{E} \cdot d\mathbf{l} = \frac{d\Phi}{dt} \tag{5.54}$$

where l_m is the circumference of the loop in C_2. Since the magnetic circuit is made of a high-permeability material, the magnitude of $d\Phi/dt$ crossing the loop is identical to that generated by C_1. Assuming that C_2 has N_2 loops, the right-hand side of Eq. (5.54) is multiplied by N_2. We obtain

$$N_2 \oint_{l_m} \mathbf{E} \cdot d\mathbf{l} = N_2 \frac{d\Phi}{dt}$$

The left-hand side of this equation represents the total circulation of the electric field intensity $\mathbf{E}$ over N_2 loops, assuming that l_m is the average circumference of the loops. Therefore, this integral represents the potential U_2 detected by the voltmeter

$$U_2 = N_2 \frac{d\Phi}{dt} \tag{5.55}$$

Dividing Eq. (5.55) by Eq. (5.53) we get

$$\frac{U_2}{U_1} = \frac{N_2}{N_1} \tag{5.56}$$

This shows that the relation between the two potentials is directly proportional to the number of turns in the two circuits as it should be for an ideal transformer.

6
Interaction between Electromagnetic and Mechanical Forces

6.1. Introduction

The majority of electromagnetic devices operate on the principle of conversion of electromagnetic energy to mechanical energy or vice versa. Examples are the vast number of electric motors and generators and innumerable applications of electromagnets (relays, electromagnetic valves, transducers, etc.). In these structures, we attempt to generate mechanical forces such that the moving parts of the device can perform work. Many of the appliances and all electric machines we use belong to this category. There are, however, other types of devices which are designed to convert one form of electric energy into another. Examples are transformers, converters, and inverters.

The interaction between electrical and mechanical quantities is a complex interaction and must be approached carefully. There is no single, general method of treatment which can, in a simple manner, define this conversion of energy. Experience has shown that for each type of device, there is a preferred method of treatment.

In this chapter we present some of the concepts involved in this interaction. Through the use of examples, we show which principles are more convenient in one class of devices or another. Among these concepts we emphasize Maxwell's force tensor. This is particularly important with the advent of modern numerical methods for computation of fields in which the principle is utilized extensively. Yet, this principle is seldom presented in electromagnetic field books even though its application is very useful for almost all electromagnetic devices with moving parts.

6.2. Force on a Conductor

We seek an expression which allows the calculation of force acting on a conductor, carrying a current I and located in a magnetic field whose magnetic flux density is **B**.

176 6. Interaction between Electromagnetic and Mechanical Forces

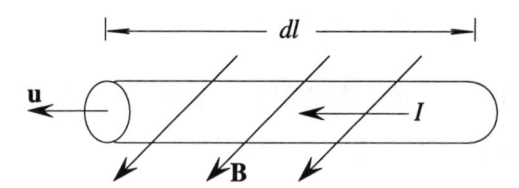

Figure 6.1.

To establish this expression, we start with a postulate for force, expressed as

$$\mathbf{f} = \mathbf{J} \times \mathbf{B} \qquad (6.1)$$

where **f** is the volumetric force density, **J** the surface current density, and **B** the magnetic flux density.

Consider a differential segment of the conductor of length dl, carrying a current I, under the influence of the vector field **B**, as shown in Figure 6.1. The cross-section of the conductor is S. A unit vector $\hat{\mathbf{u}}$ is defined in the direction of flow of the current I. With this unit vector, the vector **J** (current density) and the vector $d\mathbf{l}$ are defined as

$$\mathbf{J} = \hat{\mathbf{u}} \frac{I}{S}$$

$$d\mathbf{l} = \hat{\mathbf{u}} dl$$

Therefore, we can write

$$\hat{\mathbf{u}} = \frac{d\mathbf{l}}{dl} = \frac{\mathbf{J}}{J}$$

The force applied on this section of the conductor is

$$d\mathbf{F} = \mathbf{f} dv$$

where dv is the volume of the conductor section ($dv = Sdl$). This gives

$$d\mathbf{F} = \mathbf{J} \times \mathbf{B} S dl$$

$$d\mathbf{F} = \frac{\mathbf{J}}{J} \times \mathbf{B} J S dl = \frac{d\mathbf{l}}{dl} \times \mathbf{B} I dl$$

and finally,

Force on a Conductor   177

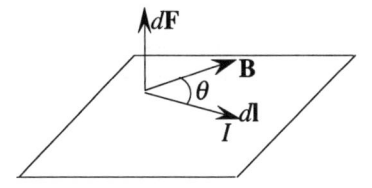

Figure 6.2. Relation between current, magnetic field intensity, and force.

$$dF = dl \times BI \tag{6.2}$$

which expresses the force on a conductor of length dl. This equation is known as Laplace's law, established experimentally by Pierre S. Laplace (1749-1827).

From the vector notation, we note that the force is perpendicular to the plane formed by the vectors dl and B (the right-hand rule) and its magnitude is

$$dF = BIdl\sin\theta \tag{6.3}$$

where θ is the angle between the vectors dl and B as in Figure 6.2.

As an example consider two, parallel, infinitely long wires, carrying a current I_1 and I_2, as shown in Figure 6.3a. In Figure 6.3b, the same two wires are shown with the currents coming out of the plane of the figure. The field intensity H_1, generated by wire number 1 at the location of wire number 2, is given by $H_1 = I_1/2\pi x$. The force F_2 is given by the vector product $dl_2 \times B_1 I_2$. The direction of the force is shown in Figure 6.3b. Its magnitude, obtained from Eq. (6.3), is

$$dF_2 = \mu_0 \frac{I_1}{2\pi x} I_2 dl_2 \sin 90°$$

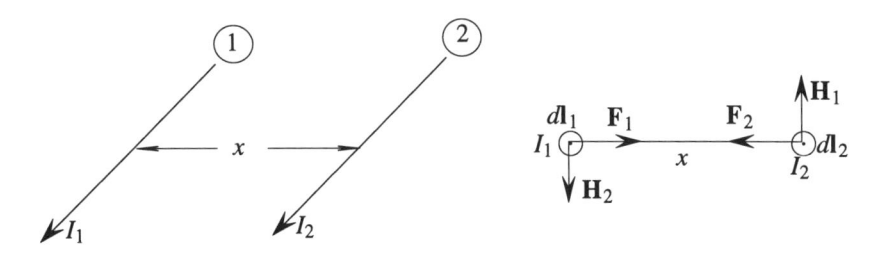

Figure 6.3a. Two parallel, infinitely long wires carrying currents I_1 and I_2.

Figure 6.3b. The two wires in cross-section showing directions of magnetic field intensities and forces.

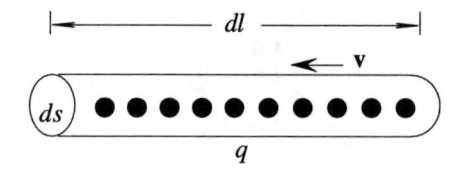

178   6. Interaction between Electromagnetic and Mechanical Forces

Figure 6.4.

The force per unit length of the wire is

$$\frac{F_2}{l_2} = \frac{\mu_0 I_1 I_2}{2\pi x}$$

Analogously, the force acting on conductor number 1 is

$$dF_1 = \mu_0 \frac{I_2}{2\pi x} I_1 dl_1 \sin 90°$$

and the force per unit length is

$$\frac{F_1}{l_1} = \frac{\mu_0 I_1 I_2}{2\pi x} = \frac{F_2}{l_2}$$

The forces exerted by two wires on each other are identical, as one would expect. Other examples to the application of Laplace's law are given at the end of this chapter.

6.3. Force on Moving Charges: The Lorentz Force

Consider a volume in space with a charge q moving at a velocity v, as shown in Figure 6.4. The current passing through this volume is

$$I = \frac{dq}{dt}$$

A current density in the cross-section ds is defined as

$$J = \frac{I}{ds}$$

Using Eq. (6.1) (that is: $\mathbf{f} = \mathbf{J} \times \mathbf{B}$), the total force on the volume is

$$d\mathbf{F} = \mathbf{J} \times \mathbf{B} ds dl$$

The vector **J**, which was defined in section 6.2, and the vector **v** are co-linear and in the same direction. The unit vector in the direction of the current is

$$\hat{\mathbf{u}} = \frac{\mathbf{J}}{J} = \frac{\mathbf{v}}{v}$$

The expression for the total force can now be written as

$$d\mathbf{F} = \frac{\mathbf{J}}{J} \times \mathbf{B} J ds dl = \frac{\mathbf{v}}{v} \times \mathbf{B} J ds dl$$

Since $Jds=I$, we have

$$d\mathbf{F} = \mathbf{v} \times \mathbf{B} I \frac{dl}{v}$$

However, **v** is the velocity at which q covers the distance dl and, therefore, $v=dl/dt$, and we obtain

$$\mathbf{dF} = \mathbf{v} \times \mathbf{B} I dt$$

The product Idt represents the amount of charge moving through the volume in a time dt. The integrated charge is q; we have

$$\mathbf{F} = q\mathbf{v} \times \mathbf{B} \qquad (6.4)$$

This is the force experienced by a charge or an assembly of charges when moving at a velocity **v**, through a region in which a magnetic flux density **B** exists. On the other hand, we know that a charge under the action of an electrostatic field **E**, also experiences a force $\mathbf{F}=q\mathbf{E}$ (see chapter 2). Therefore, the total force on a charge q, moving at a velocity **v** in a region in which both an electric field intensity **E** and a magnetic flux density **B** exist is

$$\mathbf{F} = q(\mathbf{E} + \mathbf{v} \times \mathbf{B}) \qquad (6.5)$$

This is known as the Lorentz force equation. We note that $\mathbf{v} \times \mathbf{B}$ has units of electric field intensity. The total electric field influencing the charge is, in general

$$\mathbf{E}_t = \mathbf{E} + \mathbf{v} \times \mathbf{B} \qquad (6.6)$$

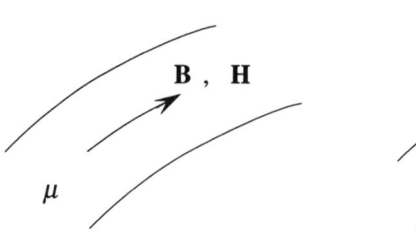

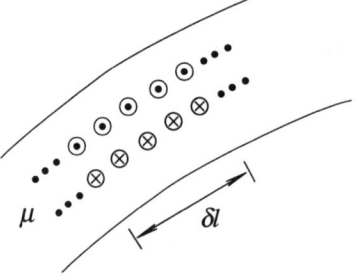

Figure 6.5a. A material with permeability µ, a magnetic field intensity **H**, and a magnetic flux density **B**.

Figure 6.5b. A solenoid that produces an equivalent magnetic field replaces the magnetic field intensity.

6.4. Energy in the Magnetic Field

If a magnetic field intensity **H** exists in a given material (and, in consequence a magnetic flux density **B**), there is a magnetic energy associated with this field. A certain amount of energy will be required to generate the flux density **B** in a given time, say between zero and t. To establish the expression for magnetic energy, consider the situation in Figure 6.5a, where a material with permeability μ, a field intensity **H**, and a flux density **B** is shown.

For the purpose of this discussion, we replace the magnetic field by a small solenoid, of length δl and cross-section δs (Figure 6.5b) such that the magnetic field intensity generated by the solenoid is equivalent to the field intensity **H**. The current is defined here as a linear current density J, expressing the current per unit length (A/m). In this case, I is equal to $J\delta l$. Writing Ampere's law, and assuming the field outside the solenoid to be zero, gives

$$\oint_C \mathbf{H} \cdot d\mathbf{l} = \int_{\delta l} H dl = I = \int_{\delta l} J dl \qquad (6.7)$$

From this, $H=J$.

In other words, the electric energy that must be supplied to the solenoid is

$$dW = \int_0^T V i dt$$

where V is the voltage applied on the solenoid. Note that the various quantities vary simultaneously as

$t \Rightarrow [0,T]$ (time)
$j \Rightarrow [0,J]$ (linear current density)
$h \Rightarrow [0,H]$ (magnetic field intensity)
$b \Rightarrow [0,B]$ (magnetic flux density)
$i \Rightarrow [0,I]$ (current in the solenoid)
$\phi \Rightarrow [0,\Phi]$ (magnetic flux)

Since $i = j\delta l$ and

$$|V| = \frac{d\phi}{dt} = \frac{d(b\delta S)}{dt} = \frac{db}{dt}\delta S$$

we get

$$\delta W = \int_0^T j\delta l \frac{db}{dt} \delta S dt = \int_0^B jdb\,\delta l\,\delta S$$

Substituting $j = h$, gives

$$\delta W = \int_0^B hdb\,\delta l\,\delta S$$

Since $\delta l\,\delta s$ is the volume of the solenoid, we obtain an expression for the volumetric energy density as

$$w_V = \int_0^B hdb \qquad (6.8)$$

In general, magnetic materials are nonlinear and the magnetic energy needed to generate a flux density B in the material, corresponds to the crosshatched area in Figure 6.6.

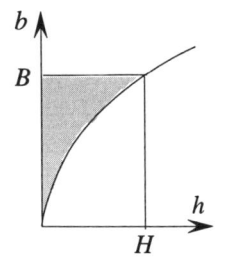

Figure 6.6. Energy needed to generate a magnetic flux density in a material with given $B(H)$ characteristics.

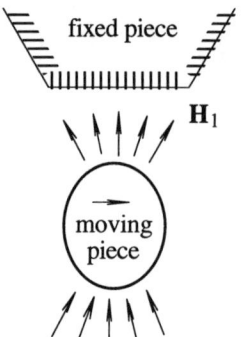

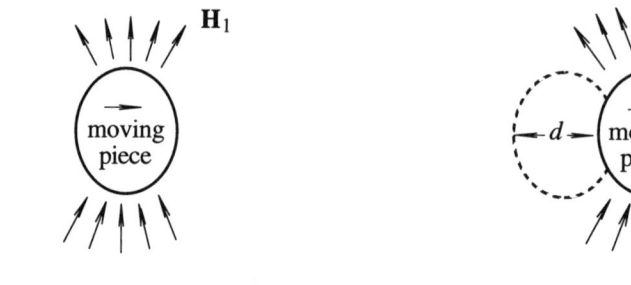

Figure 6.7a. Principle of virtual displacement. Energy in the system at a given position is calculated.

Figure 6.7b. The body is "moved" a distance d and the new energy in the system is calculated.

For materials in which μ is constant (linear materials), we have

$$w_v = \mu \int_0^H h\, dh = \mu \frac{H^2}{2}$$

Using the relation $B=\mu H$, this expression can be written in the following equivalent forms

$$w_v = \frac{\mu H^2}{2} = \frac{HB}{2} = \frac{B^2}{2\mu} \qquad (6.9)$$

6.5. Force as Variation of Energy (Virtual Work)

Calculation of the change in magnetic energy in a device in which there are moving parts is a simple method of calculating the force on moving parts in the device. This is based on the fundamental principle of virtual displacement. That is: the change in energy ΔW, of a body is equal to the force F on the body, multiplied by the displacement of the body Δd caused by the applied force ($\Delta W = F \Delta d$). Applying this principle to magnetic energy and taking its limit gives $F = \partial W/\partial x$, where F and W are, respectively, the force and the magnetic energy. The same principle can be applied to magnetic structures with moving parts. This principle is shown in Figure 6.7a and 6.7b. Suppose that the moving body is displaced a distance d as shown in Figure 6.7b. We denote the field in Figure 6.7a, H_1, and in 6.7b, H_2. We assume that the permeabilities remain constant and calculate the energy W_1 using the expression

Force as Variation of Energy (Virtual Work) 183

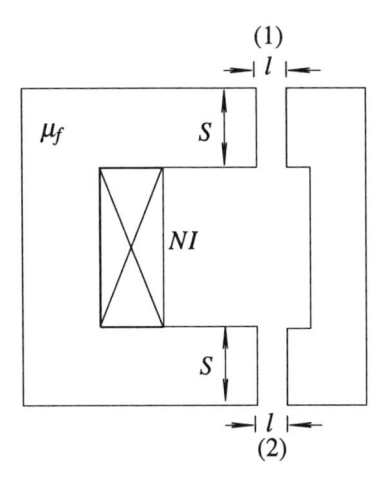

Figure 6.8. A magnetic circuit with a movable piece used to calculate force on the piece.

$$W_1 = \frac{1}{2}\int_V \mu H_1^2 dv$$

where V is the total volume of the domain under study. Analogously, we can obtain W_2. The magnetic force F acting on the body is

$$F = \frac{W_2 - W_1}{d} \qquad (6.10)$$

As an example we calculate the force exerted on the movable part of the structure shown in Figure 6.8. We calculate the force from the change in energy due to displacement between l_1 and l_2. The assumptions used are:

- The two gaps (1) and (2) are equal in length.
- $\mu \gg \mu_0$ and does not vary with changes in l.
- The field in the gaps does not leak out of the gap and is constant throughout the cross-sectional area of the gap.
- Only the fields in the gaps vary with l. (This is a good approximation in this case.)

Under these conditions, the field intensity H in the gaps, using Ampere's law is

$$H 2l = NI \quad \therefore \quad H = \frac{NI}{2l}$$

The energies W_1 and W_2 in gaps (1) and (2) are

$$W_1 = \frac{1}{2}\mu_0 H_1^2 S l_1 2 = \mu_0 \left(\frac{NI}{2l_1}\right)^2 S l_1$$

and

$$W_2 = \mu_0 \left(\frac{NI}{2l_2}\right)^2 S l_2$$

These values were multiplied by 2 since there are two gaps; therefore, since $F=(W_2-W_1)/(l_2-l_1)$, we get for the magnitude of F

$$F = \frac{\mu_0 N^2 I^2 S}{4 l_1 l_2}$$

Suppose that: $l_2=1.5$ mm; $l_1=2.5$ mm; $S=4$ cm^2; $NI=300$ At. For these values the force is

$$F = 3.02\ N$$

6.6. The Poynting Vector

First we recall that the volumetric energy density associated with the electric and magnetic fields are

$$W_{ve} = \frac{1}{2}\varepsilon E^2 \qquad (6.11)$$

and

$$W_{vm} = \frac{1}{2}\mu H^2 \qquad (6.12)$$

If the fields are time dependent, the energy of the fields also varies with time. It is important to note that if the fields are time dependent, there is a variation or transport of energy in the space occupied by these fields. This is the mechanism by which an electromagnetic wave transfers energy. If we consider a volume, a wave incident on the volume transfers energy into the volume. Similarly, if the source of the radiating wave is inside the volume, the wave transports energy to the exterior of the volume.

To understand these phenomena, we develop the expression $\nabla\cdot(\mathbf{E}\times\mathbf{H})$. Using the relation in Eq. (1.27), we can write

$$\nabla\cdot(\mathbf{E}\times\mathbf{H}) = \mathbf{H}\cdot(\nabla\times\mathbf{E}) - \mathbf{E}\cdot(\nabla\times\mathbf{H}) \qquad (6.13)$$

The Poynting Vector   185

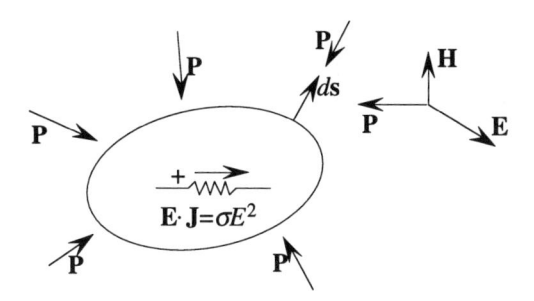

Figure 6.9. Negative energy flux represents the receiver case. Energy from outside the volume flows into the volume where it is dissipated.

Since $\nabla \times \mathbf{E} = -\partial \mathbf{B}/\partial t$ and $\nabla \times \mathbf{H} = \mathbf{J} + \partial \mathbf{D}/\partial t$, we get

$$\nabla \cdot (\mathbf{E} \times \mathbf{H}) = -\mathbf{H} \cdot \left(\frac{\partial \mathbf{B}}{\partial t}\right) - \mathbf{E} \cdot \left(\frac{\partial \mathbf{D}}{\partial t}\right) - \mathbf{E} \cdot \mathbf{J} \qquad (6.14)$$

The terms on the right-hand side are related to the volumetric energy density. To obtain the expression for energy, we integrate Eq. (6.14) over the volume v.

$$\int_v \nabla \cdot (\mathbf{E} \times \mathbf{H}) dv = -\frac{\partial}{\partial t} \int_v \left(\frac{\mathbf{H} \cdot \mathbf{B}}{2} + \frac{\mathbf{E} \cdot \mathbf{D}}{2}\right) dv - \int_v \mathbf{E} \cdot \mathbf{J} dv \qquad (6.15)$$

In this expression, use was made of the fact that if μ and ε are constant, $\partial(\mathbf{H} \cdot \mathbf{B})/\partial t = 2\mathbf{H} \cdot (\partial \mathbf{B}/\partial t)$ and $\partial(\mathbf{E} \cdot \mathbf{D})/\partial t = 2\mathbf{E} \cdot (\partial \mathbf{D}/\partial t)$.

Using the divergence theorem the left-hand side of Eq. (6.15) is transformed into a surface integral and we obtain

$$\oint_{S(v)} (\mathbf{E} \times \mathbf{H}) \cdot d\mathbf{s} = -\frac{\partial}{\partial t} \int_v \left(\frac{\mathbf{H} \cdot \mathbf{B}}{2} + \frac{\mathbf{E} \cdot \mathbf{D}}{2}\right) dv - \int_v \mathbf{E} \cdot \mathbf{J} dv \qquad (6.16)$$

where $S(v)$ is the surface enclosing the volume v.

Because $\mathbf{E}$ is given in *Volts/meter* and $\mathbf{H}$ in *Amperes/meter*, the product $\mathbf{E} \times \mathbf{H}$ has dimensions of *Watts/m²*. This is power per unit area or power density. This power density is called Poynting vector and is denoted by $\mathbf{P}$ as

$$\mathbf{P} = \mathbf{E} \times \mathbf{H} \qquad (6.17)$$

The Poynting vector is perpendicular to the plane formed by $\mathbf{E}$ and $\mathbf{H}$. From Figure 6.9, we observe that the vector $d\mathbf{s}$ points out of the volume.

In the particular case shown in Figure 6.9, the energy flux given by Poynting's vector is negative ($\mathbf{P}$ and $d\mathbf{s}$ are opposite in directions).

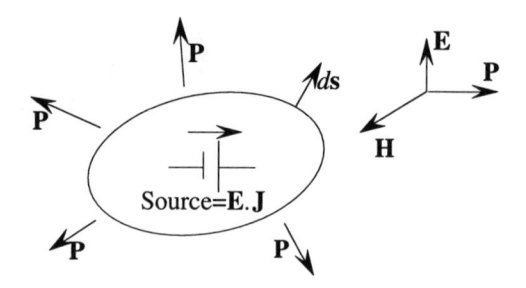

186   6. Interaction between Electromagnetic and Mechanical Forces

Figure 6.10. A source within a volume characterizes the transmitter case of the Poynting vector.

This represents a case in which the volume receives energy through a combination of electric and magnetic fields around its exterior. The first term on the right-hand side in Eq. (6.16):

$$-\frac{\partial}{\partial t}\int_v \left(\frac{\mathbf{H}\cdot\mathbf{B}}{2} + \frac{\mathbf{E}\cdot\mathbf{D}}{2}\right) dv \qquad (6.18)$$

represents the increase in the potential energy of the electric and magnetic fields in the volume. The second term in Eq. (6.16)

$$-\int_v \mathbf{E}\cdot\mathbf{J}\, dv = -\int_v \sigma E^2 dv \qquad (6.19)$$

represents that part of the energy from the exterior dissipated through ohmic losses, which may exist in the volume. In absolute values, the magnitude of the Poynting vector is equal to the sum of the two terms in Eqs. (6.18) and (6.19).

We turn now to the configuration in Figure 6.10 where the source of the energy is inside the volume v. In this case, the term

$$-\int_v \mathbf{E}\cdot\mathbf{J}\, dv \qquad (6.20)$$

represents the power supplied by the source instead of ohmic losses through Eq. (6.19). This part of the energy is utilized for the generation of the fields $\mathbf{E}$ and $\mathbf{H}$ in the interior of the volume (Eq. (6.18)). The balance of this power is flowing out of the volume according to the expression for the Poynting vector. In terms of absolute values, Eq. (6.16) can be written in the form

$$\left|\int_v \mathbf{E}\cdot\mathbf{J}\, dv\right| = \left|\frac{\partial}{\partial t}\int_v \left(\frac{\mathbf{H}\cdot\mathbf{B}}{2} + \frac{\mathbf{E}\cdot\mathbf{D}}{2}\right) dv\right| + \left|\oint_{S(v)} \mathbf{P}\cdot d\mathbf{s}\right| \qquad (6.21)$$

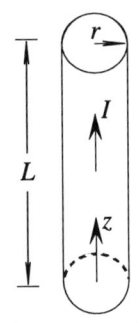

Figure 6.11. A segment of a current-carrying conductor.

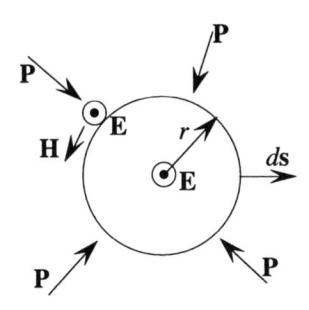

Figure 6.12. Cross-section of the conductor showing the electric and magnetic field intensities and the Poynting vector.

Even though the Poynting vector is most often used for treatment of electromagnetic waves, we calculate here, as an example, the ohmic losses in a segment of a conductor of length L and radius R, carrying a current I, uniformly distributed throughout its cross-section. The conductor is shown in Figure 6.11. In the conductor, $J=\sigma E$ in the Oz direction with $J=I/S$ and $S=\pi r^2$. Therefore, $E=I/S\sigma$, in the Oz direction. The magnetic field intensity **H** at the surface of the conductor has a magnitude $H=I/2\pi r$ and is tangential, as shown in Figure 6.12.

The vector **P**=**E**×**H** penetrates radially into the volume of the conductor. Surrounding the volume by a surface S allows the calculation of the total power entering the conductor by the expression

$$P_t = \int_S \mathbf{P} \cdot d\mathbf{s} = \int_{S_l} \mathbf{P} \cdot d\mathbf{s} + \int_{S_b} \mathbf{P} \cdot d\mathbf{s}$$

where P_t is the total power, S_l is the lateral portion of the surface, and S_b the base portion of the surface of the cylinder. On the bases of the conductor, **P**·d**s**=0. We can now write

$$P_t = \int_S \mathbf{P} \cdot d\mathbf{s} = \int_{S_l} (\mathbf{E} \times \mathbf{H}) \cdot d\mathbf{s} = -\int_{S_l} EH ds$$

$$P_t = \int_{S_l} \frac{I}{S\sigma} \frac{I}{2\pi r} ds = \frac{I^2}{S\sigma 2\pi r} 2\pi rL$$

$$P_t = I^2 \frac{L}{\sigma S} = RI^2$$

where R is the resistance $R=L/\sigma S$.

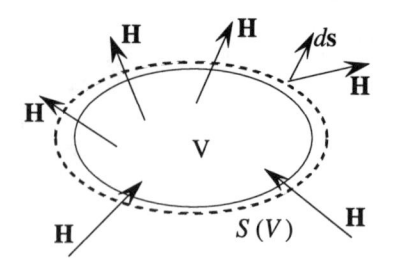

Figure 6.13. Application of Maxwell's force tensor. The force is calculated on the surface S.

It is interesting to note the physical meaning of this result. It demonstrates that the dissipated power in the conductor (due to resistance) can be obtained through the energy penetrating into the conductor by the electromagnetic fields. By use of the Poynting vector, the energy dissipated in the conductor does not enter the conductor through its connections but through the electric and magnetic fields generated by the power source (i.e., a battery) and penetrates through the surface of the conductor.

6.7. Maxwell's Force Tensor

Maxwell's force tensor is one of the most efficient general method for the calculation of forces on bodies under the influence of magnetic fields. It leads to expressions that allow the computation of forces on such diverse structures as the rotor of an electric machine or the moving piece of a relay.

The calculation of forces through Maxwell's force tensor is used extensively in computer programs for numerical computation of fields. In order for this principle to be applied, the fields must be known and the computation method used must be capable of providing these data. This is perhaps the reason why, on one hand it is extensively utilized in conjunction with numerical methods of computation and, on the other hand, it has very rarely been used during pre-computer periods when the computation of fields was difficult or impossible.

Also of importance is the fact that Maxwell's tensor facilitates understanding of the relation between the magnitude and direction of forces as functions of the magnetic fields that generate the forces.

For the practical application of Maxwell's force tensor, suppose that a body occupies a volume V and that the magnetic field intensity $\mathbf{H}$ is known on the surface $S(V)$ enclosing the body. It is also required that this body be located in air or within a material with permeability $\mu=\mu_0$. Figure 6.13 shows such a body of volume v and surface $S(V)$; ds is a differential area on $S(V)$, and its direction is out of the volume. $\hat{\mathbf{n}}$ is a unit vector normal to the surface, giving: $d\mathbf{s}=\hat{\mathbf{n}}ds$. Maxwell's tensor is given by

Maxwell's Force Tensor   189

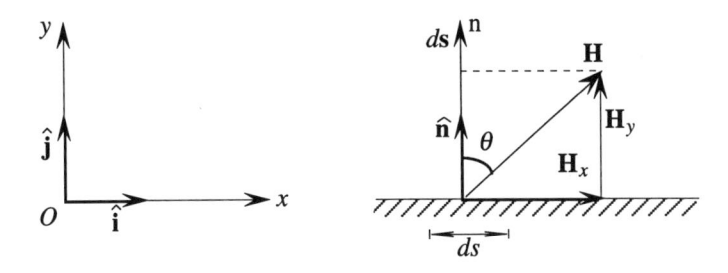

Figure 6.14.

$$d\mathbf{F} = -\frac{\mu_0}{2} H^2 \, d\mathbf{s} + \mu_0 \, (\mathbf{H} \cdot d\mathbf{s}) \, \mathbf{H} \qquad (6.22)$$

or, in the form of magnetic pressure,

$$\frac{d\mathbf{F}}{dS} = -\frac{\mu_0}{2} H^2 \, \widehat{\mathbf{n}} + \mu_0 \, (\mathbf{H} \cdot \widehat{\mathbf{n}}) \, \mathbf{H} \qquad (6.23)$$

This gives the force $d\mathbf{F}$ on a differential area, on which we know the field intensity $\mathbf{H}$. Integrating over the surface S provides the total force applied to the body. This force can be viewed as acting at the center of gravity of the body.

We analyze now a problem in the Oxy coordinate system such that the axis Ox is parallel to the differential surface $d\mathbf{s}$ of the body, as shown in Figure 6.14. Assuming the field intensity H is constant over the area ds and using the orthogonal unit vectors $\widehat{\mathbf{i}}$ and $\widehat{\mathbf{j}}$ associated with the axes Ox and Oy, we can write

$$\widehat{\mathbf{n}} = \widehat{\mathbf{j}}$$

$$\mathbf{H} = \widehat{\mathbf{i}} H_x + \widehat{\mathbf{j}} H_y$$

$$d\mathbf{F} = \widehat{\mathbf{i}} dF_x + \widehat{\mathbf{j}} dF_y$$

Substituting these expressions in Eq. (6.23) gives

$$\frac{d\mathbf{F}}{dS} = -\widehat{\mathbf{j}} \frac{\mu_0}{2} (H_x^2 + H_y^2) + \mu_0 \left[(\widehat{\mathbf{i}} H_x + \widehat{\mathbf{j}} H_y) \cdot \widehat{\mathbf{j}} \right] (\widehat{\mathbf{i}} H_x + \widehat{\mathbf{j}} H_y)$$

With some algebraic manipulation, we get

$$\frac{d\mathbf{F}}{dS} = \widehat{\mathbf{i}} \mu_0 H_x H_y + \widehat{\mathbf{j}} \frac{\mu_0}{2} (H_y^2 - H_x^2)$$

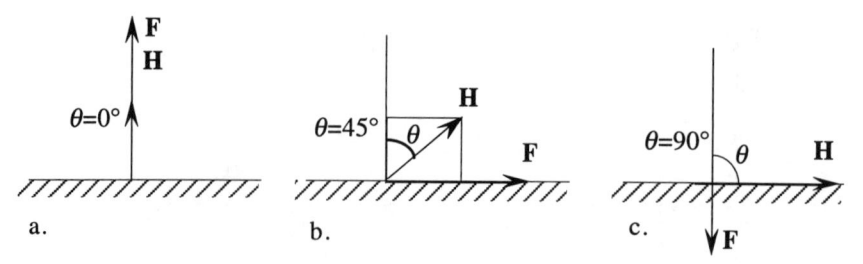

190   6. Interaction between Electromagnetic and Mechanical Forces

Figure 6.15. Various possible force components, depending on the direction of the magnetic flux density. a) Normal tension force, b) tangential force, and c) normal compression force.

or, equating components,

$$dF_x = \mu_0 H_x H_y \, ds \tag{6.24}$$

$$dF_y = \frac{\mu_0}{2}(H_y^2 - H_x^2) \, ds \tag{6.25}$$

From these expressions, the following observations are in order:

- $H_x = 0$, $H_y \neq 0$; results in $dF_x = 0$ and $dF_y \neq 0$ and, therefore, only a normal tension force to the surface exists (Figure 6.15a).
- $H_x = H_y \neq 0$; we have a case where $dF_y = 0$ and $dF_x \neq 0$; only a tangential force exists (Figure 6.15b).
- $H_x \neq 0$; $H_y = 0$; results in $dF_x = 0$ and $dF_y < 0$; a normal compression force exists (Figure 6.15c).

The figures above show that the force always forms an angle which is twice the angle θ that **H** forms with the normal. The magnitude of $d\mathbf{F}$ is given by

$$dF = \sqrt{dF_x^2 + dF_y^2}$$

or

$$dF = \sqrt{(\mu_0 H_x H_y \, ds)^2 + \left[\mu_0(H_y^2 - H_x^2)\, ds/2\right]^2}$$

which, upon simplification, becomes

$$dF = \frac{1}{2}\mu_0(H_y^2 + H_x^2) \, ds \tag{6.26}$$

or

$$dF = \frac{\mu_0}{2} H^2 ds \tag{6.27}$$

The direction of *dF* is obtained from the tangent of the angle θ, which **H** forms with the normal. This is

$$\tan\theta = \frac{H_x}{H_y} \tag{6.28}$$

The angle α, which *dF* forms with the normal, is

$$\tan\alpha = \frac{dF_x}{dF_y} \tag{6.29}$$

Using Eqs. (6.24) and (6.25) in this expression gives

$$\tan\alpha = \frac{\mu_0 H_x H_y ds}{\frac{\mu_0}{2}(H_y^2 - H_x^2)ds} = \frac{2H_x H_y}{H_y^2 - H_x^2}$$

Dividing numerator and denominator by H_y^2, we obtain

$$\tan\alpha = \frac{2\frac{H_x}{H_y}}{1 - \frac{H_x^2}{H_y^2}}$$

which, using Eq. (6.28), becomes

$$\tan\alpha = \frac{2\tan\theta}{1 - \tan^2\theta} = \tan 2\theta$$

Therefore, the angle α, which *dF* forms with the normal, is always twice the angle θ, which **H** forms with the normal. If the field intensity **H** is known, we can easily calculate the magnitude of *dF* using its components H_x and H_y. The angle θ is also known from the relation $\alpha = 2\theta$.

These practical rules demonstrate that only simple calculations are required and, in general, it is not necessary to resort to direct application of Eq. (6.23). By use of these rules, the position of **F** in Figure 6.15 can be determined.

The same idea can be applied to all differential areas *ds* comprising the surface *S*(*V*) of any arbitrary body. For each *ds*, we define a local system of coordinates *Oxy* and obtain dF_x and dF_y, as shown in Figure 6.16.

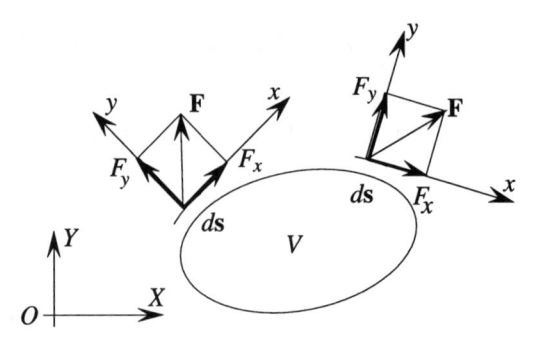

192   6. Interaction between Electromagnetic and Mechanical Forces

Figure 6.16. Application of Maxwell's force tensor to an arbitrarily shaped body. Forces are calculated for differential areas *ds* and summed over the surface enclosing the body.

It is obvious that, to evaluate the total force acting on the body, we must reference all differential forces *dFx* and *dFy* to a global system of coordinates *OXY*. This is done by simple vector summation of the differential forces. To do so, we apply the classical rotation matrix [*R*] that relates the global system *OXY* and the local system *Oxy*. The matrix is given as

$$[R] = \begin{bmatrix} cos\beta & sin\beta \\ -sin\beta & cos\beta \end{bmatrix}$$

where β is the angle between the two systems of coordinates, as indicated in Figure 6.17. Through use of the rotation matrix [*R*], the vector *d***F** is calculated as

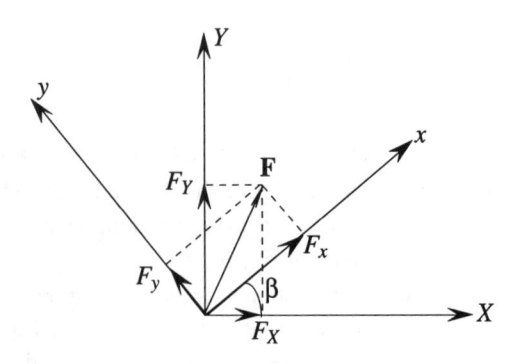

$$\{F_{xy}\} = [R]\{F_{XY}\} \qquad (6.30)$$

Figure 6.17. Relation between the local and global systems of coordinates for calculation of forces using Maxwell's force tensor.

In this case, the inverse operation is required; the force in the local system is known and needs to be transformed into the global system. This is done as

$$[dF_{XY}] = [R]^{-1}[dF_{xy}]$$

where, both sides of Eq. (6.30) were multiplied by $[R]^{-1}$. We obtain

$$\begin{bmatrix} dF_X \\ dF_Y \end{bmatrix} = \begin{bmatrix} \cos\beta & -\sin\beta \\ \sin\beta & \cos\beta \end{bmatrix} \begin{bmatrix} dF_x \\ dF_y \end{bmatrix} \quad (6.31)$$

The values of F_X and F_Y (total force components) are given by the sum of all values of dF_X and dF_Y over the surface $S(V)$.

As an example to the application of Maxwell's force tensor we solve the problem in section 6.5, where we calculated the attraction force due to the small (virtual) displacement of the magnetic circuit (Figure 6.8) by use of the change in energy. The magnetic circuit has two air gaps, 1.5 *mm* and 2.5 *mm* wide, and the average force obtained there was 3.02 N.

Here we use two gaps, each 2 *mm* (i.e., average length of the two gaps). First, the magnetic field intensity is determined using Ampere's law, based on the same assumptions as in section 6.5

$$2Hl = NI$$

$$H = \frac{NI}{2l}$$

Assuming all the magnetic flux passes through gaps (1) and (2), those parts of the surface $S(V)$ which enclose the moving piece and have nonzero forces are the surfaces of the gaps (1) and (2). The force is in the direction shown in Figure 6.18, since H forms an angle equal to zero with the normal. The force and its direction in gap (2) is the same as in gap (1). We have, therefore,

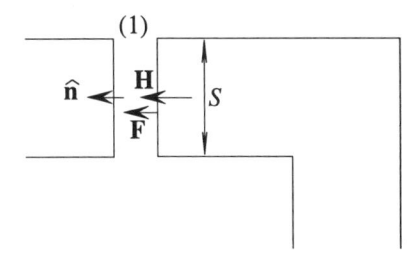

Figure 6.18. Force on the surface of the magnetic path in one of the gaps of Figure 6.8.

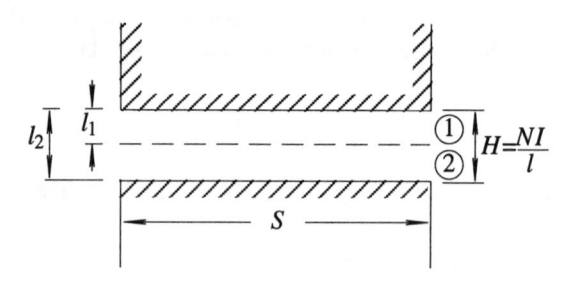

6. Interaction between Electromagnetic and Mechanical Forces

Figure 6.19. Calculation of attraction force in the gap using energy change as the gap varies from l_1 to l_2.

$$F = F_1 + F_2 = 2F_1$$

where

$$F_1 = \frac{\mu_0}{2} H^2 S$$

With the numerical values given: $l=2$ mm; $NI=300$ At; $S=4$ cm^2, we obtain

$$F = 2\frac{\mu_0}{2}\left(\frac{NI}{2l}\right)^2 S = 2.83 \ N$$

In light of the approximations made, these values for F (3.02 N from energy variation and 2.83 N from Maxwell's force tensor) are consistent. The results would be closer if l_1 and l_2 were equal to 2 mm. In general, the two calculations are close to each other and, under certain conditions, tend to the same value. To see this, we calculate the attraction force on the piece in Figure 6.19 using the concept of energy gradient, when the gap varies from l_1 to l_2. Suppose the field varies only in the gap and has only vertical components (there are no leakage fields at the edges of the gap). This gives, for the field in the gap,

$$F = \frac{W_2 - W_1}{l_2 - l_1}$$

where $V_2 = Sl_2$ is the volume of the gap. With this, we get

$$W_2 = \frac{1}{2}\mu_0 \frac{(NI)^2}{l_2^2} Sl_2 = \frac{1}{2}\mu_0 \frac{(NI)^2}{l_2} S$$

analogously for l_1,

$$W_1 = \frac{1}{2} \mu_0 \frac{(NI)^2}{l_1} S$$

and the force is

$$F = \frac{W_2 - W_1}{l_2 - l_1} = \frac{1}{2} \mu_0 (NI)^2 S \left(\frac{1}{l_2} - \frac{1}{l_1}\right) \frac{1}{(l_2 - l_1)} = -\frac{1}{2} \mu_0 \frac{(NI)^2}{l_1 l_2} S$$

Using the median between l_1 and l_2 for l such that $l = \sqrt{l_1 l_2}$, gives

$$F = -\frac{1}{2} \mu_0 \left(\frac{(NI)}{l}\right)^2 S = -\frac{1}{2} \mu_0 H^2 S$$

which is identical in magnitude to the expression obtained by the Maxwell's force tensor. The negative sign is due to the fact that the force calculated by Maxwell's tensor is a force internal to the system or, that this is the force the two pieces exert on each other. Using the energy gradient method, we get the force exerted on the pieces from outside to move them to the new location [from position (1) to position (2)]. This force must be the same as the force of attraction but in opposite direction.

We can appreciate now the considerable interest in Maxwell's force tensor. In calculating the force, we have used a single gap ($l=2$ mm), while for the energy gradient method we needed to use two calculations.

6.8. Examples

6.8.1. *Force between Two Conducting Segments*

Consider two conductors (1) and (2), of arbitrary shape, carrying currents I_1 and I_2 as shown in Figure 6.20. We wish to calculate the force that a differential length dl_1 of conductor (1) exerts on a differential length dl_2 of conductor (2).

A vector **r** is defined as $\mathbf{r} = \mathbf{P} - \mathbf{Q}$. The force $d\mathbf{F}$ at point P is given by Eq. (6.1)

$$d\mathbf{F} = \mathbf{J}_2 \times d\mathbf{B}_1 dv \tag{6.32}$$

where: $\mathbf{J}_2$ is the surface current density in conductor (2); $d\mathbf{B}_1$ is the magnetic flux density due to the segment dl_1 of conductor (1) at point P; dv is the differential volume in which we wish to calculate the force. This volume is $dv = S_2 dl_2$ where S_2 is the cross-sectional area of conductor (2).

From Biot-Savart's law, we get

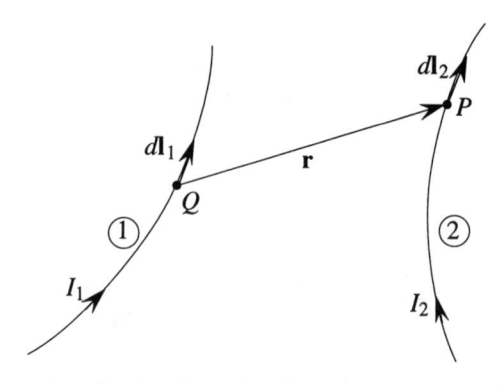

196   6. Interaction between Electromagnetic and Mechanical Forces

Figure 6.20. Calculation of forces between two arbitrary current carrying segments.

$$d\mathbf{B}_1 = \mu_0 \frac{d\mathbf{l}_1 \times \mathbf{r}}{4\pi r^3} I_1$$

Using Eq. (6.32) gives

$$d\mathbf{F} = \frac{\mu_0}{4\pi r^3} I_1 \, \mathbf{J}_2 \times (d\mathbf{l}_1 \times \mathbf{r}) S_2 dl_2 \qquad (6.33)$$

Defining the unit vector $\hat{\mathbf{u}}$ in the direction of the current I_2 at point P as $\hat{\mathbf{u}} = \mathbf{J}_2/J_2$, the vector $d\mathbf{l}_2 = \hat{\mathbf{u}} dl_2$ is defined; that is,

$$\hat{\mathbf{u}} = \frac{\mathbf{J}_2}{J_2} = \frac{d\mathbf{l}_2}{dl_2}$$

and

$$\mathbf{J}_2 = J_2 \frac{d\mathbf{l}_2}{dl_2}$$

which, when substituted in Eq. (6.33), gives

$$d\mathbf{F} = \frac{\mu_0 I_1}{4\pi r^3} I_2 \, d\mathbf{l}_2 \times (d\mathbf{l}_1 \times \mathbf{r}) \qquad (6.34)$$

where we have substituted $I_2 = J_2 S_2$.

To obtain the total force on the differential segment dl_2, we integrate Eq. (6.34) over the length of conductor (1). The integration is done on $d\mathbf{l}_1$ and results in the force, or mechanical influence, of conductor (1) on the segment dl_2 of conductor (2).

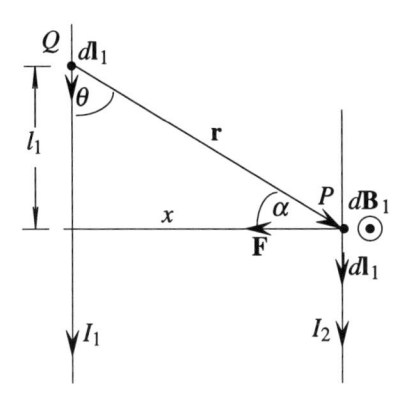

Figure 6.21. Force between two parallel wires separated a distance x.

Consider now a particular case, where conductors (1) and (2) are two parallel wires separated by a distance x. The wires carry identical currents in the same direction as shown in Figure 6.21.

First we note that the product $d\mathbf{l}_1 \times \mathbf{r}$ provides the direction of the vector $d\mathbf{B}_1$ at point P. Then, $d\mathbf{l}_2 \times (d\mathbf{l}_1 \times \mathbf{r})$ or $d\mathbf{l}_2 \times d\mathbf{B}_1$ indicates the direction of the force $d\mathbf{F}$, as shown in Figure 6.21.

The magnitude of the force per unit length of the wire can be calculated using Eq. (6.34)

$$\frac{dF}{dl_2} = \frac{\mu_0 I_1 I_2}{4\pi r^3} dl_1 r \sin\theta \qquad (6.35)$$

where the relation $|d\mathbf{l}_1 \times \mathbf{r}| = dl_1 r \sin\theta$ was used.

Since θ and α are complementary angles, the expression above can be rewritten as a function of α. We note that

$$\tan\alpha = \frac{l_1}{x} \quad \text{and} \quad dl_1 = x \sec^2\alpha \, d\alpha$$

$$\cos\alpha = \frac{x}{r} \quad \text{and} \quad r = \frac{x}{\cos\alpha}$$

Substituting these relations in Eq. (6.35) yields

$$\frac{dF}{dl_2} = \frac{\mu_0 I_1 I_2}{4\pi x} \int_{-\pi/2}^{\pi/2} \cos\alpha \, d\alpha$$

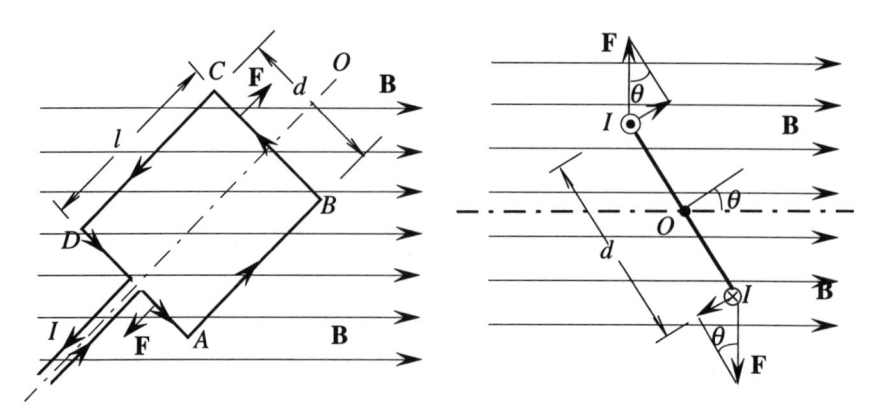

198   6. Interaction between Electromagnetic and Mechanical Forces

Figure 6.22a. A current-carrying loop in a magnetic field.

Figure 6.22b. Forces on the segments of the loop.

which, assuming the wires are infinitely long, is

$$\frac{F}{l} = \frac{\mu_0 I_1 I_2}{2\pi x} \qquad (6.36)$$

This expression is identical to that in the example in section 6.2.

6.8.2. Torque on a Loop

Consider a rectangular loop, carrying a current I, under the action of a constant magnetic field with flux density **B**, as shown in Figure 6.22a and 6.22b. Eq. (6.2) is used to calculate the force on the four segments of the loop

$$d\mathbf{F} = d\mathbf{l} \times \mathbf{B} I$$

At the outset, note that the direction of the forces on segments BC and DA are in the direction shown in Figure 6.22a. Assuming mechanical resistance does not distort the loop, the two forces cancel each other.

The force acting on segments CD and AB are indicated in Figure 6.22b. Since the vectors $d\mathbf{l}$ and **B** are perpendicular, the force **F** has the magnitude

$$F = BIl$$

Suppose the loop is isolated and is free to move about an axis, indicated as O. The force F creates a moment M about the axis, given by

$$M = 2F\frac{d}{2}\sin\theta$$

where θ is the angle the normal to the plane of the loop makes with **B**. Thus,

$$M = BIld\sin\theta \qquad (6.37)$$

Since S is the area of the loop ($S=ld$), we obtain

$$M = BIS\sin\theta \qquad (6.38)$$

If the loop is of arbitrary shape, as in Figure 6.23, the torque must be calculated in relation to each segment of the loop. A segment $d\mathbf{l}$ of the loop can be decomposed into two vector components: dl_a capable of generating torque (equivalent to the segment AB in Figure 6.22a) and dl_b, which does not generate torque (equivalent to segment BC in Figure 6.22a)

$$d\mathbf{l} = d\mathbf{l}_a + d\mathbf{l}_b$$

Substituting this in Eq. (6.2) yields

$$d\mathbf{F} = (d\mathbf{l}_a + d\mathbf{l}_b) \times \mathbf{B}I$$

or

$$d\mathbf{F} = d\mathbf{l}_a \times \mathbf{B}I + d\mathbf{l}_b \times \mathbf{B}I$$

However, the only force capable of generating torque is that related to dl_a:

$$dF = dl_a BI$$

The torque due to this force can now be calculated. The sum of torques of all segments provide the total torque on an arbitrarily shaped loop.

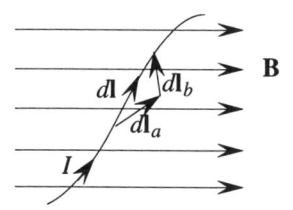

Figure 6.23. Force on an arbitrary segment of current. The segment is separated into two segments: one capable of generating torque, one incapable of generating torque.

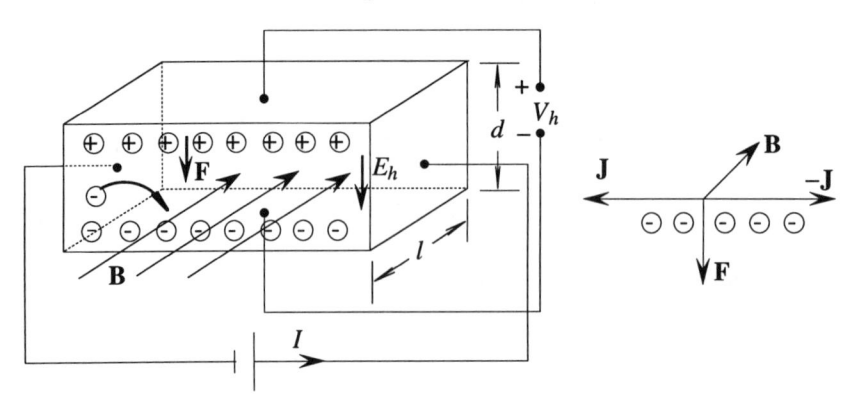

Figure 6.24a. The hall element, biased, and placed in a magnetic field.

Figure 6.24b. Force on electrons in the magnetic field.

6.8.3. The Hall Effect

Consider a conducting bar, carrying a current and subjected to the influence of a magnetic field whose flux density is **B**, as shown in Figure 6.24a. The current in the bar is, by definition, a flux of positive charges, although, in reality, we know that the current is a flux of electrons, moving in the direction −**J** shown in Figure 6.24b. Because of the movement of the electrons, there is a force that tends to move the electrons towards the lower part of the bar (**f**=**J**×**B**). At the same time, the holes vacated by the electrons appear as positive charges on the upper part of the bar. Lorentz's force [Eq. (6.5)] on the electrons is

$$\mathbf{F} = -q\mathbf{v}\times\mathbf{B}$$

where **v** is the velocity of the electrons.

Since the electrons are negatively charged, they tend to repel each other rather than concentrate on the lower part of the bar. They are forced towards the surface because of the Lorentz force. In other words, the appearance of negative charges on the lower part of the bar and positive charges on the upper part of the bar generates an electric field intensity $\mathbf{E}_h$. The force due to the field $\mathbf{E}_h$ acting on a charge $-q$ is $\mathbf{F}=-q\mathbf{E}_h$. This force is equal to the Lorentz force

$$-q\mathbf{E}_h = -q\mathbf{v}\times\mathbf{B}$$

or

$$\mathbf{E}_h = \mathbf{v}\times\mathbf{B} \qquad (6.39)$$

We now establish a relation between the surface current density J and the velocity v of the electrons.

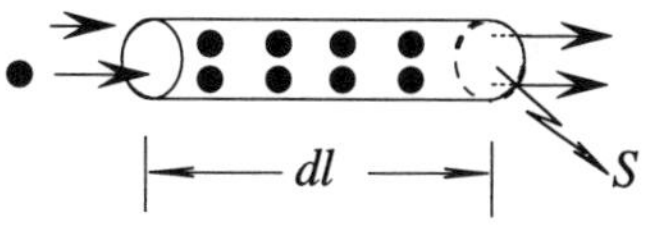

Figure 6.25.

Consider a volume V of the conductor in which there are N electrons, moving through the volume at a velocity v, as shown in Figure 6.25. If an electron takes a time dt to traverse the length dl, its velocity is

$$v = \frac{dl}{dt}$$

The current is

$$I = \frac{dq}{dt} = \frac{dq}{dl}\frac{dl}{dt} = \frac{dq}{dl}v$$

The charge dq traversing the cross-section S of the conductor in a time dt is the charge in the volume and is given by Nq (since N is the number of charges in the volume). This gives: $I=Nqv/dl$. Given the current density $J=I/S$, we obtain the relation $J=Nqv/Sdl$. Since $V=Sdl$, we have

$$J = nqv \qquad (6.40)$$

where $n=N/V$ is the number of charges per unit volume of the conductor. This number is constant and depends only on the type of material (examples: in metallic conductors, $n=8.4\times10^{28}$ m^{-3}; in Germanium, $n=10^{22}$ m^{-3}).

Using Eq. (6.40) in (6.39) gives

$$E_h = \frac{JB}{nq}\sin\theta$$

where θ is the angle between **B** and **J**.

Since V_h is the potential between the two surfaces of the bar (Figure 6.24a), the relation between V_h and E_h is given by

$$V_h = E_h d$$

and we obtain

$$V_h = \frac{JBd}{nq}\sin\theta$$

Recalling that the cross-section of the conductor, S, is $S=ld$, we can substitute $d=S/l$ in the expression above

$$V_h = \frac{JSB}{nql} \sin\theta$$

With $I=JS$, we finally have

$$V_h = \frac{IB}{nql} \sin\theta \tag{6.41}$$

We note that V_h is larger the smaller the number of free electrons n is. Thus, the reason the Hall effect is associated with semiconductors. For practical application of the effect, we wish this potential to be as large as possible.

Since n, q, l, and θ are constant for a given device, the expression in Eq. (6.41) is a relation between I, B, and V_h. The potential V_h is called the Hall potential or Hall voltage. The most common application of this device is in the Hall element sensor, which is particularly useful for measurement of the flux density B. In this case, the sensor is placed such that $\theta=90°$. From Eq. (6.41), this results in

$$B = \frac{V_h nql}{I} \tag{6.42}$$

where V_h and I are measured directly by a voltmeter and amperemeter, connected to the device, or, alternatively, a fixed current source is used for I and V_h is measured.

6.8.4. *The Linear Motor and Generator*

Consider the device shown in Figure 6.26 which functions as a linear generator. If a force **F** can be obtained by the interaction of a current I and a flux density **B**, it should also be possible to do the inverse: to apply a force on a conductor in a magnetic field and obtain a current I. This in fact defines a generator. Figure 6.26 shows this device. A conducting bar is free to move on two conducting rails. The total resistance of the bar and rails is represented by R. A uniform magnetic flux density **B** is present as shown.

If a force is applied to the bar, a potential V_r is generated on R. The generator creates a force F_m which opposes the applied force F_a. The expression for the electromotive force V_r, can be obtained as, for example, 5.6.2 (Eq. 5.43) and is equal to

$$V_r = Blv \tag{6.43}$$

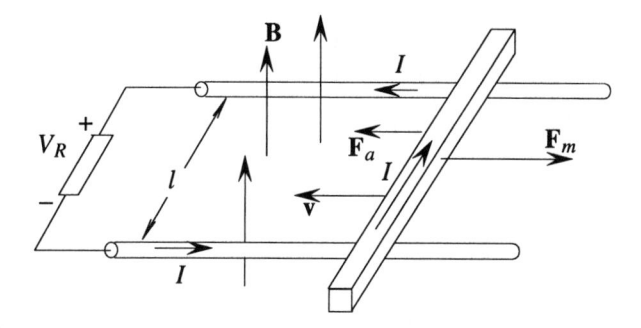

Figure 6.26. Principle of the linear generator. An *emf*, proportional to velocity, magnetic flux density, and length is generated in the sliding bar.

where v is the velocity of the bar. We note that the current is in the direction shown because the magnetic flux diminishes (the cross-section of the loop diminishes). Recalling that induced currents tend to maintain the initial flux, the existence of a current in the circuit gives rise to a force F_m whose magnitude is

$$F_m = IBl$$

and is directed in the direction opposing F_a. The induced current I is equal to $I=V_r/R=vBl/R$. To maintain the movement at constant velocity, the two forces must balance:

$$F_m = F_a = IBl = vB^2l^2/R \qquad (6.44)$$

The power P is given by $P=Fv$, giving the power of the applied force F_a as

$$P = F_a v = v^2 B^2 l^2 / R$$

This mechanical power is equal to the power dissipated through Joule's losses which, using the current $I=vBl/R$, is

$$RI^2 = v^2 B^2 l^2 / R$$

A linear motor is obtained by replacing the resistance R by a source V with polarity such that the current passes in the same direction as the current in the generator (Figure 6.27).

This device is a linear motor in which the bar moves in the direction of F_m. The equation for the potential is

$$V - V_c - IR = 0 \qquad (6.45)$$

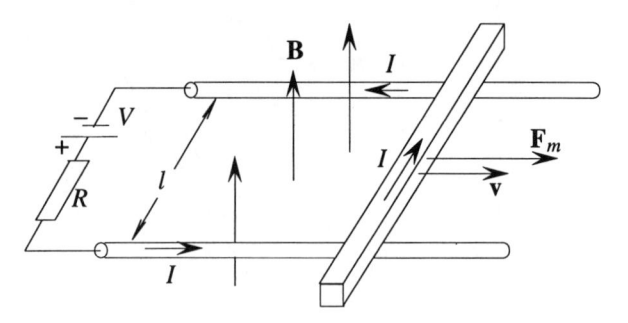

Figure 6.27. Principle of the linear motor. A current supplied by a source forces the bar to move in the direction shown.

where V_c is the induced electromotive force given by Eq. (6.43) (i.e., $V_c=vBl$). Using Eq. (6.45), the force F_m is given as

$$F_m = IBl = \frac{V - V_c}{R} Bl = \frac{V - vBl}{R} Bl \qquad (6.46)$$

This force is maximum at the start of motion when $v=0$. As the bar moves, an induced potential V_c is induced in the circuit, reducing the current in the bar. Allowing for friction between the bar and rails, a retarding force F_f, opposite in direction to F_m exists. The force by itself decreases because v increases [Eq. (6.46)]. The velocity increases until F_f and F_m is established. The dynamic operation of the device is then found from this balance of forces (Newton's second law)

$$m \frac{dv}{dt} = F_m - F_f \qquad (6.47)$$

where F_m is given in Eq. (6.46). This differential equation can be solved to characterize the movement of the bar.

The energy balance is obtained from Eq. (6.45) by multiplying the terms by I:

$$VI = V_c I + I^2 R \qquad (6.48)$$

where VI is the power supplied by the source, $V_c I$ is the power that produces mechanical work due to the movement of the bar, and $I^2 R$ is the dissipated power through resistance. We note that

$$I = F_m/Bl \quad \text{and} \quad V_c = vBl$$

and that the expression $V_c I$ turns out to be equal to $F_m v$, representing the mechanical power. We therefore have

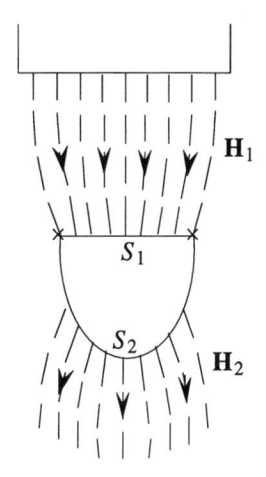

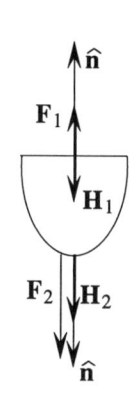

Figure 6.28. A ferromagnetic piece in the field of a permanent magnet.

Figure 6.29. The forces acting on the ferromagnetic piece.

$$VI = F_m v + I^2 R \qquad (6.49)$$

6.8.5. Attraction of a Ferromagnetic Body

Consider a body immersed in a magnetic field, generated, for example, by a permanent magnet, as in Figure 6.28. If this body is made of a ferromagnetic material, the magnetic flux tends to pass through the body, as shown schematically in Figure 6.28. To simplify calculations, we assume the flux penetrates through surface S_1 and exits through surface S_2. Further, we assume that the field intensity H_1 is constant on S_1 and field intensity H_2 is constant on S_2. First we need to identify the location and direction of the forces F_1 on S_1 and F_2 on S_2. In Figure 6.29, the angle θ (defined in Figure 6.14) is 180° on S_1; thus, $\mathbf{F}_1$ makes an angle of 360° with the normal and this force is, therefore, an attraction force. The body is attracted towards the source of the field. On surface S_2, $\theta=0°$ and $\mathbf{F}_2$ makes an angle of 0° with the normal. Force $\mathbf{F}_2$ is a repulsion force and tends to move the body away from the source. The magnitudes of these forces are

$$F_1 = \mu_0 H_1^2 S_1 / 2 \qquad F_2 = \mu_0 H_2^2 S_2 / 2$$

Since the flux is conserved, $(\Phi_1 = \Phi_2)$,

$$\mu_0 H_1 S_1 = \mu_0 H_2 S_2 \qquad \text{or:} \qquad H_1 = H_2 \frac{S_2}{S_1}$$

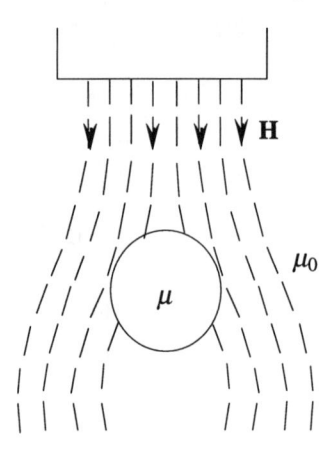

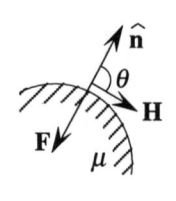

Figure 6.30. A diamagnetic body in a magnetic field.

Figure 6.31. Details of the field and force on the surface of the diamagnetic material.

The relation between F_1 and F_2 is

$$\frac{F_1}{F_2} = \frac{H_1^2 S_1}{H_2^2 S_2}$$

which, using the relation between H_1 and H_2, yields

$$\frac{F_1}{F_2} = \frac{S_2}{S_1} \quad \text{or} \quad F_1 = F_2 \frac{S_2}{S_1}$$

If $S_2 > S_1$, $F_1 > F_2$ and the body is attracted. Even though the material used in this calculation has a particular shape, the procedure used for the calculation can be applied to a body of arbitrary shape, since the part of the body closer to the source of the flux concentrates the field and, as a result, the force due to this part is dominant.

6.8.6. Repulsion of a Diamagnetic Body

In this case we consider a body for which $\mu < \mu_0$, under the influence of the field. The magnetic field behaves as shown in Figure 6.30. The flux passes mostly through air, which has a higher permeability. Figure 6.31 shows a more detailed view of the field at the interface between air and a diamagnetic material. Note that this field has a significant tangential component. The angle θ with the normal is 90° and the force is at 180° to the normal. The sum of these forces generates a total force of repulsion in relation to the source of the field.

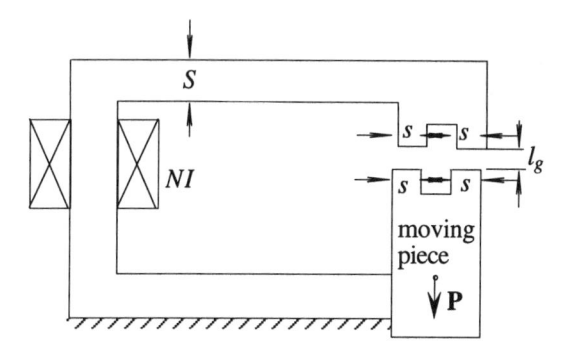

Figure 6.32. A magnetic path with a movable piece.

Having said this, we must point out that the diamagnetic effect in this discussion is exaggerated, since μ of a diamagnetic material is only slightly smaller than μ_0, and the total force of repulsion is practically imperceptible. If a large tangential force could be generated, a compression force on the surface of the body would exist.

6.8.7. Magnetic Levitation

We wish to calculate the magnetic levitation of a body in the magnetic circuit shown in Figure 6.32. The part of the magnetic circuit that contains the coil is fixed. The moving piece is free to move vertically. An attraction force **F** is created on the two poles of the piece. If this force is larger than the weight **P**, the movable piece is suspended. We assume $\mu \gg \mu_0$, such that the flux passes through the poles only and there are no edge effects. Also, the friction on the lateral part of the movable piece where it touches the fixed part of the circuit is neglected.

The numerical values for the problems are given as:

- loop $N=100$ turns
- current $I=10$ A
- cross section of the poles $s=10$ cm^2
- gap length $l_g=2$ mm
- mass $m=40$ kg
- gravity acceleration $g=9.81$ m/s^2

The magnetic field intensity between the poles of the stationary and movable piece is calculated from the circulation of the field intensity

$$hl_g = NI \quad \Rightarrow \quad h=500{,}000 \text{ At/m}$$

The force between the two parts is

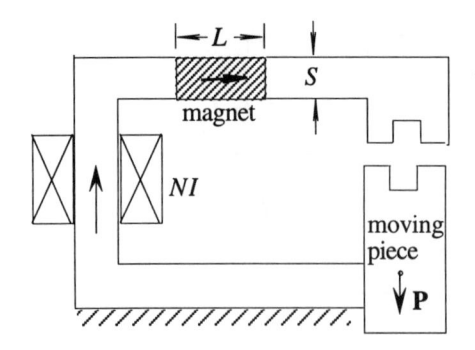

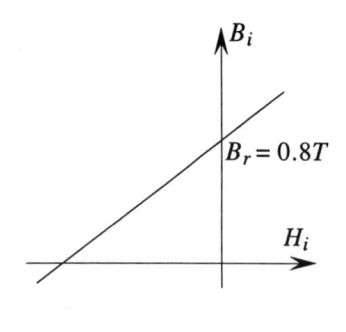

208   6. Interaction between Electromagnetic and Mechanical Forces

Figure 6.33. The magnetic path of Figure 6.32 with an added permanent magnet in the path to increase the attraction force on the moving piece.

Figure 6.34. The magnetization curve for the permanent magnet.

$$F = 2\frac{\mu_0}{2}h^2 s = 314.2 \ N \qquad (6.50)$$

The weight is $P=mg$, or: $P=392.4 \ N$. The weight is larger than the attraction force and the piece is not suspended.

In an attempt to increase the attraction force, a strong permanent magnet is placed in the magnetic circuit, with its field in the same direction as that of the coil, as shown in Figure 6.33. The length of the permanent magnet is $L=2 \ cm$; its cross-section is $S=30 \ cm^2$ and its magnetization curve is that shown in Figure 6.34. The equation corresponding to this curve is

$$B_i = \mu_0 H_i + B_r$$

The orientation of the magnet is such that the fluxes of the magnet and the coil add together. The new attraction force **F** is now calculated.

From the circulation of the magnetic field intensity

$$hl_g + H_i L = NI$$

From the conservation of flux

$$\Phi_i = \Phi_g$$

or

$$B_i S = 2\mu_0 h s$$

From the magnetization curve

$$B_i = \mu_0 H_i + 0.8$$

The system of equations to solve is

$$0.002h + 0.02H_i = 1000$$

$$30 \times 10^{-4} B_i = 2\mu_0 h \, 10 \times 10^{-4}$$

$$B_i = \mu_0 H_i + 0.8$$

The solution is $H_i = -0.396 \times 10^5$ At/m; $B_i = 0.750$ T; $h = 0.896 \times 10^6$ At/m.

The attraction force in the gap, calculated from Eq. (6.50) is $F \approx 1009$ N, sufficiently large to suspend the movable piece.

6.8.8. *The Magnetic Brake*

An interesting and very useful application of induced currents is the magnetic brake. To present the principle involved, we use the structure in Figure 6.35. In this structure, an electromagnet generates a flux density **B** in the gap. This field will be assumed to be constant. A pendulum-like piece, made of a conducting material is placed such that it can move into the gap. If the current in the electromagnet, I, is zero, the oscillation of the pendulum is not affected by the structure. If the current in the coil in not zero, the movement of the conducting plate, into the magnetic field (Figure 6.36) generates induced currents in the plate itself. The flux of the induced currents is such that it opposes the field **B**, since, initially, in the plate, the field was zero. According to Lenz's law, the induced currents tend to maintain this condition. Figure 6.36 gives the direction of the fields. The electric field due to the induced currents is given by **E**=**v**×**B** (section 6.3), and we get

$$\mathbf{J} = \sigma \mathbf{E} = \sigma \mathbf{v} \times \mathbf{B}$$

The rate at which the plate penetrates into the gap is responsible for the magnitude of the induced currents which, in this case, point upward as shown in Figure 6.37. Using Eq. (6.1) the volumetric force density **f** is

$$\mathbf{f} = \mathbf{J} \times \mathbf{B} = \sigma(\mathbf{v} \times \mathbf{B}) \times \mathbf{B}$$

If all vectors are mutually orthogonal, as is the case in this example, the force is

$$F = \sigma v B^2 Vol$$

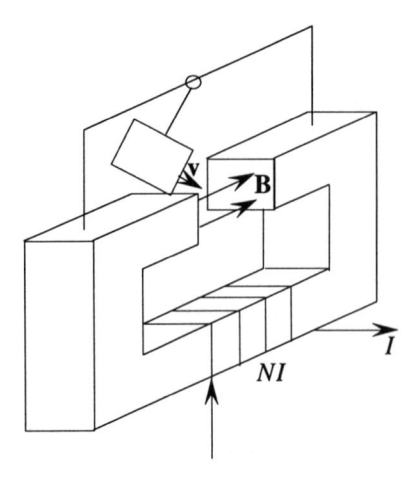

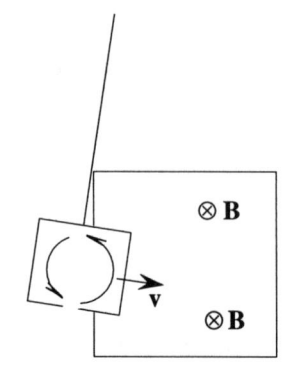

210   6. Interaction between Electromagnetic and Mechanical Forces

Figure 6.35. Principle of magnetic braking. As the moving plate penetrates into the magnetic field, eddy currents are generated in the plate.

Figure 6.36. Eddy currents in the moving plate retard its movement. Braking force is proportional to velocity, conductivity, and flux density.

and its direction, given by the cross product **J**×**B**, opposes the direction of **v**. This has the effect of damping the movement of the plate into the gap. If the conductivity of the plate were infinite, the plate would be repelled from the gap. In reality, σ is finite and the plate is decelerated as the induced currents are dissipated in the plate. The plate penetrates into the gap, decelerating, and, eventually, reaches a state of static equilibrium at the lowest point of its oscillation. In case this does not happen immediately, the plate may continue to oscillate in a highly damped motion until the pendulum has completely stopped.

This principle is used extensively on locomotives and trucks. Conducting discs are installed on the axes of the vehicle and electromagnets are placed around them such that the discs move in the gap of the electromagnets, as shown in Figure 6.38. When the mechanical brakes are applied, a current is passed through the electromagnet and the braking effects of the mechanical and magnetic brakes are added together. We note, however, that the braking effect assumes a velocity **v**. For this reason, electromagnetic brakes cannot be used to completely stop a vehicle, only to slow it down.

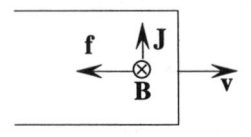

Figure 6.37. Direction of current density and force on the moving plate. The force is opposite in direction to velocity.

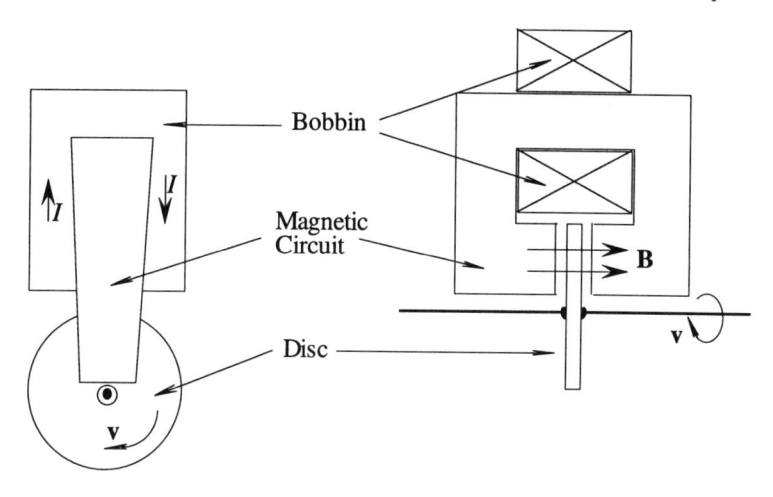

Figure 6.38. Example of an electromagnetic brake. The disk is mounted on the axle and moves inside the gap of the magnetic circuit. When current is passed through the bobbin, a braking force proportional to velocity is generated.

7
Wave Propagation and High-Frequency Electromagnetic Fields

7.1. Introduction

In Chapter 2, we introduced Maxwell's equations, in their general form, as well as in particular, useful forms. Chapter 5 discussed in some detail time-dependent phenomena, associated with induction. However, the discussion included only low-frequency applications of the time-dependent field since the displacement currents in Maxwell's equations were neglected. This chapter deals with those time-dependent equations and applications which require the inclusion of displacement currents. Both time-dependent and time-harmonic representations are discussed. Again, as in Chapter 2, the distinction between low- and high-frequency applications is based on the need to include displacement currents, rather than on frequency. In particular, the propagation of plane waves, waveguides, and cavity resonators are discussed.

The general Maxwell equations are as follows:

$$\nabla \times \mathbf{H} = \mathbf{J} + \frac{\partial \mathbf{D}}{\partial t} \qquad (7.1)$$

$$\nabla \cdot \mathbf{B} = 0 \qquad (7.2)$$

$$\nabla \times \mathbf{E} = -\frac{\partial \mathbf{B}}{\partial t} \qquad (7.3)$$

$$\nabla \cdot \mathbf{D} = \rho \qquad (7.4)$$

The electrical continuity equation was defined in section 2.2.2. by taking the divergence of both sides of Eq. (7.1) and using Eq. (7.4). The electrical continuity equation is

$$\nabla \cdot \mathbf{J} = -\frac{\partial \rho}{\partial t} \qquad (7.5)$$

Introduction 213

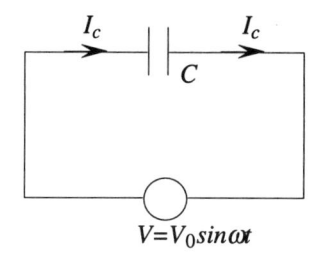

Figure 7.1. Capacitor in an ac circuit. The current through the capacitor is a displacement current.

In most low-frequency applications, $\partial \rho / \partial t$ is zero and, therefore, we normally obtain $\nabla \cdot \mathbf{J} = 0$. This indicates that the flux of the vector, or similarly, that the conduction current is conservative. In other words, the current entering a certain volume is equal to the current leaving the volume. In fact, in most electromagnetic devices, the current injected into the device is equal to the current leaving it. When this does not happen, there is an accumulation of charges in the device, or a certain amount of charge is extracted from the device. A simple device in which this occurs is the capacitor. In this case, the displacement currents are important at any frequency.

To see that displacement currents cannot always be neglected at low frequencies, consider the capacitor in Figure 7.1. The capacitor is connected to an *ac* source. A current I_c flows in the wires. This is a conduction current that cannot flow through the dielectric material between the plates. A surface, bounded by a contour, can now be drawn as in Figure 7.2. Along this contour, Ampere's law can be applied resulting in a magnetic field intensity **H**, which at the moment we do not calculate. However, the surface S is arbitrary, as long as we integrate all currents passing through the surface. A second surface can be defined as in Figure 7.2b.

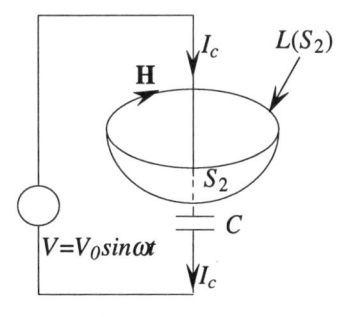

Figure 7.2a. A contour limiting a surface S through which the conduction current passes.

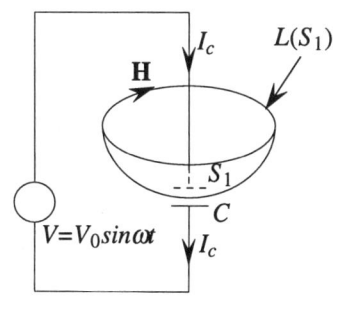

Figure 7.2b. A contour identical to that in Figure 7.2a, limiting a surface through which the displacement current passes.

214   7. Wave Propagation and High-Frequency Electromagnetic Fields

The surface is such that it passes between the plates of the capacitor, but the contour $L(S)$ is maintained the same as in Figure 7.2a. Since there is no conduction current passing through S, we first conclude that the field **H** is due to displacement currents. Similarly, since the contour in Figure 7.2b is at the same physical location as in Figure 7.2a, the displacement current I_d must be equal to the conduction current I_c. This current can be immediately calculated from circuit theory, since

$$I_c = I_d = C\frac{dV}{dt} = CV_0\omega\cos\omega t$$

For the particular case of parallel plate capacitors the displacement currents are calculated directly from the relation

$$I_d = \int_S \frac{\partial \mathbf{D}}{\partial t} \cdot d\mathbf{s} \qquad (7.6)$$

The electric flux density **D** is calculated as

$$\mathbf{D} = \varepsilon \mathbf{E}$$

and the electric field in a parallel plate capacitor is

$$E = \frac{V}{d}$$

where d is the distance between the two plates. Substitution of these into Eq. (7.6) gives

$$I_d = \int_S \frac{\partial}{\partial t}\left(\varepsilon\frac{V_0\sin\omega t}{d}\right)ds = \left(\varepsilon\frac{S}{d}\right)\omega V_0\cos\omega t$$

and using the capacitance for a parallel capacitor as $C=\varepsilon S/d$ [from Eq. (3.22)], we have

$$I_d = C\omega V_0 \cos\omega t$$

as above.

This example clearly indicates the importance of the continuity equation in general and, in fact, this example was used by Maxwell to explain the need for inclusion of the displacement currents in Ampere's law. Another aspect of displacement current is propagation of electromagnetic waves.

Finally, note that the term "displacement current" is due to the derivative $\partial \mathbf{D}/\partial t$, since $\mathbf{D}$, also called the electric displacement, is normally associated with displacement of charges.

7.2. The Wave Equation and Its Solution

Based on the introduction of the displacement currents in Ampere's law, Maxwell predicted the existence of propagating waves, a prediction that was verified experimentally in 1889, nine years after his death by Heinrich Herz. This prediction was based on the nature of the equations one obtains by using the complete set of equations. Note that in Chapter 6, we neglected the displacement currents assuming they are small compared to conduction currents. This is certainly the case in many applications, but their neglection also means that none of the wave properties of the magnetic field could be analyzed. The wave phenomena that cannot be defined without resorting to wave properties include such widely used quantities as the skin depth. The re-introduction of the displacement currents will allow us to look at these aspects of the electromagnetic field.

7.2.1. *The Time-Dependent Equations*

In Chapter 2, we introduced Maxwell's equations in various forms. This was done to facilitate solution or the explanation of concepts. As examples, in the static case we obtained either Poisson or Laplace type equations. The same process, using the general form of Maxwell's equations, yields a wave equation. It is called a wave equation because its solution is a propagating wave. We assume here that media are isotropic and linear. Under these conditions, Eqs. (7.2) and (7.4) become

$$\nabla \cdot \mathbf{H} = 0$$

and

$$\nabla \cdot \mathbf{E} = \frac{\rho}{\varepsilon}$$

To obtain a wave equation we use Eq. (7.1) and take the curl on both sides of the equation

$$\nabla \times (\nabla \times \mathbf{H}) = \nabla \times \mathbf{J} + \nabla \times \frac{\partial \mathbf{D}}{\partial t} = \nabla \times \mathbf{J} + \frac{\partial}{\partial t}(\nabla \times \mathbf{D})$$

where the time differentiation and the curl were interchanged. From Eq. (7.3) we have

$$\nabla \times D = -\varepsilon \frac{\partial B}{\partial t}$$

Using this and the relation $\nabla \times (\nabla \times H) = -\nabla^2 H + \nabla(\nabla \cdot H)$, we have

$$\nabla^2 H - \nabla(\nabla \cdot H) = -\nabla \times J + \frac{\partial}{\partial t}\left(\varepsilon \frac{\partial B}{\partial t}\right) \qquad (7.7)$$

As for the term $\nabla \times J$, we again use Eq. (7.3) and the constitutive relation $J = \sigma E$

$$\nabla \times J = \nabla \times \sigma E = -\sigma \frac{\partial B}{\partial t} = -\sigma \mu \frac{\partial H}{\partial t}$$

Substituting this in Eq. (7.7) and rearranging terms, we get

$$\nabla^2 H - \mu\varepsilon \frac{\partial^2 H}{\partial t^2} = \sigma\mu \frac{\partial H}{\partial t} \qquad (7.8)$$

This is a nonhomogeneous wave equation in terms of the magnetic field intensity. A homogeneous wave equation is obtained by removing the right-hand side term in Eq. (7.8) which is due to the current density J

$$\nabla^2 H - \mu\varepsilon \frac{\partial^2 H}{\partial t^2} = 0 \qquad (7.9)$$

The distinction between homogeneous (source free) and nonhomogeneous equations indicates the location of the source and the type of solution one expects. With the nonhomogeneous wave equations, the source is in the solution domain and is taken into account in the solution. With the source free equations, the source is outside the solution domain and we expect a solution in terms of the characteristic modes of the problem.

Instead of starting with Eq. (7.1), we can start with Eq. (7.3) and take the curl on both sides of the equation

$$\nabla \times (\nabla \times E) = -\mu \frac{\partial}{\partial t}(\nabla \times H)$$

where $B = \mu H$ was used. Using the relation $(\nabla \times (\nabla \times E) = -\nabla^2 E + \nabla(\nabla \cdot E))$ and substituting from Eq. (7.1) for $\nabla \times H$, we have

$$-\nabla^2 \mathbf{E} + \nabla(\nabla \cdot \mathbf{E}) = -\mu \frac{\partial}{\partial t}(\mathbf{J} + \frac{\partial \mathbf{D}}{\partial t}) = -\mu \frac{\partial \mathbf{J}}{\partial t} - \mu \frac{\partial^2 \mathbf{D}}{\partial t^2}$$

Substituting Eq. (7.4) for $\nabla \cdot \mathbf{E}$, $\sigma \mathbf{E}$ for $\mathbf{J}$ and $\varepsilon \mathbf{E}$ for $\mathbf{D}$, and rearranging terms, we have

$$\nabla^2 \mathbf{E} - \mu\varepsilon \frac{\partial^2 \mathbf{E}}{\partial t^2} = \nabla\left(\frac{\rho}{\varepsilon}\right) + \mu\sigma \frac{\partial \mathbf{E}}{\partial t} \qquad (7.10)$$

If the sources of the electric field do not exist, (i.e., they are outside the solution domain) we obtain the source free wave equation. In this case, both ρ and $\mathbf{J}$ must be zero and the two terms on the right-hand side of Eq. (7.10) disappear. The source free wave equation in terms of the electric field intensity is

$$\nabla^2 \mathbf{E} - \mu\varepsilon \frac{\partial^2 \mathbf{E}}{\partial t^2} = 0 \qquad (7.11)$$

Our choice of writing Maxwell's equations in these particular forms was completely arbitrary. We could just as well write them in terms of $\mathbf{B}$ or $\mathbf{D}$. In fact, by simply multiplying both sides of Eq. (7.8) and (7.9) by μ and using $\mathbf{B}=\mu\mathbf{H}$, we get the homogeneous wave equation and the source free wave equations in terms of $\mathbf{B}$ rather than $\mathbf{H}$

$$\nabla^2 \mathbf{B} - \mu\varepsilon \frac{\partial^2 \mathbf{B}}{\partial t^2} = \sigma\mu \frac{\partial \mathbf{B}}{\partial t} \qquad (7.12)$$

and

$$\nabla^2 \mathbf{B} - \mu\varepsilon \frac{\partial^2 \mathbf{B}}{\partial t^2} = 0 \qquad (7.13)$$

Similarly, by multiplying Eqs. (7.10) and (7.11) by ε and using $\mathbf{D}=\varepsilon\mathbf{E}$, we have for the homogeneous wave equation

$$\nabla^2 \mathbf{D} - \mu\varepsilon \frac{\partial^2 \mathbf{D}}{\partial t^2} = \nabla\rho + \mu\sigma \frac{\partial \mathbf{D}}{\partial t} \qquad (7.14)$$

and for the source free equation

$$\nabla^2 \mathbf{D} - \mu\varepsilon \frac{\partial^2 \mathbf{D}}{\partial t^2} = 0 \qquad (7.15)$$

In addition to using field variables we can also define potential functions and write the wave equations in terms of these functions. Potential functions may be defined in any way that is convenient and consistent with the field equations. As an example, we can define a vector potential based on the solenoidal (divergence free) nature of the magnetic field. Because

$$\nabla \cdot \mathbf{B} = 0$$

we can write

$$\nabla \cdot (\nabla \times \mathbf{A}) \equiv 0$$

for any vector **A**. Obviously, we still have to define the divergence of **A**. Thus, a magnetic vector potential can be defined as

$$\mathbf{B} = \nabla \times \mathbf{A}$$

This can now be substituted in Eq. (7.3)

$$\nabla \times \mathbf{E} = -\frac{\partial}{\partial t}(\nabla \times \mathbf{A})$$

In Chapter 3, we defined the scalar potential V based on the irrotational nature of the static electric field. We stated that if $\nabla \times \mathbf{E} = 0$, then $\mathbf{E} = -\nabla V$, where V is a scalar potential. Obviously, the time-dependent electric field is not irrotational and we cannot define the scalar potential in the same fashion. Instead, we can write the equation above as

$$\nabla \times (\mathbf{E} + \frac{\partial \mathbf{A}}{\partial t}) = 0$$

Now the expression in parentheses is irrotational and we can define the time-dependent electric scalar potential as

$$\mathbf{E} + \frac{\partial \mathbf{A}}{\partial t} = -\nabla V$$

or, the electric field as

$$\mathbf{E} = -\frac{\partial \mathbf{A}}{\partial t} - \nabla V \qquad (7.16)$$

This definition is consistent with the static case. The expression above states that the electric field has two sources. One is due to a charge distribution and exists in the static case as well as the time-dependent case ($-\nabla V$). The second is due to time-dependent currents ($\partial \mathbf{A}/\partial t$). We also note in passing that the vector potential $\mathbf{A}$ is a vector parallel to $\mathbf{E}$ and, therefore, to the current density $\mathbf{J}$: ($\mathbf{J} = \sigma \mathbf{E}$).

Returning now to Maxwell's equations, we substitute Eq. (7.16) into Eq. (7.1)

$$\nabla \times (\nabla \times \mathbf{A}) = \mu \mathbf{J} + \mu \varepsilon \frac{\partial}{\partial t}\left(-\nabla V - \frac{\partial \mathbf{A}}{\partial t}\right)$$

Using the identity for $\nabla \times (\nabla \times \mathbf{A})$ and rearranging the terms, we have

$$\nabla^2 \mathbf{A} + \mu \mathbf{J} - \mu \varepsilon \frac{\partial^2 \mathbf{A}}{\partial t^2} - \nabla\left(\nabla \cdot \mathbf{A} + \mu \varepsilon \frac{\partial V}{\partial t}\right) = 0$$

Since the divergence of $\mathbf{A}$ has not been defined, we can define it such that the last term in the equation above disappears

$$\nabla \cdot \mathbf{A} = -\mu \varepsilon \frac{\partial V}{\partial t} \qquad (7.17)$$

This condition is known as the Lorentz gage. The static form of the gage, namely $\nabla \cdot \mathbf{A} = 0$, is known as Coulomb's gage.

The nonhomogeneous wave equation for the magnetic vector potential is

$$\nabla^2 \mathbf{A} - \mu \varepsilon \frac{\partial^2 \mathbf{A}}{\partial t^2} = -\mu \mathbf{J} \qquad (7.18)$$

and the homogeneous, or source free wave equation is

$$\nabla^2 \mathbf{A} - \mu \varepsilon \frac{\partial^2 \mathbf{A}}{\partial t^2} = 0 \qquad (7.19)$$

To define a wave equation in terms of the electric scalar potential we use the relation [Eq. (7.4), in modified form]

$$\nabla \cdot \mathbf{E} = \frac{\rho}{\varepsilon}$$

Substituting for $\mathbf{E}$ from Eq. (7.16), we have

$$-\nabla \cdot \left(\frac{\partial \mathbf{A}}{\partial t} + \nabla V\right) = \frac{\rho}{\varepsilon}$$

which, after performing the divergence, is

$$\nabla^2 V + \frac{\partial}{\partial t}(\nabla \cdot \mathbf{A}) = -\frac{\rho}{\varepsilon}$$

and, again using the condition in Eq. (7.17), we get a nonhomogeneous wave equation

$$\nabla^2 V - \mu\varepsilon \frac{\partial^2 V}{\partial t^2} = -\frac{\rho}{\varepsilon} \tag{7.20}$$

The source free equation is now

$$\nabla^2 V - \mu\varepsilon \frac{\partial^2 V}{\partial t^2} = 0 \tag{7.21}$$

This last equation is a scalar wave equation while all the others defined above were vector wave equations. Before continuing, we must mention that the wave equations reduce to the well-known forms (either Poisson's or Laplace's equations) in the static case by simply setting the time derivatives to zero.

7.2.2. The Time-Harmonic Wave Equations

A particularly useful form of the wave equations is found by assuming sinusoidal fields and using the complex (phasor) notation

$$\mathbf{E} = \mathbf{E}_0 e^{j\omega t}$$

where $j = \sqrt{-1}$, $\omega = 2\pi f$, and f is the frequency of the field. Under this condition, the time-dependent form of the field is

$$\mathbf{E}(t) = Re\left(\mathbf{E}_0 e^{j\omega t}\right) = \mathbf{E}_0 \cos\omega t$$

where Re denotes the real part of the phasor.

A sinusoidal form, or any phase in the field, can be introduced in the notation as an angle ϕ

$$\mathbf{E} = \mathbf{E}_0 e^{j\phi} e^{j\omega t}$$

Under phasor conditions, the time derivatives of the fields are

$$\frac{\partial \mathbf{E}}{\partial t} = j\omega \mathbf{E}$$

and the second order derivative is

$$\frac{\partial^2 \mathbf{E}}{\partial t^2} = -\omega^2 \mathbf{E}$$

Introducing these quantities for the nonhomogeneous and source free equations above [Eqs. (7.8), (7.10), (7.12), (7.14), (7.18), or (7.20) and Eqs. (7.9), (7.11), (7.13), (7.15), (7.19), or (7.21)] we obtain the following time-harmonic wave equations:

- Nonhomogeneous wave equations

$$\nabla^2 \mathbf{H} - j\omega\sigma\mu\mathbf{H} + \omega^2\mu\varepsilon\mathbf{H} = 0 \qquad (7.22)$$

$$\nabla^2 \mathbf{E} - \nabla\left(\frac{\rho}{\varepsilon}\right) - j\omega\mu\sigma\mathbf{E} + \omega^2\mu\varepsilon\mathbf{E} = 0 \qquad (7.23)$$

$$\nabla^2 \mathbf{B} - j\omega\sigma\mu\mathbf{B} + \omega^2\mu\varepsilon\mathbf{B} = 0 \qquad (7.24)$$

$$\nabla^2 \mathbf{D} - \nabla\rho - j\omega\mu\sigma\mathbf{D} + \omega^2\mu\varepsilon\mathbf{D} = 0 \qquad (7.25)$$

$$\nabla^2 \mathbf{A} + \mu\mathbf{J} + \omega^2\mu\varepsilon\mathbf{A} = 0 \qquad (7.26)$$

$$\nabla^2 V + \omega^2\mu\varepsilon V + \frac{\rho}{\varepsilon} = 0 \qquad (7.27)$$

- Source free wave equations

$$\nabla^2 \mathbf{H} + \omega^2\mu\varepsilon\mathbf{H} = 0 \qquad (7.28)$$

$$\nabla^2 \mathbf{E} + \omega^2\mu\varepsilon\mathbf{E} = 0 \qquad (7.29)$$

$$\nabla^2 \mathbf{B} + \omega^2\mu\varepsilon\mathbf{B} = 0 \qquad (7.30)$$

$$\nabla^2 \mathbf{D} + \omega^2\mu\varepsilon\mathbf{D} = 0 \qquad (7.31)$$

$$\nabla^2 \mathbf{A} + \omega^2\mu\varepsilon\mathbf{A} = 0 \qquad (7.32)$$

$$\nabla^2 V + \omega^2\mu\varepsilon V = 0 \qquad (7.33)$$

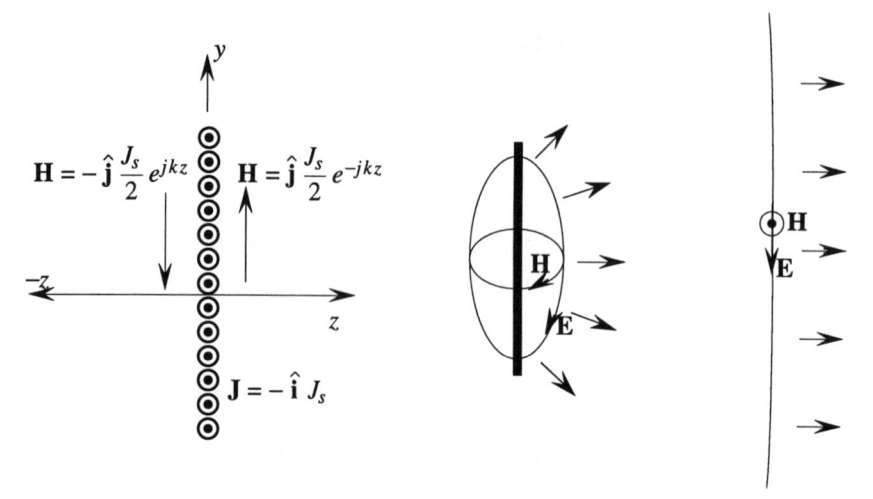

$$\mathbf{H} = -\hat{\mathbf{j}}\frac{J_s}{2}e^{jkz} \qquad \mathbf{H} = \hat{\mathbf{j}}\frac{J_s}{2}e^{-jkz}$$

$$\mathbf{J} = -\hat{\mathbf{i}}\,J_s$$

Figure 7.3a. A sheet of currents and the magnetic field intensity it generates. A plane wave propagates in the +z and −z directions.

Figure 7.3b. A short current segment generates a wave with an elliptical front. At large distances, this can be viewed as a plane wave.

7.2.3. Solution of the Wave Equation

To observe the behavior of waves we will solve the time-harmonic source free wave equation in one dimension. This may seem at first as an oversimplification, but since the homogeneous solution represents the system in which the wave propagates, it is not necessary to include the sources in the solution. The solution to a one-dimensional wave equation provides all the aspects of wave propagation without the need for complex analysis. In addition, many of the more important aspects of wave propagation are easier to define and understand in the frequency domain. Some terms that we use, such as phase, only have a meaning in the context of sinusoidal excitation. Before proceeding with the solution, we define uniform plane waves.

7.2.4. Solution for Plane Waves

A uniform plane wave is a wave (i.e., a solution to the wave equation), in which the electric and the magnetic field intensities are constant in magnitude and phase and each is in a constant direction on planes perpendicular to the direction of propagation.

The only way a uniform plane wave can be generated is by an infinite source such as an infinite current sheet as shown in Figure 7.3a. The magnetic field intensity everywhere in space is either in the positive or negative y direction and is directed parallel to the current sheet as shown. The electric field intensity is perpendicular to this in the plane of the sheet, and the wave propagates in the positive (and negative) z direction.

While this is not physically realizable for any finite source, if the observation point is remote enough from the source, the wavefront will be a very large sphere and this can be viewed as a plane for practical purposes. A situation of this kind is shown in Figure 7.3b. A finite length source (i.e., an antenna) is shown. The fields near the antenna form an elliptical surface, but far from the source they can be assumed to be on a plane. This assumption, while not perfect, allows simplification of analysis, focusing on properties rather than complex mathematics.

7.2.5. The One-Dimensional Wave Equation in Free Space and Lossless Dielectrics

To simplify the solution and discussion, we assume that the electric field has only a component in the x direction and varies only in the z direction. These assumptions imply the following conditions:

$$E_y = E_z = 0 \quad \text{and} \quad \frac{\partial E_*}{\partial x} = \frac{\partial E_*}{\partial y} = 0$$

where * denotes any component of **E**. Substitution of these into Eq. (7.29) results in

$$\frac{d^2 E_x}{dz^2} + \omega^2 \mu\varepsilon E_x = 0 \qquad (7.34)$$

where the partial derivative was replaced with the ordinary derivative because of the field dependence on z only. We denote

$$k = \sqrt{\omega^2 \mu\varepsilon}$$

Eq. (7.34) describes simple harmonic motion, therefore has the solution

$$E_x(z) = E_0^+ e^{-jkz} + E_0^- e^{jkz} \qquad (7.35)$$

where E_0^+ and E_0^- are constants to be determined from the boundary conditions of the problem. The notation (+) and (−) indicates that the first term is a propagating wave in the positive z direction and the second a propagating wave in the negative z direction, as in Figure 7.4, where arrows indicate the direction of propagation, not the direction of fields. The constants can be complex and are arbitrary. This solution can be verified by direct substitution into Eq. (7.34).

Using the phasor transformation, we can write this in the time domain as

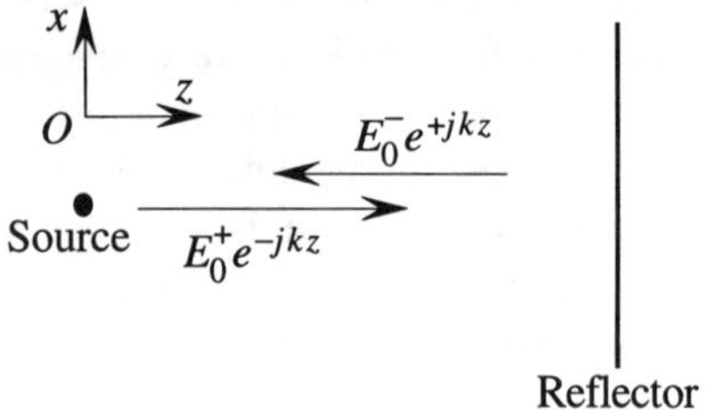

224   7. Wave Propagation and High-Frequency Electromagnetic Fields

Figure 7.4. A propagating wave. Forward and backward propagating waves are shown.

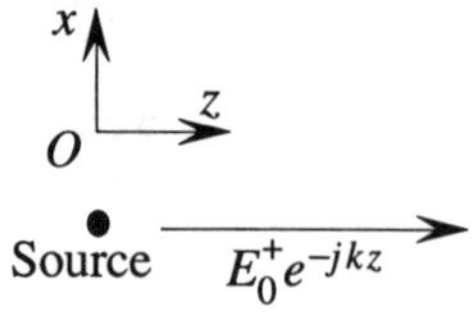

Figure 7.5. A forward propagating wave in unbounded domain.

$$E_x(z,t) = E_0^+ \cos(\omega t - kz + \phi) + E_0^- \cos(\omega t + kz + \phi)$$

where the initial (arbitrary) phase angle ϕ was added for completeness.

If we assume that the wave propagates in a boundless space, only an outward, wave exists, and E_0^- is zero (no backward propagating wave). On the other hand, if the forward propagating wave is reflected without losses (i.e., from a perfect conductor), the amplitude of the two waves is equal. Assuming only a forward propagating wave, the solution is

$$E_x(z) = E_0^+ e^{-jkz}$$

or

$$E_x(z,t) = E_0^+ \cos(\omega t - kz + \phi)$$

Figure 7.5 shows schematically a forward propagating wave without reflection. Examining these expressions, it is obvious that what changes with time is the phase of the wave. In other words, the phase of the wave "travels" at a certain velocity. To see what this velocity is we use Figure 7.6. We follow a fixed point on the wave, for which $\omega t - kz + \phi = const$. Then,

$$z = \frac{\omega t}{k} - const$$

To find the phase velocity, we write

$$\frac{dz}{dt} = v = \frac{\omega}{k} = \frac{1}{\sqrt{\mu\varepsilon}}$$

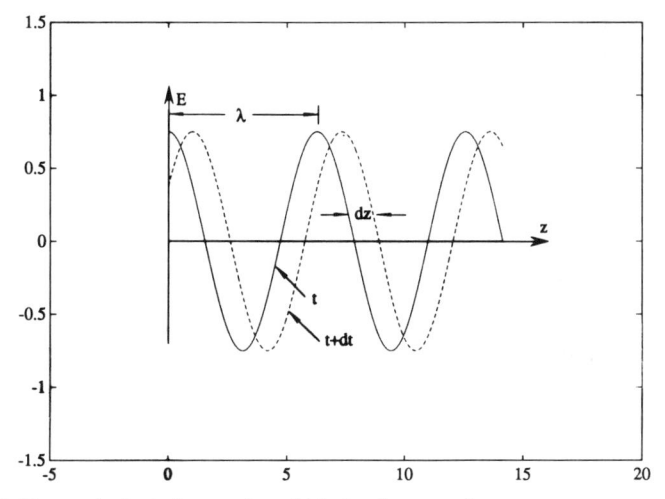

Figure 7.6. Phase velocity is the speed at which the phase travels.

In free space $\mu=\mu_0$ and $\varepsilon=\varepsilon_0$ and, therefore, the phase velocity is equal to the speed of light, or "c." The phase velocity is the speed of propagation of the electromagnetic wave in any medium. The speed of propagation in most materials is lower than c since ε_r is always equal or larger than 1, while μ_r is only smaller than 1 in diamagnetic materials. In fact, the speed of propagation in good conductors can be a small fraction of the speed of light.

Now we can also define the wavelength λ (in meters) which is that distance the wave travels in one cycle

$$\lambda = \frac{v}{f} = \frac{2\pi}{k} \quad (m) \tag{7.36}$$

Or, writing k as

$$k = \frac{2\pi}{\lambda} \tag{7.37}$$

we note that k is the number of wavelengths in one cycle. For this reason it is called the wavenumber.

As an example, consider propagation in water. The relative permittivity of water is 81. The phase velocity (speed of propagation) is 9 times slower than the speed of light. The wavelength is also 9 times shorter.

Up to this point we have treated only the electric field intensity, disregarding the magnetic field intensity. In propagation of electromagnetic waves the magnetic field intensity must be treated together with the electric field intensity. Indeed, without this coupling the propagation would not exist in the first place. To see the relation between the electric and magnetic field intensities, we use Eq. (7.3). The source free equation is

$$\nabla \times \mathbf{E} = -j\omega\mu\mathbf{H}$$

In terms of the components of the equation, this is

$$\hat{\mathbf{i}}\left(\frac{\partial E_z}{\partial y} - \frac{\partial E_y}{\partial z}\right) + \hat{\mathbf{j}}\left(\frac{\partial E_x}{\partial z} - \frac{\partial E_z}{\partial x}\right) + \hat{\mathbf{k}}\left(\frac{\partial E_y}{\partial x} - \frac{\partial E_x}{\partial y}\right) = -j\omega\mu(\hat{\mathbf{i}}H_x + \hat{\mathbf{j}}H_y + \hat{\mathbf{k}}H_z)$$

From the assumption that only the term $\partial E_x/\partial z$ exists, we have

$$H_x = H_z = 0$$

and

$$\frac{\partial E_x}{\partial z} = -j\omega\mu H_y$$

or, writing this for the forward propagating wave, we can write for H_y

$$H_y^+ = \frac{j}{\omega\mu} \frac{\partial E_x^+}{\partial z}$$

Substitution of E_x^+ from Eq. (7.35), we get

$$\frac{\partial E_x^+}{\partial z} = \frac{\partial}{\partial z}(E_0^+ e^{-jkz}) = -jk(E_0^+ e^{-jkz}) = -jkE_x^+(z)$$

This gives

$$H_y^+(z) = \frac{k}{\omega\mu} E_x^+(z)$$

The ratio between $H_y(z)$ and $E_x(z)$, has units of (1/Ohms). We define the quantity $\omega\mu/k$ as

$$\eta = \frac{\omega\mu}{k} = \sqrt{\frac{\mu}{\varepsilon}} \quad (\Omega) \qquad (7.38)$$

This quantity is an impedance and is called the "intrinsic impedance" of the material since it is only dependent on material properties. The intrinsic impedance of free space is

$$\eta_0 = \sqrt{\frac{\mu_0}{\varepsilon_0}} = 377\ \Omega$$

The intrinsic impedance is a complex number in general media. The equation for H_y^+ can now be written as

$$H_y^+(z) = \frac{1}{\eta} E_x^+(z)$$

Note here that H and E propagate in the same direction. They are orthogonal to each other and to the direction of propagation. This property makes E and H transverse electromagnetic (*TEM*) waves. The ratio between E and H is equal to the intrinsic impedance of the material in which they propagate.

The discussion above was restricted to a single component of the electric and magnetic field intensities. However, we can do the same with any other component of the electric or magnetic field, and use any other direction of propagation. The restriction to lossless media is minor, in that all that is necessary is replacement of the permittivity with the complex permittivity. This and the effect of losses are discussed next. Perhaps the most significant restriction in this discussion was the assumption of linear, isotropic materials.

7.3. Propagation of Waves in Materials

7.3.1. *Propagation of Waves in Lossy Dielectrics*

A lossy dielectric is defined as a material with small but not negligible losses. For our purpose, we assume here that a lossy dielectric is characterized by its permittivity and conductivity. Thus, we may assume that in addition to displacement currents, there are also conduction currents in the dielectric. Using Eq. (7.23), we can write the following:

$$\nabla^2 \mathbf{E} - j\omega\mu\sigma\mathbf{E} + \omega^2\mu\varepsilon\mathbf{E} = \nabla^2\mathbf{E} + \omega^2\mu\left(\varepsilon - \frac{j\sigma}{\omega}\right)\mathbf{E} = 0$$

where the term $\nabla(\rho/\varepsilon)$ was neglected, assuming there are no free charges.

Comparing this equation with the source free equation in (7.29), we may replace the term in brackets by an equivalent term ε_c

$$\varepsilon_c = \varepsilon - \frac{j\sigma}{\omega} = \varepsilon' - j\varepsilon'' \tag{7.39}$$

This is called the complex permittivity and, in general, replaces the permittivity ε in the field equations. The imaginary part of the complex permittivity is associated with conduction losses. There are also dielectric losses but we will not discuss these separately. Dielectric losses will be viewed here as part of ε'' since, in practice, the two types of losses are indistinguishable.

Now, the wave number takes the following form:

$$k = \omega\sqrt{\mu\varepsilon}\sqrt{\left(1 - \frac{j\sigma}{\omega\varepsilon}\right)}$$

which is a complex quantity. A (complex) propagation constant is defined through the following relation

$$-\gamma^2 = \omega^2\mu\varepsilon\left(1 - \frac{j\sigma}{\omega\varepsilon}\right)$$

or

$$\gamma = j\omega\sqrt{\mu\varepsilon}\sqrt{\left(1 - \frac{j\sigma}{\omega\varepsilon}\right)} \tag{7.40}$$

This term can now be substituted for jk in Eq. (7.35), using the same assumptions we made for the lossless case. The general solution has the same two wave components: one traveling in the positive z direction, the other in the negative z direction:

$$E_x(z) = E_0^+ e^{-\gamma z} + E_0^- e^{+\gamma z}$$

Similarly, assuming only an outgoing wave, we have

$$E_x(z) = E_0^+ e^{-\gamma z}$$

Writing

$$\gamma = \alpha + j\beta$$

we have

$$E_x(z) = E_0^+ e^{-(\alpha+j\beta)z} = E_0^+ e^{-\alpha z} e^{-j\beta z} \qquad (7.41)$$

In this form, the propagating wave has the same form as Eq. (7.35), where β has replaced k and the exponential term $e^{-\alpha z}$ multiplies the amplitude. This is, therefore, a wave, propagating in the positive z direction, with a phase velocity v and with an exponentially decaying amplitude. The phase velocity is now

$$v = \frac{\omega}{\beta} \qquad (7.42)$$

and the wavelength is defined as

$$\lambda = \frac{2\pi}{\beta}$$

The magnetic field intensity can be written directly from Eq. (7.41) as

$$H_y^+(z) = \frac{1}{\eta} E_x^+(z) \qquad (7.43)$$

where the intrinsic impedance is now

$$\eta = \frac{j\omega\mu}{\gamma} \qquad (7.44)$$

and is a complex number. The imaginary part of the intrinsic impedance is due to losses. The imaginary part (reactive) is normally small and is often neglected for low loss dielectrics.

The exponential term $e^{-\alpha z}$ is an attenuation term. The wave attenuates as it propagates in space. The constant α is therefore an attenuation constant and is measured in Nepers/meter. The Neper is a dimensionless constant and defines the fraction of the attenuation the wave undergoes in one meter.

7.3.2. Propagation of Plane Waves in Low-Loss Dielectrics

A low-loss dielectric is a material for which $\sigma/\omega\varepsilon<<1$ (or, equivalently, $\varepsilon''<<\varepsilon'$). The phase constant now can be approximated by

230   7. Wave Propagation and High-Frequency Electromagnetic Fields

$$\gamma = j\omega\sqrt{\mu\varepsilon}\sqrt{\left(1 - \frac{j\sigma}{\omega\varepsilon}\right)} \approx j\omega\sqrt{\mu\varepsilon}\left(1 - \frac{j\sigma}{2\omega\varepsilon}\right)$$

using the binomial expansion. The attenuation constant is therefore

$$\alpha \approx \frac{\sigma}{2}\sqrt{\frac{\mu}{\varepsilon}} \qquad (7.45)$$

and the phase constant is

$$\beta \approx \omega\sqrt{\mu\varepsilon} \qquad (7.46)$$

The phase constant for low-loss dielectrics is essentially unchanged from that for the lossless dielectric.

The intrinsic impedance in low-loss dielectrics is complex. Substituting the value of γ in Eq. (7.44), η can be approximated as

$$\eta \approx \eta_n\left(1 + \frac{j\sigma}{2\omega\varepsilon}\right) \qquad (7.47)$$

where η_n is the no-loss intrinsic impedance for the same material (i.e., with the same μ and ε')

The phase velocity remains essentially unchanged from that in the lossless case since β changes very little $(v=\omega/\beta)$.

As an example to a low loss dielectric, consider sea water at 60 Hz. The relative permittivity is 81 and the conductivity is about 4 S/m. The complex permittivity is $(81-j0.064)\varepsilon_0$. The condition for low-loss dielectric is certainly satisfied at this frequency. However, in general, water is not considered to be a low-loss dielectric, especially because of its high absorption at microwave frequencies.

7.3.3. Propagation of Plane Waves in Conductors

In highly conductive materials the losses are high and we can assume that $\sigma>>\omega\varepsilon$ or, that $\varepsilon''>>\varepsilon$. This is equivalent to ignoring the displacement currents. Note that high conductivity does not mean perfect conductivity. Perfect conductors do not have losses due to the exclusion of fields from the materials. In the case of high conductivity materials, the complex propagation constant can be approximated from Eq. (7.40) as

$$\gamma \approx j\omega\sqrt{\mu\varepsilon}\sqrt{-\frac{j\sigma}{\omega\varepsilon}} = (1+j)\sqrt{\frac{\omega\mu\sigma}{2}} \qquad (7.48)$$

where $\sqrt{j}=(1+j)/\sqrt{2}$. From this, we get

$$\alpha = \beta = \sqrt{\frac{\omega\mu\sigma}{2}} = \frac{1}{\delta} \qquad (7.49)$$

The attenuation and phase constants are equal and are very large. The wave is attenuated rapidly to the point where propagation in conducting media can only exist within short distances. The propagating wave can now be written as

$$E_x(z) = E_0^+ e^{-z/\delta} e^{-jz/\delta}$$

where the term

$$\delta = \sqrt{\frac{2}{\omega\mu\sigma}} \qquad (7.50)$$

is known as the skin depth or depth of penetration. It is defined as that distance through which the amplitude of a plane wave is attenuated to $1/e$ of its original amplitude. The skin depth in conductors is small. In the microwave range it can be of the order of a few microns (depending on material and frequency).

The phase velocity in good conductors is [from Eq. (7.42)]

$$v = \frac{\omega}{\beta} = \omega\delta = \sqrt{\frac{2\omega}{\mu\sigma}} \qquad (7.51)$$

and is obviously small compared to the phase velocity in dielectrics or free space.

The intrinsic impedance is [using Eq. (7.39)]

$$\eta = \sqrt{\frac{\mu}{\varepsilon_c}} \approx \sqrt{\frac{j\omega\mu}{\sigma}} = (1+j)\frac{1}{\sigma\delta} \qquad (7.52)$$

The phase angle of the intrinsic impedance is, therefore, 45°. This is characteristic of good conductors for which the magnetic field intensity lags behind the electric field intensity by 45°. The wavelength also changes drastically compared to free space or lossless dielectrics. It is very short and given by

$$\lambda = \frac{2\pi}{\beta} = 2\pi\delta \qquad (7.53)$$

232   7. Wave Propagation and High-Frequency Electromagnetic Fields

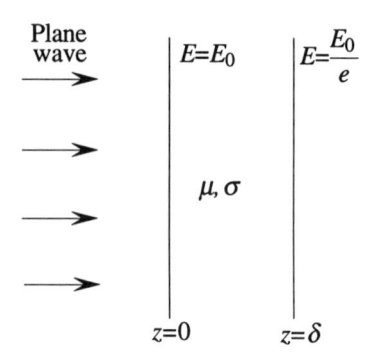

Figure 7.7. Propagation of plane electromagnetic waves in a good conductor and definition of skin depth.

7.3.4. Propagation in a Conductor: Definition of the Skin Depth.

We return now to the skin depth. This term is very useful at low-frequency applications, since it defines the range in which currents or fields can be assumed to exist. In high-frequency applications, it defines the depth in which we can have effective propagation of waves. Materials with small skin depths cannot support a propagating wave because of the large attenuation. These can be used for shielding purposes or for absorption of microwaves.

Another important point with regard to skin depth, one that is often overlooked, is that skin depth is only properly defined for plane waves and only for sinusoidal excitations. Furthermore, the formula used for skin depth only applies to good conductors with planar surfaces.

Consider the situation in Figure 7.7, where a plane wave propagates in a good conductor. As shown above, the attenuation constant for a good conductor is

$$\alpha = \sqrt{\pi f \mu \sigma}$$

If the wave propagates a distance δ, the amplitude of the electric field intensity (or magnetic field intensity) at this point is

$$E = E_0 e^{-\sqrt{\pi f \sigma \mu}\, \delta}$$

That distance within the conductor at which the amplitude has been reduced to $1/e$ of its original amplitude E_0 is defined as the skin depth, or the depth of penetration in the conductor. The condition for this is

$$\sqrt{\pi f \sigma \mu}\, \delta = 1$$

or, in a more familiar form, the depth of penetration is written as

$$\delta = \frac{1}{\sqrt{\pi f \sigma \mu}}$$

We note, that if the material in which the wave propagates is a lossy dielectric, the attenuation constant is given by Eq. (7.45)

$$\alpha = \frac{\sigma}{2}\sqrt{\frac{\mu}{\varepsilon}}$$

If we were to attempt the same definition here, the required condition is

$$\frac{\sigma}{2}\sqrt{\frac{\mu}{\varepsilon}}\delta = 1$$

and the depth of penetration is

$$\delta = \frac{2}{\sigma\sqrt{\frac{\mu}{\varepsilon}}} = \frac{2}{\sigma}\sqrt{\frac{\varepsilon}{\mu}} \qquad (7.54)$$

which, while a valid definition, is different than what we have defined as the standard depth of penetration. In particular, this is independent of frequency.

As simple numerical examples, the skin depth in copper at 1 gHz (using a conductivity of 5.81×10^7 S/m) is 2.088 μm. In sea water, which is a high loss dielectric at this frequency, the conductivity is 4 S/m, and using a relative permittivity of 81, we get a skin depth of 1.33 mm using Eq. (7.54). Calculation of the skin depth using the standard formula gives a skin depth of 7.96 mm, which is quite different.

7.4. Polarization of Plane Waves

The electric (or magnetic) field intensity of a uniform plane wave has a direction in space. This direction may either be constant or may change as the wave propagates. Polarization of a plane wave refers to the direction of the electric field intensity vector in space. Polarization is defined for the electric field vector only since the magnetic field is always obtainable from the electric field.

The electric field

$$\mathbf{E} = \hat{\mathbf{j}} E_y(z)$$

234   7. Wave Propagation and High-Frequency Electromagnetic Fields

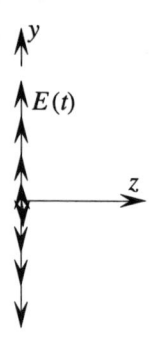

Figure 7.8. Time variation of a linearly polarized wave in the y direction.

of a wave propagating in the z direction and directed in the y direction is "linearly polarized" in the y direction. The variation of the electric field with time is shown in Figure 7.8.

An important type of polarized waves is obtained if one wave is polarized in a given direction, say the x direction, and the other at 90° to the first, say in the y direction but also lagging in phase by 90°. The electric field of the superposed wave can be written as

$$\mathbf{E} = \hat{\mathbf{i}} E_x(z) + \hat{\mathbf{j}} E_y(z) = \hat{\mathbf{i}} E_1 e^{-jkz} - \hat{\mathbf{j}} j E_2 e^{-jkz} \qquad (7.55)$$

where the amplitudes E_1 and E_2 are real and E_2 lags behind E_1 by 90°. Taking the real part of Eq. (7.55), we obtain the time domain representation as

$$\mathbf{E}(z,t) = \hat{\mathbf{i}} E_1 cos(\omega t - kz) + \hat{\mathbf{j}} E_2 cos(\omega t - kz \pm \frac{\pi}{2})$$

To examine the change of **E** at a point in space, say $z=0$, we write

$$\mathbf{E}(0,t) = \hat{\mathbf{i}} E_1 cos(\omega t) + \hat{\mathbf{j}} E_2 cos(\omega t - \frac{\pi}{2}) = \hat{\mathbf{i}} E_1 cos(\omega t) + \hat{\mathbf{j}} E_2 sin(\omega t)$$

The vector $\mathbf{E}(0,t)$ now describes an elliptical path as t changes. The elliptical path is due to the different amplitudes of the component E_1 and E_2. The rotation of the vector is in the counterclockwise direction. The wave in Eq. (7.55) is said to be elliptically polarized (see Figure 7.9a). Circular polarization is obtained if $E_1=E_2$ as in Figure 7.9b. If E_1 lags behind E_2, the rotation of the field is clockwise.

If the two fields are in phase, we obtain a simple superposition of fields and the resulting field is linearly polarized in the direction $\tan^{-1}(E_2/E_1)$. As an example, a linearly polarized wave at 45° is obtained if $E_1=E_2$.

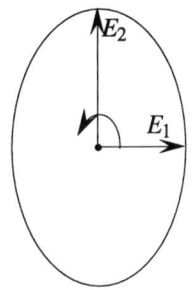

Figure 7.9a. An elliptically polarized wave in which E_2 lags behind E_1.

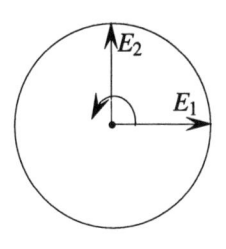

Figure 7.9b. A circularly polarized wave in which E_2 lags behind E_1.

7.5. Reflection, Refraction, and Transmission of Plane Waves

We discussed in Chapter 4 the refraction of the magnetic field at the boundaries between materials. While there our interest was in the direction of the fields at boundaries, with electromagnetic waves, we will discuss not only the direction of the fields but also their propagation properties and changes in their amplitudes as they propagate through the interfaces between materials. This aspect of propagation of waves is quite complicated and many of the properties of waves are defined by materials and their interfaces. As an example we note the reflection of waves from conducting surfaces giving rise to standing waves and the propagation through lossy (dispersive) materials which may change the nature of waves from uniform to nonuniform plane waves. By analogy to light waves, we expect similar behavior, including reflection, transmission, and refraction of waves at the interface. These are discussed next for plane waves in a number of important conditions. The various properties depend on the materials involved, on the direction of propagation, and on the polarization of the waves. While it is not possible to present all possible combinations, the principles involved are simple and will be demonstrated for a number of important interfaces.

The basic idea is to write the waves on both sides of the interface and to match the fields (electric and magnetic) on the boundary. In general, this means applying the continuity conditions of the fields at the surface. From these, the fields on both sides of the interface are deduced. We start first with normal incidence on a lossless dielectric, followed by normal incidence on a conductor. Next we look at two examples of oblique incidence at a dielectric interface and two examples of oblique incidence on conducting surfaces.

236   7. Wave Propagation and High-Frequency Electromagnetic Fields

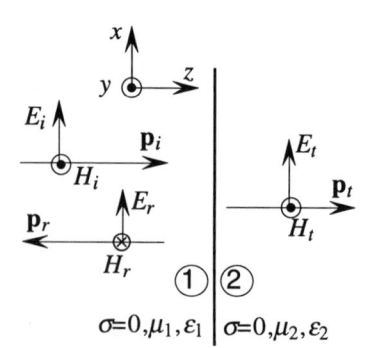

Figure 7.10. Incident, transmitted, and reflected waves at the interface between two general, lossless dielectrics.

7.5.1. Reflection and Transmission at a Lossy Dielectric Interface: Normal Incidence

Figure 7.10 shows an incident wave E_i propagating in a medium with permittivity ε_1 and permeability μ_1. The wave encounters the interface between material (1) and material (2) which is a general lossless dielectric with permittivity ε_2 and permeability μ_2. At this point, we simply assume that part of the wave is reflected and part of it is transmitted at the interface. The magnitudes of the various parts of the wave are determined next for normal incidence. We assume an incident wave propagating in the positive z direction, with amplitude E_{i1} as in Figure 7.10. The direction is denoted by the direction of the Poynting vector.

$$\mathbf{E}_i = \hat{\mathbf{i}} E_{i1} e^{-j\beta_1 z}$$

The reflected wave at the interface propagates in the negative z direction. That is,

$$\mathbf{E}_r = \hat{\mathbf{i}} E_{r1} e^{+j\beta_1 z}$$

The amplitude E_{r1} is, in general, different than E_{i1} and may be zero (no reflection at the interface). By defining a reflection coefficient, denoted by Γ, the reflected wave can be written in terms of the incident wave as

$$\mathbf{E}_{r1} = \hat{\mathbf{i}} \Gamma E_{i1} e^{+j\beta_1 z}$$

where $E_{r1} = \Gamma E_{i1}$, at point $z=0$.

Similarly, the transmitted wave can be written as

$$\mathbf{E}_{t2} = \hat{\mathbf{i}} E_{t1} e^{-j\beta_2 z}$$

where the direction of propagation is the same as for the incident wave. By defining a transmission coefficient T, we can write

$$\mathbf{E}_{t2} = \hat{\mathbf{i}} T E_{i1} e^{-j\beta_2 z}$$

where $E_{t2} = T E_{i1}$.

Since the magnetic and the electric field intensities are related through the intrinsic impedance, we can write the following for the magnetic field intensity

$$\mathbf{H}_{i1} = \hat{\mathbf{j}} \frac{E_{i1}}{\eta_1} e^{-j\beta_1 z}$$

$$\mathbf{H}_{r1} = -\hat{\mathbf{j}} \Gamma \frac{E_{i1}}{\eta_1} e^{+j\beta_1 z}$$

The negative sign indicates that the power flow, as defined by the Poynting vector ($\mathbf{P} = \mathbf{E} \times \mathbf{H}$), changes direction and now is directed towards the source. The power flow of the transmitted wave is in the same direction as the incident wave. The transmitted wave is

$$\mathbf{H}_{t2} = \hat{\mathbf{j}} T \frac{E_{i1}}{\eta_2} e^{-j\beta_2 z}$$

The transmission and reflection coefficients can now be evaluated from the relations at the boundary:

$$\mathbf{E}_i + \mathbf{E}_r = \mathbf{E}_t \quad \text{and} \quad \mathbf{H}_i + \mathbf{H}_r = \mathbf{H}_t$$

Substitution of the expressions for the fields at the interface ($z=0$) gives the following two equations:

$$1 + \Gamma = T$$

$$\frac{1}{\eta_1} - \frac{\Gamma}{\eta_1} = \frac{T}{\eta_2}$$

or

$$\Gamma = \frac{\eta_2 - \eta_1}{\eta_1 + \eta_2} \qquad (7.56)$$

and

$$T = \frac{2\eta_2}{\eta_1 + \eta_2} \tag{7.57}$$

We note here that since η_1 and η_2 are in general complex, Γ and T are also complex. Γ can be negative (depending on the relative values of the intrinsic impedances), but T is always positive. Complex transmission and reflection coefficients indicate a phase shift in the reflected and transmitted waves at the interface. Both coefficients are dimensionless.

The electric and magnetic field intensities in material (1) can be calculated as the sum of the incident and reflected waves

$$\mathbf{E}_1(z) = \hat{\mathbf{i}} E_{i1} \left(e^{-j\beta_1 z} + \Gamma e^{+j\beta_1 z} \right) = \hat{\mathbf{i}} E_{i1} \left(T e^{-j\beta_1 z} + \Gamma(j 2\sin\beta_1 z) \right) \tag{7.58}$$

where the relation $1+\Gamma=T$ was used. This equation indicates that the sum of the incident and reflected waves consists of a wave component propagating in the positive z direction and a nonpropagating wave component. The later [second term in Eq. (7.58)] is responsible for the standing waves we often encounter. We also observe that if the transmission coefficient T is zero, only a standing wave exists, while if the reflection coefficient Γ is zero, there are no standing waves. Standing waves will be discussed separately in conjunction with reflection from conducting surfaces but we point out here that standing waves may exist any time there is reflection of a wave because of the interference of the forward and backward propagating waves.

The discussion above is general and assumes two general materials. In many cases, material (1) is taken as free space. The only difference is that η_1 becomes η_0 while in the expressions for E_i and E_r, β_1 is replaced by k_0.

If material (1) is free space and material (2) is a lossless dielectric, the transmission and reflection coefficients can be written in terms of the relative permeability and relative permittivity of material (2) alone

$$\Gamma = \frac{\eta_2 - \eta_0}{\eta_0 + \eta_2} = \frac{\sqrt{\frac{\mu_{r2}}{\varepsilon_{r2}}} - 1}{\sqrt{\frac{\mu_{r2}}{\varepsilon_{r2}}} + 1}$$

$$T = \frac{2\sqrt{\frac{\mu_{r2}}{\varepsilon_{r2}}}}{\sqrt{\frac{\mu_{r2}}{\varepsilon_{r2}}} + 1}$$

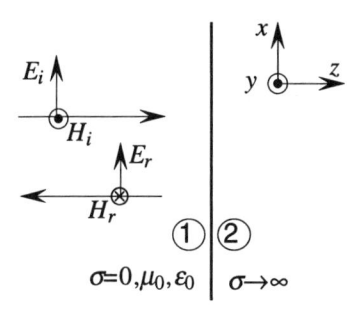

Figure 7.11. Incident and reflected waves at a conducting interface.

The electric field in free space is now

$$E_1(z) = \hat{\imath}E_{i1}\left(Te^{-jk_0z} + \Gamma(j2\sin k_0z)\right) \quad (7.59)$$

We note from these expressions that if we are interested in reducing the reflection at the interface, the ratio of μ_r and ε_r must tend to 1. In other words, the closer the materials are in their ratio, the lower the reflection. If the ratio $\mu_{r2}/\varepsilon_{r2}$ is equal to that of free space (i.e., μ_0/ε_0), there is no reflection at the interface and the whole wave is transmitted across the boundary. We can consider this to be the case of perfect impedance matching, since both materials have the same intrinsic impedance. While this situation is certainly not common in dielectrics, some ferrites can be made to closely resemble this condition by adjusting their permeability. This is a case of total transmission and is characterized by $\Gamma=0$ and $T=1$.

Similarly, if we are interested in large reflections, the permittivity must be as large as possible. In the limit, with permittivity tending to infinity, the reflection coefficient tends to 1 and the transmission coefficient tends to zero. Metals behave close to this condition and in most cases we consider a conductor's interface to be totally reflecting.

7.5.2. *Reflection and Transmission at a Conductor Interface: Normal Incidence*

We consider next the propagation of a wave from free space (or any lossless dielectric), impinging on a perfect conductor as in Figure 7.11.

From the discussion in the previous section, it is obvious that we can write directly

$$\Gamma = -1 \quad \text{and} \quad T = 0 \quad (7.60)$$

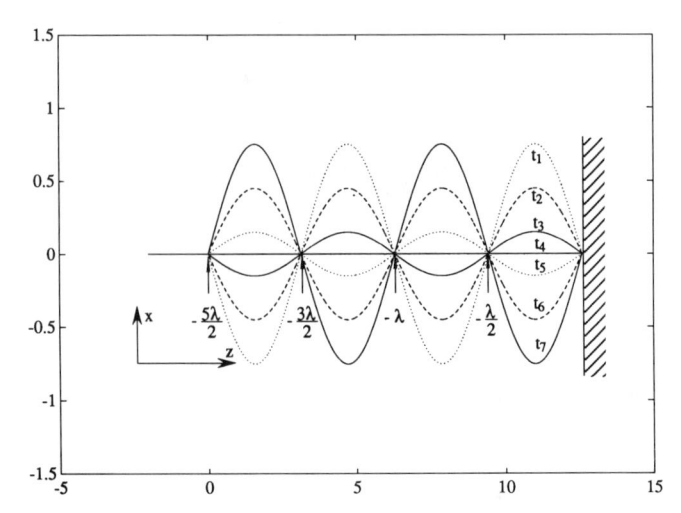

Figure 7.12. Standing wave due to reflection at a conducting interface. In this case, total reflection without attenuation is shown.

From Eq. (7.59), the electric and magnetic fields are:

$$\mathbf{E}_1(z) = -\hat{\mathbf{i}} j 2 E_{i1} \sin k_0 z$$

$$\mathbf{H}_1(z) = \hat{\mathbf{j}} 2 \frac{E_{i1}}{\eta_0} \cos k_0 z$$

The wave formed by these fields is unique in that it does not propagate. We call this a standing wave. The principle of a standing wave is shown in Figure 7.12, where the amplitude of the wave changes with time but the location of the peaks and zeroes (nodes) is constant in space. To show that this is the case, it is most convenient to calculate the Poynting vector

$$\mathbf{P} = \mathbf{E}_1(z) \times \mathbf{H}_1^*(z) = \hat{\mathbf{k}} j \frac{4}{\eta_0} |E_{i1}|^2 \sin k_0 z \, \cos k_0 z$$

where * denotes the complex conjugate vector. Since **P** is purely imaginary, no power is transferred into the conductor.

7.5.3. *Reflection and Transmission at a Finite Conductivity Conductor Interface.*

For a good but not perfect conductor we can use the relation for the intrinsic impedance of the conductor [Eq. (7.52)] to obtain the transmission and reflection coefficients

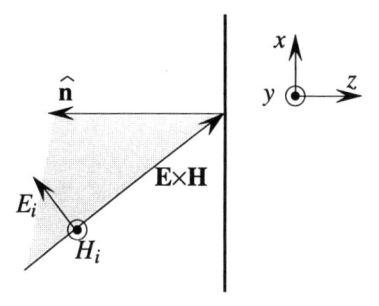

Figure 7.13. Definition of the plane of incidence.

$$\Gamma = \frac{\eta_2 - \eta_0}{\eta_0 + \eta_2} = \frac{(1+j)\dfrac{1}{\sigma\delta} - \eta_0}{(1+j)\dfrac{1}{\sigma\delta} + \eta_0} \tag{7.61}$$

$$T = \frac{2(1+j)\dfrac{1}{\sigma\delta}}{\eta_0 + (1+j)\dfrac{1}{\sigma\delta}} \tag{7.62}$$

These expression are both complex, and according to Eq. (7.52), are only approximations. The larger σ is, the closer Γ is to -1 and T to zero, as one would expect for a perfect conductor.

7.5.4. *Reflection and Transmission at an Interface: Oblique Incidence*

A plane wave, obliquely incident on an interface between two materials, undergoes changes similar to those for normal incidence. Part of the wave is transmitted and part of it reflected. In some cases, there is either only transmission or only reflection. In oblique incidence, the continuity of the field components is taken into account as a boundary condition at the interface. The behavior of the wave at the interface depends on the polarization of the wave. A plane of incidence is defined by two vectors: the direction of propagation of the incident wave (i.e., the Poynting vector), and the normal to the surface of the interface, as shown in Figure 7.13. The direction of the electric or magnetic field intensities for uniform plane waves is always normal to the direction of propagation. The electric field can be either in the plane of incidence, perpendicular to it, or at any angle to this plane. If the electric field is parallel to the plane of incidence, we refer to this as parallel polarization (sometimes called *H* polarization). Perpendicular polarization refers to the case when the electric field is perpendicular to the plane of incidence, (sometimes called *E* polarization). The general case of arbitrary polarization can always be treated by separating the electric field into its normal and parallel components as a combination of parallel and perpendicular polarization.

We treat the problem of oblique incidence by looking at propagation of waves obliquely incident at an interface for parallel, and perpendicular incidence separately for dielectrics and conductors. The case of oblique incidence for perpendicular polarization is shown explicitly, while the results for parallel polarization are given as simple modifications of the perpendicular polarization results, without the details. The case of incidence on conducting interfaces is given first, followed by incidence on lossless dielectrics. The more general case of incidence on lossy dielectrics is not treated here because it normally results in nonuniform plane waves. However the principles used here are directly applicable.

7.5.5. Oblique Incidence on a Conducting Interface: Perpendicular Polarization

The electric field is perpendicular to the plane of incidence as shown in Figure 7.14. We assume incidence from a dielectric with permittivity ε_1 on a perfect conductor plane surface.

The direction of propagation is now defined as $\hat{\mathbf{p}}_i = \hat{\mathbf{i}} sin\theta_i + \hat{\mathbf{k}} cos\theta_i$ for the incident wave and $\hat{\mathbf{p}}_r = \hat{\mathbf{i}} sin\theta_r - \hat{\mathbf{k}} cos\theta_r$ for the reflected wave. The variable z in the exponent of the wave is now replaced by $xsin\theta_i + zcos\theta_i$ for the incident wave and by $xsin\theta_r - zcos\theta_r$ for the reflected wave. The electric field propagates in the (x,z) plane and is directed perpendicular to the plane of incidence, in the y direction. Using the directions of propagation, the incident electric and magnetic fields are

$$\mathbf{E}_i(x,z) = \hat{\mathbf{j}} E_{i1} e^{-j\beta_1(xsin\theta_i + zcos\theta_i)} \tag{7.63}$$

$$\mathbf{H}_i(x,z) = \frac{E_{i1}}{\eta_1} (-\hat{\mathbf{i}} cos\theta_i + \hat{\mathbf{k}} sin\theta_i) e^{-j\beta_1(xsin\theta_i + zcos\theta_i)} \tag{7.64}$$

where the material index number was dropped as the waves only exist in material (1). The reflected fields are

$$\mathbf{E}_r(x,z) = \hat{\mathbf{j}} E_{r1} e^{-j\beta_1(xsin\theta_r - zcos\theta_r)}$$

$$\mathbf{H}_r(x,z) = \frac{E_{r1}}{\eta_1} (\hat{\mathbf{i}} cos\theta_r + \hat{\mathbf{k}} sin\theta_r) e^{-j\beta_1(xsin\theta_r - zcos\theta_r)}$$

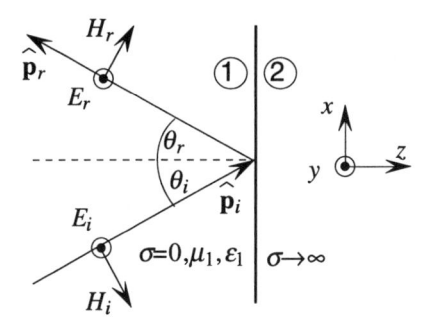

Reflection, Refraction, and Transmission of Plane Waves   243

Figure 7.14. Incident and reflected waves for oblique incidence at a dielectric-conductor interface. The electric field is polarized perpendicular to the plane of incidence.

In a perfect conductor there is no transmitted wave, therefore, at the interface ($z=0$), the total electric field intensity must be zero:

$$\mathbf{E}_i(x,0) + \mathbf{E}_r(x,0) = \hat{\mathbf{j}}\left[E_{i1}e^{-j\beta_1(x\sin\theta_r)} + E_{r1}e^{-j\beta_1(x\sin\theta_r)}\right] = 0$$

For this to be satisfied, the following must hold

$$E_{r1} = -E_{i1} \quad \text{and} \quad \theta_r = \theta_i$$

The amplitudes of the reflected and incident waves are the same and the angle of incidence and reflection are the same (Snell's law). The reflected electric and magnetic fields can be written in terms of the incident field as

$$\mathbf{E}_r(x,z) = -\hat{\mathbf{j}} E_{i1} e^{-j\beta_1(x\sin\theta_i - z\cos\theta_i)}$$

$$\mathbf{H}_r(x,z) = -\frac{E_{i1}}{\eta_1}(\hat{\mathbf{i}}\cos\theta_i + \hat{\mathbf{k}}\sin\theta_i)e^{-j\beta_1(x\sin\theta_i - z\cos\theta_i)}$$

It is now obvious that the reflection coefficient in this case is equal to -1, since the reflection coefficient is the ratio of the amplitudes of $H_r(x,z)$ and $H_i(x,z)$, and these are equal [see Eq. (7.64)].

The total field is the sum of the incident and reflected waves. The total electric and magnetic field intensities (after rearranging terms) are:

$$\mathbf{E}_1(x,z) = \hat{\mathbf{j}} E_{i1}\left[e^{-j\beta_1(z\cos\theta_i)} - e^{j\beta_1(z\cos\theta_i)}\right]e^{-j\beta_1(x\sin\theta_i)} = -\hat{\mathbf{j}} j 2E_{i1}\sin(\beta_1 z\cos\theta_i)e^{-j\beta_1(x\sin\theta_i)}$$

$$\mathbf{H}_1(x,z) = -2\frac{E_{i1}}{\eta_1}\left[\hat{\mathbf{i}}\cos\theta_i \cos(\beta_1 z\cos\theta_i) + \hat{\mathbf{k}} j\sin\theta_i \sin(\beta_1 z\cos\theta_i)\right]e^{-j\beta_1 x\sin\theta_i}$$

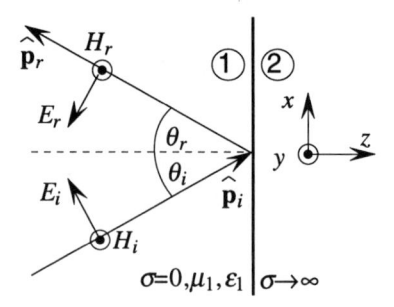

244   7. Wave Propagation and High-Frequency Electromagnetic Fields

Figure 7.15. Incident and reflected waves for oblique incidence at a dielectric-conductor interface. The electric field is polarized parallel to the plane of incidence.

where use of the following identities was made

$$j2\sin(\beta_1 z\cos\theta_i) = e^{j\beta_1 z\cos\theta_i} - e^{-j\beta_1 z\cos\theta_i}$$

$$2\cos(\beta_1 z\cos\theta_i) = e^{j\beta_1 z\cos\theta_i} + e^{-j\beta_1 z\cos\theta_i}$$

Note that the wave in the z direction is a standing wave, but the wave in the x direction is propagating.

7.5.6. Oblique Incidence on a Conducting Interface: Parallel Polarization

The discussion of the previous section applies here as well, but the components of the fields are different. The electric field now lies in the incidence plane and therefore will have x and z components while the magnetic field, is perpendicular to the plane of incidence and therefore is in the y direction (see Figure 7.15). The incident electric and magnetic fields are

$$\mathbf{E}_i(x,z) = E_{i1}(\hat{\mathbf{i}}\cos\theta_i - \hat{\mathbf{k}}\sin\theta_i)e^{-j\beta_1(x\sin\theta_i + z\cos\theta_i)}$$

$$\mathbf{H}_i(x,z) = \hat{\mathbf{j}}\frac{E_{i1}}{\eta_i}e^{-j\beta_1(x\sin\theta_i + z\cos\theta_i)}$$

Following similar considerations, we obtain for the electric and magnetic fields

$$\mathbf{E}_1(x,z) = -2E_{i1}\left[\hat{\mathbf{i}}j\cos\theta_i \sin(\beta_1 z\cos\theta_i) + \hat{\mathbf{k}}\sin\theta_i \cos(\beta_1 z\cos\theta_i)\right]e^{-j\beta_1 x\sin\theta_i}$$

$$\mathbf{H}_1(x,z) = \hat{\mathbf{j}}2\frac{E_{i1}}{\eta_1}\cos(\beta_1 z\cos\theta_i)e^{-j\beta_1 x\sin\theta_i}$$

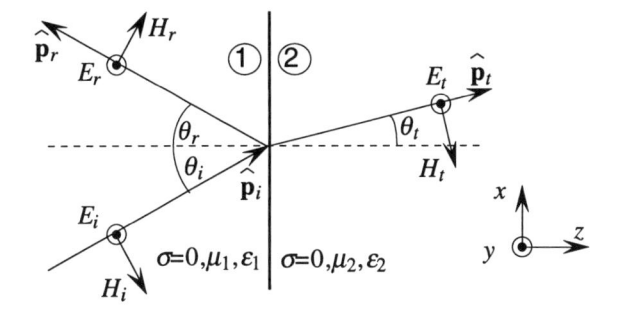

Reflection, Refraction, and Transmission of Plane Waves 245

Figure 7.16. Incident, reflected, and refracted waves for oblique incidence at a dielectric-dielectric interface. The electric field is polarized perpendicular to the plane of incidence.

As with perpendicular polarization, the wave propagating in the z direction consists of E_{1x} and H_{1y}, and these are out of phase. Therefore, we have a standing wave, oscillating exactly as for the perpendicular polarization. The wave in the x direction is a propagating wave as for perpendicular polarization.

7.5.7. Oblique Incidence on a Dielectric Interface: Perpendicular Polarization

The case considered here is shown in Figure 7.16. The two dielectrics are lossless and either one can be free space. As with normal incidence on dielectric boundaries, part of the wave will be reflected and part will be transmitted as shown in Figure 7.16. The transmitted wave is refracted as it passes into the second dielectric, depending on the relative values of the dielectric constants,

Before continuing, we recall two well-known relations from optics relating the reflection and refraction angles to the incident angles. These are

$$\theta_i = \theta_r \qquad (7.65)$$

and

$$\frac{\sin\theta_i}{\sin\theta_t} = \frac{n_2}{n_1} \qquad (7.66)$$

The first indicates that the reflection angle and the incident angle are the same (Snell's law). The second relates the refraction angle to the indices of refraction of the two materials, where the index of refraction of a material is

$$n = \sqrt{\mu_r \varepsilon_r} \qquad (7.67)$$

In the particular case in which material (1) is free space, $n_1 = 1$, and we have

$$\frac{\sin\theta_i}{\sin\theta_t} = n_2 = \sqrt{\mu_{r2}\varepsilon_{r2}}$$

The electric and magnetic fields are the same as Eqs. (7.63) and (7.64)

$$\mathbf{E}_i(x,z) = \hat{\mathbf{j}} E_{i1} e^{-j\beta_1(x\sin\theta_i + z\cos\theta_i)} \tag{7.68}$$

$$\mathbf{H}_i(x,z) = \frac{E_{i1}}{\eta_1}(-\hat{\mathbf{i}}\cos\theta_i + \hat{\mathbf{k}}\sin\theta_i)e^{-j\beta_1(x\sin\theta_i + z\cos\theta_i)} \tag{7.69}$$

where, again, the direction of propagation of the incident wave is given by $\hat{\mathbf{p}}_i = \hat{\mathbf{i}}\sin\theta_i + \hat{\mathbf{k}}\cos\theta_i$.

Analogous to the results we obtained for oblique incidence on a conducting boundary, we write the reflected fields in terms of the reflection coefficient and the incident waves as

$$\mathbf{E}_r(x,z) = \hat{\mathbf{j}} \Gamma E_{i1} e^{-j\beta_1(x\sin\theta_i - z\cos\theta_i)} \tag{7.70}$$

$$\mathbf{H}_r(x,z) = \frac{E_{i1}\Gamma}{\eta_1}(\hat{\mathbf{i}}\cos\theta_i + \hat{\mathbf{k}}\sin\theta_i)e^{-j\beta_1(x\sin\theta_i - z\cos\theta_i)} \tag{7.71}$$

where Γ remains to be determined.

Similarly, we can write the transmitted wave directly from the incident wave by means of the transmission coefficient T as

$$\mathbf{E}_t(x,z) = \hat{\mathbf{j}} T E_{i1} e^{-j\beta_2(x\sin\theta_t + z\cos\theta_t)} \tag{7.72}$$

$$\mathbf{H}_t(x,z) = \frac{E_{i1}T}{\eta_2}(-\hat{\mathbf{i}}\cos\theta_t + \hat{\mathbf{k}}\sin\theta_t)e^{-j\beta_2(x\sin\theta_t + z\cos\theta_t)} \tag{7.73}$$

To determine the transmission and reflection coefficients, we equate the tangential components of the electric field intensity and those of the magnetic field intensity on both sides of the interface (i.e., at $z=0$). From Figure 7.16, the following two relations express the continuity of the magnetic and electric fields

$$E_i + E_r = E_t \quad \text{and} \quad H_i\cos\theta_i - H_r\cos\theta_i = H_t\cos\theta_t$$

where the relation $\theta_i = \theta_r$ was used. From the relations we have

$$H_i = \frac{E_i}{\eta_1}, \quad H_r = \frac{E_r}{\eta_1}, \quad \text{and} \quad H_t = \frac{E_t}{\eta_2}$$

$$E_i + E_r = E_t \quad \text{and} \quad \frac{E_i}{\eta_1}\cos\theta_i - \frac{E_r}{\eta_1}\cos\theta_i = \frac{E_t}{\eta_2}\cos\theta_t$$

Solution of these two equations for E_r and E_t gives

$$E_r = E_i \frac{\eta_2\cos\theta_i - \eta_1\cos\theta_t}{\eta_2\cos\theta_i + \eta_1\cos\theta_t}$$

and

$$E_t = E_i \frac{2\eta_2\cos\theta_i}{\eta_2\cos\theta_i + \eta_1\cos\theta_t}$$

From which the reflection and transmission coefficients are

$$\Gamma = \frac{\eta_2\cos\theta_i - \eta_1\cos\theta_t}{\eta_2\cos\theta_i + \eta_1\cos\theta_t} \qquad (7.74)$$

$$T = \frac{2\eta_2\cos\theta_i}{\eta_2\cos\theta_i + \eta_1\cos\theta_t} \qquad (7.75)$$

Now, the total fields in each material can be written directly. In medium (2), where we only have a transmitted wave, Eqs. (7.72) and (7.73) describe the wave. In material (1), the fields are the sum of the transmitted and reflected waves [from Eqs. (7.70) and (7.68) for E_1 and from Eqs. (7.71) and (7.69) for H_1]:

$$\mathbf{E}_1(x,z) = \hat{\mathbf{j}} E_{i1} \left[e^{-j\beta_1(x\sin\theta_i + z\cos\theta_i)} + \Gamma e^{-j\beta_1(x\sin\theta_i - z\cos\theta_i)} \right]$$

$$\mathbf{H}_1(x,z) = -\hat{\mathbf{i}}\frac{E_{i1}}{\eta_1}(\cos\theta_i + \Gamma\cos\theta_i)e^{-j\beta_1(x\sin\theta_i + z\cos\theta_i)} +$$
$$\hat{\mathbf{k}}(\sin\theta_i + \Gamma\sin\theta_i)e^{-j\beta_1(x\sin\theta_i - z\cos\theta_i)}$$

E is in the y direction but H has a component in the x and z directions. As was the case for conducting interfaces, we have a propagating wave in the x direction and a standing wave in the z direction.

248   7. Wave Propagation and High-Frequency Electromagnetic Fields

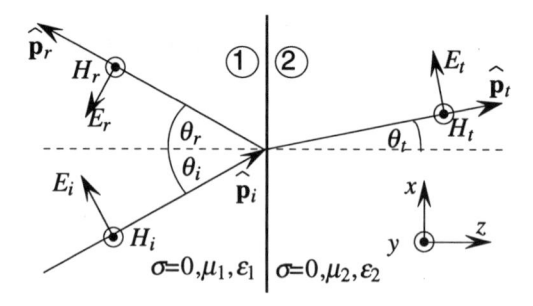

Figure 7.17. Incident, reflected, and refracted waves for oblique incidence at a dielectric-dielectric interface. The electric field is polarized parallel to the plane of incidence.

7.5.8. Oblique Incidence on a Dielectric Interface: Parallel Polarization

The situation considered here is shown in Figure 7.17. The electric field is parallel to the plane of incidence. The electric and magnetic field intensities are as given in section 7.5.6.

$$\mathbf{E}_i(x,z) = E_{i1}(\hat{\mathbf{i}}\cos\theta_i - \hat{\mathbf{k}}\sin\theta_i)e^{-j\beta_1(x\sin\theta_i + z\cos\theta_i)}$$

$$\mathbf{H}_i(x,z) = \hat{\mathbf{j}}\frac{E_{i1}}{\eta_1}e^{-j\beta_1(x\sin\theta_i + z\cos\theta_i)}$$

Following essentially the same procedures, we have for the reflection and transmission coefficients

$$\Gamma = \frac{\eta_2\cos\theta_t - \eta_1\cos\theta_i}{\eta_2\cos\theta_t + \eta_1\cos\theta_i} \tag{7.76}$$

$$T = \frac{2\eta_2\cos\theta_i}{\eta_2\cos\theta_t + \eta_1\cos\theta_i} \tag{7.77}$$

The total fields in medium (1) are calculated as in the previous section:

$$\mathbf{E}_1(x,z) = -\hat{\mathbf{i}}E_{i1}(\cos\theta_i + \Gamma\cos\theta_i)e^{-j\beta_1(x\sin\theta_i + z\cos\theta_i)} + \hat{\mathbf{k}}(\sin\theta_i + \Gamma\sin\theta_i)e^{-j\beta_1(x\sin\theta_i + z\cos\theta_i)}$$

$$\mathbf{H}_1(x,z) = \hat{\mathbf{j}}\frac{E_{i1}}{\eta_1}\left[e^{-j\beta_1(x\sin\theta_i + z\cos\theta_i)} + \Gamma e^{-j\beta_1(x\sin\theta_i + z\cos\theta_i)}\right]$$

7.6. Waveguides

The propagation of uniform plane waves as discussed above was characterized by the *TEM* nature of the wave. The electric and magnetic field intensities were perpendicular to each other and to the direction of propagation. In unbounded space, neither of the fields had a component in the direction of propagation. We will discuss now the propagation of waves in guiding structures. The purpose of these structures is to guide waves, with particular characteristics, along the structure. However, we will only discuss single-conductor waveguides, primarily as used with microwaves. Other types of guiding structures such as two-wire transmission lines or coaxial cables will not be discussed here, mainly because of the method of analysis normally employed (transmission line methods) but also because they are normally used at lower frequencies. Cavity resonators are discussed as a direct extension to waveguides.

7.6.1. TEM, TE, and, TM Waves

The uniform plane wave we used in section 7.2. was a transverse electromagnetic wave. This implied that neither the electric or the magnetic field intensity had a component in the direction of propagation. A wave with an electric field perpendicular to the direction of propagation, but with a component of the magnetic field in the direction of propagation, is called a transverse electric (*TE*) wave. Similarly, if the magnetic field is perpendicular to the direction of propagation, and the electric field has a component in the direction of propagation, this is a transverse magnetic (*TM*) wave.

Before proceeding with description of waves in waveguides, we need to define the conditions under which the three types of waves can exist. Then we use these conditions to define the possible fields in rectangular waveguides.

We start with the two curl equations and expand each into three scalar equations by equating the vector components on both sides:

$$\nabla \times \mathbf{H} = j\omega\varepsilon\mathbf{E} \quad (7.78a) \qquad \nabla \times \mathbf{E} = -j\omega\mu\mathbf{H} \quad (7.79a)$$

$$\frac{\partial H_z}{\partial y} - \frac{\partial H_y}{\partial z} = j\omega\varepsilon E_x \quad (7.78b) \qquad \frac{\partial E_z}{\partial y} - \frac{\partial E_y}{\partial z} = -j\omega\mu H_x \quad (7.79b)$$

$$\frac{\partial H_x}{\partial z} - \frac{\partial H_z}{\partial x} = j\omega\varepsilon E_y \quad (7.78c) \qquad \frac{\partial E_x}{\partial z} - \frac{\partial E_z}{\partial x} = -j\omega\mu H_y \quad (7.79c)$$

$$\frac{\partial H_y}{\partial x} - \frac{\partial H_x}{\partial y} = j\omega\varepsilon E_z \quad (7.78d) \qquad \frac{\partial E_y}{\partial x} - \frac{\partial E_x}{\partial y} = -j\omega\mu H_z \quad (7.79d)$$

Since we assume propagation in the *z* direction, with a propagation constant $\gamma = \alpha + j\beta$, the fields have the following form

250   7. Wave Propagation and High-Frequency Electromagnetic Fields

$$E_x = E_0 e^{-\gamma z}$$

where E_x was chosen as an example, and the $e^{j\omega t}$ variation is also implied. The derivatives with respect to z in Eqs. (7.78b,c) and (7.79b,c) become

$$\frac{\partial E_y}{\partial z} = -\gamma E_y, \quad \frac{\partial E_x}{\partial z} = -\gamma E_x, \quad \frac{\partial H_y}{\partial z} = -\gamma H_y, \quad \frac{\partial H_x}{\partial z} = -\gamma H_x$$

Now, Eqs. (7.78b) through (7.78d) and (7.79b) through (7.79d) are

$$\frac{\partial H_z}{\partial y} + \gamma H_y = j\omega\varepsilon E_x \quad (7.80a) \qquad \frac{\partial E_z}{\partial y} + \gamma E_y = -j\omega\mu H_x \quad (7.81a)$$

$$-\gamma H_x - \frac{\partial H_z}{\partial x} = j\omega\varepsilon E_y \quad (7.80b) \qquad -\gamma E_x - \frac{\partial E_z}{\partial x} = -j\omega\mu H_y \quad (7.81b)$$

$$\frac{\partial H_y}{\partial x} - \frac{\partial H_x}{\partial y} = j\omega\varepsilon E_z \quad (7.80c) \qquad \frac{\partial E_y}{\partial x} - \frac{\partial E_x}{\partial y} = -j\omega\mu H_z \quad (7.81c)$$

We now rewrite these equations such that the transverse components of the fields (those perpendicular to the direction of propagation), in this case E_x, E_y, H_x, and H_y, are written in terms of the longitudinal components (those components in the direction of propagation), E_z and H_z. The reason to do so is in that the longitudinal components define the type of wave. As an example, using equations (7.80a) and (7.81b), we can eliminate H_y and write the component E_x in terms of H_z and E_z. Substitution of H_y from Eq. (7.81b) in Eq. (7.80a) gives

$$\frac{\partial H_z}{\partial y} + \frac{\gamma}{j\omega\mu}\left(\gamma E_x + \frac{\partial E_z}{\partial x}\right) = j\omega\varepsilon E_x$$

Multiplying both sides by $j\omega\mu$ and rearranging terms, we get

$$E_x = \frac{1}{\gamma^2 + k^2}\left(-\gamma\frac{\partial E_z}{\partial x} - j\omega\mu\frac{\partial H_z}{\partial y}\right)$$

where $k^2 = \omega^2\mu\varepsilon$.

By repeating the process for the other three transverse components, we get a set of four equations

$$E_x = \frac{1}{\gamma^2 + k^2}\left(-\gamma\frac{\partial E_z}{\partial x} - j\omega\mu\frac{\partial H_z}{\partial y}\right) \qquad (7.82a)$$

$$E_y = \frac{1}{\gamma^2 + k^2}\left(-\gamma\frac{\partial E_z}{\partial y} + j\omega\mu\frac{\partial H_z}{\partial x}\right) \qquad (7.82b)$$

$$H_x = \frac{1}{\gamma^2 + k^2}\left(-\gamma\frac{\partial H_z}{\partial x} + j\omega\varepsilon\frac{\partial E_z}{\partial y}\right) \qquad (7.82c)$$

$$H_y = \frac{1}{\gamma^2 + k^2}\left(-\gamma\frac{\partial H_z}{\partial y} - j\omega\varepsilon\frac{\partial E_z}{\partial x}\right) \qquad (7.82d)$$

The conditions necessary for *TEM*, *TE*, and *TM* propagation of waves are defined next.

7.6.2. TEM Waves

To obtain *TEM* waves, the necessary conditions are: $E_z = H_z = 0$. This leads to the condition that all four transverse components are zero, unless $\gamma^2 + k^2 = 0$. This, in fact, indicates that the constant of propagation is

$$\gamma^2 = -k^2 \quad \text{or} \quad \gamma = j\omega\sqrt{\mu\varepsilon}$$

which is the condition we obtained for propagation of uniform plane waves in lossless media. Indeed, all the properties we obtained for propagation of plane waves in an unbounded domain, including the definition of intrinsic impedance (or wave impedance) and the speed of propagation (phase velocity) apply in this case. We only mention here that *TEM* waves cannot exist in single conductor waveguides (i.e., a rectangular or cylindrical waveguide), a property that can be shown by direct application of Ampere's law [Eq. (7.3)]. However, they can exist in coaxial cables and parallel plate waveguides. These however will not be discussed here.

7.6.3. TE Waves

For *TE* waves to exist, E_z must be zero. The transverse fields are now

$$E_x = \frac{-j\omega\mu}{\gamma^2 + k^2}\frac{\partial H_z}{\partial y} \qquad (7.83a)$$

$$E_y = \frac{j\omega\mu}{\gamma^2 + k^2}\frac{\partial H_z}{\partial x} \qquad (7.83b)$$

252   7. Wave Propagation and High-Frequency Electromagnetic Fields

$$H_x = \frac{-\gamma}{\gamma^2 + k^2} \frac{\partial H_z}{\partial x} \quad (7.83c)$$

$$H_y = \frac{-\gamma}{\gamma^2 + k^2} \frac{\partial H_z}{\partial y} \quad (7.83d)$$

Or, alternatively, we can write these as a wave equation in H_z, which is the only longitudinal component in a *TE* wave. We start with the derivative with respect to y of Eq. (7.83a) and the derivative with respect to x of Eq. (7.83b)

$$\frac{\partial E_x}{\partial y} = \frac{-j\omega\mu}{\gamma^2 + k^2} \frac{\partial^2 H_z}{\partial y^2} \quad \text{and} \quad \frac{\partial E_y}{\partial x} = \frac{j\omega\mu}{\gamma^2 + k^2} \frac{\partial^2 H_z}{\partial x^2}$$

Adding the two equations gives

$$\frac{\partial^2 H_z}{\partial x^2} + \frac{\partial^2 H_z}{\partial y^2} + \frac{\gamma^2 + k^2}{j\omega\mu}\left(-\frac{\partial E_y}{\partial x} + \frac{\partial E_x}{\partial y}\right) = 0$$

Now, using Eq. (7.79d) we have

$$\frac{\partial^2 H_z}{\partial x^2} + \frac{\partial^2 H_z}{\partial y^2} + (\gamma^2 + k^2)H_z = 0 \quad (7.84)$$

This is a wave equation in H_z alone.

7.6.4. TM Waves

A similar discussion using the condition $H_z=0$ yields the field equations for *TM* waves:

$$E_x = \frac{-\gamma}{\gamma^2 + k^2} \frac{\partial E_z}{\partial x} \quad (7.85a)$$

$$E_y = \frac{-\gamma}{\gamma^2 + k^2} \frac{\partial E_z}{\partial y} \quad (7.85b)$$

$$H_x = \frac{j\omega\varepsilon}{\gamma^2 + k^2} \frac{\partial E_z}{\partial y} \quad (7.85c)$$

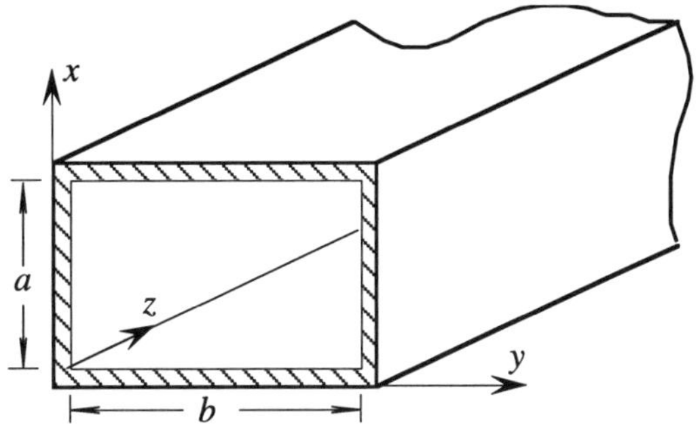

Figure 7.18. Structure and dimensions of a rectangular waveguide.

$$H_y = \frac{-j\omega\varepsilon}{\gamma^2 + k^2} \frac{\partial E_z}{\partial x} \qquad (7.85d)$$

and the wave equation equivalent to these equations is

$$\frac{\partial^2 E_z}{\partial x^2} + \frac{\partial^2 E_z}{\partial y^2} + (\gamma^2 + k^2)E_z = 0 \qquad (7.86)$$

The behavior of *TE* and *TM* modes in a rectangular waveguide is our next goal.

7.6.5. Rectangular Waveguides

A rectangular waveguide is defined as shown in Figure 7.18. The dimensions of the waveguide are assumed to be the internal dimensions and, for the moment, we will assume the walls to be perfectly conducting.

To see how the electric and magnetic fields in a waveguide of the type shown in Figure 7.18 behave, we must solve Eq. (7.84) or Eq. (7.86) subject to boundary conditions on the conducting boundaries. We start with *TM* waves because the boundary conditions are straightforward.

7.6.6. TM Modes in Waveguides

To find the fields for *TM* modes, we must solve the following equation [Eq. (7.86)]

$$\frac{\partial^2 E_z}{\partial x^2} + \frac{\partial^2 E_z}{\partial y^2} + (\gamma^2 + k^2)E_z = 0$$

subject to the condition that the electric field must vanish on the conducting boundaries. These conditions are:

$E_z(0,y)=0$ and $E_z(a,y)=0$ in the x direction

$E_z(x,0)=0$ and $E_z(x,b)=0$ in the y direction

Eq. (7.86) can be solved directly using separation of variables. This leads to two second-order ordinary differential equations, the solution to which are sinusoidal functions. Instead of proceeding in this fashion, we will simply assume a solution in terms of sinusoids, and evaluate the amplitude and spatial frequencies of the sinusoids so that it satisfies the boundary conditions. The solution we seek is of the form

$$E_z(x,y) = E_0 sin(k_x x) sin(k_y y)$$

This is, in fact, the solution one obtains through separation of variables. To satisfy the boundary conditions, k_x must be such that the *sin* function is zero at $x=0$ and $x=a$. Similarly, k_y must force the second *sin* function to zero at $y=0$ and $y=b$. These conditions yield

$$k_x = \frac{m\pi}{a} \quad \text{and} \quad k_y = \frac{n\pi}{b} \quad \text{where } m,n \text{ are integers}$$

The solution now is

$$E_z(x,y) = E_0 \sin\left(\frac{m\pi}{a}\right) x \sin\left(\frac{n\pi}{b}\right) y \tag{7.87}$$

If we now substitute this solution into Eq. (7.85a-d) and evaluate the derivative of E_z, we obtain the transverse components of E and H

$$E_x(x,y) = \frac{-\gamma}{\gamma^2 + k^2} E_0 \frac{m\pi}{a} \cos\left(\frac{m\pi}{a}\right) x \sin\left(\frac{n\pi}{b}\right) y \tag{7.88a}$$

$$E_y(x,y) = \frac{-\gamma}{\gamma^2 + k^2} E_0 \frac{n\pi}{b} \sin\left(\frac{m\pi}{a}\right) x \cos\left(\frac{n\pi}{b}\right) y \tag{7.88b}$$

$$H_x(x,y) = \frac{j\omega\varepsilon}{\gamma^2 + k^2} E_0 \frac{n\pi}{b} \sin\left(\frac{m\pi}{a}\right) x \cos\left(\frac{n\pi}{b}\right) y \tag{7.88c}$$

$$H_z(x,y) = \frac{-j\omega\varepsilon}{\gamma^2 + k^2} E_0 \frac{m\pi}{a} \cos\left(\frac{m\pi}{a}\right) x \sin\left(\frac{n\pi}{b}\right) y \qquad (7.88d)$$

To obtain a relation between the constants k_x, k_y, and γ, we substitute the general solution given in Eq. (7.87) into Eq. (7.86). This gives a relation called a dispersion relation

$$\gamma^2 + k^2 = \left(\frac{m\pi}{a}\right)^2 + \left(\frac{n\pi}{b}\right)^2$$

Or, in a more convenient form,

$$\gamma^2 = -k^2 + \left(\frac{m\pi}{a}\right)^2 + \left(\frac{n\pi}{b}\right)^2$$

Since there are no losses (perfect conductors, lossless dielectrics), we can write

$$\gamma = j\beta = j\sqrt{\omega^2 \mu\varepsilon - \left(\frac{m\pi}{a}\right)^2 - \left(\frac{n\pi}{b}\right)^2} \qquad (7.89)$$

Since m and n are discrete values, the only fields that can exist are those defined by combinations of integer values of m and n. There is an infinite number of possible combinations of the integers, and, correspondingly, an infinite number of possible fields in the waveguide. Each possible combination is called a mode. The modes defined here are for *TM* waves, therefore they are called *TM* modes. The integers m and n are the indices of the mode. Thus, if $m=1$, $n=3$, this defines a TM_{13} mode. The indices indicate the number of half-cycles in the respective direction of the waveguide.

From Eq. (7.89) we note that the propagation constant γ can be zero. This clearly indicates that the waves cannot propagate if this condition is encountered. The frequency at which this happens is called a cutoff frequency and it is obviously different for each mode. The cutoff frequency is given by the condition

$$\omega^2 \mu\varepsilon - \left[\left(\frac{m\pi}{a}\right)^2 + \left(\frac{n\pi}{b}\right)^2\right] = 0$$

or

$$f_c = \frac{1}{2\pi\sqrt{\mu\varepsilon}} \sqrt{\left(\frac{m\pi}{a}\right)^2 + \left(\frac{n\pi}{b}\right)^2} \qquad (7.90)$$

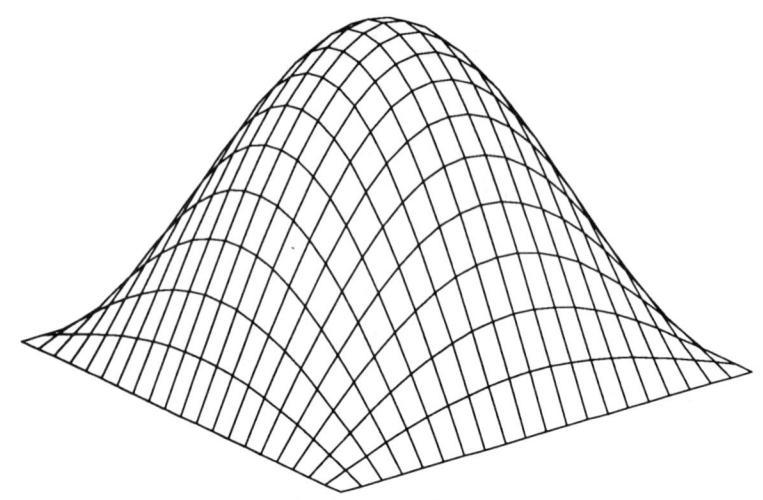

Figure 7.19. The electric field distribution in a waveguide for the TM_{11} mode.

Below this frequency there is no propagation in the waveguide.

Also, we note that neither m nor n can be zero since that would force all field components to be zero. The lowest possible mode (i.e., the one with lowest cutoff frequency) is the TM_{11} mode.

The electric field distribution (E_z) in a waveguide for the TM_{11} mode is shown in Figure 7.19. for a waveguide for which $a=2b$. The electric field is zero on the boundaries and has a half-cycle variation in each transverse direction as required.

7.6.7. TE Modes in Waveguides

For *TE* waves to exist, E_z must be zero. The wave equation to solve is now

$$\frac{\partial^2 H_z}{\partial x^2} + \frac{\partial^2 H_z}{\partial y^2} + (\gamma^2 + k^2)H_z = 0$$

subject to the condition that the x and y components of the electric field must vanish on the conducting boundaries. These conditions can be written in terms of the magnetic field intensity H_z as

$$\frac{\partial H_z}{\partial x}(0,y) = 0 \quad \text{and} \quad \frac{\partial H_z}{\partial x}(a,y) = 0 \quad \text{in the } x \text{ direction}$$

$$\frac{\partial H_z}{\partial y}(x,0) = 0 \quad \text{and} \quad \frac{\partial H_z}{\partial y}(x,b) = 0 \quad \text{in the } y \text{ direction}$$

The solution is obtained in an analogous manner to that for *TM* waves and is

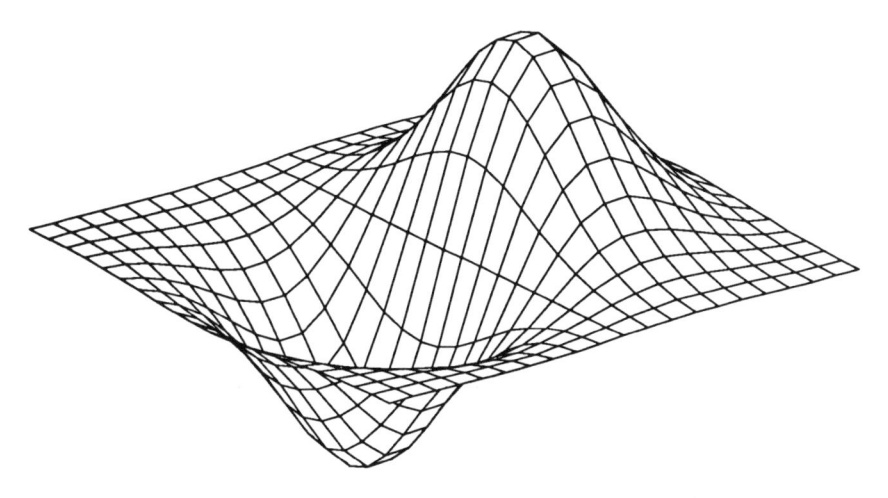

Figure 7.20. The magnetic field distribution in a waveguide for the TE_{21} mode.

$$H_z(x,y) = H_0 \cos\left(\frac{m\pi}{a}\right)x \cos\left(\frac{n\pi}{b}\right)y \qquad (7.91)$$

Substitution of this solution in Eq. (7.83a-d) provides the transverse components of the fields as:

$$E_x(x,y) = \frac{j\omega\mu}{\gamma^2 + k^2} H_0 \frac{n\pi}{b} \cos\left(\frac{m\pi}{a}\right)x \sin\left(\frac{n\pi}{b}\right)y \qquad (7.92a)$$

$$E_y(x,y) = \frac{-j\omega\mu}{\gamma^2 + k^2} H_0 \frac{m\pi}{a} \sin\left(\frac{m\pi}{a}\right)x \cos\left(\frac{n\pi}{b}\right)y \qquad (7.92b)$$

$$H_x(x,y) = \frac{\gamma}{\gamma^2 + k^2} H_0 \frac{m\pi}{a} \sin\left(\frac{m\pi}{a}\right)x \cos\left(\frac{n\pi}{b}\right)y \qquad (7.92c)$$

$$H_y(x,y) = \frac{\gamma}{\gamma^2 + k^2} H_0 \frac{n\pi}{b} \cos\left(\frac{m\pi}{a}\right)x \sin\left(\frac{n\pi}{b}\right)y \qquad (7.92d)$$

The dispersion relation for *TE* waves is obtained by substituting the solution in Eq. (7.91) into the wave equation [Eq. (7.84)] and is identical to that for the *TM* modes. The cutoff frequency is also the same as that for *TM* modes.

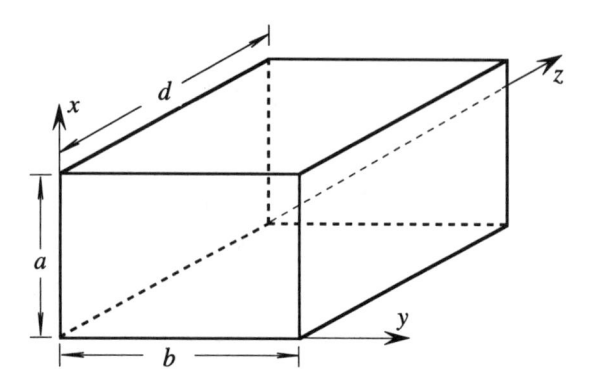

Figure 7.21. Structure and dimensions of a rectangular cavity resonator.

However, unlike *TM* modes, in *TE* modes either m or n can be zero. This indicates that the lower resonant mode is a TE_{0n} or TE_{m0}, depending on the dimensions a and b. If $a>b$, the lowest cutoff frequency is for a TE_{10} mode. Also, *TM* and *TE* modes with the same indices have the same frequency.

The magnetic field (H_z) distribution in a waveguide with $a=2b$, for the TE_{21} mode is shown in Figure 7.20. The field is maximum on the surfaces and has two half-cycle variations in the x direction and half-cycle variation in the y direction.

The discussion on waveguides was limited here to rectangular cross-section waveguides. Other shapes can be analyzed in a similar fashion. However, since the analysis entails the solution of the wave equation, only simple shapes can be solved exactly. In particular, cylindrical waveguides can be analyzed using steps identical to those presented above.

7.7. Cavity Resonators

A cavity is built by adding two walls inside a waveguide. A rectangular cavity is built by adding two conducting walls at $z=0$ and $z=d$, as shown in Figure 7.21. This can be viewed as a modified waveguide, in which, there are standing waves in the z direction as well as in the x and y directions. The main difference between cavities and waveguides is that in cavities, the z direction imposes additional boundary conditions and there is no propagation of waves as in waveguides. The cavity acts as a resonant structure in which there is exchange of energy between the electric and magnetic field at given (resonant) frequencies. This is equivalent to resonant *LC* circuits in the case of lossless cavities and to *RLC* circuits in the case of lossy cavities.

The analysis of fields in a cavity require the solution of the full three-dimensional wave equation with the required boundary conditions. The procedure here will be to take the *TM* and *TE* waves we have already defined and to modify them to satisfy the additional boundary conditions imposed by the additional conducting walls.

The *TM* fields satisfy the following equation

$$\frac{\partial^2 E_z}{\partial x^2} + \frac{\partial^2 E_z}{\partial y^2} + \frac{\partial^2 E_z}{\partial z^2} + (\gamma^2 + k^2)E_z = 0 \qquad (7.93)$$

while the *TE* fields satisfy

$$\frac{\partial^2 H_z}{\partial x^2} + \frac{\partial^2 H_z}{\partial y^2} + \frac{\partial^2 H_z}{\partial z^2} + (\gamma^2 + k^2)H_z = 0 \qquad (7.94)$$

where the term in brackets $(\gamma^2 + k^2)$ is also known as the cutoff wave number.

Comparing these with Eqs. (7.86) and (7.84), it is obvious that the fields E_z and H_z are those in Eqs. (7.87) and (7.91) multiplied by an additional product due to the z dependence of the field.

7.7.1. TM and TE Modes in Cavity Resonators

For *TM* modes, the additional boundary conditions are:

$$E_x(x,y,0) = 0 \quad E_x(x,y,d) = 0 \quad E_y(x,y,0) = 0 \quad E_y(x,y,d) = 0$$

The electric field in the z direction that satisfies Eq. (7.93) and these boundary conditions is

$$E_z(x,y,z) = E_0 \sin\left(\frac{m\pi}{a}\right) x \sin\left(\frac{n\pi}{b}\right) y \cos\left(\frac{p\pi}{d}\right) z \qquad (7.95)$$

Now we can substitute this into Eq. (7.85) to obtain

$$E_x(x,y,z) = \frac{-\gamma}{\gamma^2 + k^2} E_0 \frac{m\pi}{a} \cos\left(\frac{m\pi}{a}\right) x \sin\left(\frac{n\pi}{b}\right) y \cos\left(\frac{p\pi}{d}\right) z \quad (7.96a)$$

$$E_y(x,y,z) = \frac{-\gamma}{\gamma^2 + k^2} E_0 \frac{n\pi}{b} \sin\left(\frac{m\pi}{a}\right) x \cos\left(\frac{n\pi}{b}\right) y \cos\left(\frac{p\pi}{d}\right) z \quad (7.96b)$$

$$H_x(x,y,z) = \frac{j\omega\varepsilon}{\gamma^2 + k^2} E_0 \frac{n\pi}{b} \sin\left(\frac{m\pi}{a}\right) x \cos\left(\frac{n\pi}{b}\right) y \cos\left(\frac{p\pi}{d}\right) z \quad (7.96c)$$

260   7. Wave Propagation and High-Frequency Electromagnetic Fields

$$H_y(x,y,z) = \frac{-j\omega\varepsilon}{\gamma^2 + k^2} E_0 \frac{m\pi}{a} \cos\left(\frac{m\pi}{a}\right) x \sin\left(\frac{n\pi}{b}\right) y \cos\left(\frac{p\pi}{d}\right) z \quad (7.96d)$$

The first of these equation can be written, in general terms, as

$$E_x = E_0 e^{-\gamma z}$$

Therefore,

$$\frac{\partial E_x}{\partial z} = -\gamma E_x \quad (7.97)$$

Using this derivative in Eq. (7.96a) and (7.96b), the fields are now

$$E_x(x,y,z) = \frac{-1}{\gamma^2 + k^2} E_0 \frac{p\pi}{d} \frac{m\pi}{a} \cos\left(\frac{m\pi}{a}\right) x \sin\left(\frac{n\pi}{b}\right) y \sin\left(\frac{p\pi}{d}\right) z \quad (7.98a)$$

$$E_y(x,y,z) = \frac{-1}{\gamma^2 + k^2} E_0 \frac{p\pi}{d} \frac{n\pi}{b} \sin\left(\frac{m\pi}{a}\right) x \cos\left(\frac{n\pi}{b}\right) y \sin\left(\frac{p\pi}{d}\right) z \quad (7.98b)$$

$$H_x(x,y,z) = \frac{j\omega\varepsilon}{\gamma^2 + k^2} E_0 \frac{n\pi}{b} \sin\left(\frac{m\pi}{a}\right) x \cos\left(\frac{n\pi}{b}\right) y \cos\left(\frac{p\pi}{d}\right) z \quad (7.98c)$$

$$H_y(x,y,z) = \frac{-j\omega\varepsilon}{\gamma^2 + k^2} E_0 \frac{m\pi}{a} \cos\left(\frac{m\pi}{a}\right) x \sin\left(\frac{n\pi}{b}\right) y \cos\left(\frac{p\pi}{d}\right) z \quad (7.98d)$$

Similarly, we can write the dispersion relation as

$$k^2 = \left(\frac{m\pi}{a}\right)^2 + \left(\frac{n\pi}{b}\right)^2 + \left(\frac{p\pi}{d}\right)^2 \quad (7.99)$$

or

$$f_{mnp} = \frac{1}{2\pi\sqrt{\mu\varepsilon}} \sqrt{\left(\frac{m\pi}{a}\right)^2 + \left(\frac{n\pi}{b}\right)^2 + \left(\frac{p\pi}{d}\right)^2} \quad (7.100)$$

where the indices m,n,p, indicate the mode at which the cavity resonates.

7.7.2. TE Modes in a Cavity

A discussion similar to that for the *TM* waves can be followed here. The z component of the magnetic field is now

$$H_z(x,y,z) = H_0 \cos\left(\frac{m\pi}{a}\right)x \cos\left(\frac{n\pi}{b}\right)y \sin\left(\frac{p\pi}{d}\right)z \qquad (7.101)$$

and substitution in the *TE* field equation (7.83) gives

$$E_x(x,y,z) = \frac{j\omega\mu}{\gamma^2 + k^2} H_0 \frac{n\pi}{b} \cos\left(\frac{m\pi}{a}\right)x \sin\left(\frac{n\pi}{b}\right)y \sin\left(\frac{p\pi}{d}\right)z \quad (7.102a)$$

$$E_y(x,y,z) = \frac{-j\omega\mu}{\gamma^2 + k^2} H_0 \frac{m\pi}{a} \sin\left(\frac{m\pi}{a}\right)x \cos\left(\frac{n\pi}{b}\right)y \sin\left(\frac{p\pi}{d}\right)z \quad (7.102b)$$

$$H_x(x,y,z) = \frac{\gamma}{\gamma^2 + k^2} H_0 \frac{m\pi}{a} \sin\left(\frac{m\pi}{a}\right)x \cos\left(\frac{n\pi}{b}\right)y \cos\left(\frac{p\pi}{d}\right)z \quad (7.102c)$$

$$H_y(x,y,z) = \frac{\gamma}{\gamma^2 + k^2} H_0 \frac{n\pi}{b} \cos\left(\frac{m\pi}{a}\right)x \sin\left(\frac{n\pi}{b}\right)y \cos\left(\frac{p\pi}{d}\right)z \quad (7.102d)$$

The dispersion relation is the same as in Eq. (7.99) and the resonant frequency for *TE* modes is also given by Eq. (7.100).

From the equations for *TM* and *TE* modes we note that for *TM* modes, neither m and n can be zero while p can be zero. For *TE* modes, either m or n can be zero (but not both) while p must be nonzero.

The various modes possible are defined by m,n, and p. Some of the modes may have the same resonant frequency even though they are different modes. As an example, for a cubic cavity ($a=b=d$), TM_{110}, TE_{011}, TE_{101} all have the same frequency. These are called degenerate modes. The mode with lowest frequency (again, it depends on a, b, and d) is called the dominant mode.

7.7.3. Energy in a Cavity

Power and energy relations in a cavity are defined by the Poynting theorem. Since there is a certain amount of energy stored in the fields of a cavity, the calculation of this power is an important aspect of analysis. This is particularly obvious if we recall that in a resonant device, these relations change dramatically at or near resonance. This was true with resonant circuits and is certainly true with resonant cavities. The stored and dissipated energy in a cavity defines the basic qualities of the cavity. A lossless cavity is not practically realizable, therefore, we also define a

262   7. Wave Propagation and High-Frequency Electromagnetic Fields

quantity called "quality factor" of the cavity, which is a measure of losses in the cavity. A shift in the resonant frequency of the cavity can also be described in terms of energy. These relations can then be used to characterize a cavity and for measurements in the cavity.

The Poynting vector was defined in section 6.6. as

$$\mathbf{P} = \mathbf{E} \times \mathbf{H}$$

where $\mathbf{E}$ and $\mathbf{H}$ are the electric and magnetic field intensities in the cavity. $\mathbf{P}$ is the Poynting vector. Its direction indicates the direction of propagation of energy and its magnitude is the power density. For our purposes here it is more convenient to write the total power through a surface

$$\int_s (\mathbf{E} \times \mathbf{H}) \cdot ds = -\frac{\partial}{\partial t} \int_v \left(\frac{1}{2} \varepsilon E^2 + \frac{1}{2} \mu H^2 \right) dv - \int_v \sigma E^2 dv$$

Or, in a simplified form, as

$$\int_s (\mathbf{E} \times \mathbf{H}) \cdot ds = -\frac{\partial}{\partial t} \int_v (w_e + w_m) dv - \int_v p_d dv \qquad (7.103)$$

where w_e is the stored electric energy density, w_m the stored magnetic energy density, and p_d the dissipated power density. The average power density is

$$P_{av} = \frac{1}{2} Re(\mathbf{E} \times \mathbf{H})$$

The stored electric and magnetic energy in the cavity can then be written as

$$W_0 = \int_v \left(\frac{\varepsilon |E|^2}{4} + \frac{\mu |H|^2}{4} \right) dv \qquad (7.104)$$

where E and H are the fields in the cavity and v the volume of the cavity. This relation is correct at any frequency regardless of resonance. The dissipated power is

$$P_d = \frac{1}{2} \int_v \sigma |E|^2 dv \qquad (7.105)$$

where σ may represent conducting materials in the cavity, including lossy dielectrics or, more often, conductivity of the walls.

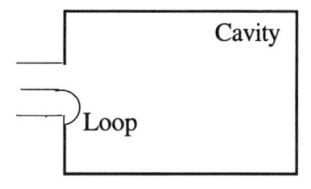

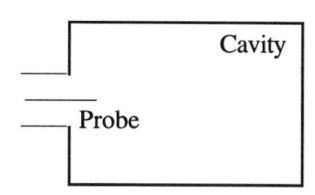

Figure 7.22a. Coupling to a cavity by a small loop in the cavity.

Figure 7.22b. Coupling to a cavity by a small probe in the cavity.

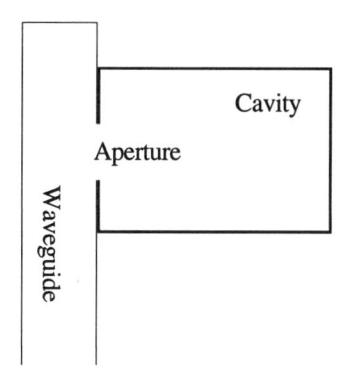

Figure 7.22c. Coupling to a cavity by a small aperture in the wall of the cavity.

7.7.4. Quality Factor of a Cavity Resonator

The quality factor of the cavity resonator is defined as the ratio between the stored energy in the cavity and the dissipated power per unit time

$$Q = \frac{\omega_0 W_0}{P_d} \qquad (7.106)$$

where ω_0 is the resonant frequency. Since the higher the Q factor, the more selective the cavity is, Q is a measure of the bandwidth of the cavity. It also defines, indirectly, the amount of energy needed to couple into the cavity to maintain an energy balance. Ideal cavities have an infinite quality factor.

7.7.5. Coupling to Cavities

The coupling of electromagnetic energy to cavities has not been addressed specifically in the discussion above. If the cavity were to be ideal, the fields within the cavity would be of infinite amplitude, provided that the necessary modes can be excited. In realistic cavities, there are always some losses but these are usually small. The fields are large and the amount of stored energy is also large.

However, the small amount of power dissipated has to be compensated for by external sources otherwise the cavity would cease to oscillate. This is done by coupling energy into the cavity. A cavity for which the lost energy is balanced is called a critically coupled cavity.

The introduction of energy into the cavity can be done in a number of ways. The most obvious of these is to have within the cavity a source that generates the necessary fields. A small loop [Figure (7.22a)] or a simple probe excitation [Figure (7.22b)] can be used. Similarly, the cavity can be coupled through a small aperture through which a small amount of energy "leaks" into the cavity [Figure (7.22c)]. The only requirement of these methods of coupling is that they excite the required modes. The three coupling methods in Figure 7.22 excite different modes.

8
Introduction to the Finite Element Method

8.1. Introduction

This chapter introduces the finite element method (FEM) as a tool for solution of classical electromagnetic problems. Although we discuss the main points in the application of the finite element method to electromagnetic design, including formulation and implementation, those who seek deeper understanding of the finite element method should consult some of the works listed in the bibliography section.

The evolution of the finite element method is intimately linked to developments in the engineering and computer sciences. Its application in a variety of areas, especially in the nuclear, aeronautic, and transportation industries, is testimony to the high degree of accuracy the method is capable of, as well as to its ability to model complex problems.

Engineering science allows us to describe the physical behavior of a system using partial differential equations while the finite element method is the method of choice in solving many of these equations efficiently and accurately. In fact, 2D and 3D problems involving linear and nonlinear conditions, as well as other practical conditions and situations, can be successfully solved using the finite element method.

The development and use of the finite element method requires the joint use of three distinct sciences:

- Engineering sciences – to establish the governing equations related to the physical problem.
- Numerical methods – needed to apply the finite element concepts and solve the system.
- Computer science – to build and maintain efficient software tools needed for implementation.

Approximately 60 years ago, some mechanical systems, related to stress analysis, were already being solved by numerical techniques. These early attempts can be viewed as the initial inspiration for the development of the FEM. In 1956, the concept of finite elements was introduced in the *Journal of Aeronautical Science* by Turner, Clough, Martin, and Topp (see bibliography section).

After 1960, the FEM developed rapidly; residual and variational methods were widely applied and high-precision elements were developed. Beginning with the late 1960s, the first books on the method appeared, describing the method and its application in diverse engineering areas. The first significant application in electrical engineering was in 1970, when P. P. Silvester and M. V. K. Chari published the paper "Finite Elements Solution of Saturable Magnetic Fields Problems" proposing the use of this method for electromagnetic problems, including in its formulation the solution of nonlinear problems. The methods used before 1970 were not completely satisfactory, in particular when the structure had a complex geometry or when nonlinearity in ferromagnetic materials was considered. The aforementioned paper was not the first to be published in this area, but we consider its appearance as the opening of a new age for applied electromagnetism, which was in need of reliable calculation methods.

The main goal of this chapter is to describe the finite element method as a numerical technique. We use electrostatic equations in developing the various concepts so as to aid in understanding, while more complex and more general electromagnetic phenomena will be presented in the following chapters.

8.2. The Galerkin Method – Basic Concepts

Because we are describing the finite element method in a relatively brief text, we chose to describe it using a simple example with known solution – the parallel plate capacitor. The various, successive steps are pointed out during the procedure of applying the technique. In this section, we introduce the basic concepts of the Galerkin method.

a. *Step 1* – The establishment of the physical equations.

Consider a parallel plate capacitor with a uniform dielectric of permittivity ε between the plates, as shown in Figure 8.1.

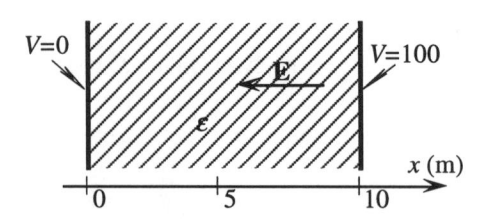

Figure 8.1. The parallel plate capacitor.

Assuming the electric field has only an x component, the problem is one-dimensional. The potential between the plates varies linearly from $V=0$ at $x=0$ and $V=100V$ at $x=10m$.

The electric field intensity **E** is related to the scalar potential V as

$$\mathbf{E} = -\nabla V \qquad (8.1)$$

In this case, the solution is known (see section 3.7.) and is

$$\mathbf{E} = -\hat{\mathbf{i}}\frac{\partial V}{\partial x} = -\hat{\mathbf{i}}\, 10\, \frac{V}{m}$$

Our intention here is to show how the same solution is found using the finite element method.

The equation to be solved is

$$\nabla \cdot \mathbf{D} = \rho$$

Although there are no charge densities in this problem, we will maintain the term ρ for completeness in presentation. Using $\mathbf{D}=\varepsilon\mathbf{E}$ and Eq. (8.1), we obtain

$$\nabla \cdot \varepsilon \mathbf{E} = \nabla \cdot \varepsilon(-\nabla V) = 0$$

or

$$\nabla \cdot \varepsilon(-\nabla V) = \rho \qquad \text{or} \qquad \frac{\partial}{\partial x}\left(\varepsilon \frac{\partial V}{\partial x}\right) = -\rho \qquad (8.2)$$

This equation needs to be solved subject to the boundary conditions

$$V(0) = 0 \qquad \text{and} \qquad V(10) = 100\ V$$

b. *Step 2* – **Discretization and finite element modeling**

In the FEM, the solution domain is subdivided or "discretized." In 1D applications, the line representing the solution domain is cut into small segments and each segment is called a "finite element." The points defining the element are called "nodes" or "degrees of freedom." For instance, in 2D applications (as we shall see shortly), the domain can be discretized into finite area patches such as triangles.

In our example, the domain is divided into N segments as in Figure 8.2a. A generic element "n" is isolated in Figure 8.2b.

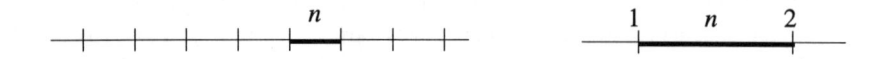

Figure 8.2a. The discretized domain.

Figure 8.2b. The finite element "n."

The finite element is bounded by "nodes." In this example, the nodes are numbered as "1" and "2," called an "internal" or "local" numbering. The assembly of elements is called a "mesh" and the node numbers of the element in the mesh can be different. The local numbering sequence is only used to perform some algebraic calculations on element "n."

We assume the potential varies linearly inside the element as

$$V(x) = a_1 + a_2 x \qquad (8.3)$$

This gives at node No. 1

$$x = x_1 \quad \text{and} \quad V(x_1) = V_1$$

and at node No. 2

$$x = x_2 \quad \text{and} \quad V(x_2) = V_2$$

With these relations, the constants a_1 and a_2 in Eq. (8.3) can be calculated by solving the following system of equations:

$$\begin{aligned} V_1 &= a_1 + a_2 x_1 \\ V_2 &= a_1 + a_2 x_2 \end{aligned} \qquad (8.4)$$

Solving for a_1 gives

$$a_1 = \frac{\begin{vmatrix} V_1 & x_1 \\ V_2 & x_2 \end{vmatrix}}{\begin{vmatrix} 1 & x_1 \\ 1 & x_2 \end{vmatrix}} = \frac{V_1 x_2 - V_2 x_1}{x_2 - x_1}$$

And, analogously, for a_2

$$a_2 = \frac{V_2 - V_1}{x_2 - x_1}$$

Denoting the length of the element as $L = x_2 - x_1$, we have

$$a_1 = \frac{1}{L}(V_1 x_2 - V_2 x_1) \qquad (8.5a)$$

$$a_2 = \frac{1}{L}(V_2 - V_1) \qquad (8.5b)$$

Substituting Eq. (8.5a) and (8.5b) into Eq. (8.3), we obtain

$$V(x) = \frac{1}{L}\left[V_1 x_2 - V_2 x_1 + (V_2 - V_1)x\right] \qquad (8.6)$$

which can be written as

$$V(x) = \frac{1}{L}\left[(x_2 - 1x)V_1 + (-x_1 + 1x)V_2\right]$$

or

$$V(x) = \frac{1}{L}\sum_{i=1,2}(p_i + q_i x)V_i \qquad (8.7)$$

where

$$\begin{array}{ll} p_1 = x_2 & p_2 = -x_1 \\ q_1 = -1 & q_2 = 1 \end{array} \qquad (8.8)$$

The expression in Eq. (8.7) is now

$$V(x) = \frac{1}{L}(p_1 + q_1 x)V_1 + \frac{1}{L}(p_2 + q_2 x)V_2$$

We define the functions $\phi_1(x)$ and $\phi_2(x)$ as

$$\phi_1(x) = \frac{1}{L}(p_1 + q_1 x) \qquad (8.9a)$$

$$\phi_2(x) = \frac{1}{L}(p_2 + q_2 x) \qquad (8.9b)$$

With these, Eq. (8.7) becomes

$$V(x) = \phi_1(x)V_1 + \phi_2(x)V_2 \qquad (8.10)$$

We observe that setting $x = x_1$ in Eq. (8.10) gives

$$V(x_1) = 1 V_1 + 0 V_2 = V_1$$

since

$$\phi_1(x_1) = \frac{1}{L}(p_1 + q_1 x_1) = \frac{1}{L}(x_2 - x_1) = 1$$

$$\phi_2(x_1) = \frac{1}{L}(p_2 + q_2 x_1) = \frac{1}{L}(x_1 - x_1) = 0$$

Analogously,

$$V(x_2) = 0V_1 + 1V_2 = V_2$$

since

$$\phi_1(x_2) = 0 \quad \text{and} \quad \phi_2(x_2) = 1$$

Also, if x is placed between x_1 and x_2, $V(x)$ is given as a combination of V_1 and V_2 as, for example, $V = 0.7V_1 + 0.3V_2$, where $\phi_1 = 0.7$ and $\phi_2 = 0.3$.

c. *Step 3* – **Application of the weighted residual method.**

Now it is time to distinguish between the "exact solution," V_e, and the solution obtained with the finite element method, "V."

For the exact solution, we have from Eq. (8.2):

$$\nabla \cdot (\varepsilon \nabla V_e) + \rho = 0$$

However, the solution we obtain using the FEM is an approximation and differs from the exact solution. When substituting this solution into Eq. (8.2), it generates a "residual" R:

$$\nabla \cdot (\varepsilon \nabla V) + \rho = R \qquad (8.11)$$

To establish a numerical procedure, we force R to be zero using the following operation:

$$\int_\Omega WR \, d\Omega = 0 \qquad (8.12)$$

where W is a "weighting function" and Ω represents the domain in which the condition is enforced. In our case, the expression in Eq. (8.12) is

$$\int_\Omega W\big[\nabla \cdot (\varepsilon \nabla V) + \rho\big] d\Omega = 0 \qquad (8.13)$$

Using Eq. (1.26), we get

$$U\nabla \cdot \mathbf{A} = \nabla \cdot U\mathbf{A} - \mathbf{A} \cdot \nabla U$$

and

$$\int_\Omega U\nabla \cdot \mathbf{A} d\Omega = \int_\Omega \nabla \cdot U\mathbf{A} d\Omega - \int_\Omega \mathbf{A} \cdot \nabla U d\Omega$$

Applying the divergence theorem (section 1.5.2) to the first term on the right-hand side, we obtain

$$\int_\Omega U\nabla \cdot \mathbf{A} d\Omega = \oint_{S(\Omega)} U\mathbf{A} \cdot d\mathbf{s} - \int_\Omega \mathbf{A} \cdot \nabla U d\Omega$$

where $S(\Omega)$ is the surface enclosing the domain Ω. Returning now to Eq. (8.13), we substitute $U=W$ and $\mathbf{A}=\varepsilon\nabla V$. With these, Eq. (8.13) becomes

$$\int_\Omega W\left[\nabla \cdot (\varepsilon\nabla V) + \rho\right] d\Omega = \oint_{S(\Omega)} W\varepsilon\nabla V \cdot d\mathbf{s} - \int_\Omega \varepsilon\nabla V \cdot \nabla W d\Omega + \int_\Omega W\rho d\Omega \quad (8.14)$$

This is set to zero. The first term on the right hand side is related to the boundary conditions of the problem and will be discussed later, separately. We merely comment here that since Eq. (8.14) is set to zero the integral on $S(\Omega)$ and the sum of integrals on Ω on the right hand side must be zero.

Eq. (8.14) is commonly called the "weak form" of the formulation. The origin of this terminology is in the fact that in Eq. (8.13) there are second-order derivatives, while in Eq. (8.14) there are only first-order derivatives, resulting in a "weaker" order of derivation which is easier to handle in terms of numerical techniques.

Next, the concept of discretization will be linked with the weighted residual method. Equation (8.12) for the discretized domain Ω is written as:

$$\sum_{k=1,K} \int_{\Omega_k} W_k R_k d\Omega = 0 \quad (8.15)$$

where W_k is the weighting function for node k, K is the total number of unknown nodes, and Ω_k is the partial domain to which node k belongs. This corresponds to K equations, for K unknown potential values at the K nodes in the solution domain.

272 8. Introduction to the Finite Element Method

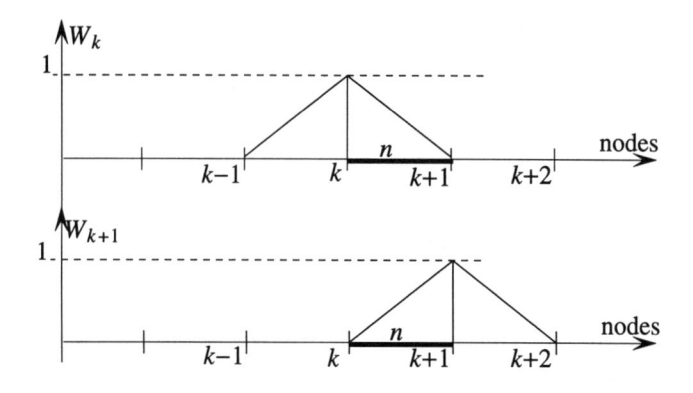

Figure 8.3. Weighting functions W_k and W_{k+1}.

The weighting functions are established as shown in Figure 8.3, where the weighting functions W_k and W_{k+1} corresponding to the nodes k and $k+1$ in element n are shown. From Figure 8.3, we note that the weighting function W_k acts strongly at node k where it equals 1 and decreases linearly away from the node, becoming zero at nodes $k-1$ and $k+1$. Similarly, W_{k+1} equals 1 at node $k+1$ and decreases to zero at nodes k and $k+2$.

Since Eq. (8.15) represents a sum only on element n in Figure 8.3, the situation in Figure 8.4 arises. That is, the sum of the weighting functions in the element is evaluated. From Figure 8.4, we observe that the functions W_k and W_{k+1}, in element "n," are identical to the functions $\phi_1(x)$ and $\phi_2(x)$ which we defined in Eq. (8.9), if we equate node 1 with node k and node 2 with node $k+1$. Therefore, instead of performing the integration node by node [as suggested by Eq. (8.15)], we can integrate element by element. Furthermore, we can use the functions ϕ_i as the weighting functions. When we do so, the method is called the "Galerkin method." This represents a particular choice of weighting functions and therefore a particular weighted residual method. The Galerkin method is widely used in electromagnetics, while the more general weighted residual method is seldom used. For this reason, we will consider here only the Galerkin method.

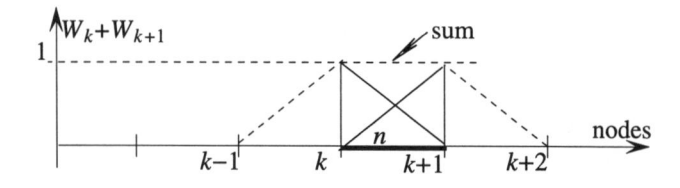

Figure 8.4. Sum of the weighting functions for element "n."

d. Step 4 – Application of the finite element method and solution.

The integrals on Ω in Eq. (8.14) for the discretized domain become

$$\sum_{n=1,N} \int_{\Omega_n} [\varepsilon \nabla V \cdot \nabla \phi_n - \rho \phi_n] d\Omega = 0 \tag{8.16}$$

where n represents a generic element and N is the number of elements, or segments, in the solution domain. The evaluation of the integral in Eq. (8.16) for an element n follows. Expanding the expression in Eq. (8.7), we obtain

$$V(x) = \frac{1}{L}(p_1 + q_1 x)V_1 + \frac{1}{L}(p_2 + q_2 x)V_2$$

From this, ∇V is

$$\nabla V(x) = \hat{\mathbf{i}} \frac{\partial}{\partial x}\left(\sum_{i=1,2} \frac{1}{L}(p_i + q_i x)V_i\right) = \hat{\mathbf{i}} \sum_{i=1,2} \frac{1}{L} q_i V_i \tag{8.17}$$

or

$$\nabla V(x) = \hat{\mathbf{i}}\left(\frac{1}{L} q_1 V_1 + \frac{1}{L} q_2 V_2\right) \tag{8.18}$$

Now we calculate $\nabla \phi_n$, noting that ϕ_n is the combination of ϕ_1 and ϕ_2 (see Figure 8.4). For ϕ_1,

$$\nabla \phi_1 = \hat{\mathbf{i}} \frac{\partial}{\partial x}\left(\frac{1}{L}(p_1 + q_1 x)\right)$$

or

$$\nabla \phi_1 = \hat{\mathbf{i}} \frac{1}{L} q_1 \tag{8.19}$$

Substituting Eqs. (8.18) and (8.19) into Eq. (8.16), we get

$$\int_L \varepsilon \hat{\mathbf{i}}\left(\frac{1}{L} q_1 V_1 + \frac{1}{L} q_2 V_2\right)\cdot\left(\hat{\mathbf{i}} \frac{1}{L} q_1\right) dl$$

This is

274   8. Introduction to the Finite Element Method

$$\int_L \varepsilon \left(\frac{1}{L^2} q_1 q_1 V_1 + \frac{1}{L^2} q_1 q_2 V_2 \right) dx$$

Since the integrand does not depend on x, we get

$$\frac{\varepsilon}{L}(q_1 q_1 V_1 + q_1 q_2 V_2)$$

or, in matrix form,

$$\frac{\varepsilon}{L}[\begin{matrix} q_1 q_1 & q_1 q_2 \end{matrix}] \begin{bmatrix} V_1 \\ V_2 \end{bmatrix} \qquad (8.20)$$

Performing the same calculations for the second node, we obtain for ϕ_2:

$$\nabla \phi_2 = \hat{\mathbf{i}} \frac{1}{L} q_2 \qquad (8.21)$$

and, in a manner analogous to Eq. (8.20), we have

$$\frac{\varepsilon}{L}[\begin{matrix} q_2 q_1 & q_2 q_2 \end{matrix}] \begin{bmatrix} V_1 \\ V_2 \end{bmatrix} \qquad (8.22)$$

Taking into account both nodes, we get the following system:

$$\frac{\varepsilon}{L} \begin{bmatrix} q_1 q_1 & q_1 q_2 \\ q_2 q_1 & q_2 q_2 \end{bmatrix} \begin{bmatrix} V_1 \\ V_2 \end{bmatrix} \qquad (8.23)$$

Now we evaluate the second term of Eq. (8.16); considering ρ as constant in element n, we have

$$-\rho \int_{\Omega_n} \phi_n dx$$

Taking, for example, $\phi_n = \phi_1$, and observing the slope of ϕ_1 in Figure 8.3 (W_k is now ϕ_1 for element n), this integral represents the evaluation of the surface of the triangle with basis L and height "1;" therefore, for nodes 1 and 2 it yields the matrix contribution

$$-\frac{\rho L}{2} \begin{bmatrix} 1 \\ 1 \end{bmatrix} \qquad (8.24)$$

This contribution to the system does not depend on the unknown potentials and, therefore, is assembled onto the right-hand side of the global system of equations, at locations corresponding to the numbers of the two nodes. Equation (8.24) is also called the "source" term, since it generates electric fields in addition to those generated by the so called Dirichlet boundary conditions. If $\rho=0$, the electric field is generated only by the Dirichlet conditions on the boundary. Dirichlet boundary conditions will be discussed shortly.

Suppose now that the solution domain is divided into two finite elements as shown in Figure 8.5 (in this example $\rho=0$). In this representation, there is only one unknown value at $x=7$, which is at node No. 2. The potentials at nodes No. 1 and 3 are known boundary conditions.

Calculating the coefficients in the matrix in Eq. (8.23) for element No. 1 gives:

$$\frac{\varepsilon}{7}\begin{bmatrix} (-1)(-1) & (-1)(1) \\ (1)(-1) & (1)(1) \end{bmatrix}\begin{bmatrix} V_1 \\ V_2 \end{bmatrix}$$

where $L=7$, $q_1=-1$, $q_2=1$ were used.

Because there are three nodes in the solution domain, the matrix above, also called an "elemental matrix," must be placed or assembled into a larger "global matrix" which includes all the nodes in the system. Doing so results in the following global matrix:

$$\varepsilon \begin{bmatrix} \frac{1}{7} & -\frac{1}{7} & 0 \\ -\frac{1}{7} & \frac{1}{7} & 0 \\ 0 & 0 & 0 \end{bmatrix} \begin{bmatrix} V_1 \\ V_2 \\ V_3 \end{bmatrix} \quad (8.25)$$

Evaluating now element No. 2 (nodes No. 2 and 3) results in the following elemental matrix:

$$\varepsilon \begin{bmatrix} \frac{1}{3} & -\frac{1}{3} \\ -\frac{1}{3} & \frac{1}{3} \end{bmatrix} \begin{bmatrix} V_2 \\ V_3 \end{bmatrix}$$

This is added, or assembled into the global system in Eq. (8.25):

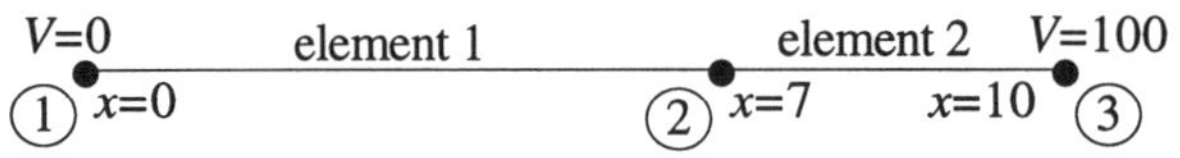

Figure 8.5. A discretized *1D* domain.

$$\varepsilon \begin{bmatrix} \frac{1}{7} & -\frac{1}{7} & 0 \\ -\frac{1}{7} & \frac{1}{7}+\frac{1}{3} & -\frac{1}{3} \\ 0 & -\frac{1}{3} & \frac{1}{3} \end{bmatrix} \begin{bmatrix} V_1 \\ V_2 \\ V_3 \end{bmatrix} = \begin{bmatrix} 0 \\ 0 \\ 0 \end{bmatrix} \qquad (8.26)$$

The right-hand side has been set to zero as required in Eq. (8.16).
Because the values of V_1 and V_3 are equal to 0 and 100, respectively (boundary conditions), the system in Eq. (8.26) is modified as

$$\begin{bmatrix} 1 & 0 & 0 \\ -\frac{1}{7} & \frac{10}{21} & -\frac{1}{3} \\ 0 & 0 & 1 \end{bmatrix} \begin{bmatrix} V_1 \\ V_2 \\ V_3 \end{bmatrix} = \begin{bmatrix} 0 \\ 0 \\ 100 \end{bmatrix} \qquad (8.27)$$

This is the same as writing the second equation alone:

$$-\frac{1}{7}V_1 + \frac{10}{21}V_2 - \frac{1}{3}V_3 = 0$$

With $V_1=0$ and $V_3=100$, we obtain

$$V_2 = 70 \ V$$

as expected. Finally, we can also obtain the value of the electric field intensity as

$$\mathbf{E} = -\nabla V(x) = -\hat{\mathbf{i}} \frac{\partial}{\partial x} \left(\sum_{i=1,2} \frac{1}{L}(p_i + q_i x) V_i \right)$$

or

$$\mathbf{E} = -\hat{\mathbf{i}} \sum_{i=1,2} \frac{1}{L} q_i V_i = -\hat{\mathbf{i}} \frac{1}{L}(q_1 V_1 + q_2 V_2) \qquad (8.28)$$

For element No. 1, this gives

$$\mathbf{E}_1 = -\hat{\mathbf{i}} \frac{1}{7}\left(-1 \times 0 + 1 \times 70\right) = -\hat{\mathbf{i}} 10 \ \frac{V}{m}$$

For the second element, we have

$$\mathbf{E}_2 = -\hat{\mathbf{i}} \frac{1}{3}\left(-1 \times 70 + 1 \times 100\right) = -\hat{\mathbf{i}} 10 \ \frac{V}{m}$$

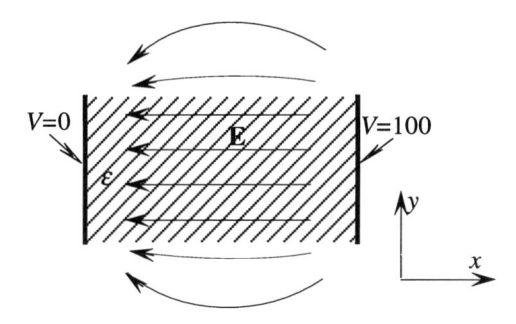

Figure 8.6. 2D capacitor. The electric field intensity is shown.

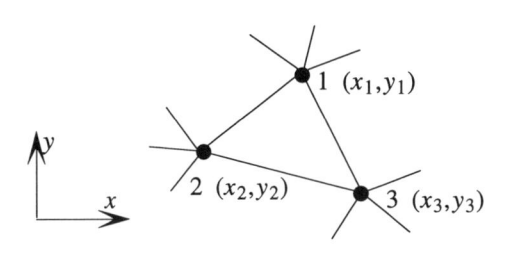

Figure 8.7. An element in a triangular element mesh.

8.3. The Galerkin Method – Extension to 2D

Suppose now we wish to consider the parallel plate capacitor of the previous section but also allow the electric field to have a component in the y direction. That is, we wish to take into account the edge effects due to the finite plate size as shown in Figure 8.6.

In this case, two dimensional finite elements must be defined. We use triangular elements. Suppose a mesh of triangular elements has been defined over a surface. A general element in this mesh is shown in Figure 8.7.

If the potential varies linearly within the triangle, the element is known as a "first-order element." For this type of element, the expansion of the potential is

$$V(x,y) = a_1 + a_2 x + a_3 y \qquad (8.29)$$

As for the 1D element, this relation must hold at the nodes of the element. For the nodes in Figure 8.7, we get

$$V_1 = a_1 + a_2 x_1 + a_3 y_1$$
$$V_2 = a_1 + a_2 x_2 + a_3 y_2$$
$$V_3 = a_1 + a_2 x_3 + a_3 y_3$$

278   8. Introduction to the Finite Element Method

From these three equations, we determine the required values of a_1, a_2, and a_3 by calculating the determinants in the following:

$$a_1 = \frac{1}{D}\begin{vmatrix} V_1 & x_1 & y_1 \\ V_2 & x_2 & y_2 \\ V_3 & x_3 & y_3 \end{vmatrix} \qquad a_2 = \frac{1}{D}\begin{vmatrix} 1 & V_1 & y_1 \\ 1 & V_2 & y_2 \\ 1 & V_3 & y_3 \end{vmatrix}$$

$$a_3 = \frac{1}{D}\begin{vmatrix} 1 & x_1 & V_1 \\ 1 & x_2 & V_2 \\ 1 & x_3 & V_3 \end{vmatrix} \quad \text{with} \quad D = \begin{vmatrix} 1 & x_1 & y_1 \\ 1 & x_2 & y_2 \\ 1 & x_3 & y_3 \end{vmatrix}$$

(8.30)

The value of D equals twice the area of the element as can be verified directly. Substituting the values of a_1, a_2, and a_3 in Eq. (8.29) and simplifying the expressions gives

$$V(x,y) = \sum_{i=1}^{3} \frac{1}{D}(p_i + q_i x + r_i y) V_i \qquad (8.31)$$

where

$$p_1 = x_2 y_3 - x_3 y_2 \qquad q_1 = y_2 - y_3 \qquad r_1 = x_3 - x_2 \qquad (8.32)$$

while the remaining terms: p_2, q_2, r_2, p_3, q_3, and r_3 are obtained by cyclical permutation of the indices.

Because $\mathbf{E} = -\nabla V$, we have

$$\mathbf{E} = \hat{\mathbf{i}} E_x + \hat{\mathbf{j}} E_y = -\hat{\mathbf{i}}\frac{\partial V}{\partial x} - \hat{\mathbf{j}}\frac{\partial V}{\partial y} \qquad (8.33a)$$

and

$$E_x = -\frac{1}{D}\sum_{i=1}^{3} q_i V_i \qquad E_y = -\frac{1}{D}\sum_{i=1}^{3} r_i V_i \qquad (8.33b)$$

8.3.1. *The Boundary Conditions*

When treating two-dimensional problems, we need to examine the boundary conditions in some detail. The boundary conditions are related to the first integral on the right-hand side of Eq. (8.14), which is:

$$\oint_{L(S)} W \varepsilon \nabla V \cdot d\mathbf{s} = 0 \qquad (8.34)$$

There are two types of boundary conditions we need to contend with:

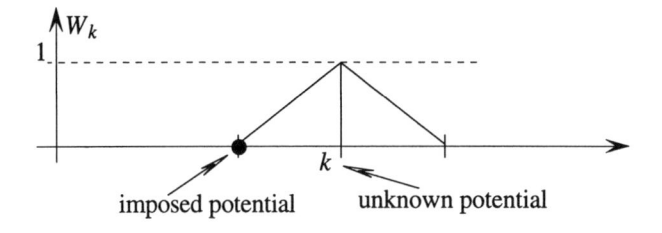

Figure 8.8. Dirichlet boundary condition scheme.

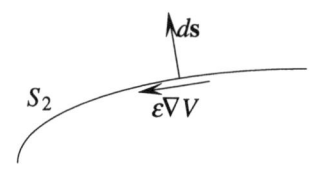

Figure 8.9. Neumann boundary condition scheme.

a. Dirichlet boundary condition – imposed potential.

Consider a physical configuration in which the potentials are known on part S_1 of the boundary. This is called a "Dirichlet boundary condition." When we come to write the equations for the unknowns at the nodes of the mesh, the weighting functions W_k are only needed for the internal nodes of the mesh. At the Dirichlet boundary nodes the weighting functions are zero (see section 8.2 and Figure 8.8). This condition assumes that Eq. (8.34) is satisfied.

b. Neumann condition – unknown nodal values on the boundary.

In certain cases, on part of the boundary $S_2 = L(S) - S_1$, the values of the potential are unknown. On this part of the boundary, Eq. (8.14) must be written and the weighting function in Eq. (8.34) is not zero. Moreover, because the integral in Eq. (8.34) is set to zero, we have

$$\varepsilon \nabla V \cdot d\mathbf{s} = 0 \qquad (8.35)$$

Examining this expression, and taking into account the scalar product, we conclude that the electric field intensity must be tangential to the boundary S_2 as shown in Figure 8.9.

8.3.2. *Calculation of the 2D Elemental Matrix*

To construct the elemental matrix, we must evaluate the expression in Eq. (8.16). To do so, we write Eq. (8.31) as follows:

8. Introduction to the Finite Element Method

$$V(x,y) = \phi_1(x,y)V_1 + \phi_2(x,y)V_2 + \phi_3(x,y)V_3 \quad (8.36)$$

where

$$\phi_1(x,y) = \frac{1}{D}(p_1 + q_1 x + r_1 y) \quad (8.37a)$$

$$\phi_2(x,y) = \frac{1}{D}(p_2 + q_2 x + r_2 y) \quad (8.37b)$$

$$\phi_3(x,y) = \frac{1}{D}(p_3 + q_3 x + r_3 y) \quad (8.37c)$$

Similarly, we can write

$$\nabla \phi_1 = \hat{i}\frac{1}{D} q_1 + \hat{j}\frac{1}{D} r_1 \quad (8.38a)$$

$$\nabla \phi_2 = \hat{i}\frac{1}{D} q_2 + \hat{j}\frac{1}{D} r_2 \quad (8.38b)$$

$$\nabla \phi_3 = \hat{i}\frac{1}{D} q_3 + \hat{j}\frac{1}{D} r_3 \quad (8.38c)$$

Also,

$$\nabla V = \hat{i}\frac{1}{D}(q_1 V_1 + q_2 V_2 + q_3 V_3) + \hat{j}\frac{1}{D}(r_1 V_1 + r_2 V_2 + r_3 V_3) \quad (8.39)$$

Since Eqs. (8.38) and (8.39) are constants, the first term of Eq. (8.16) becomes

$$\frac{\varepsilon}{D^2}(\hat{i} q_1 + \hat{j} r_1) \cdot [\hat{i}(q_1 V_1 + q_2 V_2 + q_3 V_3) + \hat{j}(r_1 V_1 + r_2 V_2 + r_3 V_3)] \int_{S_n} ds$$

and, noting that the integral on S_n equals the area of element n (that is, it equals $D/2$), we get in matrix form

$$\frac{\varepsilon}{2D}[(q_1 q_1 + r_1 r_1) \quad (q_1 q_2 + r_1 r_2) \quad (q_1 q_3 + r_1 r_3)] \begin{bmatrix} V_1 \\ V_2 \\ V_3 \end{bmatrix}$$

Extending the integral for $n=2$ and $n=3$, we obtain the elemental stiffness matrix

$$\frac{\varepsilon}{2D}\begin{bmatrix} q_1 q_1 + r_1 r_1 & q_1 q_2 + r_1 r_2 & q_1 q_3 + r_1 r_3 \\ \text{Symmetric} & q_2 q_2 + r_2 r_2 & q_2 q_3 + r_2 r_3 \\ \text{Symmetric} & \text{Symmetric} & q_3 q_3 + r_3 r_3 \end{bmatrix} \begin{bmatrix} V_1 \\ V_2 \\ V_3 \end{bmatrix} \quad (8.40)$$

Assembly of the elemental matrices into a global matrix and solution of the resulting system of equations is similar to that for the 1D element in the previous section.

Similarly to the 1D element, we evaluate now the second term of Eq. (8.16)

$$\int_{S_n} \phi_n \rho \, ds \qquad (8.41)$$

As for the 1D element, each of the functions ϕ_i equals 1 at node i and decreases to zero at all other nodes of the element. For example, ϕ_1 equals 1 at node No. 1 and zero at nodes No. 2 and 3, as shown in Figure 8.10.

The evaluation of the integral in Eq. (8.41) corresponds to calculating the volume of the pyramid of height 1 shown in Figure 8.10. This gives

$$\frac{1}{3}\rho\frac{D}{2}$$

This, however, is only due to the function ϕ_1. Performing identical calculations on ϕ_2 and ϕ_3, we obtain the contribution due to charge density as

$$\frac{\rho D}{6}\begin{bmatrix}1\\1\\1\end{bmatrix} \qquad (8.42)$$

which is the source term to be assembled on the right-hand side of the matrix system.

8.4. The Variational Method – Basic Concepts

The variational method is a different way of applying the finite element method. The method has been employed extensively in electromagnetic formulations in the past, but recently Galerkin's method has most often been employed.

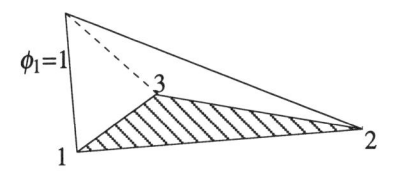

Figure 8.10. Function ϕ_1 for a triangular element.

8. Introduction to the Finite Element Method

With the variational method, instead of solving directly the governing equations, we minimize the "energy functional" (or, more often, an "energy related functional") corresponding to the governing equations. Needless to say, this can only be done if the energy functional is known. Using a very simple and not entirely rigorous analogy, the mathematical idea behind this method can be described by the following example. Suppose that the equation "$ax+b=0$" must be solved; instead of solving it directly, we can minimize the function

$$f(x) = \frac{ax^2}{2} + bx + c$$

Assuming that this function has a minimum, this minimum corresponds to the point where $df/dx=0$, or

$$\frac{df}{dx} = ax + b = 0$$

which is the equation we wish to solve.

In physical terms, the state for which the energy functional is minimum, corresponds to an equilibrium state of any physical system possessing potential energy. In the example chosen here, which is described by the equation $\nabla \cdot \mathbf{D} = \rho$, the energy functional in a dielectric medium is

$$F = \int_\Omega \left[\frac{1}{2} \varepsilon E^2 - \rho V \right] d\Omega \qquad (8.43)$$

The discretized functional for the 1D problem, with N elements (or segments), is

$$F = \sum_{n=1}^{N} \int_L \left[\frac{1}{2} \varepsilon E_n^2 - \rho V \right] dl \qquad (8.44)$$

To minimize this function, we need to find the minimum with respect to each unknown potential value in the solution domain

$$\frac{\partial F}{\partial V_k} = \sum_{n=1}^{N} \frac{\partial F_n}{\partial V_k} = 0$$

which results in k equations. Therefore, we need to evaluate $\partial F_n/\partial V_k$ for all elements. For a generic element, the first integral of Eq. (8.44) is

$$\frac{\partial F_n}{\partial V_k} = \frac{\partial}{\partial V_k} \int_L \frac{1}{2} \varepsilon E_n^2 dl$$

The Variational Method - Basic Concepts   283

With the adopted approximation for $V(x)$ [Eq. (8.3)], the electric field intensity is constant in the element and we can write

$$\frac{\partial F_n}{\partial V_k} = \frac{\partial}{\partial V_k}\left[\frac{1}{2}\varepsilon E^2\right]\int_L dl = \frac{1}{2}\varepsilon \frac{\partial E^2}{\partial V_k} L \qquad (8.45)$$

From Eq. (8.28),

$$\mathbf{E} = -\hat{\mathbf{i}}\frac{1}{L}\sum_{i=1,2} q_i V_i = -\hat{\mathbf{i}}\frac{1}{L}(q_1 V_1 + q_2 V_2)$$

and

$$E^2 = \frac{1}{L^2}(q_1 V_1 + q_2 V_2)^2$$

For $k=1$, this gives

$$\frac{\partial E^2}{\partial V_1} = \frac{2}{L^2}(q_1 V_1 + q_2 V_2)q_1$$

and, substituting this in Eq. (8.45)

$$\frac{\partial F_n}{\partial V_1} = \frac{1}{2}\varepsilon L \frac{2}{L^2}(q_1 q_1 V_1 + q_1 q_2 V_2)$$

or

$$\frac{\partial F_n}{\partial V_1} = \frac{\varepsilon}{L}[\, q_1 q_1 \quad q_1 q_2 \,]\begin{bmatrix} V_1 \\ V_2 \end{bmatrix}$$

Performing similar operations for the second node, we obtain

$$\begin{bmatrix} \dfrac{\partial F_n}{\partial V_1} \\ \dfrac{\partial F_n}{\partial V_2} \end{bmatrix} = \frac{\varepsilon}{L}\begin{bmatrix} q_1 q_1 & q_1 q_2 \\ q_2 q_1 & q_2 q_2 \end{bmatrix}\begin{bmatrix} V_1 \\ V_2 \end{bmatrix} \qquad (8.46)$$

This is the stiffness or contribution matrix for the element and is identical to Eq. (8.23) which was obtained using the Galerkin method.

For the second term of Eq. (8.44), we can write:

$$\frac{\partial F_n}{\partial V_k} = -\frac{\partial}{\partial V_k}\int_L \rho V dl$$

Noting that $V(x)=\phi_1(x)V_1+\phi_2(x)V_2$ and assuming that $V_k=V_1$, we have for the expression above

$$-\int_L \rho\phi_1 dl$$

which has already been calculated. For nodes 1 and 2, we obtain the source term

$$-\frac{\rho L}{2}\begin{bmatrix}1\\1\end{bmatrix}$$

This is identical to the expression in Eq. (8.24).

The assembly of the elemental matrices and source terms of all elements in the solution domain into a global matrix, and solution of the matrix for the unknown values of potential are identical to those performed in the previous sections.

8.5. The Variational Method – Extension to 2D

In order to extend the variational formulation of the previous section to two dimensional applications, it is necessary first to dwell on the fundamentals of the variational method since important statements such as the "Euler equation" and boundary conditions arise from the method. This is done next.

8.5.1. *The Variational Formulation*

Suppose that a functional F exists, and that it is a function of a variable P and of its partial derivatives $P'_x=\partial P/\partial x$, $P'_y=\partial P/\partial y$, and $P'_z=\partial P/\partial z$. Then:

$$F = \int_V f(P, P'_x, P'_y, P'_z)dv$$

where V is the volume in which F is defined.

The necessary and sufficient condition for F to be stationary (minimum condition) is that for any small variation δP of the independent variable P, the corresponding variation δF is zero. This is written formally as

The Variational Method - Extension to 2D 285

$$\delta F = \int_V \left(\frac{\partial f}{\partial P} \delta P + \frac{\partial f}{\partial P'_x} \delta P'_x + \frac{\partial f}{\partial P'_y} \delta P'_y + \frac{\partial f}{\partial P'_z} \delta P'_z \right) dv = 0 \qquad (8.47)$$

Note that $\delta P'_x = \delta(\partial P/\partial x) = \partial/\partial x (\delta P)$ (and similarly for $\delta P'_y$ and $\delta P'_z$). The following can now be defined:

$$\hat{i} \frac{\partial}{\partial x}(\delta P) + \hat{j} \frac{\partial}{\partial y}(\delta P) + \hat{k} \frac{\partial}{\partial z}(\delta P) = \nabla(\delta P) \qquad (8.48)$$

and also,

$$\hat{i} \frac{\partial f}{\partial P'_x} + \hat{j} \frac{\partial f}{\partial P'_y} + \hat{k} \frac{\partial f}{\partial P'_z} = \mathbf{g} \qquad (8.49)$$

The last three terms in the brackets in Eq. (8.47) can be written in the following form, using Eqs. (8.48) and (8.49)

$$\frac{\partial f}{\partial P'_x} \delta P'_x + \frac{\partial f}{\partial P'_y} \delta P'_y + \frac{\partial f}{\partial P'_z} \delta P'_z = \mathbf{g} \cdot \nabla(\delta P)$$

and that part of Eq. (8.47) relating to these terms is, therefore,

$$\int_V \mathbf{g} \cdot \nabla(\delta P) dv \qquad (8.50)$$

The scalar product between the ∇ operator and a general vector function $c\mathbf{d}$ (c is a scalar, $\mathbf{d}$ is a vector) is

$$\nabla \cdot (c\mathbf{d}) = c\nabla \cdot \mathbf{d} + \mathbf{d} \cdot \nabla c$$

Therefore, Eq. (8.50) can be written as

$$\int_V \mathbf{g} \cdot \nabla(\delta P) dv = \int_V \nabla \cdot (\delta P \mathbf{g}) dv - \int_V \delta P (\nabla \cdot \mathbf{g}) dv$$

Applying the divergence theorem, we get

$$\int_V \mathbf{g} \cdot \nabla(\delta P) dv = \oint_S \delta P \mathbf{g} \cdot \hat{\mathbf{n}} ds - \int_V \delta P (\nabla \cdot \mathbf{g}) dv \qquad (8.51)$$

where S is the surface enclosing the volume V and $\hat{\mathbf{n}}$ is the normal unit vector at any point on S. Substituting the terms of Eq. (8.51) into Eq. (8.47), we get

$$\delta F = \int_V \left(\frac{\partial f}{\partial P} - \nabla \cdot \mathbf{g}\right)\delta P dv + \oint_S \delta P \mathbf{g} \cdot \hat{\mathbf{n}} ds = 0$$

Because the two integrals are independent, for their sum to be zero the following must be satisfied:

$$\int_V \left(\frac{\partial f}{\partial P} - \nabla \cdot \mathbf{g}\right)\delta P dv = 0 \tag{8.52}$$

and

$$\oint_S \delta P \mathbf{g} \cdot \hat{\mathbf{n}} ds = 0 \tag{8.53}$$

Since δP is arbitrary, and since Eq. (8.52) must be zero, we get

$$\left(\frac{\partial f}{\partial P} - \nabla \cdot \mathbf{g}\right) = 0$$

or

$$\frac{\partial f}{\partial P} - \frac{\partial}{\partial x}\left(\frac{\partial f}{\partial P'_x}\right) - \frac{\partial}{\partial y}\left(\frac{\partial f}{\partial P'_y}\right) - \frac{\partial}{\partial z}\left(\frac{\partial f}{\partial P'_z}\right) = 0 \tag{8.54}$$

This equation is known as "Euler's equation." For the functional F to be valid, it must have an integrand that satisfies Eq. (8.54). Equation (8.53), which relates to the integration on the surface S, is now examined separately. For this equation to be zero, either δP or "$\mathbf{g} \cdot \hat{\mathbf{n}}$" is zero. We can, a priori, divide the surface S into two parts S_1 and S_2 (one of these surfaces can be nonexistent, or have zero area):

- S_1 is that part of the boundary on which $\delta P=0$; on this part the potential is imposed as a boundary condition, called a "Dirichlet condition." Since P is fixed, its variation δP is zero.
- $S_2=S-S_1$ is the part of the boundary on which $\mathbf{g} \cdot \hat{\mathbf{n}}=0$; on this boundary we have the so called "Neumann boundary condition;" on this boundary, no condition is imposed. The Neumann boundary condition is natural and is as an intrinsic property of this formulation. On S_2, we can write

$$\mathbf{g} \cdot \hat{\mathbf{n}} = \frac{\partial f}{\partial P'_x} n_x + \frac{\partial f}{\partial P'_y} n_y + \frac{\partial f}{\partial P'_z} n_z = 0 \qquad (8.55)$$

where

$$\hat{\mathbf{n}} = \hat{\mathbf{i}} n_x + \hat{\mathbf{j}} n_y + \hat{\mathbf{k}} n_z$$

In the following paragraphs we will examine the physical meaning of "Euler's equation" [Eq. (8.54)] and the boundary conditions, especially Eq. (8.55), as applied to the cases above.

Going back to the energy functional, we can say that to establish the value of P minimizing the functional F entails:

- Establishment of the functional F whose integrand satisfies Euler's equation.
- Application of boundary conditions, pertinent to the problem, where either P is fixed (Dirichlet condition), or Eq. (8.55) (Neumann condition) is satisfied.

In our example, assuming the charge density ρ is not zero, the electrostatic field functional for a dielectric medium is

$$F = \int_V f(V, V'_x, V'_y) \, dv = \int_V \left(\frac{1}{2} \varepsilon E^2 - \rho V \right) dv \qquad (8.56)$$

where the integrand in the first term is the volume energy density, as described in Chapter 3. The Euler equation for this 2D case is

$$\frac{\partial f}{\partial V} - \frac{\partial}{\partial x}\left(\frac{\partial f}{\partial V'_x}\right) - \frac{\partial}{\partial y}\left(\frac{\partial f}{\partial V'_y}\right) = 0 \qquad (8.57)$$

where the variable P in Eq. (8.54) is the scalar potential V. Calculating the different terms of this expression, we get

$$\frac{\partial f}{\partial V} = \frac{\partial}{\partial V}\left(\frac{1}{2}\varepsilon E^2 - \rho V\right) = -\rho \qquad (8.58)$$

Noting that

$$V'_x = \frac{\partial V}{\partial x} = -E_x$$

we get

$$\frac{\partial f}{\partial V'_x} = -\frac{\partial f}{\partial E_x} = -\frac{\partial}{\partial E_x}\left(\frac{1}{2}\varepsilon E^2 - \rho V\right) = -\frac{\partial}{\partial E_x}\left(\frac{1}{2}\varepsilon E^2\right)$$

Because E depends on E_x,

$$-\frac{\partial}{\partial E_x}\left(\frac{1}{2}\varepsilon E^2\right) = -\frac{\partial}{\partial E}\left(\frac{1}{2}\varepsilon E^2\right)\frac{\partial E}{\partial E_x} = -\varepsilon E\frac{\partial E}{\partial E_x}$$

Because $E=(E_x^2+E_y^2)^{1/2}$, after some algebraic operations we get

$$\frac{\partial f}{\partial V_x'} = -\varepsilon E_x \qquad (8.59)$$

and identically for the second term,

$$\frac{\partial f}{\partial V_y'} = -\varepsilon E_y \qquad (8.60)$$

substituting Eqs. (8.58), (8.59), and (8.60) in Eq. (8.57) gives

$$-\rho + \frac{\partial}{\partial x}(\varepsilon E_x) + \frac{\partial}{\partial y}(\varepsilon E_y) = 0$$

which can be written in the following form

$$(\hat{\mathbf{i}}\frac{\partial}{\partial x}+\hat{\mathbf{j}}\frac{\partial}{\partial y})\cdot(\hat{\mathbf{i}}\varepsilon E_x + \hat{\mathbf{j}}\varepsilon E_y) = \rho$$

Using $\mathbf{D}=\varepsilon\mathbf{E}$, we get, by applying Euler's equation

$$\nabla\cdot\mathbf{D} = \rho$$

This is Maxwell's equation for our problem, and therefore Euler's equation is valid. As for the Neumann boundary condition in two dimensions, we have from Eq. (8.55),

$$\frac{\partial f}{\partial V_x'}n_x + \frac{\partial f}{\partial V_y'}n_y = 0$$

Using Eqs. (8.59) and (8.60), this expression yields

$$D_x n_x + D_y n_y = \mathbf{D}\cdot\hat{\mathbf{n}} = 0$$

In Figure 8.11a the electrical field $\mathbf{E}$ is parallel to the boundary S_2. Thus, for the boundary on which no condition is specified, the field is tangent to the boundary.

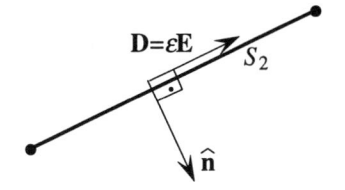

Figure 8.11a. Neumann boundary condition. If no condition is specified on a boundary, the electric field intensity is parallel to that boundary.

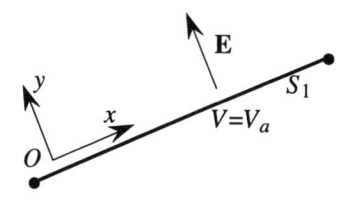

Figure 8.11b. Dirichlet boundary condition. The electric field intensity is perpendicular to a Dirichlet boundary.

As for the Dirichlet condition, examine Figure 8.11b, where the potential $V=V_a$ is imposed on boundary S_1. We define a local coordinate system Oxy with the direction Ox on line S_1. In this case, since $V=V_a$ is constant in the direction Ox, $dV/dx=0$ and, therefore, $E_x=0$. This limitation does not exist in the Oy direction, and the field $E_y \neq 0$, corresponding to the situation shown in Figure. 8.11b.

8.5.2. Calculation of the 2D Elemental Matrix

To derive the 2D elemental matrix, we minimize the functional in Eq. (8.56) as follows:

$$\frac{\partial F_n}{\partial V_k} = \frac{\partial}{\partial V_k} \int_{S_n} \left(\frac{1}{2} \varepsilon E_n^2 - \rho V \right) ds \tag{8.61}$$

The first term of Eq. (8.61) gives

$$\frac{1}{2} \varepsilon \frac{\partial E_n^2}{\partial V_k} \frac{D}{2} \tag{8.62}$$

where, using Eqs. (8.33a) and (8.33b)

$$E^2 = \frac{1}{D^2}\left[(q_1 V_1 + q_2 V_2 + q_3 V_3)^2 + (r_1 V_1 + r_2 V_2 + r_3 V_3)^2\right]$$

The derivative with respect to V_1 is

$$\frac{\partial E^2}{\partial V_1} = \frac{2}{D^2}\left[(q_1 V_1 + q_2 V_2 + q_3 V_3)q_1 + (r_1 V_1 + r_2 V_2 + r_3 V_3)r_1\right]$$

which, when substituted in Eq. (8.62), gives

$$\frac{\varepsilon}{2D}[(q_1q_1 + r_1r_1)V_1 + (q_1q_2 + r_1r_2)V_2 + (q_1q_3 + r_1r_3)V_3]$$

Extending this calculation to the three nodes of the 2D element and writing them in matrix form, we get

$$\frac{\varepsilon}{2D}\begin{bmatrix} q_1q_1 + r_1r_1 & q_1q_2 + r_1r_2 & q_1q_3 + r_1r_3 \\ Symmetric & q_2q_2 + r_2r_2 & q_2q_3 + r_2r_3 \\ Symmetric & Symmetric & q_3q_3 + r_3r_3 \end{bmatrix}\begin{bmatrix} V_1 \\ V_2 \\ V_3 \end{bmatrix} \quad (8.63)$$

This system is identical to the expression in Eq. (8.40) which was obtained using Galerkin's method.

The second term of Eq. (8.61) is

$$-\rho \frac{\partial}{\partial V_k} \int_{S_n} V ds = -\rho \int_{S_n} \frac{\partial V}{\partial V_k} ds$$

Because

$$V(x,y) = \phi_1(x,y)V_1 + \phi_2(x,y)V_2 + \phi_3(x,y)V_3$$

the derivative of $V(x,y)$ with respect to V_1 equals ϕ_1, with respect to V_2 equals ϕ_2, and with respect to V_3, equals ϕ_3. Thus, the integration for $k=1$ is

$$-\rho \int_{S_n} \frac{\partial V}{\partial V_1} ds = -\rho \int_{S_n} \phi_1 ds = -\frac{\rho D}{6}$$

Repeating this step for $k=2$ and $k=3$, and writing the result in matrix form we obtain

$$\begin{bmatrix} \dfrac{\partial F_n}{\partial V_1} \\ \dfrac{\partial F_n}{\partial V_2} \\ \dfrac{\partial F_n}{\partial V_3} \end{bmatrix} = \frac{\varepsilon}{2D}\begin{bmatrix} q_1q_1 + r_1r_1 & q_1q_2 + r_1r_2 & q_1q_3 + r_1r_3 \\ Symmetric & q_2q_2 + r_2r_2 & q_2q_3 + r_2r_3 \\ Symmetric & Symmetric & q_3q_3 + r_3r_3 \end{bmatrix}\begin{bmatrix} V_1 \\ V_2 \\ V_3 \end{bmatrix} - \frac{D}{6}\begin{bmatrix} \rho \\ \rho \\ \rho \end{bmatrix} \quad (8.64)$$

This is, of course, identical to the result obtained from the Galerkin formulation. As in the previous case, the contribution due to the charge density must be assembled into the right-hand side of the system of equations since it is independent of the potential.

8.6. Generalization of the Finite Element Method

The finite elements described in the previous sections were very simple, primarily because we allowed only linear variation between the nodes of the elements. There are more accurate finite elements, but their introductions requires some concepts which we will introduce in the following section.

First, it is worth mentioning here that for a 1D element, with a linear variation of the potential between the nodes, two nodes are necessary, as we have seen in section 8.2. The potential varies as

$$V(x) = a_1 + a_2 x$$

We used the two nodes to evaluate the two constants a_1 and a_2 by satisfying this equation at the location of the two nodes:

$$V_1 = a_1 + a_2 x_1$$
$$V_2 = a_1 + a_2 x_2$$

That is, the approximation for the potential is a first-order polynomial approximation. If we wish to obtain better accuracy, we can use quadratic elements which have the following variation for potential:

$$V(x) = a_1 + a_2 x + a_3 x^2 \qquad (8.65)$$

This approximation requires three nodes to determine the constants a_1, a_2, and a_3. Assuming an element with three nodes is given, the constants are evaluated from the following:

$$V_1 = a_1 + a_2 x_1 + a_3 x_1^2$$
$$V_2 = a_1 + a_2 x_2 + a_3 x_2^2$$
$$V_3 = a_1 + a_2 x_3 + a_3 x_3^2$$

where V_1, V_2, and V_3 are the unknown potentials at the coordinates (x_1,y_1), (x_2,y_2), and (x_3,y_3).

The main point in the discussion above is that there is a relationship between the order of the approximation and the number of nodes defining the element. Although we used a 1D element, this is also true in 2D and 3D elements. For example, a quadratic variation of the potential in 2D elements is

$$V(x,y) = a_1 + a_2 x + a_3 y + a_4 xy + a_5 x^2 + a_6 y^2 \qquad (8.66)$$

This requires six nodes, such as a six-node triangular element. This will be discussed shortly.

8.6.1. *High-Order Finite Elements: General*

Figure 8.12 shows some of the most commonly used finite elements in one, two, and three dimensions. The higher order (second- or third-order) elements are also called high-precision elements.

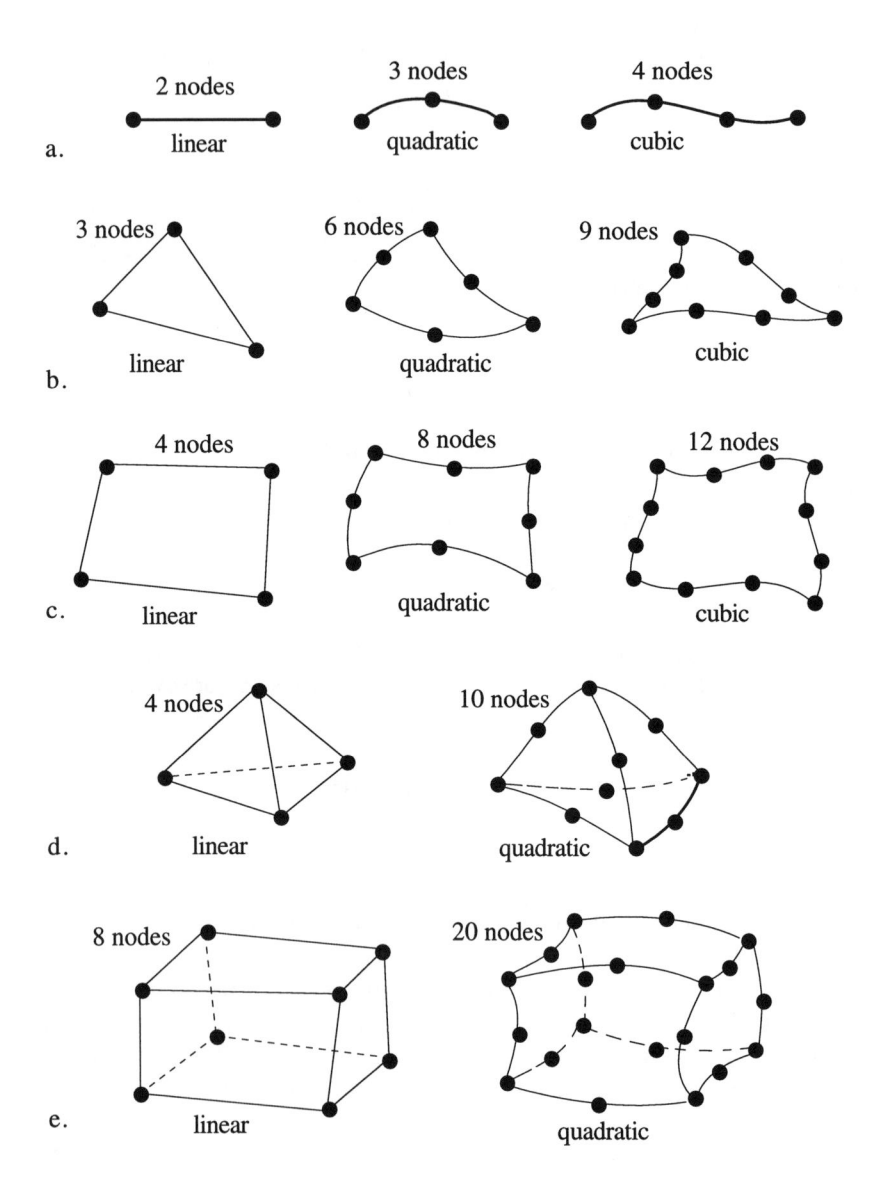

Figure 8.12. a. 1D elements. b. Triangular 2D elements. c. Quadrilateral 2D elements. d. Tetrahedral 3D elements. e. Hexahedral 3D elements.

There are many other finite elements but the elements shown in Figure 8.12 are the most commonly used in electromagnetic applications. Information about additional elements can be found in references in the bibliography section.

To apply the finite elements shown here, it is first necessary to introduce some notation and relations, which we do in the following section.

8.6.2. High-Order Finite Elements: Notation

To facilitate the definition of various finite elements, we introduce the idea of a "reference" or "local" element and the reference or local system of coordinates or space. Figure 8.13 shows an example and the relationship between the local and global systems of coordinates. The various relations needed to define an element are generated in the local system of coordinates because it is easier to do so. Then, a unique transformation is established which transforms the element from the local coordinate system to the global coordinate system. This transformation is accomplished by the so called "geometric transformation functions" or "mapping functions" or "shape functions" which express the real coordinates x, y in terms of the local coordinates u, v.

In Figure 8.13, the triangle in local coordinates is defined as

$$u \geq 0 \qquad v \geq 0 \qquad u + v \leq 1 \tag{8.67}$$

The approximation within the triangle can be written in terms of the shape functions $N(u,v)$ as

$$x(u,v) = N_1(u,v)x_1 + N_2(u,v)x_2 + N_3(u,v)x_3 \tag{8.68}$$

or, in matrix form as

$$x(u,v) = \begin{bmatrix} N_1(u,v) & N_2(u,v) & N_3(u,v) \end{bmatrix} \begin{bmatrix} x_1 \\ x_2 \\ x_3 \end{bmatrix} \tag{8.69}$$

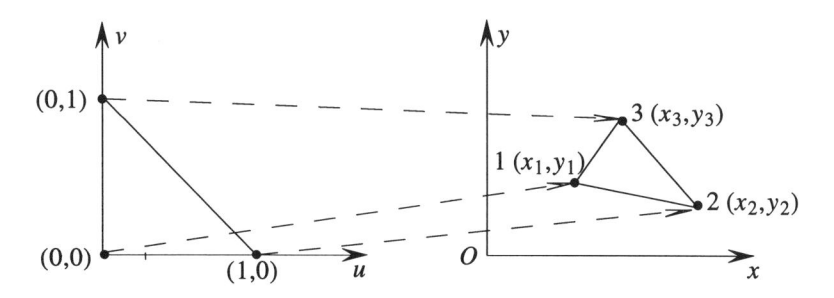

Figure 8.13. A finite element defined in a local system of coordinates and mapping to the global system of coordinates.

8. Introduction to the Finite Element Method

For first-order triangles, the shape functions in local coordinates are

$$N_1(u,v) = 1 - u - v \qquad N_2(u,v) = u \qquad N_3(u,v) = v$$

And Eq. (8.69) becomes

$$x(u,v) = \begin{bmatrix} 1-u-v & u & v \end{bmatrix} \begin{bmatrix} x_1 \\ x_2 \\ x_3 \end{bmatrix} \qquad (8.70)$$

For the node at the origin of the local system of coordinates ($u=0$, $v=0$) we get

$$x(0,0) = \begin{bmatrix} 1-0-0 & 0 & 0 \end{bmatrix} \begin{bmatrix} x_1 \\ x_2 \\ x_3 \end{bmatrix} = x_1$$

For the node at ($u=1$, $v=0$) we have

$$x(1,0) = \begin{bmatrix} 1-1-0 & 1 & 0 \end{bmatrix} \begin{bmatrix} x_1 \\ x_2 \\ x_3 \end{bmatrix} = x_2$$

and, similarly, for the node at ($u=0$, $v=1$)

$$x(0,1) = \begin{bmatrix} 1-0-1 & 0 & 1 \end{bmatrix} \begin{bmatrix} x_1 \\ x_2 \\ x_3 \end{bmatrix} = x_3$$

Identical transformations apply to the y coordinates:

$$y(u,v) = \begin{bmatrix} 1-u-v & u & v \end{bmatrix} \begin{bmatrix} y_1 \\ y_2 \\ y_3 \end{bmatrix} \qquad (8.71)$$

This means that the functions N_1, N_2, and N_3 are valid for x and y. The net effect is that node ($u=0$, $v=0$) is mapped onto (x_1, y_1), node ($u=1$, $v=0$) is mapped onto (x_2, y_2), and the node at ($u=0$, $v=1$) is mapped onto (x_3, y_3). As an example, suppose we map the centroid of the triangle which, in the local element, is located at $u=1/3$, $v=1/3$. In the global element these become

$$x(1/3, 1/3) = \begin{bmatrix} 1 - \frac{1}{3} - \frac{1}{3} & \frac{1}{3} & \frac{1}{3} \end{bmatrix} \begin{bmatrix} x_1 \\ x_2 \\ x_3 \end{bmatrix}$$

or

$$x(1/3, 1/3) = (x_1 + x_2 + x_3)/3$$

and similarly for the y coordinate,

$$y(1/3, 1/3) = (y_1 + y_2 + y_3)/3$$

Thus, for any point (u,v), there corresponds a unique point (x,y).

An important property related to the coordinates transformation is the "Jacobian" matrix $[J]$:

$$[J] = \begin{bmatrix} \dfrac{\partial x}{\partial u} & \dfrac{\partial y}{\partial u} \\ \dfrac{\partial x}{\partial v} & \dfrac{\partial y}{\partial v} \end{bmatrix} \qquad (8.72)$$

The transformation is only possible if this matrix is not singular. To evaluate the Jacobian, we calculate the terms of the matrix using Eqs. (8.70) and (8.71):

$$\frac{\partial x}{\partial u} = \begin{bmatrix} -1 & 1 & 0 \end{bmatrix} \begin{bmatrix} x_1 \\ x_2 \\ x_3 \end{bmatrix} \qquad \frac{\partial x}{\partial v} = \begin{bmatrix} -1 & 0 & 1 \end{bmatrix} \begin{bmatrix} x_1 \\ x_2 \\ x_3 \end{bmatrix}$$

or

$$\frac{\partial x}{\partial u} = x_2 - x_1 \qquad \frac{\partial x}{\partial v} = x_3 - x_1 \qquad (8.73a)$$

and, analogously,

$$\frac{\partial y}{\partial u} = y_2 - y_1 \qquad \frac{\partial y}{\partial v} = y_3 - y_1 \qquad (8.73b)$$

With these, the Jacobian is

$$[J] = \begin{bmatrix} x_2 - x_1 & y_2 - y_1 \\ x_3 - x_1 & y_3 - y_1 \end{bmatrix}$$

and its determinant is

$$det[J] = (x_2 - x_1)(y_3 - y_1) - (x_3 - x_1)(y_2 - y_1) \qquad (8.74)$$

The determinant equals twice the area of the triangle. The Jacobian is zero if the area of the triangle is zero (the three nodes of the triangle are on a single line). This is obviously not an acceptable finite element and should be avoided.

8.6.3. *High-Order Finite Elements: Implementation*

The 2D approximation for potential in a first-order triangle is [from Eq. (8.36)]

$$V(x,y) = \phi_1(x,y)V_1 + \phi_2(x,y)V_2 + \phi_3(x,y)V_3$$

or, in matrix form,

$$V(x,y) = \begin{bmatrix} \phi_1(x,y) & \phi_2(x,y) & \phi_3(x,y) \end{bmatrix} \begin{bmatrix} V_1 \\ V_2 \\ V_3 \end{bmatrix} \quad (8.75)$$

where

$$\phi_1(x,y) = (p_1 + q_1 x + r_1 y)/D$$
$$\phi_2(x,y) = (p_2 + q_2 x + r_2 y)/D$$
$$\phi_3(x,y) = (p_3 + q_3 x + r_3 y)/D$$

with the following values:

$$\phi_1(x_1,y_1) = 1 \qquad \phi_2(x_1,y_1) = 0 \qquad \phi_3(x_1,y_1) = 0$$
$$\phi_1(x_2,y_2) = 0 \qquad \phi_2(x_2,y_2) = 1 \qquad \phi_3(x_2,y_2) = 0$$
$$\phi_1(x_3,y_3) = 0 \qquad \phi_2(x_3,y_3) = 0 \qquad \phi_3(x_3,y_3) = 1$$

For any internal point in the triangle, the values of ϕ_1, ϕ_2, and ϕ_3 vary depending on the location of the point in relation to the three nodes. The functions ϕ_1, ϕ_2, and ϕ_3 are called "interpolation functions." These interpolation functions are written in global coordinates. Interpolation functions in global coordinates are often used for simple elements such as the first-order triangular element described above. For most elements, but in particular for higher order elements, it is much easier to define first the interpolation functions in local coordinates and then map these using mapping functions into the global domain. The interpolation functions in local coordinates will be denoted here $N^*(u,v)$. Theoretically, the mapping functions, denoted $N(u,v)$, are different than the interpolation functions $N^*(u,v)$. However, in practice they are most often chosen to be the same, defining an "isoparametric" mapping process. From now on, we will refer to both functions as $N(u,v)$. In the example above, we have

$$V(u,v) = \begin{bmatrix} 1-u-v & u & v \end{bmatrix} \begin{bmatrix} V_1 \\ V_2 \\ V_3 \end{bmatrix} \quad (8.76)$$

The value we obtain for $V(u,v)$ or $V(x,y)$ for any specific values of u and v or the corresponding values of x and y are the same as will be shown next. Suppose $u=1/4$, $v=1/2$. By direct evaluation, we get

$$V(u,v) = \left[1 - \frac{1}{4} - \frac{1}{2} \quad \frac{1}{4} \quad \frac{1}{2}\right] \begin{bmatrix} V_1 \\ V_2 \\ V_3 \end{bmatrix} = \frac{1}{4}(V_1 + V_2 + 2V_3) \quad (8.77)$$

The coordinates are transformed as

$$x(1/4,1/2) = \left[1 - \frac{1}{4} - \frac{1}{2} \quad \frac{1}{4} \quad \frac{1}{2}\right] \begin{bmatrix} x_1 \\ x_2 \\ x_3 \end{bmatrix} = \frac{1}{4}(x_1 + x_2 + 2x_3) \quad (8.78a)$$

and

$$y(1/4,1/2) = \frac{1}{4}(y_1 + y_2 + 2y_3) \quad (8.78b)$$

The interpolation function ϕ_1 is

$$\phi_1(x,y) = (p_1 + q_1 x + r_1 y)/D$$

Replacing the value of p_1, q_1, and r_1 as defined in Eq. (8.32) and using the coordinates x, y from Eqs. (8.78a) and (8.78b), gives

$$\phi_1(x,y) = \frac{1}{D}\left[x_2 y_3 - x_3 y_2 + (y_2 - y_3)x + (x_3 - x_2)y\right]$$
$$= \frac{1}{D}\left[x_2 y_3 - x_3 y_2 + (y_2 - y_1)\frac{1}{4}(x_1 + x_2 + 2x_3) + (x_3 - x_2)\frac{1}{4}(y_1 + y_2 + 2y_3)\right]$$

which, after some algebra, gives

$$\phi_1(x,y) = \frac{1}{4}$$

Performing similar calculations for ϕ_2 and ϕ_3, we get

$$\phi_2(x,y) = \frac{1}{4} \quad \text{and} \quad \phi_3(x,y) = \frac{2}{4}$$

And, finally, we get

$$V(x(1/4,1/2); y(1/4,1/2)) = \frac{1}{4}(V_1 + V_2 + 2V_3) \quad (8.79)$$

a result which we obtained in Eq. (8.77) in a much simpler way. We note that

$$V(u=1/4, v=1/2) = V(x(1/4,1/2), y(1/4,1/2))$$

Even though this is only an example and cannot be viewed as general proof, this property is always valid.

8.6.4. Continuity of Finite Elements

An element is said to have a C^0 continuity if the variable approximated over the element [in this case, $V(x,y)$] is continuous across the interface between adjacent elements. These elements are also called Lagrange elements. An element is said to have C^1 continuity if both the variable and its first derivative are continuous across inter element interfaces. Similarly, a C^n continuous element means that the variable and its first n derivatives are continuous across element interfaces. For such elements, the derivatives at the nodes are also unknowns, representing additional degrees of freedom.

In electromagnetics, derivatives are often discontinuous at the interface between two different materials. For example, in the discussion above we used the potential as the variable. Its derivatives are the electric field intensity ($\mathbf{E}=-\nabla V$). It is therefore appropriate to use C^0 elements to ensure continuity of the potential V and allow discontinuity of the electric field intensity at interfaces between two dielectric materials with different permittivity. Because the need for discontinuous derivatives is common in electromagnetics, the Lagrange (C^0) elements are most commonly used. It is possible to use C^1 elements in domains without material discontinuities, but the complication in defining these types of elements all but precludes their use. For this reason, we will use only C^0 elements in this work.

As mentioned before, a finite element is called "isoparametric" when the shape functions (the geometric interpolation functions) are identical to the interpolation functions. The isoparametric elements are the most commonly used elements in FE codes. However, if we decide to use, for example, triangular elements with straight edges, the element can be mapped with linear functions while, for reasons of accuracy, we may wish to use quadratic functions for the interpolation functions. Because the order of the mapping function is lower than that of the interpolating function, the element is "subparametric." The opposite can also be implemented: mapping functions with order higher than the interpolation function in which case the element becomes "overparametric or hyperparametric." While subparametric mapping is common, overparametric mapping is not.

8.6.5. Polynomial Basis

Another fundamental characteristic of finite elements is their polynomial basis. To see what the polynomial base of a finite element is, recall that for a linear triangle, the approximation for $V(x,y)$ is

$$V(x,y) = a_1 + a_2 x + a_3 y \qquad (8.80)$$

The same relation holds in the local coordinates

$$V(u,v) = a_1 + a_2 u + a_3 v \qquad (8.81)$$

Generalization of the Finite Element Method 299

with the following property:

$$V(x(u,v), y(u,v)) = V(u,v)$$

which means that V has the same value when calculated through Eq. (8.29) or (8.81) for any corresponding point, as was shown in section 8.6.3.

The expression in Eq. (8.81) can be written as

$$V(u,v) = \begin{bmatrix} 1 & u & v \end{bmatrix} \begin{bmatrix} a_1 \\ a_2 \\ a_3 \end{bmatrix} \quad (8.82)$$

and the vector

$$\begin{bmatrix} 1 & u & v \end{bmatrix}$$

is called the "polynomial basis" of the element.

We will consider now the six-node quadratic triangle shown in Figure 8.12b, for which the potential $V(x,y)$ is given in Eq. (8.66). The corresponding equation in local coordinates is

$$V(u,v) = a_1 + a_2 u + a_3 v + a_4 u^2 + a_5 uv + a_6 v^2 \quad (8.83)$$

and the polynomial basis for this element is

$$\begin{bmatrix} 1 & u & v & u^2 & uv & v^2 \end{bmatrix} \quad (8.84)$$

This basis has six terms and is called a "complete basis" since all combinations of u and v are present in the expansion.

Consider now the first-order rectangular element in Figure 8.12c. The element has four nodes and the approximation in local coordinates is

$$V(u,v) = a_1 + a_2 u + a_3 v + a_4 uv \quad (8.85)$$

The polynomial basis is

$$\begin{bmatrix} 1 & u & v & uv \end{bmatrix} \quad (8.86)$$

Now the basis has only four terms and is obviously an incomplete basis since the terms u^2 and v^2 are absent.

In conclusion, we point out that each finite element has its own polynomial basis.

8.6.6. Transformation of Quantities – the Jacobian

In actual solution of a problem, the derivatives $\partial V/\partial x$, $\partial V/\partial y$, and $\partial V/\partial z$ are also required, in addition to the potentials V. The normal method of obtaining these derivatives is to calculate the Jacobian in the local system of coordinates and then to transform the derivatives into the global system of coordinates. This transformation is facilitated by the Jacobian through the following relation:

$$\begin{bmatrix} \dfrac{\partial}{\partial u} \\ \dfrac{\partial}{\partial v} \\ \dfrac{\partial}{\partial p} \end{bmatrix} = \begin{bmatrix} \dfrac{\partial x}{\partial u} & \dfrac{\partial y}{\partial u} & \dfrac{\partial z}{\partial u} \\ \dfrac{\partial x}{\partial v} & \dfrac{\partial y}{\partial v} & \dfrac{\partial z}{\partial v} \\ \dfrac{\partial x}{\partial p} & \dfrac{\partial y}{\partial p} & \dfrac{\partial z}{\partial p} \end{bmatrix} \begin{bmatrix} \dfrac{\partial}{\partial x} \\ \dfrac{\partial}{\partial y} \\ \dfrac{\partial}{\partial z} \end{bmatrix} \qquad (8.87)$$

This is denoted in short form as

$$\partial u = J \partial x \qquad (8.88)$$

On the other hand, we can also write

$$\begin{bmatrix} \dfrac{\partial}{\partial x} \\ \dfrac{\partial}{\partial y} \\ \dfrac{\partial}{\partial z} \end{bmatrix} = \begin{bmatrix} \dfrac{\partial u}{\partial x} & \dfrac{\partial v}{\partial x} & \dfrac{\partial p}{\partial x} \\ \dfrac{\partial u}{\partial y} & \dfrac{\partial v}{\partial y} & \dfrac{\partial p}{\partial y} \\ \dfrac{\partial u}{\partial z} & \dfrac{\partial v}{\partial z} & \dfrac{\partial p}{\partial z} \end{bmatrix} \begin{bmatrix} \dfrac{\partial}{\partial u} \\ \dfrac{\partial}{\partial v} \\ \dfrac{\partial}{\partial p} \end{bmatrix} \qquad (8.89)$$

which is denoted as

$$\partial x = J^{-1} \partial u \qquad (8.90)$$

In 2D calculations we write

$$[J] = \begin{bmatrix} J_{11} & J_{12} \\ J_{21} & J_{22} \end{bmatrix} \quad \text{and} \quad J^{-1} = \frac{1}{\det [J]} \begin{bmatrix} J_{11} & -J_{12} \\ -J_{21} & J_{22} \end{bmatrix} \qquad (8.91)$$

The Jacobian will prove to be a useful tool and we will make much use of it.
As an example, we calculate the Jacobian $[J]$ for a first-order triangle:

Generalization of the Finite Element Method   301

$$J = \begin{bmatrix} \dfrac{\partial x}{\partial u} & \dfrac{\partial y}{\partial u} \\ \dfrac{\partial x}{\partial v} & \dfrac{\partial y}{\partial v} \end{bmatrix} = \begin{bmatrix} \dfrac{\partial}{\partial u} \\ \dfrac{\partial}{\partial v} \end{bmatrix} [\ x(u,v)\ \ y(u,v)\ ] \qquad (8.92)$$

Taking into account the shape functions as in Eq. (8.69), we get

$$J = \begin{bmatrix} \dfrac{\partial}{\partial u} \\ \dfrac{\partial}{\partial v} \end{bmatrix} [\ N_1(u,v)\ \ N_2(u,v)\ \ N_3(u,v)\ ] \begin{bmatrix} x_1 & y_1 \\ x_2 & y_2 \\ x_3 & y_3 \end{bmatrix}$$

We obtain

$$J = \begin{bmatrix} \dfrac{\partial N_1}{\partial u} & \dfrac{\partial N_2}{\partial u} & \dfrac{\partial N_3}{\partial u} \\ \dfrac{\partial N_1}{\partial v} & \dfrac{\partial N_2}{\partial v} & \dfrac{\partial N_3}{\partial v} \end{bmatrix} \begin{bmatrix} x_1 & y_1 \\ x_2 & y_2 \\ x_3 & y_3 \end{bmatrix} \qquad (8.93)$$

or

$$J = \begin{bmatrix} \dfrac{\partial(1-u-v)}{\partial u} & \dfrac{\partial(u)}{\partial u} & \dfrac{\partial(v)}{\partial u} \\ \dfrac{\partial(1-u-v)}{\partial v} & \dfrac{\partial(u)}{\partial v} & \dfrac{\partial(v)}{\partial v} \end{bmatrix} \begin{bmatrix} x_1 & y_1 \\ x_2 & y_2 \\ x_3 & y_3 \end{bmatrix} \qquad (8.94)$$

This gives

$$J = \begin{bmatrix} -1 & 1 & 0 \\ -1 & 0 & 1 \end{bmatrix} \begin{bmatrix} x_1 & y_1 \\ x_2 & y_2 \\ x_3 & y_3 \end{bmatrix} = \begin{bmatrix} x_2 - x_1 & y_2 - y_1 \\ x_3 - x_1 & y_3 - y_1 \end{bmatrix} \qquad (8.95)$$

and

$$J^{-1} = \dfrac{1}{det[J]} \begin{bmatrix} y_3 - y_1 & y_1 - y_2 \\ x_1 - x_3 & x_2 - x_1 \end{bmatrix} \qquad (8.96)$$

where $det[J]$ equals twice the area of the triangle as mentioned before.

(0,10)

(10,0)

(0,0)

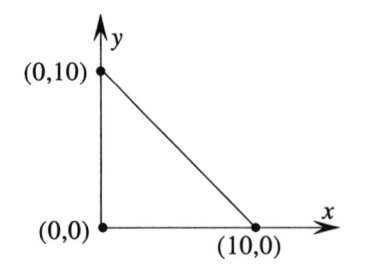

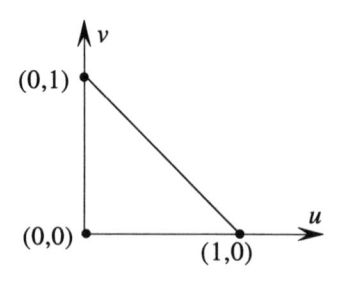

Figure 8.14. Triangle in global coordinates.

(0,1)

(1,0)

(0,0)

Figure 8.15. Triangle in local coordinates.

8.6.7. Evaluation of the Integrals

The change of variables makes the integration relatively easy. In fact, in some cases it would be next to impossible to perform the integration without the use of local coordinates and transformation to global coordinates. The integration of a function $f(x)$ over an element in global coordinates, obtained through the use of the local element, is

$$\int_{V_x} f(x) dx dy dz = \int_{V_u} f(x(u,v,p)) det[J] du dv dp \qquad (8.97)$$

where for simplicity, the function $f(x,y,z)$ is denoted as $f(x)$.

A very simple example is now shown. Suppose that the function $f(x,y)=1$ has to be integrated on the triangle in Figure 8.14.

In global coordinates, we have

$$\int_{V_x} 1 \, dx dy = 1 \times \frac{10 \times 10}{2} = 50$$

Using the local coordinates in Figure 8.15, the integral is

$$\int_{V_u} 1 \, det[J] du dv$$

As discussed above, $det[J]$ equals twice the area of the triangle, and we write

$$100 \int_0^1 \int_0^{1-v} 1 \, du dv = 100 \int_0^1 [u]_0^{1-v} dv = 100 \int_0^1 (1-v) dv = \frac{1}{2} 100 = 50$$

Even though this is a particular, and very simple example, Eq. (8.97) is a valid method of evaluating the integral through use of local coordinates.

To clarify some of the concepts discussed above we will evaluate now the stiffness and source matrices for the electrostatic problem introduced at the beginning of the chapter. The electrostatic governing equation is

$$\nabla \cdot \mathbf{D} = \rho$$

or

$$\nabla \cdot (\varepsilon \nabla V) + \rho = 0 \qquad (8.98)$$

The problem is considered in two dimensions using first-order triangular elements. First, to conform to the common notation in finite element calculations, we denote:

$$N(x,y) = \phi(x,y)$$

That is, $N_1(x,y) = \phi_1(x,y)$, $N_2(x,y) = \phi_2(x,y)$, and $N_3(x,y) = \phi_3(x,y)$. With this notation we can write

$$\nabla N(x,y) = \begin{bmatrix} \dfrac{\partial N}{\partial x} \\ \dfrac{\partial N}{\partial y} \end{bmatrix}$$

Noting that

$$\begin{bmatrix} \dfrac{\partial N}{\partial x} \\ \dfrac{\partial N}{\partial y} \end{bmatrix} = [J]^{-1} \begin{bmatrix} \dfrac{\partial N}{\partial u} \\ \dfrac{\partial N}{\partial v} \end{bmatrix} \qquad (8.99)$$

and from Eq. (8.94),

$$\begin{bmatrix} \dfrac{\partial N}{\partial u} \\ \dfrac{\partial N}{\partial v} \end{bmatrix} = \begin{bmatrix} -1 & 1 & 0 \\ -1 & 0 & 1 \end{bmatrix}$$

and with $[J]^{-1}$ given in Eq. (8.96), we obtain

$$\begin{bmatrix} \dfrac{\partial N}{\partial x} \\ \dfrac{\partial N}{\partial y} \end{bmatrix} = \dfrac{1}{D} \begin{bmatrix} y_2 - y_3 & y_3 - y_1 & y_1 - y_2 \\ x_3 - x_2 & x_1 - x_3 & x_2 - x_1 \end{bmatrix} \qquad (8.100)$$

Using the notation in Eq. (8.32), we have

$$\begin{bmatrix} \dfrac{\partial N}{\partial x} \\ \dfrac{\partial N}{\partial y} \end{bmatrix} = \dfrac{1}{D} \begin{bmatrix} q_1 & q_2 & q_3 \\ r_1 & r_2 & r_3 \end{bmatrix} \qquad (8.101)$$

Applying the Galerkin method we calculate the elemental stiffens matrix [see Eq. (8.16)] using

$$\int_{S_i} \nabla N^t \cdot \varepsilon \nabla V ds \qquad (8.102)$$

where S_i is the area of element i. Using the interpolation functions (which are identical to the shape functions), we have

$$V(x,y) = N(x,y)V \qquad (8.103)$$

which is a compact notation for

$$V(x,y) = \begin{bmatrix} N_1(x,y) & N_2(x,y) & N_3(x,y) \end{bmatrix} \begin{bmatrix} V_1 \\ V_2 \\ V_3 \end{bmatrix}$$

With these, $\nabla V(x,y)$ becomes

$$\nabla V(x,y) = \nabla N(x,y)V = \begin{bmatrix} \dfrac{\partial N}{\partial x} \\ \dfrac{\partial N}{\partial y} \end{bmatrix} V \qquad (8.104)$$

Using Eq. (8.101), we get

$$\nabla V(x,y) = \dfrac{1}{D} \begin{bmatrix} q_1 & q_2 & q_3 \\ r_1 & r_2 & r_3 \end{bmatrix} \begin{bmatrix} V_1 \\ V_2 \\ V_3 \end{bmatrix} \qquad (8.105)$$

This is in accordance with Eq. (8.39). Now substituting Eq. (8.101) and (8.105) in Eq. (8.102), we get

$$\int_0^1 \int_0^{1-v} \nabla N^t \cdot \varepsilon \nabla N V \, det[J] \, dudv \qquad (8.106)$$

or

$$\int_0^1 \int_0^{1-v} \frac{\varepsilon}{D^2} \begin{bmatrix} q_1 & r_1 \\ q_2 & r_2 \\ q_3 & r_3 \end{bmatrix} \begin{bmatrix} q_1 & q_2 & q_3 \\ r_1 & r_2 & r_3 \end{bmatrix} det[J] \, dudv \begin{bmatrix} V_1 \\ V_2 \\ V_3 \end{bmatrix}$$

Since $det[J] = D$, we have

$$\frac{\varepsilon}{D} \begin{bmatrix} q_1 & r_1 \\ q_2 & r_2 \\ q_3 & r_3 \end{bmatrix} \begin{bmatrix} q_1 & q_2 & q_3 \\ r_1 & r_2 & r_3 \end{bmatrix} \begin{bmatrix} V_1 \\ V_2 \\ V_3 \end{bmatrix} \int_0^1 \int_0^{1-v} dudv$$

The integrals give 1/2 and the final result is

$$\frac{\varepsilon}{2D} \begin{bmatrix} q_1 q_1 + r_1 r_1 & q_1 q_2 + r_1 r_2 & q_1 q_3 + r_1 r_3 \\ Symmetric & q_2 q_2 + r_2 r_2 & q_2 q_3 + r_2 r_3 \\ Symmetric & Symmetric & q_3 q_3 + r_3 r_3 \end{bmatrix} \begin{bmatrix} V_1 \\ V_2 \\ V_3 \end{bmatrix} \qquad (8.107)$$

This, of course, is identical to the elemental stiffness matrix we obtained in Eq. (8.40).

Now that the elemental stiffness matrix has been calculated, we consider the source term which is

$$\int_{S_i} N^t \rho \, dxdy = \int_0^1 \int_0^{1-v} \begin{bmatrix} 1-u-v \\ u \\ v \end{bmatrix} \rho \, det[J] \, dudv \qquad (8.108)$$

We have

$$\int_{S_i} N^t \rho \, dxdy = \rho D \int_0^1 \int_0^{1-v} \begin{bmatrix} 1-u-v \\ u \\ v \end{bmatrix} dudv = \frac{\rho D}{6} \begin{bmatrix} 1 \\ 1 \\ 1 \end{bmatrix} \qquad (8.109)$$

This is again identical to the expression in Eq. (8.42). This source term is independent of the potentials and therefore is part of the right-hand side of the system of equations.

8.7. Numerical Integration

When calculating, for example, the first term of Eq. (8.109), it is necessary to evaluate the following integral:

$$\rho D \int_0^1 \int_0^{1-v} (1 - u - v)\, du\, dv = \rho D \int_0^1 \left(u - \frac{u^2}{2} - uv \right)_0^{1-v} dv =$$

$$\rho D \int_0^1 \left(1 - v - \frac{(1-v)^2}{2} - v(1-v) \right) dv = \frac{\rho D}{6}$$

which is the result shown in Eq. (8.109).

It is clear from the example above that even for the very simple 2D element, the integration requires much work. This is in spite of the fact the $\det[J]$ is a constant. For second-order elements, the shape functions are much more complex and $\det[J]$ is not necessarily constant. Therefore, in general, it is practically impossible to evaluate the integrals analytically.

Because of these reasons, the application of finite element codes is normally associated with numerical integration, and efficient integration algorithms feature prominently in fast and efficient finite element codes.

8.7.1. Evaluation of the Integrals

There are some integrals that can be performed analytically. These are presented next.

a. 1D element.

$$\int_{-1}^1 u^i du = \begin{cases} 0 & i \text{ odd} \\ \dfrac{2}{i+1} & i \text{ even} \end{cases} \quad (8.110)$$

b. 2D quadrilateral element.

$$\int_{-1}^1 \int_{-1}^1 u^i v^j du\, dv = \begin{cases} 0 & i \text{ or } j \text{ odd} \\ \dfrac{4}{(i+1)(j+1)} & i \text{ and } j \text{ even} \end{cases} \quad (8.111)$$

c. 2D triangular element.

$$\int_0^1 \int_0^{1-u} u^i v^j du\, dv = \frac{i!\, j!}{(i+j+2)!} \quad (8.112)$$

d. 3D cubic element.

$$\int_{-1}^{1}\int_{-1}^{1}\int_{-1}^{1} u^i v^j p^k du dv\, dp = \begin{cases} 0 & i\ or\ j\ or\ k\ odd \\ \dfrac{8}{(i+1)(j+1)(k+1)} & i, j,\ and\ k\ even \end{cases} \quad (8.113)$$

e. 3D tetrahedral element.

$$\int_{0}^{1}\int_{0}^{1-u}\int_{0}^{1-u-v} u^i v^j p^k du dv dp = \frac{i!\, j!\, k!}{(i+j+k+3)!} \quad (8.114)$$

8.7.2. Basic Principles of Numerical Integration

Although any integral required for finite element calculations may be performed using the expressions of the previous section, it is not practical to calculate the integrals in this fashion. It is more common and more practical to use numerical integration methods of the type:

$$\int K(u)du = \sum_{i=1}^{r} w_i K(u_i) \quad (8.115)$$

This means that the integrand K is not modified and therein lies the most attractive feature of these methods. In this form, r is the number of integration points, u_i the coordinates of the integration points, and w_i weights associated with the integration points. The integral is reduced to a sum over a relatively small number of values as we shall see shortly.

In general, integration over each type of finite element can be performed by different numbers of integration points r. The number of points defines the accuracy of the integration with larger number of points providing better accuracy. We will discuss this point later including the optimal number of points required for integration.

Next we show the calculation of the weights w and the integration points for the Gauss method of integration, the most-used method of integration in finite elements for electromagnetic application.

In Gauss' method, the r coefficients w_i and the integration points u_i are determined so as to integrate the polynomial of order $m \leq 2r-1$ exactly:

$$\int_{-1}^{+1} Y(u)du = w_1 Y(u_1) + w_2 Y(u_2) + \ldots + w_r Y(u_r) \quad (8.116)$$

or, written as a sum,

308   8. Introduction to the Finite Element Method

$$\int_{-1}^{+1} Y(u)du = \sum_{i=1}^{r} w_i Y(u_i)$$

To evaluate $Y(u)$, we write it as a polynomial with $2r$ coefficients as follows:

$$Y(u) = a_1 + a_2 u + a_3 u^2 + \ldots + a_{2r} u^{2r-1} \qquad (8.117)$$

This provides an exact evaluation of $Y(u)$. Using the expression in Eq. (8.116), we get

$$a_1 \int_{-1}^{+1} du + a_2 \int_{-1}^{+1} u\,du + a_3 \int_{-1}^{+1} u^2 du + \ldots + a_{2r} \int_{-1}^{+1} u^{2r-1} du =$$
$$w_1(a_1 + a_2 u_1 + a_3 u_1^2 + \ldots + a_{2r} u_1^{2r-1}) +$$
$$w_2(a_1 + a_2 u_2 + a_3 u_2^2 + \ldots + a_{2r} u_2^{2r-1}) + \qquad (8.118a)$$
$$\vdots$$
$$w_r(a_1 + a_2 u_r + a_2 u_r^2 + \ldots + a_{2r} u_r^{2r-1})$$

The right-hand side can be rearranged as

$$a_1(w_1 + w_2 + w_3 + \ldots + w_r) +$$
$$a_2(w_1 u_1 + w_2 u_2 + w_3 u_3 + \ldots + w_r u_r) +$$
$$\vdots \qquad (8.118b)$$
$$a_{2r}(w_1 u_1^{2r-1} + w_2 u_2^{2r-1} + a_3 u_3^{2r-1} + \ldots + w_r u_r^{2r-1})$$

From Eq. (8.118a) and (8.118b), we have

$$\int_{-1}^{+1} du = w_1 + w_2 + w_3 + \ldots + w_r$$

$$\int_{-1}^{+1} u\,du = w_1 u_1 + w_2 u_2 + w_3 u_3 + \ldots + w_r u_r$$

$$\int_{-1}^{+1} u^2 du = w_1 u_1^2 + w_2 u_2^2 + w_3 u_3^2 + \ldots + w_r u_r^2$$

$$\vdots$$

$$\int_{-1}^{+1} u^{2r-1} du = w_1 u_1^{2r-1} + w_2 u_2^{2r-1} + w_3 u_3^{2r-1} + \ldots + w_r u_r^{2r-1}$$

The left-hand side integrals in the equations above may be evaluated analytically using Eq. (8.110) and we obtain a system of $2r$ equations:

$$2 = w_1 + w_2 + w_3 + \ldots + w_r$$
$$0 = w_1 u_1 + w_2 u_2 + w_3 u_3 + \ldots + w_r u_r$$
$$\frac{2}{3} = w_1 u_1^2 + w_2 u_2^2 + w_3 u_3^2 + \ldots + w_r u_r^2$$
$$\vdots \qquad (8.119)$$
$$0 = w_1 u_1^{2r-1} + w_2 u_2^{2r-1} + w_3 u_3^{2r-1} + \ldots + w_r u_r^{2r-1}$$

These equations are linear in w_i and nonlinear in u_i, and can be solved for the $2r$ parameters (w_i and u_i) under the conditions

$$w_i > 0, \quad -1 < u_i < 1, \quad \text{for } i=1,2,3,\ldots r$$

As a simple example, consider the Gauss method with two integration points ($r=2$). The integral is

$$\int_{-1}^{+1} Y(u) du = w_1 Y(u_1) + w_2 Y(u_2)$$

The polynomial that can be integrated exactly is of order $2r-1=3$:

$$Y(u) = a_1 + a_2 u + a_3 u^2 + a_4 u^3$$

Applying the method above, we get from the system in Eq. (8.119):

$$2 = w_1 + w_2$$
$$0 = w_1 u_1 + w_2 u_2$$
$$\frac{2}{3} = w_1 u_1^2 + w_2 u_2^2$$
$$0 = w_1 u_1^3 + w_2 u_2^3$$

The solution to this is

$$w_1 = w_2 = 1$$

and

$$u_1 = \frac{1}{\sqrt{3}} \qquad u_2 = -\frac{1}{\sqrt{3}}$$

There are other integration methods, some of which are very similar to the method above. Additional information on integration methods may be found in the bibliography section. Also, it is worth mentioning that the weights and points

8. Introduction to the Finite Element Method

calculated above exist as tabulated values for all practical applications. The integration for 2D and 3D elements follows rules similar to those discussed above.

As an example, for triangular elements with a single integration point ($r=1$), we get

$$u_1 = \frac{1}{3} \qquad v_1 = \frac{1}{3} \qquad w_1 = \frac{1}{2}$$

Suppose that we wish to evaluate the first term in Eq. (8.108)

$$\int_0^1 \int_0^{1-v} (1-u-v)\rho \det[J] \, du \, dv = \rho \det[J] \int_0^1 \int_0^{1-v} (1-u-v) \, du \, dv$$

In this case, the integrand is

$$f(u,v) = 1 - u - v$$

and with $r=1$, we get

$$\int_0^1 \int_0^{1-v} (1-u-v)\rho \det[J] \, du \, dv = \rho \det[J] \int_0^1 \int_0^{1-v} (1-u-v) \, du \, dv$$

This is an exact result because the polynomial approximation over the element is first order ($2r-1=1$).

Suppose now that we choose $r=3$; that is, we choose to integrate using three integration points (see Figure 8.16). In this case the points and weights are

$$u_1 = \frac{1}{2} \qquad u_2 = 0 \qquad u_3 = \frac{1}{2}$$
$$v_1 = \frac{1}{2} \qquad v_2 = \frac{1}{2} \qquad v_3 = 0$$
$$w_1 = \frac{1}{6} \qquad w_2 = \frac{1}{6} \qquad w_3 = \frac{1}{6}$$

and we obtain

$$\rho \det[J] \int_0^1 \int_0^{1-v} f(u,v) \, du \, dv =$$

$$\rho D \left[\frac{1}{6}\left(1 - \frac{1}{2} - \frac{1}{2}\right) + \frac{1}{6}\left(1 - 0 - \frac{1}{2}\right) + \frac{1}{6}\left(1 - \frac{1}{2} - 0\right) \right] = \frac{\rho D}{6}$$

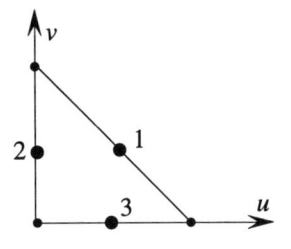

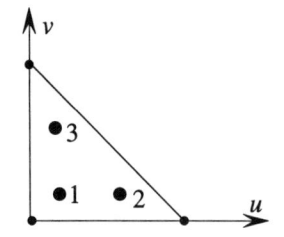

Figure 8.16. Three integration points for a triangle (one possible choice).

Figure 8.17. Three integration points for a triangle (a second choice).

For $r=3$, it is possible to use a different set of integration points (and associated weights) as shown in Figure 8.17:

$$u_1 = \frac{1}{6} \quad u_2 = \frac{2}{3} \quad u_3 = \frac{1}{6}$$
$$v_1 = \frac{1}{6} \quad v_2 = \frac{1}{6} \quad v_3 = \frac{2}{3}$$
$$w_1 = \frac{1}{6} \quad w_2 = \frac{1}{6} \quad w_3 = \frac{1}{6}$$

These points and weights give the same result.

8.7.3. Accuracy and Errors in Numerical Integration

It was mentioned above that the Gauss method of integration can integrate a polynomial of order $m \leq 2r-1$ exactly. Thus, it is important to choose the number of points r so as to perform exact integration to the desired approximation. It is also worth mentioning that the method is very efficient as it requires very few integration points (one or two are usually sufficient) in comparison to other methods.
The error in integration using the Gauss method is given as

$$e = \frac{2^{2r-1}(r\,!)^4}{(2r-1)[(2r)!]^3} \frac{d^{2r}Y}{du^{2r}} \quad (8.120)$$

For example, suppose we wish to integrate the polynomial

$$Y(u) = 1 + u^2 + u^3 + u^4 \quad (8.121)$$

using two integration points. For these, the points and weights are

$$u_1 = \frac{1}{\sqrt{3}} \quad u_2 = -\frac{1}{\sqrt{3}} \quad w_1 = w_2 = 1$$

8. Introduction to the Finite Element Method

With two integration points, a third-order polynomial can be integrated exactly. Using the Gauss integration formula, we get

$$I = \int_{-1}^{1} (1 + u^2 + u^3 + u^4)\,du = 1\left(1 + \frac{1}{3} + \frac{1}{3\sqrt{3}} + \frac{1}{9}\right) + 1\left(1 + \frac{1}{3} - \frac{1}{3\sqrt{3}} + \frac{1}{9}\right) = \frac{130}{45}$$

The exact integration gives

$$I_{ex} = \int_{-1}^{1} (1 + u^2 + u^3 + u^4)\,du = \frac{138}{45}$$

The error is, therefore, $8/45$. The formula in Eq. (8.120) gives

$$e = \frac{2^{4-1}(2!)^4}{(4-1)[4!]^3} \frac{d^4 Y}{du^4} = \frac{24}{135} = \frac{8}{45}$$

From Eq. (8.120), we note that for three integration points, the derivative $d^{2r}Y/du^{2r} = d^6 Y/du^6 = 0$ and the integration would be exact.

In 2D problems, we first consider integration over quadrilateral elements. Suppose we use r_1 integration points in the u direction and r_2 points in the v direction. The integration will be exact in the u direction for polynomials with highest order terms m_1 where $m_1 = 2r_1 - 1$. Similarly, for the v direction. As a practical rule, the product $r = r_1 r_2$ gives the number of points that will integrate exactly the monomials

$$u^i v^j \quad \text{for:} \quad \begin{cases} 0 \le i \le 2r_1 - 1 \\ 0 \le j \le 2r_2 - 1 \end{cases}$$

For triangular elements, the "direct" method is preferred (the method is also based on the Gauss method) because it integrates exactly the monomials of order m such as

$$u^i v^j \quad \text{for:} \quad i+j \le m$$

The relationship between the order m and the number of integration points is shown in Table 8.1.

In concluding this section, it is worth reiterating that the integration points and weights for any order elements are available in tables and need merely be applied. There is rarely any need to find these points and weights.

Table 8.1. Number of integration points for triangular elements.

Order m	Integration points r
1	1
2	3
3	4
4	6
5	7
6	12

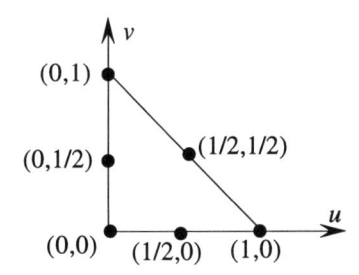

Figure 8.18. Nodes of a quadratic triangular element.

8.8. Some Specific Finite Elements

Putting together the concepts described in the previous sections, we can summarize the characteristics of a finite element. The finite element is described by:

- The shape of the element (triangular, quadrilateral, etc.).
- The coordinates of its geometric nodes (see, for example, Figure 8.18).
- The number of unknowns (or degrees of freedom). The element in Figure 8.18 has six nodes and therefore six unknowns.
- The nodal variable (V in the example presented at the beginning of the chapter).
- The polynomial basis of the element. For Figure 8.18, the polynomial basis is

$$\begin{bmatrix} 1 & u & v & u^2 & uv & v^2 \end{bmatrix}$$

- The class or type of continuity: here only C^0 continuity is considered.
- The shape or mapping functions $N(u,v)$ and its derivatives $\partial N/\partial u$, $\partial N/\partial v$ (in 3D also $\partial N/\partial p$). Also, the interpolation functions, which in this text are the same as the shape functions (isoparametric elements).
- The numerical integration table, indicating how the element is integrated. Conceptually, this step is independent of the finite element but, in practice, each finite element has its integration table and may be integrated differently.

8.8.1. 1D Elements

a. Linear 1D element.

Figure 8.18a. 1D element in local and global coordinates.

Table 8.2. Shape functions and derivatives for linear 1D elements.

Node	$[N]$	$[\partial N/\partial u]$
1	$(1-u)/2$	$-1/2$
2	$(1+u)/2$	$1/2$

Polynomial basis:

$$\begin{bmatrix} 1 & u \end{bmatrix}$$

b. Quadratic 1D element.

Figure 8.19. Quadratic 1D element in local and global coordinates.

Table 8.3. Shape functions and derivatives for linear 1D elements.

Node	$[N]$	$[\partial N/\partial u]$
1	$-u(1-u)/2$	$(-1+2u)/2$
2	$(1-u^2)/2$	$-2u$
3	$u(1+u)/2$	$(1+2u)/2$

Polynomial basis:

$$\begin{bmatrix} 1 & u & u^2 \end{bmatrix}$$

Note: For 1D elements, analytical integration is easily performed and should always be used.

8.8.2. 2D Elements

a. First-order triangular element.

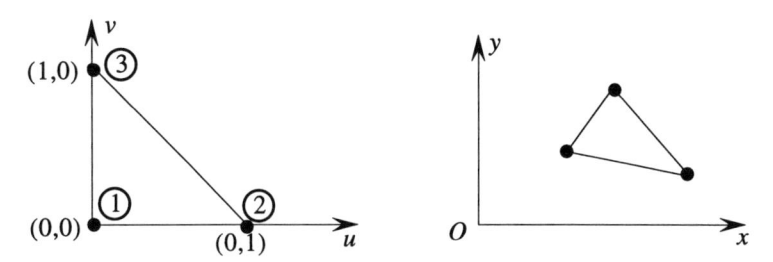

Figure 8.20. Triangular element in local and global coordinates.

Table 8.4. Shape functions and their derivatives for the triangular element in Figure 8.20.

Node	$[N]$	$[\partial N/\partial u]$	$[\partial N/\partial v]$
1	$1 - u - v$	-1	-1
2	u	1	0
3	v	0	1

Polynomial basis:

$$[\begin{array}{ccc} 1 & u & v \end{array}]$$

Note: analytical integration is both possible and recommended in this element.

b. Second-order triangular element.

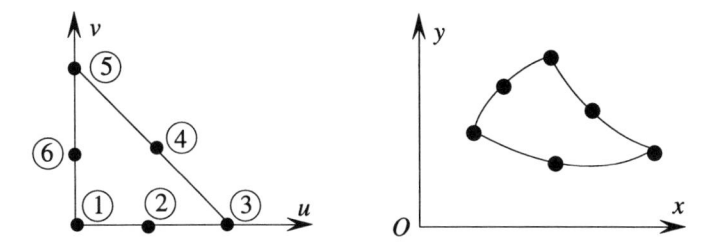

Figure 8.21. Second-order triangular element in local and global coordinates.

The coordinates of the nodes in the local coordinates are (0,0; 1/2,0; 1,0; 1/2,1/2; 0,1; 0,1/2). The shape functions for this element are given in Table 8.5

and, because these are second-order, the element edges in the global coordinate system may be curved as shown in Figure 8.21.

Table 8.5. Shape functions and their derivatives for 6-node, isoparametric triangular elements.

Node	$[N]$	$[\partial N/\partial u]$	$[\partial N/\partial v]$
1	$-t(1-2t)$	$1-4t$	$1-4t$
2	$4ut$	$4(t-u)$	$-4u$
3	$-u(1-2u)$	$-1+4u$	0
4	$4uv$	$4v$	$4u$
5	$-v(1-2v)$	0	$-1+4v$
6	$4vt$	$-4v$	$4(t-v)$

where: t=1−u−v

Polynomial basis:

$$\begin{bmatrix} 1 & u & v & u^2 & uv & v^2 \end{bmatrix}$$

The Jacobian $[J]$ is given as

$$[J] = \begin{bmatrix} \dfrac{\partial N}{\partial u} \\ \dfrac{\partial N}{\partial v} \end{bmatrix} [\, x \ \ y \,] = \begin{bmatrix} 1-4t & 4(t-u) & \cdots \\ 1-4t & -4u & \cdots \end{bmatrix} \begin{bmatrix} x_1 & y_1 \\ x_2 & y_2 \\ \vdots & \vdots \\ x_6 & y_6 \end{bmatrix} \quad (8.122)$$

Table 8.6. Integration points and corresponding weights for one, three, and seven point Gauss-Legendre integration.

Order m	Integration Points	u_i	v_i	W_i
1	1	1/3	1/3	1/2
2	3	1/6	1/6	1/6
		2/3	1/6	1/6
		1/6	2/3	1/6
5	7	1/3	1/3	9/80
		a	a	0.066197076
		1−2a	a	0.066197076
		a	1−2a	0.066197076
		b	b	0.062969590
		1−2b	b	0.062969590
		b	1−2b	0.062969590

where: a=0.470142064; b=0.101286507

c. Quadrilateral bi-linear element.

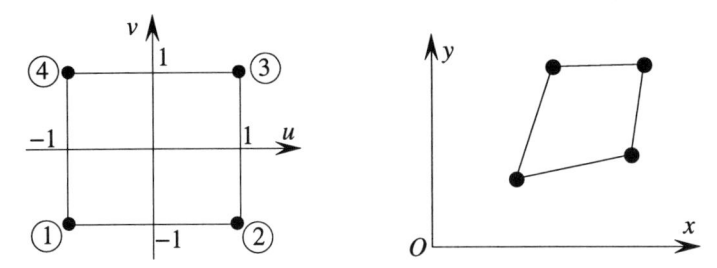

Figure 8.22. Quadrilateral bi-linear element in local and global coordinates.

Table 8.7. Shape functions and their derivatives for the quadrilateral bi-linear element.

Node	$[N]$	$[\partial N/\partial u]$	$[\partial N/\partial v]$
1	$(1-u)(1-v)/4$	$(-1+v)/4$	$(-1+u)/4$
2	$(1+u)(1-v)/4$	$(1-v)/4$	$(-1-u)/4$
3	$(1+u)(1+v)/4$	$(1+v)/4$	$(1+u)/4$
4	$(1-u)(1+v)/4$	$(-1-v)/4$	$(1-u)/4$

Polynomial basis:

$$[\quad 1 \quad u \quad v \quad uv \quad]$$

For numerical integration, it is recommended that four integration points ($r=2$) be used. This assumes exact integration up to a third-degree polynomial. The integration points are

$$(u_i, v_i) = \left(\pm \frac{1}{\sqrt{3}}, \pm \frac{1}{\sqrt{3}}\right), \qquad w_i = 1$$

d. Quadrilateral quadratic element.

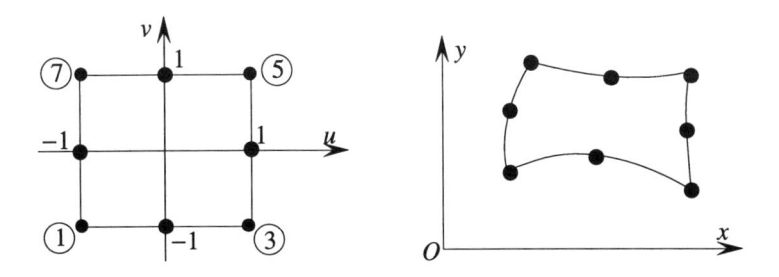

Figure 8.23. A quadrilateral, quadratic finite element in local and global coordinates.

318   8. Introduction to the Finite Element Method

Because this element is quadratic, curvilinear edges in the global system of coordinates can be modeled.

Table 8.8. Shape functions and their derivatives for the quadratic quadrilateral element in Figure 8.23.

Node	$[N]$	$[\partial N/\partial u]$	$[\partial N/\partial v]$
1	$-(1-u)(1-v)(1+u+v)/4$	$(1-v)(2u+v)/4$	$(1-u)(u+2v)/4$
2	$(1-u^2)(1-v)/2$	$-(1-v)u$	$-(1-u^2)/2$
3	$-(1+u)(1-v)(1-u+v)/4$	$(1-v)(2u-v)/4$	$-(1+u)(u-2v)/4$
4	$(1+u)(1-v^2)/2$	$(1-v^2)/2$	$-(1+u)v$
5	$-(1+u)(1+v)(1-u-v)/4$	$(1+v)(2u+v)/4$	$(1+u)(u+2v)/4$
6	$(1-u^2)(1+v)/2$	$-(1+v)u$	$(1-u^2)/2$
7	$-(1-u)(1+v)(1+u-v)/4$	$(1+v)(2u-v)/4$	$-(1-u)(u-2v)/4$
8	$(1-u)(1-v^2)/2$	$-(1-v^2)/2$	$-(1-u)v$

Polynomial basis:

$$\begin{bmatrix} 1 & u & v & u^2 & uv & v^2 & u^2v & uv^2 \end{bmatrix}$$

To perform integration over this element the procedure used for the bi-linear element may be used here as well.

The quadratic quadrilateral element is an "incomplete" element because its polynomial basis is missing the ninth term (u^2v^2) that would make the polynomial expansion complete. The complete element requires an additional node at $u=0$, $v=0$. However, the element as shown here is actually more often used than the "complete" element.

8.8.3. 3D Elements

a. Linear tetrahedral element.

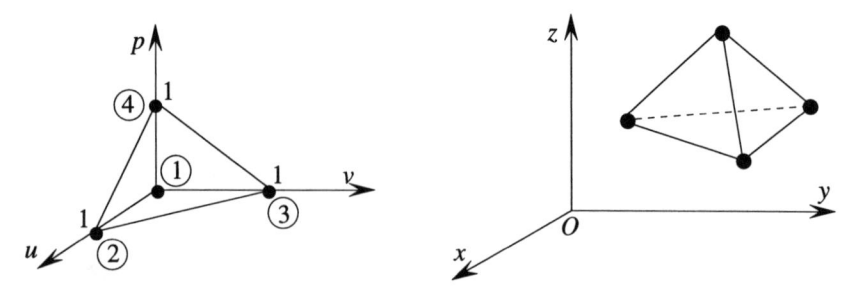

Figure 8.24. Linear tetrahedral element in local and global coordinates.

Table 8.9. Shape functions and their derivatives for the linear tetrahedral element.

Node	$[N]$	$[\partial N/\partial u]$	$[\partial N/\partial v]$	$[\partial N/\partial p]$
1	$1 - u - v - p$	-1	-1	-1
2	u	1	0	0
3	v	0	1	0
4	p	0	0	1

Polynomial basis:

$$[\ 1 \quad u \quad v \quad p\ ]$$

This element is very similar in terms of evaluation of parameters and application to the triangular element. Integration is performed analytically and the Jacobian matrix is calculated analogously to that in section 8.6.6. The latter is

$$[J] = \begin{bmatrix} x_2 - x_1 & y_2 - y_1 & z_2 - z_1 \\ x_3 - x_1 & y_3 - y_1 & z_3 - z_1 \\ x_4 - x_1 & y_4 - y_1 & z_4 - z_1 \end{bmatrix} \qquad (8.123)$$

The determinant of the Jacobian equals six times the volume of the tetrahedral element.

b. Quadratic tetrahedral element.

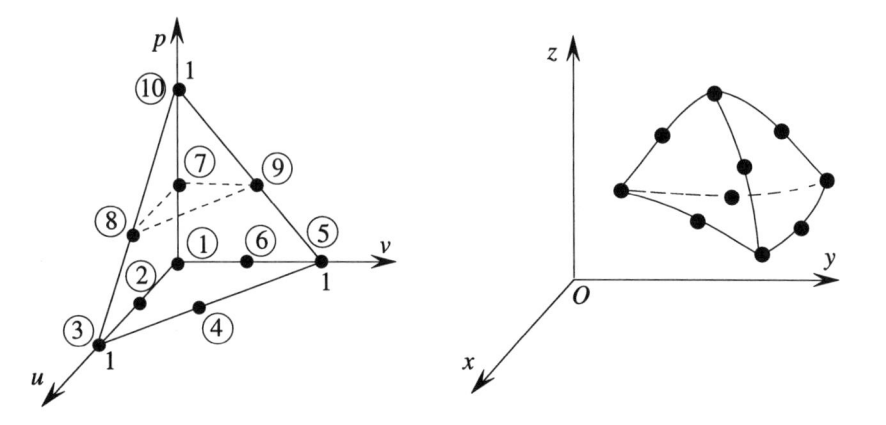

Figure 8.25. Quadratic tetrahedral finite element in local and global coordinates.

As with all quadratic elements, the element in global coordinates may have curvilinear edges.

320   8. Introduction to the Finite Element Method

Table 8.10. Shape functions and their derivatives for the quadratic tetrahedral element.

Node	$[N]$	$[\partial N/\partial u]$	$[\partial N/\partial v]$	$[\partial N/\partial p]$
1	$-t(1-2t)$	$1-4t$	$1-4t$	$1-4t$
2	$4ut$	$4(t-u)$	$-4u$	$-4u$
3	$-u(1-2u)$	$-1+4u$	0	0
4	$4uv$	$4v$	$4u$	0
5	$-v(1-2v)$	0	$-1+4v$	0
6	$4vt$	$-4v$	$4(t-v)$	$-4v$
7	$4pt$	$-4p$	$-4p$	$4(t-p)$
8	$4up$	$4p$	0	$4u$
9	$4vp$	0	$4p$	$4v$
10	$-p(1-2p)$	0	0	$-1+4p$

where $t = 1-u-v-p$

Polynomial basis:

$$\begin{bmatrix} 1 & u & v & p & u^2 & uv & v^2 & vp & p^2 & up \end{bmatrix}$$

The Jacobian for the element is given by

$$[J] = \begin{bmatrix} \dfrac{\partial N}{\partial u} \\[4pt] \dfrac{\partial N}{\partial v} \\[4pt] \dfrac{\partial N}{\partial p} \end{bmatrix} \begin{bmatrix} x & y & z \end{bmatrix} = \begin{bmatrix} 1-4t & 4(t-u) & \cdots \\ 1-4t & -4u & \cdots \\ 1-4t & -4u & \cdots \end{bmatrix} \begin{bmatrix} x_1 & y_1 & z_1 \\ x_2 & y_2 & z_2 \\ \vdots & \vdots & \vdots \\ x_{10} & y_{10} & z_{10} \end{bmatrix} \quad (8.124)$$

Table 8.11. Integration points and corresponding weights for one, four and five point Gauss-Legendre integration over the tetrahedral element.

Integration points	Order m	W_i	u_i	v_i	p_i
1 point	1	1/6	1/4	1/4	1/4
4 points: $a=0.1381966$ $b=0.5854102$	2	1/24 1/24 1/24 1/24	a a a b	a a b a	a b a a
5 points: $a=1/4$ $b=1/6$ $c=1/2$	3	$-2/15$ 3/40 3/40 3/40 3/40	a b b b c	a b b c b	a b c b b

b. Tri-linear hexahedral element.

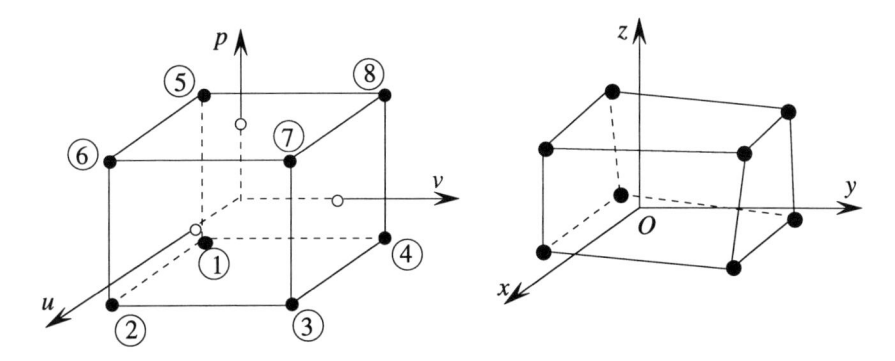

Figure 8.26. The tri-linear hexahedral element in local and global coordinates.

Table 8.12. Shape functions and their derivatives for the first order hexahedral element.

Node	$[N]$	$[\partial N/\partial u]$	$[\partial N/\partial v]$	$[\partial N/\partial p]$
1	$a_2 b_2 c_2/8$	$-b_2 c_2/8$	$-a_2 c_2/8$	$-a_2 b_2/8$
2	$a_1 b_2 c_2/8$	$b_2 c_2/8$	$-a_1 c_2/8$	$-a_1 b_2/8$
3	$a_1 b_1 c_2/8$	$b_1 c_2/8$	$a_1 c_2/8$	$-a_1 b_1/8$
4	$a_2 b_1 c_2/8$	$-b_1 c_2/8$	$a_2 c_2/8$	$-a_2 b_1/8$
5	$a_2 b_2 c_1/8$	$-b_2 c_1/8$	$-a_2 c_1/8$	$a_2 b_2/8$
6	$a_1 b_2 c_1/8$	$b_2 c_1/8$	$-a_1 c_1/8$	$a_1 b_2/8$
7	$a_1 b_1 c_1/8$	$b_1 c_1/8$	$a_1 c_1/8$	$a_1 b_1/8$
8	$a_2 b_1 c_1/8$	$-b_1 c_1/8$	$a_2 c_1/8$	$a_2 b_1/8$

where: $a_1=(1+u)$ $a_2=1-u$ $b_1=1+v$ $b_2=1-v$ $c_1=1+p$ $c_2=1-p$

Polynomial basis:

$$\begin{bmatrix} 1 & u & v & p & uv & vp & pu & uvp \end{bmatrix}$$

For the numerical integration the following eight points and weights are used:

$$(u_i, v_i, p_i) = \left(\pm \frac{1}{\sqrt{3}}, \pm \frac{1}{\sqrt{3}}, \pm \frac{1}{\sqrt{3}} \right), \qquad w_i = 1$$

d. Quadratic hexahedral element.

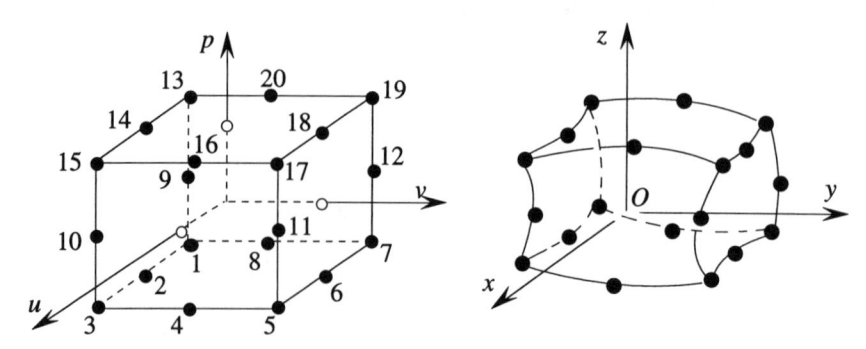

Figure 8.27. Quadratic hexahedral finite element in local and global coordinates.

The shape functions and their derivatives are summarized in Table 8.13a and 8.13b.

Table 8.13a. Shape functions and their derivatives with respect to u for the twenty-node hexahedral element.

Node	$[N_i]$	$[\partial N_i/\partial u]$
1	$(1-u)(1-v)(1-p)(-2-u-v-p)/8$	$-u(1-v)(1-p)(-1-2u-v-p)/8$
2	$(1-u^2)(1-v)(1-p)/4$	$-u(1-v)(1-p)/2$
3	$(1+u)(1-v)(1-p)(-2+u-v-p)/8$	$u(1-v)(1-p)(-1+2u-v-p)/8$
4	$(1+u)(1-v^2)(1-p)/4$	$u(1-v^2)(1-p)/4$
5	$(1+u)(1+v)(1-p)(-2+u+v-p)/8$	$u(1+v)(1-p)(-1+2u+v-p)/8$
6	$(1-u^2)(1+v)(1-p)/4$	$-u(1+v)(1-p)/2$
7	$(1-u)(1+v)(1-p)(-2-u+v-p)/8$	$-u(1+v)(1-p)(-1-2u+v-p)/8$
8	$(1-u)(1-v^2)(1-p)/4$	$-u(1-v^2)(1-p)/4$
9	$(1-u)(1-v)(1-p^2)/4$	$-u(1-v)(1-p^2)/4$
10	$(1+u)(1-v)(1-p^2)/4$	$-u(1-v)(1-p^2)/4$
11	$(1+u)(1+v)(1-p^2)/4$	$-u(1-v)(1-p^2)/4$
12	$(1-u)(1+v)(1-p^2)/4$	$-u(1-v)(1-p^2)/4$
13	$(1-u)(1-v)(1+p)(-2-u-v+p)/8$	$-u(1-v)(1+p)(-1-2u-v+p)/8$
14	$(1-u^2)(1-v)(1+p)/4$	$-u(1-v)(1+p)/2$
15	$(1+u)(1-v)(1+p)(-2+u-v+p)/8$	$u(1-v)(1+p)(-1+2u-v+p)/8$
16	$(1+u)(1-v^2)(1+p)/4$	$u(1-v^2)(1+p)/4$
17	$(1+u)(1+v)(1+p)(-2+u+v+p)/8$	$u(1+v)(1+p)(-1+2u+v+p)/8$
18	$(1-u^2)(1+v)(1+p)/4$	$-u(1+v)(1+p)/2$
19	$(1-u)(1+v)(1+p)(-2-u+v+p)/8$	$-u(1+v)(1+p)(-1-2u+v+p)/8$
20	$(1-u)(1-v^2)(1+p)/4$	$-u(1-v^2)(1+p)/4$

Table 8.13b. Derivatives of the shape functions with respect to v and p for the twenty-node hexahedral element.

Node	$[\partial N_i/\partial u]$	$[\partial N_i/\partial u]$
1	$-v(1-u)(1-p)(-1-u-2v-p)/8$	$-p(1-u)(1-v)(-1-u-v-2p)/8$
2	$-v(1-u^2)(1-p)/4$	$-p(1-u^2)(1-v)/4$
3	$-v(1+u)(1-p)(-1+u-2v-p)/8$	$-p(1+u)(1-v)(-1+u-v-2p)/8$
4	$-v(1+u)(1-p)/2$	$-p(1+u)(1-v^2)/4$
5	$v(1+u)(1-p)(-1+u+2v-p)/8$	$p(1+u)(1+v)(-1+u+v-2p)/8$
6	$v(1-u^2)(1-p)/4$	$-p(1-u^2)(1+v)/4$
7	$v(1-u)(1-p)(-1-u+2v-p)/8$	$-p(1-u)(1+v)(-1-u+v-2p)/8$
8	$-v(1-u)(1-p)/2$	$-p(1-u)(1-v^2)/4$
9	$-v(1-u)(1-p^2)/4$	$-p(1+u)(1-v)/2$
10	$-v(1+u)(1-p^2)/4$	$-p(1+u)(1-v)/2$
11	$v(1+u)(1-p^2)/4$	$-p(1+u)(1+v)/2$
12	$v(1-u)(1-p^2)/4$	$-p(1-u)(1+v)/2$
13	$-v(1-u)(1+p)(-1-u-2v+p)/8$	$p(1-u)(1-v)(-1-u-v+2p)/8$
14	$-v(1-u^2)(1+p)/4$	$p(1-u^2)(1-v)/4$
15	$-v(1+u)(1+p)(-1+u-2v+p)/8$	$p(1+u)(1-v)(-1+u-v+2p)/8$
16	$-v(1+u)(1+p)/2$	$p(1+u)(1-v^2)/4$
17	$v(1+u)(1+p)(-1+u+2v+p)/8$	$p(1+u)(1+v)(-1+u+v+2p)/8$
18	$v(1-u^2)(1+p)/4$	$p(1-u^2)(1+v)/4$
19	$v(1-u)(1+p)(-1-u+2v+p)/8$	$p(1-u)(1+v)(-1-u+v+2p)/8$
20	$-v(1-u)(1+p)/2$	$p(1-u)(1-v^2)/4$

Polynomial basis:

$$\begin{bmatrix} 1 & u & v & p & u^2 & uv & v^2 & vp & p^2 & up & u^2v \\ uv^2 & v^2p & vp^2 & up^2 & u^2p & uvp & u^2vp & uv^2p & uvp^2 \end{bmatrix}$$

Numerical integration for this element can be accomplished as described in the previous section.

8.9. Coupling Different Finite Elements; Infinite Elements

8.9.1. Coupling Different Types of Finite Elements

During the discretization process of a physical domain it is possible to use more than one type of element in the same domain, provided the continuity between

324   8. Introduction to the Finite Element Method

elements is maintained. Figure 8.28 shows the use of triangular and quadrilateral elements.

For each of the elements, the regular shape functions and integration are used separately without modifications. The coupling is done through the common node designation. This method is quite useful, especially when the geometry can be better described with a combination of elements.

Another type of coupling between different elements is shown in Figure 8.29a. This however requires some attention because element (1) is a linear element while element (2) is a modified quadratic element.

To match the two elements in Figure 8.29a, we start with element (2) shown in local coordinates in Figure 8.29b. Node No. 8 must be removed and, therefore, the quadratic element will be modified, since there are now only seven shape functions. To do so, we start with the eight shape functions of the regular quadratic element. These are

$$N = [\ N_1 \quad N_2 \quad N_3 \quad N_4 \quad N_5 \quad N_6 \quad N_7 \quad N_8\ ]$$

Now, node No. 8 is assumed to be dependent and its potential is forced to equal the average between the potential at node No. 1 and node No. 7. This is accomplished by modifying the shape functions of the element as follows:

$$N = [N_1 + N_8/2 \quad N_2 \quad N_3 \quad N_4 \quad N_5 \quad N_6 \quad N_7 + N_8/2]$$

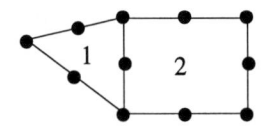

Figure 8.28. Coupling of a triangular and a quadrilateral element.

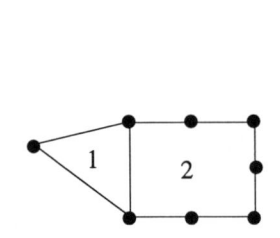

Figure 8.29a. Coupling of linear and quadratic elements.

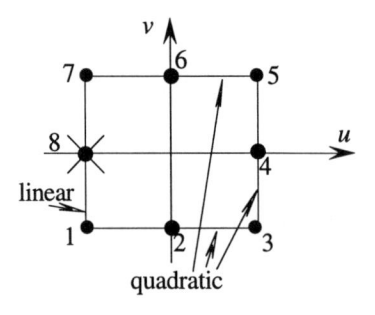

Figure 8.29b. Modifying the quadratic element.

Assuming now that in the process of defining the mesh, node No. 2 must also be eliminated (because another first-order triangular element is connected to the edge on which node No. 2 is placed), the shape functions become

$$N = [N_1 + N_8/2 + N_2/2 \quad N_3 + N_2/2 \quad N_4 \quad N_5 \quad N_6 \quad N_7 + N_8/2]$$

This method is commonly used when using the so-called adaptive methods of mesh generation, a topic we will take up later. The main purpose of this seemingly more complicated method of discretization is to improve solution while decreasing the amount of computation needed, by means of coupling of different finite elements, to better match local conditions.

8.9.2. Infinite Elements

In this section, we discuss briefly the concept of infinite elements, using a 1D element as a simple example.

Suppose that we wish to construct a one-dimensional element of infinite length in the x direction. This is useful when the solution domain extends to infinity, that is, when the quantity we solve for, only tends to zero at infinity and therefore normal Dirichlet boundary conditions cannot be used. The coordinate x of the infinite element may be expressed as

$$x = x_1 + \alpha \frac{1+u}{1-u} \qquad (8.125)$$

The infinite element in local and global coordinates is shown in Figure 8.30. Because the potential is zero at infinity, and using the regular interpolation function (equal to the shape function), we can write

$$V(u) = \begin{bmatrix} \dfrac{1-u}{2} & \dfrac{1+u}{2} \end{bmatrix} \begin{bmatrix} V_1 \\ 0 \end{bmatrix}$$

and

$$V(u) = \frac{1-u}{2} V_1 \qquad (8.126)$$

Figure 8.30. 1D infinite element in local and global coordinates.

Rewriting Eq. (8.125) as

$$u = \frac{(x - x_1) - \alpha}{(x - x_1) + \alpha} \tag{8.127}$$

Substituting this in Eq. (8.126), we get

$$V(x) = \frac{1}{1 + (x - x_1)/\alpha} V_1$$

which, for $x=x_1$ gives $V(x_1)=V_1$ and for $x=\infty$ gives $V(\infty)=0$.

Another method of defining an infinite element and its shape functions is to modify the shape function

$$N_1(u) = \frac{1-u}{2}$$

by multiplying it by a function $f(u)$ which is equal to zero for $u=1$. One such function is

$$f(u) = e^{-[\beta(1+u)/(1-u)]}$$

This gives the approximation over the element:

$$V(u) = \left[\left(\frac{1-u}{2}\right) e^{-[\beta(1+u)/(1-u)]} \quad \frac{1+u}{2} \right] \begin{bmatrix} V_1 \\ 0 \end{bmatrix}$$

or

$$V(u) = V_1 \left(\frac{1-u}{2}\right) e^{-[\beta(1+u)/(1-u)]} \tag{8.128}$$

Substituting Eq. (8.127) into Eq. (8.128), we get

$$V(x) = e^{-(\beta/\alpha)(x-x_1)} \left(\frac{1}{1 + (x-x_1)/\alpha}\right) V_1 \tag{8.129}$$

Similar techniques apply to two- and three-dimensional elements.

8.10. Calculation of Some Terms in Poisson's Equation

In this section, we evaluate some terms that appear in electromagnetic field equations. These will be used in the following chapters when we explore physical phenomena.

In support of the following chapters, we use second-order triangular elements which were discussed in section 8.8.2.

Consider the hypothetical general equation:

$$\nabla \cdot (a\nabla U) + bU + c\frac{\partial U}{\partial x} + s = 0 \qquad (8.130)$$

where U is the unknown quantity, a, b, and c are constants and s is the source term. Applying Galerkin's method, the four terms of the equation above are transformed into matrices which will be assembled into a global matrix system before solution can take place.

8.10.1. *The Stiffness Matrix*

After applying the Galerkin method to Eq. (8.130) and separating the stiffness matrix from the boundary conditions [see section 8.2., step c, and Eq. (8.16)] the first term in Eq. (8.130) results in

$$\int_{S_i} a\nabla N^T \cdot \nabla N \; U \; dxdy \qquad (8.131)$$

To perform the integration in the local coordinate system we write from Eq. (8.131)

$$\int_{S_{local}} a\nabla N^T \cdot \nabla N \; det[J] \; dudv \; U \qquad (8.132)$$

The integral in Eq. (8.132) is calculated numerically; its integrand is evaluated as follows. The Jacobian is

$$[J] = \begin{bmatrix} \dfrac{\partial N}{\partial u} \\ \dfrac{\partial N}{\partial v} \end{bmatrix} [\; x \quad y \;]$$

328   8. Introduction to the Finite Element Method

For the quadratic triangular element [using Table 8.5 and Eq. (8.122)], this is

$$[J] = \begin{bmatrix} 1-4t & 4(t-u)-1+4u & 4v & 0 & -4v \\ 1-4t & -4u & 0 & 4u & -1+4v & 4(t-v) \end{bmatrix} \begin{bmatrix} x_1 & y_1 \\ x_2 & y_2 \\ x_3 & y_3 \\ x_4 & y_4 \\ x_5 & y_5 \\ x_6 & y_6 \end{bmatrix} \quad (8.133)$$

The product between the two matrices above gives the 2×2 Jacobian matrix. The determinant of the Jacobian is then easily obtained. The Jacobian is denoted as

$$[J] = \begin{bmatrix} J_{11} & J_{12} \\ J_{21} & J_{22} \end{bmatrix} \quad (8.134)$$

Next we calculate the term ∇N. This is (see section 8.6.7)

$$\nabla N = [J]^{-1} \begin{bmatrix} \dfrac{\partial N}{\partial u} \\ \dfrac{\partial N}{\partial v} \end{bmatrix}$$

With the Jacobian above, we get

$$\nabla N = \frac{1}{\det [J]} \begin{bmatrix} J_{22} & -J_{12} \\ -J_{21} & J_{11} \end{bmatrix} \begin{bmatrix} 1-4t & 4(t-u) & \cdots & -4v \\ 1-4t & -4u & \cdots & 4(t-v) \end{bmatrix} \quad (8.135)$$

This gives a (2×6) matrix denoted as

$$\nabla N = \begin{bmatrix} dnx_1 & dnx_2 & \cdots & dnx_6 \\ dny_1 & dny_2 & \cdots & dny_6 \end{bmatrix} \quad (8.136)$$

and Eq. (8.132) becomes

$$\int_{S_{local}} \begin{bmatrix} dnx_1 & dny_1 \\ dnx_2 & dny_2 \\ \vdots & \vdots \\ dnx_6 & dny_6 \end{bmatrix} \begin{bmatrix} dnx_1 & dnx_2 & \cdots & dnx_6 \\ dny_1 & dny_2 & \cdots & dny_6 \end{bmatrix} a \, det[J] \, du \, dv \begin{bmatrix} U_1 \\ U_2 \\ U_3 \\ \vdots \\ U_6 \end{bmatrix} \quad (8.137)$$

This results in a (6×6) symmetric system of equations. To evaluate the expressions, we note that both ∇N and $\det [J]$ depend on u and v and, therefore, we use the integration points (u_i, v_i) for both expressions (see Table 8.6). After integration, we obtain the stiffness matrix.

8.10.2. *Evaluation of the Second Term in Eq. (8.130)*

Application of Galerkin's method to the second term in Eq. (8.130) gives

$$\int_{S_{local}} N^t \cdot N \, U \, b \, det[J] \, dudv \qquad (8.138)$$

Noting that $U(u,v) = N(u,v)U$ we get

$$U(u,v) = [N_1(u,v) \quad \cdots \quad N_6(u,v)] \begin{bmatrix} U_1 \\ U_2 \\ \vdots \\ U_6 \end{bmatrix}$$

With the aid of Table 8.5, this gives

$$\int_{S_{local}} b \begin{bmatrix} -t(1-2t) \\ 4ut \\ \vdots \\ 4vt \end{bmatrix} [-t(1-2t) \quad 4ut \quad \cdots \quad 4vt] det[J] \, dudv \begin{bmatrix} U_1 \\ U_2 \\ \vdots \\ U_6 \end{bmatrix} \qquad (8.139)$$

This is also a 6×6 symmetric system. The procedure for integration is the same as indicated in the previous section including the recommended number of integration points (seven).

8.10.3. *Evaluation of the Third Term in Eq. (8.130)*

The third term in Eq. (8.130) results in the following:

$$\int_{S_{local}} N^t \cdot \nabla_x N \, U \, c \, det[J] \, dudv \qquad (8.140)$$

where $\nabla_x NU$ denotes the derivative of U in the x direction. Using Eq. (8.136) in Eq. (8.140) gives

$$\int_{S_{local}} \begin{bmatrix} -t(1-2t) \\ 4ut \\ \vdots \\ 4vt \end{bmatrix} [dnx_1 \quad dnx_2 \quad \cdots \quad dnx_6] c \, det[J] \, dudv \begin{bmatrix} U_1 \\ U_2 \\ \vdots \\ U_6 \end{bmatrix} \qquad (8.141)$$

Unlike the previous matrices, this matrix is nonsymmetric, but integration is done as in the previous two sections.

8.10.4. Evaluation of the Source Term

The source term is represented by the fourth term in Eq. (8.130). This term gives

$$\int_{S_{local}} N^t s \, det[J] \, dudv \qquad (8.142)$$

or

$$\int_{S_{local}} \begin{bmatrix} -t(1-2t) \\ 4ut \\ \vdots \\ 4vt \end{bmatrix} s \, det[J] \, dudv \qquad (8.143)$$

This vector does not depend on the unknowns and is therefore part of the right-hand side of the system of equations for the element.

8.11. A Simplified 2D Second-Order Finite Element Program

Now that the basic concepts necessary for finite element calculations were introduced, we present an example using a simple finite element program to bring together the different concepts discussed before.

8.11.1. The Problem to Be Solved

Starting with Eq. (8.130) and retaining only the first and fourth term, we get

$$\nabla \cdot (a \nabla U) + s = 0 \qquad (8.144)$$

where the first term is the term leading to the stiffness matrix and the second is the source term. This equation may represent many physical situation but in our case it may represent electrostatic applications:

$$\nabla \cdot (\varepsilon \nabla V) + \rho = 0 \qquad (8.145)$$

This equation was used at the beginning of the chapter. However, rather than solving this equation again, we introduce now, briefly, another electromagnetic equation which is similar to Eq. (8.145). This equation describes magnetostatic problems in 2D applications and is written as

$$\nabla \cdot (\nu \nabla A) + J = 0 \qquad (8.146)$$

where ν is the reluctivity of the material ($\nu = 1/\mu$), A is the magnitude of the z component of the magnetic vector potential $\mathbf{A}$ (in this approach, the magnetic vector potential has only a z component). J is the magnitude of the z component of current density (which also has a single component in the z direction). $\mathbf{A}$ and $\mathbf{J}$ are perpendicular to the x-y plane. $\mathbf{A}$ is related to $\mathbf{B}$ as

$$\mathbf{B} = \nabla \times \mathbf{A} \qquad (8.147)$$

$$\mathbf{B} = \begin{vmatrix} \hat{\mathbf{i}} & \hat{\mathbf{j}} & \hat{\mathbf{k}} \\ \dfrac{\partial}{\partial x} & \dfrac{\partial}{\partial y} & \dfrac{\partial}{\partial z} \\ 0 & 0 & A \end{vmatrix} = \hat{\mathbf{i}} \dfrac{\partial A}{\partial y} - \hat{\mathbf{j}} \dfrac{\partial A}{\partial x} \qquad (8.148)$$

The governing equation is

$$\nabla \times \mathbf{H} = \mathbf{J}$$

With $\mathbf{H} = \nu \mathbf{B}$, we get

$$\nabla \times (\nu \mathbf{B}) = \mathbf{J}$$

Substituting Eq. (8.148) into this equation, we obtain Eq. (8.146). Much more information and discussion on this equation and its solution will be given in the following chapter. For now, we simply view this as an example.

The example to be solved here is shown in Figure 8.31.

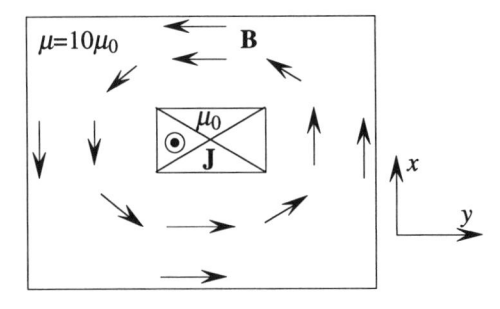

Figure 8.31. Conducting bar carrying a current density $\mathbf{J}$.

332   8. Introduction to the Finite Element Method

In this case, the magnetic flux (related to **B**) flows in a block of material with permeability equal to $10\mu_0$. The conductor carrying the current (related to **J**) has permeability μ_0 (such as copper or aluminum).

Solving Eq. (8.146) for the magnetic vector potential A, we obtain the components of the magnetic flux density by finding the derivatives $\partial A/\partial x$ and $\partial A/\partial y$ [see Eq. (8.148)].

8.11.2. *The Discretized Domain*

The discretized domain is shown in Figure 8.32.

The bar is 60×60 *mm* and the nodes are numbered horizontally starting on the bottom as shown. There are a total of seventy two elements out of which eight are source elements (elements No. 29, 30, 31, 32, 41, 42, 43, and 44). The current density is $J=10^6$ A/m^2 and is perpendicular to the plane. Second-order triangular elements are used throughout with the node numbers as shown. For example, element No. 1 is defined by nodes No. 1, 2, 3, 16, 29, and 15. Nodes on the edges of the elements are placed midway between corner nodes. For example, $x_2=(x_3+x_1)/2$ and $y_2=(y_3+y_1)/2$.

The boundary conditions in this case are $A=0$ on the outer boundary, which means that the magnetic flux density has no normal components on the outer boundaries. The following nodes represent boundary lines and will be assigned zero magnetic vector potential values:

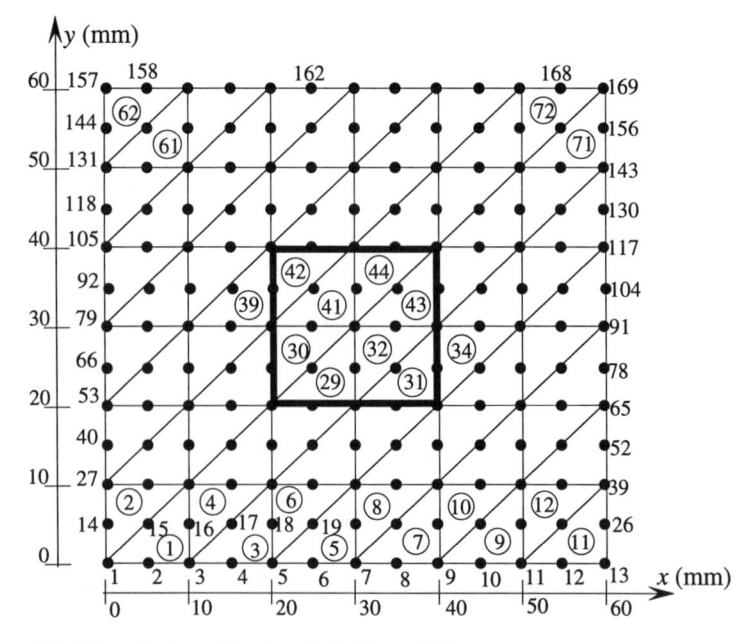

Figure 8.32. Discretization of the domain in Figure 8.31.

[1, 2, 3, 4, 5, 6, 7, 8, 9, 10, 11, 12, 13]
[14, 27, 40, 53, 66, 79, 92, 105, 118, 131, 144]
[157, 158, 159, 160, 161, 162, 163, 164, 165, 166, 167, 168, 169]
[26, 39, 52, 65, 78, 91, 104, 117, 130, 143, 156]

8.11.3. *The Finite Element Program*

A listing of the second order finite element program, written in standard FORTRAN is shown below. Reading of the program is advised so that the reader may understand the implementation of the concepts described above.

The main arrays and variables are:

- NNO – Number of nodes.
- NEL – Number of elements.
- NF – Number of lines with imposed boundary conditions.
- KT(NEL,6) – Node numbers in the element.
- NMAT(NEL) – Material index for the element.
- SOURCEL(NEL) – Value of the source in the element.
- X(NNO), Y(NNO) – Coordinates of the nodes.
- NUMBMAT – Number of different materials.
- PROP(6) – Property index of the materials.
- AA(NNO,NNO), BB(NNO) – Left- and right-side global system of the equations.

Program Listing

```
      program Secdemo
c--------------- Call Data to read the problem data
      call Data(nno,nel,nf)
c--------------- Call Zero to set matrices to zero
      call Zero(nno)
c--------------- Call Form to create the matrix system
      call Form(nel)
c--------------- Call Condi to insert Dirichlet boundary conditions
      call Condi(nf,nno)
c--------------- Call Elim to solve the matr. system by Gauss Elim. method
      call Elim(nno)
c--------------- Call Sort to furnish results
      call Sort(nno,nel)
      stop
      end
c----------------------------------------------------------------------
      subroutine Data(nno,nel,nf)
      common /mesh/ nmat(200),kt(200,6),ncd(10,20),sourcel(200)
      common /realx/ vi(10),prop(6),x(300),y(300)
      dimension source(6),nss(6,20)
      character arq*20
c--------------- Data reading
      write(6,'(a$)')' File Name='
      read(5,'(a20)')arq
      open (9,file=arq,form='formatted')
c--------------- nno-number of nodes; nel-number of elements
      read(9,'(i3)')nno
```

334   8. Introduction to the Finite Element Method

```fortran
      read(9,'(i3)')nel
      do i=1,nel
c-------------- kt-element nodes; nmat-material; sourcel-source
      read(9,'(6i4,i3,e12.4)')(kt(i,j),j=1,6),nmat(i),sourcel(i)
      enddo
      do i=1,nno
c-------------- x,y-coords. of the nodes
      read(9,'(2f8.4)')x(i),y(i)
      enddo
c-------------- nf-number of boundary (Dirichlet) conditions
      read(9,'(i3)')nf
      do i=1,nf
c-------------- vi-imposed potential; ncd-nodes with this potential
      read(9,'(f7.2)')vi(i)
      read(9,'(20i4)')(ncd(i,j),j=1,20)
      enddo
c-------------- numbmat-number of materials; prop-constit. property value
      read(9,'(i3)')numbmat
      do i=1,numbmat
      read(9,'(e9.4)')prop(i)
      enddo
      return
      end
c------------------------------------------------------------------
      subroutine Zero(nno)
      common /matrix1/ aa(200,200),vv(200),bb(200)
      do i=1,nno
      bb(i)=0.
      vv(i)=0.
      do j=1,nno
      aa(i,j)=0.
      enddo
      enddo
      return
      end
c------------------------------------------------------------------
      subroutine Form(nel)
      common /mesh/ nmat(200),kt(200,6),ncd(10,20),sourcel(200)
      common /realx/ vi(10),prop(6),x(300),y(300)
      common /matrix1/ aa(200,200),vv(200),bb(200)
      common /matrix2/ u7(7),v7(7),w7(7),dnu(6),dnv(6),xp(6),yp(6),
     *                 u3(3),v3(3),w3(3),xj(2,2),rn(6),dnx(6),dny(6)
      dimension n(6),r(6,6),s(6)
c-------------- Data for 7 integration points (table 8.6)
      a=0.470142064
      b=0.101286507
      w1=0.066197076
      w2=0.062969590
      u7(1)=1./3.
      u7(2)=a
      u7(3)=1.-2.*a
      u7(4)=a
      u7(5)=b
      u7(6)=1.-2.*b
      u7(7)=b
      v7(1)=1./3.
      v7(2)=a
      v7(3)=a
      v7(4)=1.-2.*a
      v7(5)=b
      v7(6)=b
      v7(7)=1.-2.*b
      w7(1)=9./80.
      w7(2)=w1
      w7(3)=w1
      w7(4)=w1
      w7(5)=w2
      w7(6)=w2
      w7(7)=w2
c-------------- Loop on elements
      do 1 i=1,nel
c-------------- Arrays n, xp and yp created as internal numbering
c-------------- of the element i
      do j=1,6
```

```
              n(j)=kt(i,j)
           enddo
           nm=nmat(i)
           xprop=prop(nm)
           ss=sourcel(i)
           do j=1,6
              xp(j)=x(n(j))
              yp(j)=y(n(j))
           enddo
c--------------- Stiffness r(6,6) and source s(6) elemental
c--------------- matrices set to zero
           do j=1,6
              s(j)=0.
              do k=1,6
                 r(j,k)=0.
              enddo
           enddo
c--------------- Integration points loop
           do 2 it=1,7
              w=w7(it)
              u=u7(it)
              v=v7(it)
c--------------- Call Dnxy to calculate grad N, and det(J)
              call Dnxy(u,v,detj)
              do j=1,6
                 s(j)=s(j)+rn(j)*ss*detj*w
                 do k=1,6
c--------------- Calculation of the Stiffness matrix
                    r(j,k)=r(j,k)+(dnx(j)*dnx(k)+dny(j)*dny(k))*xprop*detj*w
                 enddo
              enddo
 2         continue
c--------------- Assembling the global system
           do j=1,6
              jj=n(j)
c--------------- Source vector on the right side
              bb(jj)=bb(jj)+s(j)
              do k=1,6
                 kk=n(k)
c--------------- Stiffness matrix on the left side
                 aa(jj,kk)=aa(jj,kk)+r(j,k)
              enddo
           enddo
 1      continue
        return
        end
c-------------------------------------------------------------------
        subroutine Dnxy(u,v,detj)
        common /matrix2/ u7(7),v7(7),w7(7),dnu(6),dnv(6),xp(6),yp(6),
     *                   u3(3),v3(3),w3(3),xj(2,2),rn(6),dnx(6),dny(6)
c--------------- Calculation of dN/du and dN/dv  (table 8.5)
        t=1.-u-v
        rn(1)=-t*(1.-2.*t)
        rn(2)=4.*u*t
        rn(3)=-u*(1.-2.*u)
        rn(4)=4.*u*v
        rn(5)=-v*(1.-2.*v)
        rn(6)=4.*v*t
        dnu(1)=1.-4.*t
        dnu(2)=4.*(t-u)
        dnu(3)=-1.+4.*u
        dnu(4)=4*v
        dnu(5)=0.
        dnu(6)=-4.*v
        dnv(1)=1.-4.*t
        dnv(2)=-4.*u
        dnv(3)=0.
        dnv(4)=4.*u
        dnv(5)=-1.+4.*v
        dnv(6)=4.*(t-v)
c--------------- Calculation of Jacobian terms
        xj(1,1)=0.
        xj(1,2)=0.
        xj(2,1)=0.
```

```
      xj(2,2)=0.
      do i=1,6
        xj(1,1)=xj(1,1)+dnu(i)*xp(i)
        xj(1,2)=xj(1,2)+dnu(i)*yp(i)
        xj(2,1)=xj(2,1)+dnv(i)*xp(i)
        xj(2,2)=xj(2,2)+dnv(i)*yp(i)
      enddo
c--------------- Calculation of det(J)
      detj=xj(1,1)*xj(2,2)-xj(2,1)*xj(1,2)
c--------------- Calculation of derivatives DN/dx and DN/dy
      do i=1,6
        dnx(i)=(xj(2,2)*dnu(i)-xj(1,2)*dnv(i))/detj
        dny(i)=(-xj(2,1)*dnu(i)+xj(1,1)*dnv(i))/detj
      enddo
      return
      end
c------------------------------------------------------------------
      subroutine Condi(nf,nno)
      common /mesh/ nmat(200),kt(200,6),ncd(10,20),source1(200)
      common /realx/ vi(10),prop(6),x(300),y(300)
      common /matrix1/ aa(200,200),vv(200),bb(200)
      do 1 i=1,nf
        do 2 j=1,20
          noc=ncd(i,j)
          if(noc.eq.0)goto 1
c--------------- Zero is placed in all terms of the line
          do k=1,nno
            aa(noc,k)=0.
          enddo
c--------------- 1. in diagonal and imposed value on the right side vector
          aa(noc,noc)=1.
          bb(noc)=vi(i)
 2      continue
 1    continue
      return
      end
c------------------------------------------------------------------
      subroutine Elim(nno)
      common /matrix1/ aa(200,200),vv(200),bb(200)
c--------------- Applic. of Gauss Elimination method to solve AA*vv=bb
      nn=nno-1
      do 1 i=1,nn
        do 1 m=i+1,nno
          fat=aa(m,i)/aa(i,i)
          bb(m)=bb(m)-bb(i)*fat
          do 5 j=i+1,nno
 5          aa(m,j)=aa(m,j)-aa(i,j)*fat
 1    continue
      vv(nno)=bb(nno)/aa(nno,nno)
      do 7 i=nn,1,-1
        som=0.
        do 8 j=i+1,nno
 8        som=som+aa(i,j)*vv(j)
        vv(i)=(bb(i)-som)/aa(i,i)
 7    continue
      return
      end
c------------------------------------------------------------------
      subroutine Sort(nno,nel)
      character q*1
      common /mesh/ nmat(200),kt(200,6),ncd(10,20),source1(200)
      common /realx/ vi(10),prop(6),x(300),y(300)
      common /matrix1/ aa(200,200),vv(200),bb(200)
      common /matrix2/ u7(7),v7(7),w7(7),dnu(6),dnv(6),xp(6),yp(6),
     *                 u3(3),v3(3),w3(3),xj(2,2),rn(6),dnx(6),dny(6)
      dimension n(6)
      write(6,'(a)')' Set Printer'
      read(5,'(a)')q
c--------------- Potentials printing
      do i=1,nno
        write(6,'(a,i3,a,e10.4)')' Node-',i,'    Potential=',vv(i)
      enddo
c--------------- Printing of potentials and derivatives
c---------------    in the element centroids
```

```
      ub=1./3.
      vb=1./3.
      do i=1,nel
        do j=1,6
          n(j)=kt(i,j)
          xp(j)=x(n(j))
          yp(j)=y(n(j))
          dnx(j)=0.
          dny(j)=0.
        enddo
        call Dnxy(ub,vb,detj)
        potcen=0.
        dnxv=0.
        dnyv=0.
        do j=1,6
          potcen=potcen+rn(j)*vv(n(j))
          dnxv=dnxv+dnx(j)*vv(n(j))
          dnyv=dnyv+dny(j)*vv(n(j))
        enddo
        write(6,'(a,i3,a,e10.4,a,e10.4,a,e10.4)')
     *'  Elem-',i,'  Pot=',potcen,'     dV/dx=',dnxv,'    dV/dy=',dnyv
      enddo
      return
      end
```

Some remarks are in order here:

- The program listed above is a very simple, elementary program intended to demonstrate some of the concepts discussed in this chapter such as shape functions, numerical integration, matrix assembly, and the like. For this purpose, the two most important subroutines are FORM and DNXY.
- In subroutine FORM, the loops on the elements and numerical integration are performed. Assembly of the stiffness and source matrices are also done in FORM.
- Subroutine DNXY calculates the Jacobian, its inverse, its determinant, and the derivatives $\partial N/\partial x$ and $\partial N/\partial y$.
- It would be possible to simplify this program further by using subparametric mapping instead of the isoparametric mapping used in the current program. This would make the edges of the elements in global coordinates, straight lines (rather than curvilinear) and, because the Jacobian is simplified, the evaluation of matrices would also be simpler.
- Insertion of boundary conditions is performed after the global matrix has been fully assembled.
- Solution of the system of equations is done using a standard Gaussian elimination method.
- The matrix AA(NNO,NNO) is assumed to be fully populated and nonsymmetric. The solution here takes no advantage of either the sparse or the symmetric nature of the matrix. In more elaborate finite element programs both symmetry and sparsness are exploited to increase speed and efficiency.
- The results listed by SORT are the potentials at the nodes of the mesh, and the derivatives $\partial A/\partial x$ and $\partial A/\partial y$ at the centroids of the elements. The potentials at the centroids are also calculated.

The data entry for program SECDEMO is listed below (see subroutine DATA to follow the sequence of input data).

Input Data Listing

```
169
72
  1   2   3  16  29  15 1  .0000E+00
 29  28  27  14   1  15 1  .0000E+00
  3   4   5  18  31  17 1  .0000E+00
 31  30  29  16   3  17 1  .0000E+00
  5   6   7  20  33  19 1  .0000E+00
 33  32  31  18   5  19 1  .0000E+00
  7   8   9  22  35  21 1  .0000E+00
 35  34  33  20   7  21 1  .0000E+00
  9  10  11  24  37  23 1  .0000E+00
 37  36  35  22   9  23 1  .0000E+00
 11  12  13  26  39  25 1  .0000E+00
 39  38  37  24  11  25 1  .0000E+00
 27  28  29  42  55  41 1  .0000E+00
 55  54  53  40  27  41 1  .0000E+00
 29  30  31  44  57  43 1  .0000E+00
 57  56  55  42  29  43 1  .0000E+00
 31  32  33  46  59  45 1  .0000E+00
 59  58  57  44  31  45 1  .0000E+00
 33  34  35  48  61  47 1  .0000E+00
 61  60  59  46  33  47 1  .0000E+00
 35  36  37  50  63  49 1  .0000E+00
 63  62  61  48  35  49 1  .0000E+00
 37  38  39  52  65  51 1  .0000E+00
 65  64  63  50  37  51 1  .0000E+00
 53  54  55  68  81  67 1  .0000E+00
 81  80  79  66  53  67 1  .0000E+00
 55  56  57  70  83  69 1  .0000E+00
 83  82  81  68  55  69 1  .0000E+00
 57  58  59  72  85  71 2  .1000E+07
 85  84  83  70  57  71 2  .1000E+07
 59  60  61  74  87  73 2  .1000E+07
 87  86  85  72  59  73 2  .1000E+07
 61  62  63  76  89  75 1  .0000E+00
 89  88  87  74  61  75 1  .0000E+00
 63  64  65  78  91  77 1  .0000E+00
 91  90  89  76  63  77 1  .0000E+00
 79  80  81  94 107  93 1  .0000E+00
107 106 105  92  79  93 1  .0000E+00
 81  82  83  96 109  95 1  .0000E+00
109 108 107  94  81  95 1  .0000E+00
 83  84  85  98 111  97 2  .1000E+07
111 110 109  96  83  97 2  .1000E+07
 85  86  87 100 113  99 2  .1000E+07
113 112 111  98  85  99 2  .1000E+07
 87  88  89 102 115 101 1  .0000E+00
115 114 113 100  87 101 1  .0000E+00
 89  90  91 104 117 103 1  .0000E+00
117 116 115 102  89 103 1  .0000E+00
105 106 107 120 133 119 1  .0000E+00
133 132 131 118 105 119 1  .0000E+00
107 108 109 122 135 121 1  .0000E+00
135 134 133 120 107 121 1  .0000E+00
109 110 111 124 137 123 1  .0000E+00
137 136 135 122 109 123 1  .0000E+00
111 112 113 126 139 125 1  .0000E+00
139 138 137 124 111 125 1  .0000E+00
113 114 115 128 141 127 1  .0000E+00
141 140 139 126 113 127 1  .0000E+00
115 116 117 130 143 129 1  .0000E+00
143 142 141 128 115 129 1  .0000E+00
131 132 133 146 159 145 1  .0000E+00
159 158 157 144 131 145 1  .0000E+00
133 134 135 148 161 147 1  .0000E+00
161 160 159 146 133 147 1  .0000E+00
135 136 137 150 163 149 1  .0000E+00
163 162 161 148 135 149 1  .0000E+00
137 138 139 152 165 151 1  .0000E+00
165 164 163 150 137 151 1  .0000E+00
139 140 141 154 167 153 1  .0000E+00
167 166 165 152 139 153 1  .0000E+00
141 142 143 156 169 155 1  .0000E+00
169 168 167 154 141 155 1  .0000E+00
.0000  .0000
.0050  .0000
.0100  .0000
.0150  .0000
.0200  .0000
.0250  .0000
.0300  .0000
.0350  .0000
.0400  .0000
.0450  .0000
.0500  .0000
.0550  .0000
.0600  .0000
.0000  .0050
.0050  .0050
.0100  .0050
.0150  .0050
.0200  .0050
.0250  .0050
.0300  .0050
.0350  .0050
.0400  .0050
.0450  .0050
.0500  .0050
.0550  .0050
.0600  .0050
.0000  .0100
.0050  .0100
.0100  .0100
.0150  .0100
.0200  .0100
.0250  .0100
.0300  .0100
.0350  .0100
.0400  .0100
.0450  .0100
.0500  .0100
.0550  .0100
.0600  .0100
.0000  .0150
.0050  .0150
.0100  .0150
.0150  .0150
.0200  .0150
.0250  .0150
.0300  .0150
.0350  .0150
.0400  .0150
.0450  .0150
.0500  .0150
.0550  .0150
```

A 2D Second-Order Finite Element Program

.0600	.0150	.0350	.0400
.0000	.0200	.0400	.0400
.0050	.0200	.0450	.0400
.0100	.0200	.0500	.0400
.0150	.0200	.0550	.0400
.0200	.0200	.0600	.0400
.0250	.0200	.0000	.0450
.0300	.0200	.0050	.0450
.0350	.0200	.0100	.0450
.0400	.0200	.0150	.0450
.0450	.0200	.0200	.0450
.0500	.0200	.0250	.0450
.0550	.0200	.0300	.0450
.0600	.0200	.0350	.0450
.0000	.0250	.0400	.0450
.0050	.0250	.0450	.0450
.0100	.0250	.0500	.0450
.0150	.0250	.0550	.0450
.0200	.0250	.0600	.0450
.0250	.0250	.0000	.0500
.0300	.0250	.0050	.0500
.0350	.0250	.0100	.0500
.0400	.0250	.0150	.0500
.0450	.0250	.0200	.0500
.0500	.0250	.0250	.0500
.0550	.0250	.0300	.0500
.0600	.0250	.0350	.0500
.0000	.0300	.0400	.0500
.0050	.0300	.0450	.0500
.0100	.0300	.0500	.0500
.0150	.0300	.0550	.0500
.0200	.0300	.0600	.0500
.0250	.0300	.0000	.0550
.0300	.0300	.0050	.0550
.0350	.0300	.0100	.0550
.0400	.0300	.0150	.0550
.0450	.0300	.0200	.0550
.0500	.0300	.0250	.0550
.0550	.0300	.0300	.0550
.0600	.0300	.0350	.0550
.0000	.0350	.0400	.0550
.0050	.0350	.0450	.0550
.0100	.0350	.0500	.0550
.0150	.0350	.0550	.0550
.0200	.0350	.0600	.0550
.0250	.0350	.0000	.0600
.0300	.0350	.0050	.0600
.0350	.0350	.0100	.0600
.0400	.0350	.0150	.0600
.0450	.0350	.0200	.0600
.0500	.0350	.0250	.0600
.0550	.0350	.0300	.0600
.0600	.0350	.0350	.0600
.0000	.0400	.0400	.0600
.0050	.0400	.0450	.0600
.0100	.0400	.0500	.0600
.0150	.0400	.0550	.0600
.0200	.0400	.0600	.0600
.0250	.0400		
.0300	.0400		

4
.00
1 2 3 4 5 6 7 8 9 10 11 12 13 0 0 0 0 0 0 0
.00
157 158 159 160 161 162 163 164 165 166 167 168 169 0 0 0 0 0 0 0
.00
14 27 40 53 66 79 92 105 118 131 144 0 0 0 0 0 0 0 0
.00
26 39 52 65 78 91 104 117 130 143 156 0 0 0 0 0 0 0 0
2
.7958E+05
.7958E+06

8. Introduction to the Finite Element Method

After solution we obtain the following output. Note that the derivatives printed, $\partial V/\partial x$ and $\partial V/\partial y$ correspond directly to $-B_y$ and B_x, respectively [Eq. (8.148)]. If we were solving Eq. (8.145) instead, the derivatives would correspond to $-E_x$ and $-E_y$.

Output Data Listing

Node	Potential	Node	Potential
Node- 1	.0000E+00	Node- 62	.4985E-03
Node- 2	.0000E+00	Node- 63	.3015E-03
Node- 3	.0000E+00	Node- 64	.1458E-03
Node- 4	.0000E+00	Node- 65	.0000E+00
Node- 5	.0000E+00	Node- 66	.0000E+00
Node- 6	.0000E+00	Node- 67	.1666E-03
Node- 7	.0000E+00	Node- 68	.3469E-03
Node- 8	.0000E+00	Node- 69	.5575E-03
Node- 9	.0000E+00	Node- 70	.8095E-03
Node- 10	.0000E+00	Node- 71	.8342E-03
Node- 11	.0000E+00	Node- 72	.8433E-03
Node- 12	.0000E+00	Node- 73	.8342E-03
Node- 13	.0000E+00	Node- 74	.8095E-03
Node- 14	.0000E+00	Node- 75	.5575E-03
Node- 15	.3913E-04	Node- 76	.3469E-03
Node- 16	.7826E-04	Node- 77	.1666E-03
Node- 17	.1149E-03	Node- 78	.0000E+00
Node- 18	.1458E-03	Node- 79	.0000E+00
Node- 19	.1666E-03	Node- 80	.1738E-03
Node- 20	.1738E-03	Node- 81	.3621E-03
Node- 21	.1666E-03	Node- 82	.5751E-03
Node- 22	.1458E-03	Node- 83	.8234E-03
Node- 23	.1149E-03	Node- 84	.8433E-03
Node- 24	.7826E-04	Node- 85	.8499E-03
Node- 25	.3913E-04	Node- 86	.8433E-03
Node- 26	.0000E+00	Node- 87	.8234E-03
Node- 27	.0000E+00	Node- 88	.5751E-03
Node- 28	.7826E-04	Node- 89	.3621E-03
Node- 29	.1590E-03	Node- 90	.1738E-03
Node- 30	.2356E-03	Node- 91	.0000E+00
Node- 31	.3015E-03	Node- 92	.0000E+00
Node- 32	.3469E-03	Node- 93	.1666E-03
Node- 33	.3621E-03	Node- 94	.3469E-03
Node- 34	.3469E-03	Node- 95	.5575E-03
Node- 35	.3015E-03	Node- 96	.8095E-03
Node- 36	.2356E-03	Node- 97	.8342E-03
Node- 37	.1590E-03	Node- 98	.8433E-03
Node- 38	.7826E-04	Node- 99	.8342E-03
Node- 39	.0000E+00	Node-100	.8095E-03
Node- 40	.0000E+00	Node-101	.5575E-03
Node- 41	.1149E-03	Node-102	.3469E-03
Node- 42	.2356E-03	Node-103	.1666E-03
Node- 43	.3670E-03	Node-104	.0000E+00
Node- 44	.4985E-03	Node-105	.0000E+00
Node- 45	.5575E-03	Node-106	.1458E-03
Node- 46	.5751E-03	Node-107	.3015E-03
Node- 47	.5575E-03	Node-108	.4985E-03
Node- 48	.4985E-03	Node-109	.7678E-03
Node- 49	.3670E-03	Node-110	.8095E-03
Node- 50	.2356E-03	Node-111	.8234E-03
Node- 51	.1149E-03	Node-112	.8095E-03
Node- 52	.0000E+00	Node-113	.7678E-03
Node- 53	.0000E+00	Node-114	.4985E-03
Node- 54	.1458E-03	Node-115	.3015E-03
Node- 55	.3015E-03	Node-116	.1458E-03
Node- 56	.4985E-03	Node-117	.0000E+00
Node- 57	.7678E-03	Node-118	.0000E+00
Node- 58	.8095E-03	Node-119	.1149E-03
Node- 59	.8234E-03	Node-120	.2356E-03
Node- 60	.8095E-03	Node-121	.3670E-03
Node- 61	.7678E-03	Node-122	.4985E-03

A 2D Second-Order Finite Element Program 341

Node-123	Potential= .5575E-03		Node-147	Potential= .1149E-03
Node-124	Potential= .5751E-03		Node-148	Potential= .1458E-03
Node-125	Potential= .5575E-03		Node-149	Potential= .1666E-03
Node-126	Potential= .4985E-03		Node-150	Potential= .1738E-03
Node-127	Potential= .3670E-03		Node-151	Potential= .1666E-03
Node-128	Potential= .2356E-03		Node-152	Potential= .1458E-03
Node-129	Potential= .1149E-03		Node-153	Potential= .1149E-03
Node-130	Potential= .0000E+00		Node-154	Potential= .7826E-04
Node-131	Potential= .0000E+00		Node-155	Potential= .3913E-04
Node-132	Potential= .7826E-04		Node-156	Potential= .0000E+00
Node-133	Potential= .1590E-03		Node-157	Potential= .0000E+00
Node-134	Potential= .2356E-03		Node-158	Potential= .0000E+00
Node-135	Potential= .3015E-03		Node-159	Potential= .0000E+00
Node-136	Potential= .3469E-03		Node-160	Potential= .0000E+00
Node-137	Potential= .3621E-03		Node-161	Potential= .0000E+00
Node-138	Potential= .3469E-03		Node-162	Potential= .0000E+00
Node-139	Potential= .3015E-03		Node-163	Potential= .0000E+00
Node-140	Potential= .2356E-03		Node-164	Potential= .0000E+00
Node-141	Potential= .1590E-03		Node-165	Potential= .0000E+00
Node-142	Potential= .7826E-04		Node-166	Potential= .0000E+00
Node-143	Potential= .0000E+00		Node-167	Potential= .0000E+00
Node-144	Potential= .0000E+00		Node-168	Potential= .0000E+00
Node-145	Potential= .3913E-04		Node-169	Potential= .0000E+00
Node-146	Potential= .7826E-04			

Elem- 1	Pot= .3451E-04	dV/dx= .5217E-02	dV/dy= .1052E-01
Elem- 2	Pot= .3451E-04	dV/dx= .1052E-01	dV/dy= .5217E-02
Elem- 3	Pot= .8235E-04	dV/dx= .4114E-02	dV/dy= .2537E-01
Elem- 4	Pot= .1394E-03	dV/dx= .9637E-02	dV/dy= .2139E-01
Elem- 5	Pot= .1111E-03	dV/dx= .9600E-03	dV/dy= .3429E-01
Elem- 6	Pot= .2193E-03	dV/dx= .4800E-02	dV/dy= .3409E-01
Elem- 7	Pot= .1053E-03	dV/dx=-.2782E-02	dV/dy= .3227E-01
Elem- 8	Pot= .2318E-03	dV/dx=-.2978E-02	dV/dy= .3611E-01
Elem- 9	Pot= .6819E-04	dV/dx=-.4887E-02	dV/dy= .2062E-01
Elem- 10	Pot= .1694E-03	dV/dx=-.8865E-02	dV/dy= .2614E-01
Elem- 11	Pot= .1739E-04	dV/dx=-.5217E-02	dV/dy= .5217E-02
Elem- 12	Pot= .6929E-04	dV/dx=-.1052E-01	dV/dy= .1052E-01
Elem- 13	Pot= .1394E-03	dV/dx= .2139E-01	dV/dy= .9637E-02
Elem- 14	Pot= .8235E-04	dV/dx= .2537E-01	dV/dy= .4114E-02
Elem- 15	Pot= .3529E-03	dV/dx= .2227E-01	dV/dy= .3307E-01
Elem- 16	Pot= .3529E-03	dV/dx= .3307E-01	dV/dy= .2227E-01
Elem- 17	Pot= .4923E-03	dV/dx= .4366E-02	dV/dy= .4345E-01
Elem- 18	Pot= .6188E-03	dV/dx= .9722E-02	dV/dy= .4914E-01
Elem- 19	Pot= .4645E-03	dV/dx=-.9888E-02	dV/dy= .4362E-01
Elem- 20	Pot= .6461E-03	dV/dx=-.4200E-02	dV/dy= .4897E-01
Elem- 21	Pot= .2879E-03	dV/dx=-.2227E-01	dV/dy= .2227E-01
Elem- 22	Pot= .4539E-03	dV/dx=-.3307E-01	dV/dy= .3307E-01
Elem- 23	Pot= .6819E-04	dV/dx=-.2062E-01	dV/dy= .4887E-02
Elem- 24	Pot= .1694E-03	dV/dx=-.2614E-01	dV/dy= .8865E-02
Elem- 25	Pot= .2193E-03	dV/dx= .3409E-01	dV/dy= .4800E-02
Elem- 26	Pot= .1111E-03	dV/dx= .3429E-01	dV/dy= .9600E-03
Elem- 27	Pot= .6188E-03	dV/dx= .4914E-01	dV/dy= .9722E-02
Elem- 28	Pot= .4923E-03	dV/dx= .4345E-01	dV/dy= .4366E-02
Elem- 29	Pot= .8341E-03	dV/dx= .3060E-02	dV/dy= .4189E-02
Elem- 30	Pot= .8341E-03	dV/dx= .4189E-02	dV/dy= .3060E-02
Elem- 31	Pot= .8220E-03	dV/dx=-.5154E-02	dV/dy= .5154E-02
Elem- 32	Pot= .8430E-03	dV/dx=-.2094E-02	dV/dy= .2094E-02
Elem- 33	Pot= .4645E-03	dV/dx=-.4362E-01	dV/dy= .9888E-02
Elem- 34	Pot= .6461E-03	dV/dx=-.4897E-01	dV/dy= .4200E-02
Elem- 35	Pot= .1053E-03	dV/dx=-.3227E-01	dV/dy= .2782E-02
Elem- 36	Pot= .2318E-03	dV/dx=-.3611E-01	dV/dy= .2978E-02
Elem- 37	Pot= .2318E-03	dV/dx= .3611E-01	dV/dy=-.2978E-02
Elem- 38	Pot= .1053E-03	dV/dx= .3227E-01	dV/dy=-.2782E-02
Elem- 39	Pot= .6461E-03	dV/dx= .4897E-01	dV/dy=-.4200E-02
Elem- 40	Pot= .4645E-03	dV/dx= .4362E-01	dV/dy=-.9888E-02
Elem- 41	Pot= .8430E-03	dV/dx= .2094E-02	dV/dy=-.2094E-02
Elem- 42	Pot= .8220E-03	dV/dx= .5154E-02	dV/dy=-.5154E-02
Elem- 43	Pot= .8341E-03	dV/dx=-.4189E-02	dV/dy=-.3060E-02
Elem- 44	Pot= .8341E-03	dV/dx=-.3060E-02	dV/dy=-.4189E-02
Elem- 45	Pot= .4923E-03	dV/dx=-.4345E-01	dV/dy=-.4366E-02
Elem- 46	Pot= .6188E-03	dV/dx=-.4914E-01	dV/dy=-.9722E-02
Elem- 47	Pot= .1111E-03	dV/dx=-.3429E-01	dV/dy=-.9600E-03
Elem- 48	Pot= .2193E-03	dV/dx=-.3409E-01	dV/dy=-.4800E-02
Elem- 49	Pot= .1694E-03	dV/dx= .2614E-01	dV/dy=-.8865E-02
Elem- 50	Pot= .6819E-04	dV/dx= .2062E-01	dV/dy=-.4887E-02

342   8. Introduction to the Finite Element Method

```
Elem- 51   Pot= .4539E-03   dV/dx= .3307E-01    dV/dy=-.3307E-01
Elem- 52   Pot= .2879E-03   dV/dx= .2227E-01    dV/dy=-.2227E-01
Elem- 53   Pot= .6461E-03   dV/dx= .4200E-02    dV/dy=-.4897E-01
Elem- 54   Pot= .4645E-03   dV/dx= .9888E-02    dV/dy=-.4362E-01
Elem- 55   Pot= .6188E-03   dV/dx=-.9722E-02    dV/dy=-.4914E-01
Elem- 56   Pot= .4923E-03   dV/dx=-.4366E-02    dV/dy=-.4345E-01
Elem- 57   Pot= .3529E-03   dV/dx=-.3307E-01    dV/dy=-.2227E-01
Elem- 58   Pot= .3529E-03   dV/dx=-.2227E-01    dV/dy=-.3307E-01
Elem- 59   Pot= .8235E-04   dV/dx=-.2537E-01    dV/dy=-.4114E-02
Elem- 60   Pot= .1394E-03   dV/dx=-.2139E-01    dV/dy=-.9637E-02
Elem- 61   Pot= .6929E-04   dV/dx= .1052E-01    dV/dy=-.1052E-01
Elem- 62   Pot= .1739E-04   dV/dx= .5217E-02    dV/dy=-.5217E-02
Elem- 63   Pot= .1694E-03   dV/dx= .8865E-02    dV/dy=-.2614E-01
Elem- 64   Pot= .6819E-04   dV/dx= .4887E-02    dV/dy=-.2062E-01
Elem- 65   Pot= .2318E-03   dV/dx= .2978E-02    dV/dy=-.3611E-01
Elem- 66   Pot= .1053E-03   dV/dx= .2782E-02    dV/dy=-.3227E-01
Elem- 67   Pot= .2193E-03   dV/dx=-.4800E-02    dV/dy=-.3409E-01
Elem- 68   Pot= .1111E-03   dV/dx=-.9600E-03    dV/dy=-.3429E-01
Elem- 69   Pot= .1394E-03   dV/dx=-.9637E-02    dV/dy=-.2139E-01
Elem- 70   Pot= .8235E-04   dV/dx=-.4114E-02    dV/dy=-.2537E-01
Elem- 71   Pot= .3451E-04   dV/dx=-.1052E-01    dV/dy=-.5217E-02
Elem- 72   Pot= .3451E-04   dV/dx=-.5217E-02    dV/dy=-.1052E-01
```

From these results, we note the following:

- The central node of the mesh (in the middle of the conductor, number 85), has the highest potential value in the solution domain as expected.
- Symmetrically located nodes have identical potential values.
- Symmetrically located elements (such as No. 34 and No. 39) have the same value of derivatives (fields).

These observations are simple but useful indications that the software is working properly. Whenever a code is used, it is advisable that at least simple verification as indicated here be performed.

The results from SECDEMO were verified with results from EFCAD which has been used satisfactorily for many years. The plot of constant potentials provided by EFCAD is shown in Figure 8.33.

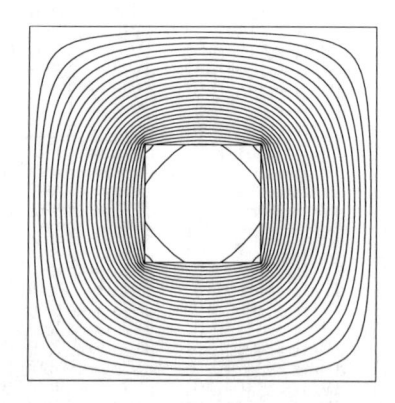

Figure 8.33. Plot of equal potential lines provided by EFCAD.

9
The Variational Finite Element Method: Some Static Applications

9.1. Introduction

The finite element method was presented as a numerical tool in the previous chapter. Here we apply the method to what may be considered classical electromagnetic problems. The understanding of the solution of these important problems may be seen as a natural extension of classical electromagnetism.

The finite element implementation is done, in general, using either a variational approach or the Galerkin method.

In this chapter, we introduce the variational method applied to some cases normally associated with the variational approach. The following chapter will discus Galerkin's method.

9.2. Some Static Applications

The cases presented here are considered as static in time. Even though, at first glance, this might appear an important limitation (because most devices have moving parts, or operate under time dependent conditions), there are many situations where a dynamic problem can be studied as a composition of static cases. Here we establish, using Maxwell's equations, the second-order partial differential equations (Laplace's and Poisson's equations) associated with specific physical phenomena. Instead of specifying the equations in terms of field variables, we use the scalar and vector potentials as primary variables.

9.2.1. *Electrostatic Fields: Dielectric Materials*

Consider the situation shown in Figure 9.1, where the domain under study contains several dielectric media with different permittivities "ε." On lines A and B potentials V_a and V_b are imposed. Assume that a static charge q defined by a volume charge density ρ exists in the interior of the domain. Since $\nabla \times \mathbf{E} = 0$, a scalar electrical potential V *(Volts)*, linked to the electrical field intensity $\mathbf{E}$ by

344   9. The Variational Finite Element Method: Some Static Applications

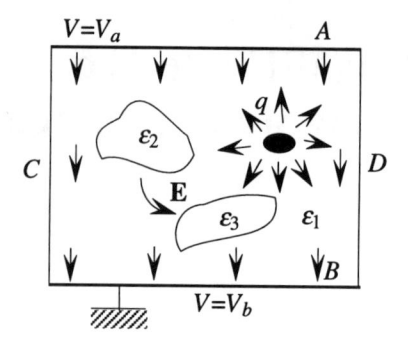

Figure 9.1. A problem solved using the electric scalar potential. The solution domain includes dielectric materials and charges with specified voltages on boundaries A and B.

$$\mathbf{E} = -\nabla V \qquad (9.1)$$

is defined. Also, because

$$\nabla \cdot \mathbf{D} = \rho \qquad (9.2)$$

and using the constitutive relation

$$\mathbf{D} = \varepsilon \mathbf{E} \qquad (9.3)$$

we have

$$\nabla \cdot \varepsilon \mathbf{E} = \rho$$

or, in terms of the potential,

$$\nabla \cdot \varepsilon (-\nabla V) = \rho$$

which, in explicit form in two dimensions (2D), is

$$\frac{\partial}{\partial x} \varepsilon \frac{\partial V}{\partial x} + \frac{\partial}{\partial y} \varepsilon \frac{\partial V}{\partial y} = -\rho \qquad (9.4)$$

If the domain contains only one material, the permittivity can be considered constant, and we get

$$\frac{\partial^2 V}{\partial x^2} + \frac{\partial^2 V}{\partial y^2} = -\frac{\rho}{\varepsilon} \qquad (9.5)$$

This is Poisson's equation which describes the potential distribution in a the domain.

In practical electrical engineering problems, the charge density is frequently zero. In this case, Eq. (9.4) is transformed into Laplace's equation:

$$\frac{\partial}{\partial x}\varepsilon\frac{\partial V}{\partial x} + \frac{\partial}{\partial y}\varepsilon\frac{\partial V}{\partial y} = 0 \qquad (9.6)$$

Here, the sources of the field are the boundary conditions imposed as potentials (on lines A and B). In contrast, in Eq. (9.5), the field can be generated by the charge density ρ alone.

Equations (9.4), (9.5), and (9.6) can be solved for V using the finite element method. This is equivalent to solving the original Maxwell equations since knowledge of V provides the electric field intensity **E**, using Eq. (9.1).

9.2.2. Stationary Currents: Conducting Materials

The physical situation considered here is shown in Figure 9.2. The potential difference $(V_a - V_b)$ establishes the currents in the composite medium with conductivities σ_1, σ_2, and σ_3.

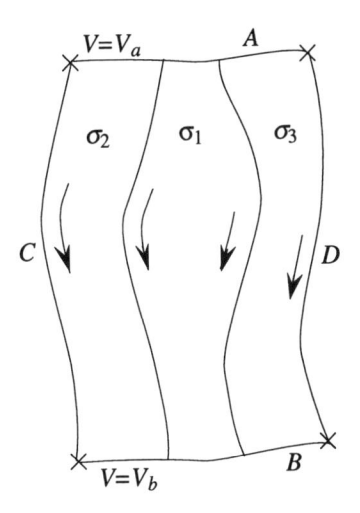

Figure 9.2. Currents in a domain made of materials with different conductivities. One appropriate method of solution is the electric scalar potential.

To define an equation for this situation we use the relation $\mathbf{E}=-\nabla V$, and the electrical continuity equation, presented in Chapter 2

$$\nabla \cdot \mathbf{J} = 0 \qquad (9.7)$$

With the point form of Ohm's law, $\mathbf{J}=\sigma\mathbf{E}$, we have

$$\nabla \cdot \sigma \mathbf{E} = \nabla \cdot \sigma(-\nabla V) = 0$$

which, in explicit form, is

$$\frac{\partial}{\partial x}\sigma\frac{\partial V}{\partial x} + \frac{\partial}{\partial y}\sigma\frac{\partial V}{\partial y} = 0 \qquad (9.8)$$

If there is only one material in the domain, with constant conductivity, Eq. (9.8) becomes

$$\frac{\partial^2 V}{\partial x^2} + \frac{\partial^2 V}{\partial y^2} = 0 \qquad (9.9)$$

Solving for V, allows the calculation of $\mathbf{E}$ and, using Ohm's law, allows the calculation of $\mathbf{J}$.

9.2.3. Magnetic Fields: Scalar Potential

The geometry relating to this situation is shown in Figure 9.3a and 9.3b. We assume that the magnetomotive force of the coil (Figure 9.3a) is established between lines A and B. This is a good approximation if the permeability of the magnetic circuit is high.

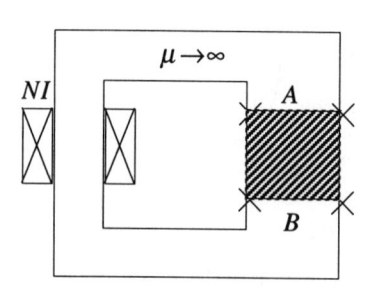

Figure 9.3a. A magnetic circuit with a gap between lines A and B. Because of the high permeability, only the gap is considered.

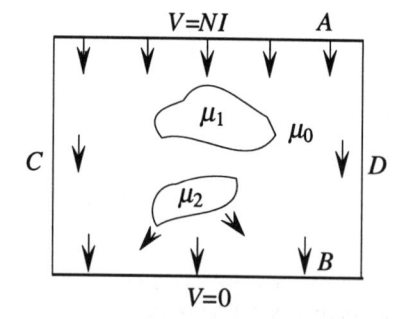

Figure 9.3b. The gap with magnetic scalar potentials enforced on lines A and B. Lines C and D are Neumann boundary conditions.

Thus, we only need to consider the air-gap domain shown in Figure 9.3b where the magnetic field is generated by the potential difference on the boundaries.

There are no currents in the domain (**J**=0), therefore, it is possible to define a scalar magnetic potential ψ (the units for this scalar potential are *Ampere.turns*) related to the magnetic field **H** by

$$\mathbf{H} = -\nabla \psi \qquad (9.10)$$

The validity of this definition is established by the following:

$$\nabla \times \mathbf{H} = 0$$

and, therefore,

$$\nabla \times (-\nabla \psi) = 0$$

Note that if **J**≠0, this relation is inconsistent, since $\nabla \times (-\nabla \psi)$ is always equal to 0. Using the following:

$$\mathbf{B} = \mu \mathbf{H} \qquad (9.11)$$

$$\nabla \cdot \mathbf{B} = 0 \qquad (9.12)$$

we get

$$\nabla \cdot (\mu \mathbf{H}) = \nabla \cdot \mu(-\nabla \psi) = 0$$

which can be written as

$$\frac{\partial}{\partial x} \mu \frac{\partial \psi}{\partial x} + \frac{\partial}{\partial y} \mu \frac{\partial \psi}{\partial y} = 0 \qquad (9.13)$$

The permeability μ, in general, is not constant because almost all electrical devices have magnetic paths that contain nonlinear materials.

9.2.4. *The Magnetic Field: Vector Potential*

If we wish to calculate the field in a domain with current sources, (as in the example of Figure 9.3a, including the coil and the magnetic circuit), the scalar potential formulation shown above is not applicable because, using this formulation, **J** is required to be equal to 0. If **J**≠0, it is possible to use the vector potential **A**, related to **B** by

$$\mathbf{B} = \nabla \times \mathbf{A} \qquad (9.14)$$

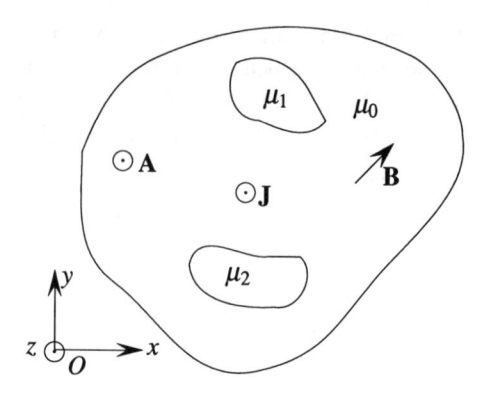

9. The Variational Finite Element Method: Some Static Applications

Figure 9.4. Definition of a two-dimensional domain. Current densities and the magnetic vector potential are perpendicular to the plane. The magnetic flux density is parallel to the plane.

In 2D problems, the vectors **J** and **A** have only one component, perpendicular to the plane Oxy in Figure 9.4. Denoting $\hat{\mathbf{i}}$, $\hat{\mathbf{j}}$, and $\hat{\mathbf{k}}$ the orthogonal unit vectors in directions Ox, Oy, and Oz, we have

$$\mathbf{A} = \hat{\mathbf{k}}A$$

$$\mathbf{J} = \hat{\mathbf{k}}J$$

$$\mathbf{B} = \hat{\mathbf{i}}B_x + \hat{\mathbf{j}}B_y$$

To establish the formulation for this case, we use the equations

$$\nabla \times \mathbf{H} = \mathbf{J} \tag{9.15}$$

$$\mathbf{H} = \nu \mathbf{B} \tag{9.16}$$

where ν is the magnetic reluctivity ($\nu = 1/\mu$). We have now

$$\nabla \times \nu \mathbf{B} = \nabla \times \nu \nabla \times \mathbf{A} = \mathbf{J} \tag{9.17}$$

where $\mathbf{B} = \nabla \times \mathbf{A}$ is, in the 2D case,

$$\mathbf{B} = \det \begin{bmatrix} \hat{\mathbf{i}} & \hat{\mathbf{j}} & \hat{\mathbf{k}} \\ \dfrac{\partial}{\partial x} & \dfrac{\partial}{\partial y} & \dfrac{\partial}{\partial z} \\ 0 & 0 & A \end{bmatrix}$$

With this, the magnetic flux density is written as

$$\mathbf{B} = \hat{\mathbf{i}}B_x + \hat{\mathbf{j}}B_y = \hat{\mathbf{i}}\frac{\partial A}{\partial y} - \hat{\mathbf{j}}\frac{\partial A}{\partial x} \qquad (9.18)$$

or

$$B_x = \frac{\partial A}{\partial y} \quad \text{and} \quad B_y = -\frac{\partial A}{\partial x}$$

Substituting Eq. (9.18) in (9.17), and since in 2D cases there are no variations in the Oz direction, we have

$$\det\begin{bmatrix} \hat{\mathbf{i}} & \hat{\mathbf{j}} & \hat{\mathbf{k}} \\ \dfrac{\partial}{\partial x} & \dfrac{\partial}{\partial y} & \dfrac{\partial}{\partial z} \\ \dfrac{v\partial A}{\partial y} & -\dfrac{v\partial A}{\partial x} & 0 \end{bmatrix} = \hat{\mathbf{k}}J$$

or, equating terms for the Oz component

$$\frac{\partial}{\partial x}v\frac{\partial A}{\partial x} + \frac{\partial}{\partial y}v\frac{\partial A}{\partial y} = -J \qquad (9.19)$$

This is Poisson's equation for the two-dimensional magnetic vector potential.

Now consider the line in Figure 9.5, where we assume that the potential **A** does not vary along the line. Defining a local system of coordinates Oxy, and assuming that the direction Ox coincides with the line, we have $\partial A/\partial x=0$, since A is constant. Therefore, $B_y=-\partial A/\partial x=0$. However, in the direction Oy, A can vary and $B_x=\partial A/\partial y$ can be nonzero. From this, we conclude that the magnetic field is parallel to the equipotential lines of A.

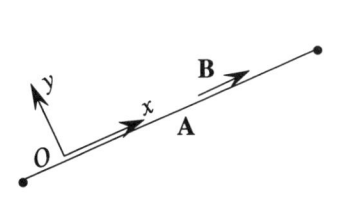

Figure 9.5. The magnetic flux density is parallel to equipotentials of **A**.

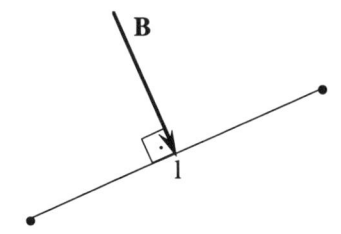

Figure 9.6. In two-dimensional geometries, magnetic flux is given per unit depth.

In 2D fields, it is possible to attribute to the vector potential a very important physical meaning. First, note that, because the structure is two-dimensional, the magnetic flux is given per unit depth of the structure, as shown in Figure 9.6. The flux per unit depth is $\Phi = Bl$ (Wb/m). If the actual flux is needed, it is necessary to multiply this value by the depth of the structure.

An example is shown in Figure 9.7, where the xy plane and the orthogonal surface S of the actual device are indicated. At point (1), $A=A_1$ and at point (2), $A=A_2$, both in the Oz direction. Therefore,

$$\Phi = \int_S \mathbf{B} \cdot d\mathbf{s} = \int_S \nabla \times \mathbf{A} \cdot d\mathbf{s}$$

using Stokes' theorem, we obtain

$$\Phi = \oint_C \mathbf{A} \cdot d\mathbf{l}$$

We calculate the circulation of $\mathbf{A}$ on contour C, in two parts. On the parts of the contour where $\mathbf{A}$ is perpendicular to $d\mathbf{l}$ the circulation is zero; on the others, assuming that $\mathbf{A}$ is constant and parallel to $d\mathbf{l}$ (2D approach), we have

$$\Phi = A_1 P - A_2 P$$

where P is the depth of the device. The difference between the potentials A_1 and A_2 gives the magnetic flux (in Wb/m) because

$$A_1 - A_2 = \frac{\Phi}{P}$$

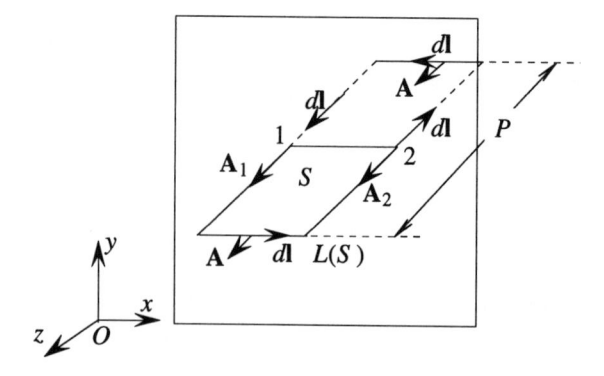

Figure 9.7. A contour in a 2D geometry used to show the interpretation of the magnetic vector potential as the difference in flux per unit depth.

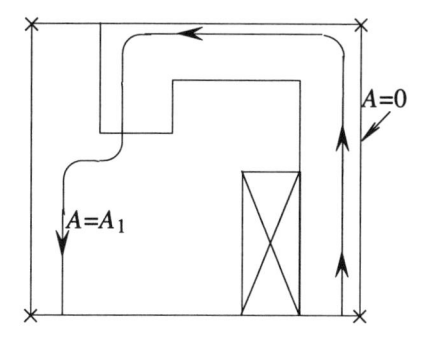

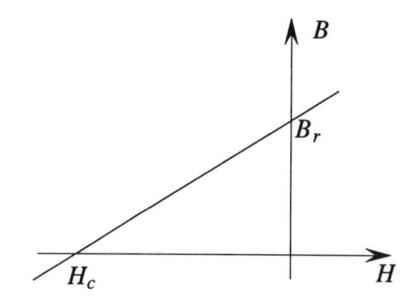

Figure 9.8. A geometry with two equipotential lines. The region between the two lines defines a flux tube.

Figure 9.9. Magnetization curve for a permanent magnet.

Note that the flux is defined by the potential difference and not by the absolute values of A, meaning that, as was the case in the example of Figure 3.5, it is necessary to fix one potential as a reference value and only then the remaining value is defined. Consider now the example of Figure 9.8, where a potential $A=0$ is imposed on the domain boundary.

After computing the potential distribution in the domain, the potential A is known everywhere in the domain and we can obtain the equipotential line $A=A_1$, as indicated in Figure 9.8. Because this line is a field line, the region included between the boundary $(A=0)$ and this equipotential line $(A=A_1)$ is a flux tube. Thus, the drawing of a number of equipotential lines provides an efficient method for visualization of the magnetic field distribution in the domain.

Suppose now that the domain contains permanent magnets (Figure 9.9) and that their magnetic characteristic curves are given by

$$\mathbf{B} = \mu\mathbf{H} + \mathbf{B}_r \qquad (9.20)$$

where μ is the magnet permeability and $\mathbf{B}_r$ is the remnant flux density. $\mathbf{B}_r$ is related to the coercive field intensity $\mathbf{H}_c$ by

$$\mathbf{B}_r = \mu\mathbf{H}_c$$

This is a good approximation for the commonly used magnets, (such as ferrite and Samarium-Cobalt magnets) whose relative permeability is close to 1.

Writing **H** in Eq. (9.20) explicitly, we have

$$\mathbf{H} = \frac{1}{\mu}(\mathbf{B} - \mathbf{B}_r) \qquad (9.21)$$

which, together with the relation $\nabla\times\mathbf{H}=0$ gives

$$\nabla \times \frac{1}{\mu}(\mathbf{B} - \mathbf{B}_r) = 0 \qquad (9.22)$$

since, for a permanent magnet, $\mathbf{J}=0$. This type of problem can also be solved by the finite element method, using the vector potential as a variable.

9.2.5. The Electric Vector Potential

In analogy to the magnetic vector potential, we can define the electric vector potential $\mathbf{T}$ related to the current density $\mathbf{J}$ by

$$\mathbf{J} = \nabla \times \mathbf{T} \qquad (9.23)$$

Assuming $\mathbf{E}$ to be time independent, $\nabla \times \mathbf{E} = 0$ and, with $\mathbf{E}=\mathbf{J}/\sigma$, we now have

$$\nabla \times \frac{1}{\sigma} \nabla \times \mathbf{T} = 0 \qquad (9.24)$$

Comparing this with the formulation presented in the previous paragraph, the following equivalent relationships can be written:

$$\mathbf{A} \Leftrightarrow \mathbf{T}$$

$$\mathbf{H} \Leftrightarrow \mathbf{E}$$

$$\mathbf{B} = \mu \mathbf{H} \Leftrightarrow \mathbf{J} = \sigma \mathbf{E}$$

$$B_x = \frac{\partial A}{\partial y}; \; B_y = -\frac{\partial A}{\partial x} \Leftrightarrow J_x = \frac{\partial T}{\partial y}; \; J_y = -\frac{\partial T}{\partial x}$$

$$\Phi = \int_S \mathbf{B} \cdot d\mathbf{s} \Leftrightarrow I = \int_S \mathbf{J} \cdot d\mathbf{s}$$

$$A_1 - A_2 = \Phi/P \Leftrightarrow T_1 - T_2 = I/P$$

where $\mathbf{T}$ is given in A/m and P is the depth of the device.

Using this notation, the difference between T_1 and T_2 is, in this case, the current between the lines on which $T=T_1$ and $T=T_2$, as indicated in Figure 9.10. In the preceding case $\mathbf{B}$ was parallel to the equipotential lines. Here, $\mathbf{J}$ is parallel to the equipotential lines.

This formulation is very useful when, instead of the potential difference between lines C and D (section 9.2.2) the current difference between lines A and B is known. This current is imposed as boundary conditions (i.e., $T=T_1$ on line A and $T=T_2$ on line B), making $T_1-T_2=I/P$.

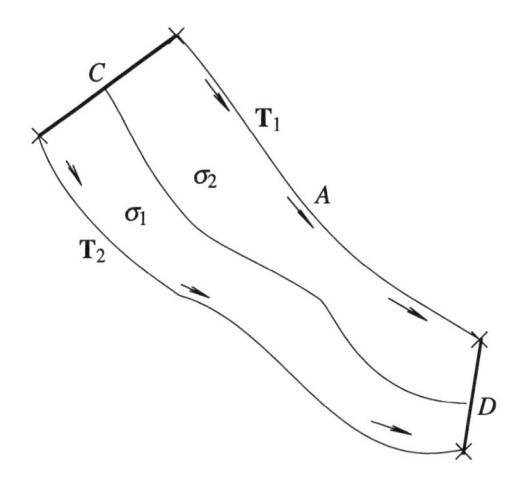

Figure 9.10. A problem that can be solved using the electric vector potential. The electric vector potential is specified on lines A and B as Dirichlet boundary conditions.

Note that in the magnetic case the magnetic field can be generated by currents (as in Figure 9.8) because the basic equation is $\nabla \times \mathbf{H} = \mathbf{J}$. In electric cases, the field can only be generated by a potential difference imposed on the boundaries, since the formulation is based on the equation $\nabla \times \mathbf{E} = 0$. Laplace's equation for this problem is similar to Eq. (9.19),

$$\frac{\partial}{\partial x}\frac{1}{\sigma}\frac{\partial T}{\partial x} + \frac{\partial}{\partial y}\frac{1}{\sigma}\frac{\partial T}{\partial y} = 0 \qquad (9.25)$$

9.3. The Variational Method

The variational method was presented in Chapter 8. Here we present only those aspects having direct bearing on the cases discussed previously. First, we should note that the variational method is a mathematical method which, a priori, does not have any relation to the finite element method. The finite element method is a numerical technique, that, associated with the variational method results in a calculation procedure normally referred to as a "variational finite element formulation."

With the variational method, instead of solving directly the governing equations [Eqs. (9.4), (9.8), (9.13), (9.19) or (9.25)], the corresponding energy functional is minimized.

9.3.1. *The Variational Formulation*

Suppose that a functional F exists, and that it is a function of a variable P and of its partial derivatives $P'_x = \partial P/\partial x$, $P'_y = \partial P/\partial y$, and $P'_z = \partial P/\partial z$. Then:

$$F = \int_V f(P, P'_x, P'_y, P'_z) dv$$

where V is the volume in which F is defined.

The necessary and sufficient condition for F to be stationary (minimum condition) is that for any small variation δP of the independent variable P, the corresponding variation δF is zero. This is written formally as

$$\delta F = \int_V \left(\frac{\partial f}{\partial P} \delta P + \frac{\partial f}{\partial P'_x} \delta P'_x + \frac{\partial f}{\partial P'_y} \delta P'_y + \frac{\partial f}{\partial P'_z} \delta P'_z \right) dv = 0 \qquad (9.26)$$

From the expression in Eq. (8.52) in section 8.5, we conclude that the following equation must hold

$$\frac{\partial f}{\partial P} - \nabla \cdot \mathbf{g} = 0 \qquad (9.27)$$

where $\mathbf{g}$ is defined as

$$\mathbf{g} = \hat{\mathbf{i}} \frac{\partial f}{\partial P'_x} + \hat{\mathbf{j}} \frac{\partial f}{\partial P'_y} + \hat{\mathbf{k}} \frac{\partial f}{\partial P'_z} \qquad (9.28)$$

Thus, Eq. (9.27) becomes

$$\frac{\partial f}{\partial P} - \frac{\partial}{\partial x}\left(\frac{\partial f}{\partial P'_x}\right) - \frac{\partial}{\partial y}\left(\frac{\partial f}{\partial P'_y}\right) - \frac{\partial}{\partial z}\left(\frac{\partial f}{\partial P'_z}\right) = 0 \qquad (9.29)$$

This equation is known as "Euler's equation." For the functional F to be valid, it must have an integrand that satisfies Eq. (9.29). To treat the boundary conditions, we divide the boundary S into two parts (see section 8.5.1):
- S_1 is that part of the boundary on which the potential is imposed as a boundary condition, called a "Dirichlet boundary condition."
- $S_2 = S - S_1$, is the part of the boundary on which $\mathbf{g} \cdot \hat{\mathbf{n}} = 0$; on this boundary we have the so called "Neumann boundary condition;" on this boundary, no condition is imposed. The Neumann boundary condition is natural and is as an intrinsic property of this formulation. On S_2, we can write

$$\mathbf{g} \cdot \hat{\mathbf{n}} = \frac{\partial f}{\partial P'_x} n_x + \frac{\partial f}{\partial P'_y} n_y + \frac{\partial f}{\partial P'_z} n_z = 0 \qquad (9.30)$$

where

$$\hat{\mathbf{n}} = \hat{\mathbf{i}} n_x + \hat{\mathbf{j}} n_y + \hat{\mathbf{k}} n_z \qquad (9.31)$$

In the following paragraphs we will examine the physical meaning of "Euler's equation" [Eq. (9.29)] and the boundary conditions, especially Eq. (9.30), as applied to the cases above.

Going back to the energy functional, we can say that to establish the value of P minimizing the functional F entails:

- Establishment of the functional F whose integrand satisfies Euler's equation.
- Application of boundary conditions, pertinent to the problem, where either P is fixed (Dirichlet condition), or Eq. (9.30) (Neumann condition), is satisfied.

9.3.2. Functionals Involving Scalar Potentials

The electrostatic field functional for a dielectric medium is

$$F = \int_v f(V, V'_x, V'_y) \, dv = \int_v \left(\frac{1}{2} \varepsilon E^2 - \rho V \right) dv \qquad (9.32)$$

where the integrand in the first term is the volume energy density, as described in Chapter 3. The Euler equation for this 2D case is

$$\frac{\partial f}{\partial V} - \frac{\partial}{\partial x}\left(\frac{\partial f}{\partial V'_x}\right) - \frac{\partial}{\partial y}\left(\frac{\partial f}{\partial V'_y}\right) = 0 \qquad (9.33)$$

where the variable P in Eq. (9.29) is the scalar potential V. Calculating the different terms of this expression, we get

$$\frac{\partial f}{\partial V} = \frac{\partial}{\partial V}\left(\frac{1}{2}\varepsilon E^2 - \rho V\right) = -\rho \qquad (9.34)$$

noting that

$$V'_x = \frac{\partial V}{\partial x} = -E_x$$

Eq. (9.34) becomes

$$\frac{\partial f}{\partial V_x'} = -\frac{\partial f}{\partial E_x} = -\frac{\partial}{\partial E_x}\left(\frac{1}{2}\varepsilon E^2 - \rho V\right) = -\frac{\partial}{\partial E_x}\left(\frac{1}{2}\varepsilon E^2\right) \qquad (9.35)$$

since E depends on E_x,

$$-\frac{\partial}{\partial E_x}\left(\frac{1}{2}\varepsilon E^2\right) = -\frac{\partial}{\partial E}\left(\frac{1}{2}\varepsilon E^2\right)\frac{\partial E}{\partial E_x} = -\varepsilon E\frac{\partial E}{\partial E_x} \qquad (9.36)$$

Because $E=(E_x^2+E_y^2)^{1/2}$, after some algebraic operations we get

$$\frac{\partial f}{\partial V_x'} = -\varepsilon E_x \qquad (9.37)$$

and identically for the second term,

$$\frac{\partial f}{\partial V_y'} = -\varepsilon E_y \qquad (9.38)$$

substituting Eqs. (9.34), (9.37), and (9.38) in Eq. (9.33), we get

$$-\rho + \frac{\partial}{\partial x}(\varepsilon E_x) + \frac{\partial}{\partial y}(\varepsilon E_y) = 0$$

which can be written in the following form

$$(\hat{\mathbf{i}}\frac{\partial}{\partial x} + \hat{\mathbf{j}}\frac{\partial}{\partial y})\cdot(\hat{\mathbf{i}}\varepsilon E_x + \hat{\mathbf{j}}\varepsilon E_y) = \rho \qquad (9.39)$$

using $\mathbf{D}=\varepsilon\mathbf{E}$, we get, by applying Euler's equation

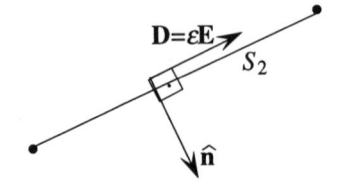

$$\nabla\cdot\mathbf{D} = \rho$$

Figure 9.11a. Neumann boundary condition. If no condition is specified on a boundary, the electric field intensity is parallel to that boundary.

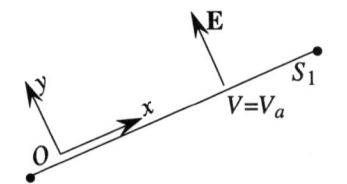

Figure 9.11b. Dirichlet boundary condition. The electric field intensity is perpendicular to a Dirichlet boundary.

This is Maxwell's equation for this problem, and therefore Euler's equation is valid. As for the Neumann boundary condition, we have from Eq. (9.30)

$$\frac{\partial f}{\partial V'_x} n_x + \frac{\partial f}{\partial V'_y} n_y = 0$$

Using Eqs. (9.37) and (9.38), we get

$$D_x n_x + D_y n_y = \mathbf{D} \cdot \hat{\mathbf{n}} = 0$$

In Figure 9.11a the electrical field $\mathbf{E}$ is parallel to the boundary S_2. Thus, for the boundary on which no condition is specified, the field is tangent to the boundary. For the cases of Figures 9.1, 9.2, and 9.3 (all scalar potential cases), the field is tangent to the boundaries C and D.

As for the Dirichlet condition, examine Figure 9.11b, where the potential $V=V_a$ is imposed on boundary S_1. We define a local coordinate system Oxy with the direction Ox on line S_1. In this case, since $V=V_a$ is constant in the direction Ox, $dV/dx=0$ and therefore $E_x=0$. This limitation does not exist in the Oy direction, and the field $E_y \neq 0$, corresponding to the situation shown in Figure. 9.11b.

In a similar manner, for the problem of stationary currents (section 9.2.2) the energy functional is

$$F = \int_v \frac{1}{2} \sigma E^2 dv \qquad (9.40)$$

we have

$$\frac{\partial f}{\partial V'_x} = -\sigma E_x = -J_x$$

and similarly for the y component

$$\frac{\partial f}{\partial V'_y} = -\sigma E_y = -J_y$$

which, when substituted in the Euler equation yields

$$\frac{\partial}{\partial x}(-J_x) + \frac{\partial}{\partial y}(-J_y) = 0$$

or, in a more compact form,

$$\nabla \cdot \mathbf{J} = 0$$

9. The Variational Finite Element Method: Some Static Applications

In magnetic scalar potential applications, the energy functional must take into account the fact that the ferromagnetic material is nonlinear [saturation in the **B(H)** curve]. The energy functional is

$$F = \int_V \left[\int_0^H B\, dH \right] dv \qquad (9.41)$$

The term

$$\frac{\partial f}{\partial V'_y} = -\frac{\partial f}{\partial H_x} = -\frac{\partial f}{\partial H}\frac{\partial H}{\partial H_x} \qquad (9.42)$$

is calculated as

$$\frac{\partial f}{\partial H} = \frac{\partial}{\partial H}\int_0^H B\, dH = B$$

and

$$\frac{\partial H}{\partial H_x} = \frac{\partial}{\partial H_x}\sqrt{H_x^2 + H_y^2} = \frac{1}{2}\left(H_x^2 + H_y^2\right)^{-1/2} 2H_x = \frac{H_x}{H}$$

The expression in Eq. (9.42) is

$$\frac{\partial f}{\partial V'_x} = -\mu H_x = -B_x$$

In complete analogy with the previous formulations, we have

$$\frac{\partial f}{\partial V'_y} = -B_y$$

The Euler equation is then

$$\frac{\partial}{\partial x}(-B_x) + \frac{\partial}{\partial y}(-B_y) = 0$$

or, in compact form,

$$\nabla \cdot \mathbf{B} = 0$$

The principles above apply without modifications for the boundary conditions.

9.3.3. *The Vector Potential Functionals*

Since $\nabla \times \mathbf{H} = \mathbf{J}$ is a point-form equation, we present first the functional for domains with current densities $\mathbf{J}$ and nonlinear permeability μ. Then, regions with permanent magnets will be calculated by modifying the nonlinear energy functional. The functional for the first case is

$$F = \int_V \left[\int_0^B H dB - JA \right] dv \qquad (9.43)$$

where $\mathbf{H} = \mathbf{B}/\mu$.

We now calculate the terms of Euler's equation, observing that here $P = A$, $P'_x = A'_x$ and $P'_y = A'_y$.

$$\frac{\partial f}{\partial A} = -J \qquad (9.44)$$

Since

$$A'_x = \frac{\partial A}{\partial x} = -B_y \quad \text{and} \quad A'_y = \frac{\partial A}{\partial y} = B_x$$

we have

$$\frac{\partial f}{\partial A'_x} = -\frac{\partial}{\partial B} \left[\int_0^B H dB - JA \right] \frac{\partial B}{\partial B_y} = -H \frac{\partial}{\partial B_y} \left(B_x^2 + B_y^2 \right)^{1/2}$$

or, after evaluation,

$$\frac{\partial f}{\partial A'_x} = -H \frac{B_y}{B} = -\frac{B_y}{\mu} = -H_y \qquad (9.45a)$$

and similarly for the x component of H,

$$\frac{\partial f}{\partial A'_y} = \frac{\partial f}{\partial B_x} = \frac{\partial f}{\partial B} \frac{\partial B}{\partial B_x} = H_x \qquad (9.45b)$$

Substituting Eqs. (9.44), (9.45a), and (9.45b) into Euler's Eq. [Eq. (9.29)] gives

$$-J - \frac{\partial}{\partial x}(-H_y) + \frac{\partial}{\partial y}(H_x) = 0$$

360   9. The Variational Finite Element Method: Some Static Applications

This is equivalent to

$$\nabla \times \mathbf{H} = \mathbf{J}$$

which is the equation to be solved. To incorporate the Dirichlet boundary condition, we use the fact that on S_1, A is imposed and constant, and the field is tangent to the surface (Figures. 9.5 and 9.12). For the Neumann boundary condition, Eq. (9.30) must be satisfied

$$\mathbf{g} \cdot \mathbf{n} = \frac{\partial f}{\partial A'_x} n_x + \frac{\partial f}{\partial A'_y} n_y = 0$$

Using Eqs. (9.45a) and (9.45b), we get

$$-H_y n_x + H_x n_y = 0$$

or, in general terms,

$$\mathbf{H} \times \hat{\mathbf{n}} = 0$$

meaning that the vectors $\mathbf{H}$ and $\hat{\mathbf{n}}$ are co-linear, as indicated in Figure 9.12.

For applications that include permanent magnets, the energy functional is

$$F = \int_v \frac{1}{2\mu} (\mathbf{B} - \mathbf{B}_r) \cdot (\mathbf{B} - \mathbf{B}_r) dv \qquad (9.46)$$

To find Euler's equation, we calculate

$$\frac{\partial f}{\partial A'_x} = -\frac{\partial f}{\partial B_y} = -\frac{\partial f}{\partial B_i} \cdot \frac{\partial B_i}{\partial B_y}$$

denoting $\mathbf{B}_i = \mathbf{B} - \mathbf{B}_r$, we have

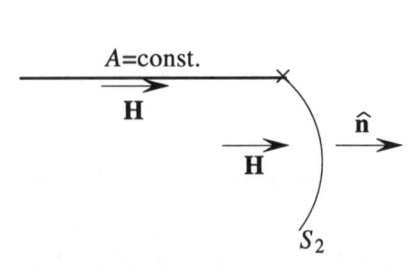

Figure 9.12. Boundary conditions for the magnetic vector potential. The magnetic field intensity is parallel to a constant vector potential boundary.

$$f = \frac{B_i^2}{2\mu}$$

and

$$\frac{\partial f}{\partial A_x'} = -\frac{\partial}{\partial B_i}\left(\frac{B_i^2}{2\mu}\right)\frac{\partial B_i}{\partial B_y} = -\frac{B_i}{\mu}\frac{\partial B_i}{\partial B_y} = -\frac{1}{2\mu}\frac{\partial B_i^2}{\partial B_y}$$

or

$$\frac{\partial f}{\partial A_x'} = -\frac{1}{2\mu}\frac{\partial(B_{ix}^2 + B_{iy}^2)}{\partial B_y} = -\frac{1}{2\mu}\frac{\partial B_{iy}^2}{\partial B_y} = -\frac{1}{2\mu}\frac{\partial}{\partial B_y}(B_y - B_{ry})^2$$

where $B_y - B_{ry}$ is the Oy component of $\mathbf{B} - \mathbf{B}_r$, that is:

$$\frac{\partial f}{\partial A_x'} = -\frac{1}{\mu}(B_y - B_{ry}) = -\frac{B_{iy}}{\mu}$$

and, analogously,

$$\frac{\partial f}{\partial A_y'} = \frac{B_{ix}}{\mu}$$

Using these results in Euler's equation, we have

$$-\frac{\partial}{\partial x}\left(-\frac{B_{iy}}{\mu}\right) - \frac{\partial}{\partial y}\left(\frac{B_{ix}}{\mu}\right) = 0$$

or

$$\nabla \times \frac{\mathbf{B}_i}{\mu} = 0$$

and, substituting for $\mathbf{B}_i$:

$$\nabla \times \frac{1}{\mu}(\mathbf{B} - \mathbf{B}_r) = 0$$

These are equal to Eq. (9.22), for the field in the permanent magnet region.

For electric vector potential applications (section 9.2.5) the energy functional is

$$F = \int_v \frac{1}{2}\frac{J^2}{\sigma}dv \qquad (9.47)$$

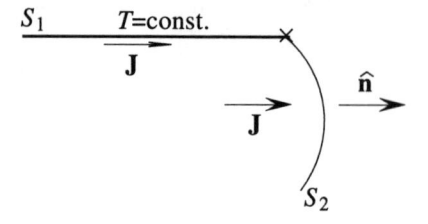

Figure 9.13. Boundary conditions for the electric vector potential. The current density is parallel to a constant potential boundary.

therefore,

$$\frac{\partial f}{\partial T'_x} = -\frac{\partial f}{\partial J_y} = -\frac{\partial f}{\partial J} \cdot \frac{\partial J}{\partial J_y} = -\frac{J_y}{\sigma} = -E_y$$

and

$$\frac{\partial f}{\partial T'_y} = E_x$$

which, when substituted into Euler's equation, gives

$$\frac{\partial}{\partial x}(-E_y) + \frac{\partial}{\partial y}(E_x) = \nabla \times \mathbf{E} = 0$$

describing the physical phenomenon for this case. For the boundary conditions there is a direct analogy with the magnetic case, as shown in Figure 9.13.

9.4. The Finite Element Method

The solutions to Laplace's and Poisson's equations for the problems described above, [Eqs. (9.4), (9.8), (9.13), (9.19) and (9.25)] are, for most practical geometries, impossible to obtain by analytical methods.

The finite element method is based on discretization of the solution domain into small regions called "finite elements" (extensive discussion of the finite element method may be found in Chapter 8). These elements form a mesh over the solution domain. A convenient formulation is applied to these elements, from which a system of equations is obtained.

The solution to this system provides the solution to the problem. It is clear that the smaller these elements are (or the denser the mesh) the closer the numerical solution will be to the exact, or continuous solution.

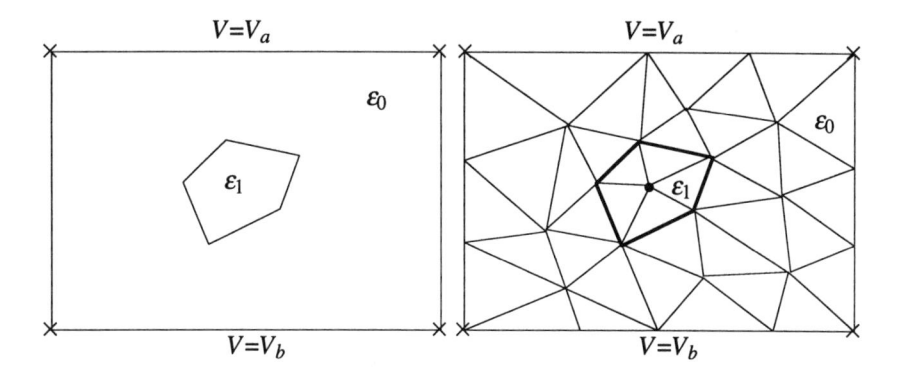

Figure 9.14a. A problem with two dielectric materials and applied potentials on two boundaries.

Figure 9.14b. A finite element mesh over the solution domain in Figure 9.14a.

Consider the problem of Figure 9.14a, where a dielectric material fills the domain between two planes at potentials V_a and V_b. A second dielectric material is located within the domain as shown. To apply the finite element method to this problem, it is necessary to discretize the solution domain into small regions. In this case, the finite elements are of triangular shape, as indicated in Figure 9.14b.

Note that a finite element can not contain two different materials. The boundary between two materials must also be a boundary between finite elements. Also, the mesh, or the assembly of finite elements, must conform to the form shown in Figure 9.15a.

Situations like the one shown in fig 9.15b are called "nonconforming." We define an "element" as the region inside a finite element and a "node" as the meeting point between the sides of the triangle (vertex). In the "nonconforming" mesh there are nodes that do not coincide with the triangle vertices, as indicated in Figure 9.15b. These situations must be avoided, or the finite element method will produce incorrect results.

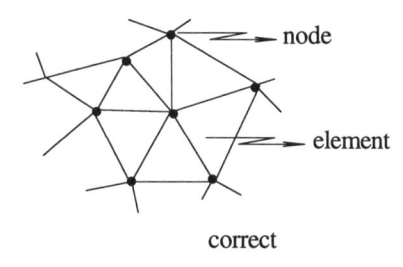

Figure 9.15a. Correct finite element discretization. All elements are conforming.

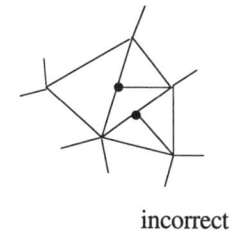

Figure 9.15b. Incorrect discretization. Some elements are not conforming.

9. The Variational Finite Element Method: Some Static Applications

In this chapter we make use primarily of the so called "first order finite elements." In this type of elements the potential varies linearly inside the element, as:

$$V(x,y) = a_1 + a_2 x + a_3 y \qquad (9.48)$$

There are more sophisticated elements in which, for example, the potential has a quadratic variation within the element. Actually, there are several types of finite elements, with even more complex structures. Theoretically, these elements should produce more accurate results.

However, experience shows that first-order elements, when correctly applied provide excellent results, and, being both efficient and simple, they are used more frequently. Complex problems, such as determination of fields in narrow air-gaps in electrical machines, are not easily solved, in spite of the use of more sophisticated elements.

It is worth remarking here that electromagnetic problems are, generally, hard to solve (nonlinearity, induced currents) and the application of simple elements simplifies the development of efficient software packages without the need for very complex algorithms. Also, in this chapter we emphasize the physical aspects of the problems rather than the finite element method itself.

Figure 9.16 shows an element in a mesh, whose nodes are P_1, P_2, and P_3, locally numbered as 1, 2, and 3. However, in the global mesh this numbering can be different. Equation 9.48 must be satisfied for the three nodes of the element:

$$V_1 = a_1 + a_2 x_1 + a_3 y_1$$
$$V_2 = a_1 + a_2 x_2 + a_3 y_2$$
$$V_3 = a_1 + a_2 x_3 + a_3 y_3$$

The solution of this system of equations provides the values of a_1, a_2, and a_3 as a function of the coordinates and potentials of the nodes. After substituting the values of a_1, a_2, and a_3 in Eq. (9.48), and after some algebraic operations, we can write this equation as the following sum

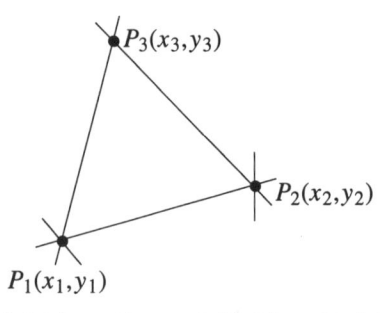

Figure 9.16. A triangular finite element in a mesh. The element is defined by the coordinates of the three nodes.

$$V(x,y) = \frac{1}{D}\sum_{l=1}^{3}(p_l + q_l x + r_l y)V_l \qquad (9.49)$$

where D is equal to twice the surface of the triangle. The terms p_1, q_1, and r_1 are

$$p_1 = x_2 y_3 - y_2 x_3$$
$$q_1 = y_2 - y_3$$
$$r_1 = x_3 - x_2$$

and the other terms, p_2, p_3, q_2, q_3, r_2, and r_3, are obtained by cyclic permutation of the index; for example,

$$p_2 = x_3 y_1 - y_3 x_1$$

The partial derivatives of the potential give the field values, and as an example, for the scalar potential we get

$$\mathbf{H} = -\nabla V = -\left(\hat{\mathbf{i}}\frac{\partial}{\partial x} + \hat{\mathbf{j}}\frac{\partial}{\partial y}\right)V$$

and, therefore,

$$H_x = -\frac{\partial V}{\partial x} \quad \text{and} \quad H_y = -\frac{\partial V}{\partial y}$$

From Eq. (9.48), we note that

$$H_x = -\frac{\partial}{\partial x}(a_1 + a_2 x + a_3 y) = -a_2$$

and

$$H_y = -a_3$$

are constant values in the triangle. Using Eq. (9.49), we have

$$\mathbf{H} = -\nabla V = -\frac{1}{D}\sum_{l=1}^{3}(\hat{\mathbf{i}}q_l + \hat{\mathbf{j}}r_l)V_l \qquad (9.50)$$

and, therefore,

$$H_x = -\frac{1}{D}(q_1 V_1 + q_2 V_2 + q_3 V_3) \qquad (9.51a)$$
$$H_y = -\frac{1}{D}(r_1 V_1 + r_2 V_2 + r_3 V_3) \qquad (9.51b)$$

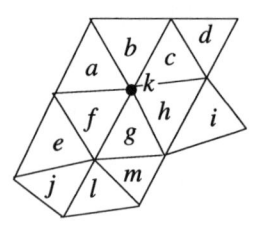

Figure 9.17. The functional for each element is zero outside its own element. The derivatives of the functional with respect to k depend only on elements a, b, c, f, g, and h.

9.5. Application of Finite Elements with the Variational Method

The variational finite element method is the combination of the variational method with the finite element technique. It has been shown previously that the functionals presented are valid. For discretized domains we can write

$$F = \sum_i F_i$$

where F_i functionals are summed over the i elements of the mesh, or the discretized domain. Assuming that the mesh has K nodes, the functional must be minimized for the V_k node potential:

$$\frac{\partial F}{\partial V_k} = \sum_i \frac{\partial F_i}{\partial V_k} = 0 \qquad (9.52)$$

Note that, even though the sum is over all i elements of the mesh, F_i is nonzero only in element i and zero outside this element. Therefore, F_i depends only on the nodes of element i. For example, in Figure 9.17, when calculating the sum in (9.52) for node k, this sum is limited to

$$\sum_i \frac{\partial F_i}{\partial V_k} = \frac{\partial F_a}{\partial V_k} + \frac{\partial F_b}{\partial V_k} + \frac{\partial F_c}{\partial V_k} + \frac{\partial F_f}{\partial V_k} + \frac{\partial F_g}{\partial V_k} + \frac{\partial F_h}{\partial V_k}$$

because, for example, $\partial F_j/\partial V_k=0$, since the partial derivative of the functional of triangle j does not depend on node k in element i.

The important point in establishing a numerical procedure of calculation is to obtain the generic term $\partial F_i/\partial V_k$ in the sum of Eq. (9.52) for the different problems presented above.

9.5.1. *Application to the Electrostatic Field*

The discretized functional for the electrostatic field case is

$$F_i = \int_{S_i} \left(\tfrac{1}{2}\varepsilon E^2 - \rho V\right) ds \tag{9.53}$$

where, in 2D, the volume is equal to the numerical value of the area S_i of element i (the depth of the element is unity). We calculate $\partial F_j/\partial V_k$, assuming that V_k is the potential of one of the three nodes 1, 2, or 3 of element i

$$\frac{\partial F_i}{\partial V_k} = \frac{\partial}{\partial V_k} \int_{S_i} \left(\tfrac{1}{2}\varepsilon E^2 - \rho V\right) ds \tag{9.54}$$

First, we calculate the first term of the functional. Using Eq. (9.50) but with the electric field intensity

$$\mathbf{E} = -\frac{1}{D}\sum_{l=1}^{3} (\hat{\mathbf{i}} q_l + \hat{\mathbf{j}} r_l) V_l \tag{9.55}$$

This expression does not depend on the x or y coordinate ($\mathbf{E}$ is constant in the element). The integration is, therefore, simple:

$$\frac{\partial}{\partial V_k} \int_{S_i} \tfrac{1}{2}\varepsilon E^2 ds = \frac{\partial}{\partial V_k} \tfrac{1}{2}\varepsilon E^2 \int_{S_i} ds = \tfrac{1}{2}\varepsilon \frac{D}{2}\frac{\partial E^2}{\partial V_k} \tag{9.56}$$

Because D is twice the area of the triangle and $E^2 = \mathbf{E}\cdot\mathbf{E}$, we get

$$\frac{\partial E^2}{\partial V_k} = \frac{\partial}{\partial V_k}\frac{1}{D^2}\left[\sum_{l=1}^{3}(\hat{\mathbf{i}}q_l + \hat{\mathbf{j}}r_l)V_l\right]^2 = \frac{\partial}{\partial V_k}\frac{1}{D^2}\left[\left(\sum_{l=1}^{3}q_l V_l\right)^2 + \left(\sum_{l=1}^{3}r_l V_l\right)^2\right]$$

where Eq. (9.55) was substituted for $\mathbf{E}$. The relation is

$$\frac{\partial E^2}{\partial V_k} = \frac{1}{D^2}\frac{\partial}{\partial V_k}\left[\left(\sum_{l=1}^{3}q_l V_l\right)^2 + \left(\sum_{l=1}^{3}r_l V_l\right)^2\right]$$

Since k is one of the three nodes in the sum, we have

$$\frac{\partial E^2}{\partial V_k} = \frac{1}{D^2}\left[2\sum_{l=1}^{3}(q_l V_l)q_k + 2\sum_{l=1}^{3}(r_l V_l)r_k\right] \tag{9.57}$$

Substituting Eq. (9.57) into Eq. (9.56) yields

$$\frac{\varepsilon}{2D}\left[(q_1V_1 + q_2V_2 + q_3V_3)q_k + (r_1V_1 + r_2V_2 + r_3V_3)r_k\right]$$

or

$$\frac{\varepsilon}{2D}\left[(q_1q_k + r_1r_k)V_1 + (q_2q_k + r_2r_k)V_2 + (q_3q_k + r_3r_k)V_3\right]$$

or, using the matrix form,

$$\frac{\varepsilon}{2D}\left[(q_1q_k + r_1r_k)\ (q_2q_k + r_2r_k)\ (q_3q_k + r_3r_k)\right]\begin{bmatrix}V_1\\V_2\\V_3\end{bmatrix}$$

In the numerical procedure, when evaluating an element i, the matrix expression above is obtained for all three nodes; k assumes the values 1, 2, and 3. The complete matrix for the element is

$$\frac{\varepsilon}{2D}\begin{bmatrix}q_1q_1 + r_1r_1 & q_1q_2 + r_1r_2 & q_1q_3 + r_1r_3\\q_2q_1 + r_2r_1 & q_2q_2 + r_2r_2 & q_2q_3 + r_2r_3\\q_3q_1 + r_3r_1 & q_3q_2 + r_3r_2 & q_3q_3 + r_3r_3\end{bmatrix}\begin{bmatrix}V_1\\V_2\\V_3\end{bmatrix} \quad (9.58)$$

The matrix of coefficients is symmetric. It is called the "elemental matrix," is normally denoted as $S(3,3)$ and its generic term is

$$S(n,k) = \frac{\varepsilon}{2D}(q_nq_k + r_nr_k) \quad (9.59)$$

where n and k are the rows and columns of $S(3,3)$. This matrix is often called a "stiffness matrix," because of its origin in stress analysis of structures.

We now calculate the second term of the functional, assuming that ρ is constant in element i

$$\frac{\partial}{\partial V_k}\int_{S_i} -\rho V\,ds = -\rho\frac{\partial}{\partial V_k}\int_{S_i} V\,ds$$

First, note that the potential $V(x,y)$, given by Eq. (9.49) is a sum of three terms

$$V(x,y) = \sum_{l=1}^{3}(p_l + q_lx + r_ly)V_l$$

Denoting α_l the coefficient multiplying the potentials V_l,

$$\alpha_l(x,y) = \frac{1}{D}(p_l + q_l x + r_l y)$$

we have

$$V(x,y) = \sum_{l=1}^{3} \alpha_l V_l = \alpha_1 V_1 + \alpha_2 V_2 + \alpha_3 V_3$$

If, for example, $x=x_1$ and $y=y_1$, it follows that $V(x_1,y_1)=V_1$, and, therefore,

$$V(x_1,y_1) = 1 V_1 + 0 V_2 + 0 V_3$$

and the coefficients are:

$$\alpha_1(x_1,y_1) = 1 \qquad \alpha_2(x_1,y_1) = 0 \qquad \alpha_3(x_1,y_1) = 0$$

Analogously, for $x=x_2$ and $y=y_2$, we have $\alpha_1=0$, $\alpha_2=1$, and $\alpha_3=0$. Finally, for $x=x_3$ and $y=y_3$, $\alpha_1=0$, $\alpha_2=0$, and $\alpha_3=1$. Thus α_j is a function equal to 1 for $x=x_j$ and $y=y_j$ and zero at all other nodes. α_1 is shown graphically in Figure 9.18. The graphical representation of α_2 and α_3 is similar, at their corresponding nodes.

Returning now to the expression we wish to calculate, we have

$$-\rho \frac{\partial}{\partial V_k} \int_{S_i} V\, ds = -\rho \int_{S_i} \frac{\partial V}{\partial V_k}\, ds = -\rho \int_{S_i} \frac{\partial}{\partial V_k}(\alpha_1 V_1 + \alpha_2 V_2 + \alpha_3 V_3)\, ds$$

or

$$-\rho \int_{S_i} \alpha_k\, ds$$

where k is a node of the triangle.

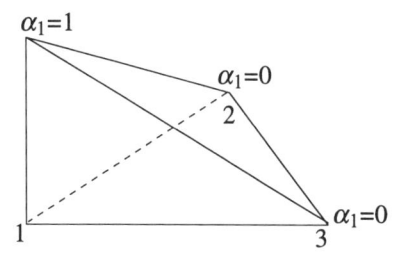

Figure 9.18. Graphical representation of one of the coefficients of the linear approximation for a triangle. The coefficient is equal to one at its corresponding node and zero at all other nodes.

Taking into account the form of α, the integral gives the volume of the tetrahedron in Figure 9.18. Because the height of the tetrahedron is unity, we get

$$-\rho \int_{S_i} \alpha_k \, ds = -\rho \frac{D}{2} \frac{1}{3} = -\frac{1}{6} \rho D$$

Performing similar calculations for the three nodes of the element and adding to the matrix [Eq. (9.58)], we finally obtain

$$\begin{bmatrix} \frac{\partial F_i}{\partial V_1} \\ \frac{\partial F_i}{\partial V_2} \\ \frac{\partial F_i}{\partial V_3} \end{bmatrix} = \begin{bmatrix} S_{1,1} & S_{1,2} & S_{1,3} \\ S_{2,1} & S_{2,2} & S_{2,3} \\ S_{3,1} & S_{3,2} & S_{3,3} \end{bmatrix} \begin{bmatrix} V_1 \\ V_2 \\ V_3 \end{bmatrix} - \frac{D}{6} \begin{bmatrix} \rho \\ \rho \\ \rho \end{bmatrix} \qquad (9.60)$$

Eq. (9.60) is the basic building block of a finite element solution. The method of forming the global system of equations based on the elemental matrix in Eq. (9.60) will be presented shortly.

9.5.2. Application to the Case of Stationary Currents

The physical situation was presented in section 9.2.2. Its behavior is described by Eq. (9.8), and it has the following discretized functional

$$F_i = \int_{S_i} \frac{1}{2} \sigma E^2 \, ds \qquad (9.61)$$

This expression is similar to the first term of the integral in Eq. (9.53). By analogy, the calculation of $\partial F_i / \partial V_k$ gives a matrix system equivalent to Eq. (9.58). The generic term of this system, $S(n,k)$ is

$$S(n,k) = \frac{\sigma}{2D} (q_n q_k + r_n r_k) \qquad (9.62)$$

9.5.3. Application to the Magnetic Field: Scalar Potential

This case was discussed in section 9.2.3. Equation (9.13) has the following, equivalent functional

$$F_i = \int_{S_i} \left[\int_0^H B \, dH \right] ds \qquad (9.63)$$

Calculating the term $\partial F_i/\partial V_k$:

$$\frac{\partial F_i}{\partial V_k} = \int_{S_i} \frac{\partial}{\partial V_k}\left[\int_0^H BdH\right]ds$$

and, since H depends on V_k, we have

$$\frac{\partial F_i}{\partial V_k} = \int_{S_i} \frac{\partial}{\partial H}\left[\int_0^H BdH\right]\frac{\partial H}{\partial V_k}ds = \int_{S_i} B \frac{\partial H}{\partial V_k}ds$$

For the finite elements used here, B and H are constant within the element:

$$\frac{\partial F_i}{\partial V_k} = \frac{D}{2} B \frac{\partial H}{\partial V_k} = \frac{D}{2} \mu H \frac{\partial H}{\partial V_k} = \frac{\mu D}{4} \frac{\partial H^2}{\partial V_k} \tag{9.64}$$

The calculation of $\partial H^2/\partial V_k$ provides an expression equivalent to Eq. (9.57), which, when placed in Eq. (9.64), results in a matrix system analogous to Eq.(9.58), with the generic term

$$S(n,k) = \frac{\mu}{2D}(q_n q_k + r_n r_k) \tag{9.65}$$

9.5.4. *Application to the Magnetic Field: Vector Potential*

Equation (9.19), in section 9.2.4, is the description of this formulation. The discretized functional is

$$F_i = \int_{S_i}\left[\int_0^B HdB - JA\right]ds \tag{9.66}$$

The term $\partial F_i/\partial A_k$ is obtained by the sum of the partial results of the two terms of Eq. (9.66). The first term is analogous to the term calculated in Eq. (9.63). Noting that $\mathbf{H}=\nu\mathbf{B}$, where $\nu=1/\mu$, we have

$$S(n,k) = \frac{\nu}{2D}(q_n q_k + r_n r_k) \tag{9.67}$$

The term

$$\frac{\partial}{\partial A_k}\int_{S_i} -JAds$$

is similar to the term

$$\frac{\partial}{\partial V_k} \int_{S_i} -\rho V ds$$

discussed in section 9.5.1. With the result of section 9.5.1, we have for $\partial F_i/\partial V_k$ the following matrix expression

$$\begin{bmatrix} \frac{\partial F_i}{\partial A_1} \\ \frac{\partial F_i}{\partial A_2} \\ \frac{\partial F_i}{\partial A_3} \end{bmatrix} = \begin{bmatrix} S_{1,1} & S_{1,2} & S_{1,3} \\ S_{2,1} & S_{2,2} & S_{2,3} \\ S_{3,1} & S_{3,2} & S_{3,3} \end{bmatrix} \begin{bmatrix} A_1 \\ A_2 \\ A_3 \end{bmatrix} - \frac{D}{6} \begin{bmatrix} J \\ J \\ J \end{bmatrix} \qquad (9.68)$$

We examine now the permanent magnet case, where, assuming that the magnet has constant permeability, the corresponding functional [Eq. (9.46)], is

$$F_i = \int_{S_i} \frac{v}{2}(B - B_r)^2 ds = \int_{S_i} \frac{v}{2} B_i^2 ds \qquad (9.69)$$

We also assume that, in element i, which is interior the magnet, there are no currents. Therefore,

$$\frac{\partial F_i}{\partial A_k} = \frac{\partial F_i}{\partial B_i} \frac{\partial B_i}{\partial A_k} = \frac{D}{2} \frac{v}{2} \frac{\partial B_i^2}{\partial B_i} \frac{\partial B_i}{\partial A_k}$$

$$\frac{\partial F_i}{\partial A_k} = \frac{Dv}{4} 2B_i \frac{\partial B_i}{\partial A_k} \qquad (9.70)$$

Recall that, even though **A** is a vector, in the 2D formulation only the perpendicular component is used. Considering its scalar magnitude, we can write

$$B_i \frac{\partial B_i}{\partial A_k} = \mathbf{B}_i \cdot \frac{\partial \mathbf{B}_i}{\partial A_k}$$

because the vectors B_i and $\partial B_i/\partial A_k$ (its derivative with respect to a scalar) have the same direction. Noting that $\mathbf{B}_i$ is

$$\mathbf{B}_i = \nabla \times \mathbf{A} - B_r = \frac{1}{D} \sum_{l=1}^{3} (\hat{\mathbf{i}} r_l - \hat{\mathbf{j}} q_l) A_l - \hat{\mathbf{i}} B_{rx} - \hat{\mathbf{j}} B_{ry}$$

where $\mathbf{B}_r$ is given by its components $\mathbf{B}_r = \hat{\mathbf{i}} B_{rx} + \hat{\mathbf{j}} B_{ry}$, we obtain

$$\frac{\partial \mathbf{B}_i}{\partial A_k} = \frac{1}{D}(\hat{\mathbf{i}} r_k - \hat{\mathbf{j}} q_k)$$

Using Eq. (9.70), this becomes

$$\frac{\partial F_i}{\partial A_k} = \frac{Dv}{2}\left[\frac{1}{D}\sum_{l=1}^{3}(\hat{\mathbf{i}} r_l - \hat{\mathbf{j}} q_l)A_l - \hat{\mathbf{i}} B_{rx} - \hat{\mathbf{j}} B_{ry}\right] \cdot \left[\frac{1}{D}(\hat{\mathbf{i}} r_k - \hat{\mathbf{j}} q_k)\right]$$

After some algebra, we get

$$\frac{\partial F_i}{\partial A_k} = \frac{v}{2D}\sum_{l=1}^{3}(q_k q_l + r_k r_l)A_l + \frac{v}{2}(B_{ry}q_k - B_{rx}r_k) \qquad (9.71)$$

In matrix form, this expression can be written as

$$\begin{bmatrix}\dfrac{\partial F_i}{\partial A_1}\\ \dfrac{\partial F_i}{\partial A_2}\\ \dfrac{\partial F_i}{\partial A_3}\end{bmatrix} = \begin{bmatrix} S_{1,1} & S_{1,2} & S_{1,3}\\ S_{2,1} & S_{2,2} & S_{2,3}\\ S_{3,1} & S_{3,2} & S_{3,3}\end{bmatrix}\begin{bmatrix}A_1\\ A_2\\ A_3\end{bmatrix} + \frac{v}{2}\begin{bmatrix}B_{ry}q_1 - B_{rx}r_1\\ B_{ry}q_2 - B_{rx}r_2\\ B_{ry}q_3 - B_{rx}r_3\end{bmatrix} \qquad (9.72)$$

9.5.5. Application to the Electric Vector Potential

This case, described in section 9.2.5, has the following functional

$$F_i = \int_{S_i} \frac{J^2}{2\sigma} ds \qquad (9.73)$$

The calculation of $\partial F_i/\partial T_k$, where T_k is the electric vector potential at node k, can be obtained similarly to the magnetic vector potential, considering μ as a constant, and without currents in the solution domain. This gives

$$\int_{S_i}\left[\int_0^B H dB\right] ds = \int_{S_i}\frac{B^2}{2\mu} ds$$

This expression is similar to Eq. (9.73). Therefore, the generic term $S(n,k)$ for Eq. (9.73) is analogous to that obtained for Eq. (9.67), that is:

$$S(n,k) = \frac{1}{2D\sigma}(q_n q_k + r_n r_k) \qquad (9.74)$$

9.6. Assembly of the Matrix System

The minimization of the functionals in section 9.5 gave the elemental matrix for the various physical situations. The terms $\partial F_i/\partial V_k$ are matrix equations and can be written in the following general form:

$$\begin{bmatrix} \dfrac{\partial F_i}{\partial V_1} \\ \dfrac{\partial F_i}{\partial V_2} \\ \dfrac{\partial F_i}{\partial V_3} \end{bmatrix} = \begin{bmatrix} S_{1,1} & S_{1,2} & S_{1,3} \\ S_{2,1} & S_{2,2} & S_{2,3} \\ S_{3,1} & S_{3,2} & S_{3,3} \end{bmatrix} \begin{bmatrix} V_1 \\ V_2 \\ V_3 \end{bmatrix} + \begin{bmatrix} Q_1 \\ Q_2 \\ Q_3 \end{bmatrix} \qquad (9.75)$$

The vector $[Q]$ is the so-called source vector of the problem, which depending on the case, can be electric charges (Eq. 9.60), currents (Eq. 9.68), or permanent magnets (Eq. 9.72). This vector is called a source vector because it describes the influence of the sources in the element (if any). Generally, these are the sources of the fields – either electric or magnetic, depending on the case.

Eq. (9.75), calculated for an element i, is part of the sum

$$\sum_{i=1}^{n} \frac{\partial F_i}{\partial V_k} = 0$$

where n is the number of elements in the finite element mesh. To calculate this sum we assemble the contribution of all elements in a global matrix system in which all the nodes of the mesh are considered. Each element contributes a system similar to that in Eq. (9.75). The K nodes of the mesh are numbered 1 to K.

local numbering	global numbering
1	7
2	10
3	14

The local numbering for each element (1, 2, and 3) corresponds to some global mesh numbers. For example, suppose that for a triangle the local and global numbers are as indicated bellow. The assembly is done as shown in Figure 9.19.

Initially, all the terms of the global matrix system $SS(K,K)$ are set to zero. Then, each element i is evaluated, and its contribution $S(3,3)$ [Eq. (9.75)] is added to the $SS(K,K)$ terms as shown above. This operation describes the sum

$$\sum_{i=1}^{n} \frac{\partial F_i}{\partial V_k} = 0$$

$$
\begin{array}{c}
 \\ 1 \\ 2 \\ \vdots \\ 7 \\ \vdots \\ 10 \\ \vdots \\ 14 \\ \vdots \\ K
\end{array}
\begin{array}{c}
\begin{array}{ccccccc} 1 & 2\ldots 7 & \ldots & 10 & \ldots & 14 & \ldots\ K \end{array} \\
\left[\begin{array}{ccccc}
 & | & | & | & \\
 & | & | & | & \\
 & | & | & | & \\
\text{———} & S_{1,1} & S_{1,2} & S_{1,3} & \text{——} \\
 & & & & \\
\text{———} & S_{2,1} & S_{2,2} & S_{2,3} & \text{——} \\
 & & & & \\
\text{———} & S_{3,1} & S_{3,2} & S_{3,3} & \text{——} \\
 & | & | & | & \\
 & | & | & | &
\end{array}\right]
\end{array}
\left[\begin{array}{c} V_1 \\ V_2 \\ \vdots \\ V_7 \\ \vdots \\ V_{10} \\ \vdots \\ V_{14} \\ \vdots \\ V_K \end{array}\right]
=
\left[\begin{array}{c} \\ \\ \\ -Q_1 \\ \\ -Q_2 \\ \\ -Q_3 \\ \\ \\ \end{array}\right]
\begin{array}{c} 1 \\ 2 \\ \vdots \\ 7 \\ \vdots \\ 10 \\ \vdots \\ 14 \\ \vdots \\ K \end{array}
$$

Figure 9.19. Assembly of elemental contributions into the global matrix. Assembly of an element with node numbers *7*, *10*, and *14* is shown.

After all n elements are calculated and assembled we have formed the complete, or global system

$$[SS][V] = \{Q\} \tag{9.76}$$

Before proceeding with the solution, it is necessary to impose the Dirichlet boundary conditions. This can be done in a very simple manner using a method based on the following rule: if a value V_m is to be imposed at node m, the diagonal position of row m of the matrix is set to one. All other coefficients in the row are set to zero. In row m of vector $[Q]$, the value of V_m is entered. Performing the product of the vector $[V]$ with row m of $[SS]$ gives

$$0\ V_1 + 0\ V_2 + \ldots + 1\ V_m + 0\ V_{m+1} + \ldots + 0\ V_K = V_m$$

which gives the value V_m for node m.

After the system $[SS][V]=[Q]$ is assembled and the Dirichlet conditions are imposed, the system can be solved using any of a number of standard methods of solution. This step provides the values of the unknown potentials.

9.7. Axi-Symmetric Applications

In electrical engineering, there are many important geometries with axial (or rotational) symmetry. This includes many applications of solenoids (relays, valves). The formulations presented above were all based on rectangular coordinates. The use of two-dimensional description was based on the assumption that the structure has no geometric variation in the direction perpendicular to the plane of study and that the fields are constant in this direction. While this treatment is a good approximation to many problems, in the case of solenoids and other circular geometries, it is not a convenient approach.

In fact, these problems are three dimensional in nature. However, if the geometry possesses an axial (or rotational) symmetry, the problem can be treated as two dimensional, using the methods described in the previous sections, provided some simple modifications in the numerical procedures are made. These problems are called "axi-symmetric," and are studied in the cross-sectional plane P, shown in Figure 9.20a, using cylindrical coordinates.

The solenoid in Figure 9.20a is shown in a cross-section for visualization purposes. Figure 9.20b indicates the cross-sectional plane and the axis of symmetry used to define the problem. The same cross-sectional plane, in Cartesian coordinates, would describe an infinitely long, conducting bar with current I, as shown in Figure 9.21. While a valid geometry in itself, this is very different from the solenoid problem in Figure 9.20.

Plane P of Figure 9.20b is shown in the Figure 9.22, together with the cylindrical coordinates (r,ϕ,z). An infinitesimal volume dv is, in this case, $dv=rdrd\phi dz$, with sides dr, $rd\phi$ and dz as indicated in Figure 9.22.

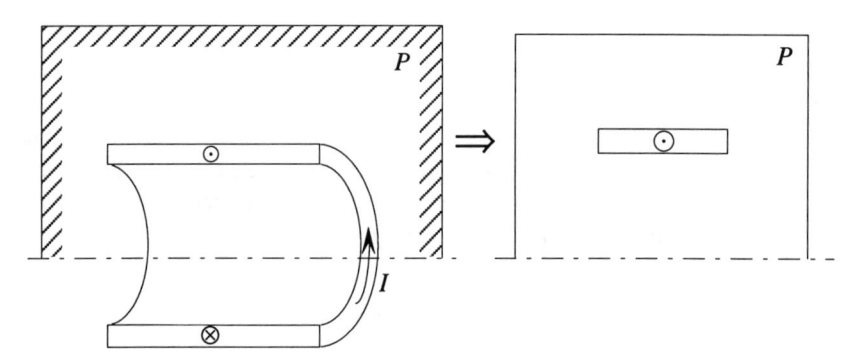

Figure 9.20a. Analysis of an axi-symmetric problem. A short solenoid is shown in cross-section.

Figure 9.20b. The solution domain is an axial cross-section through the axi-symmetric geometry.

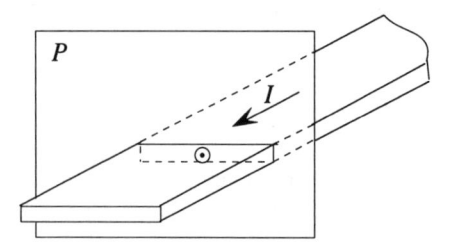

Figure 9.21. Equivalent cross-section in Cartesian coordinates. Unlike the axi-symmetric geometry, this implies an infinite, straight current layer.

Axi-Symmetric Applications 377

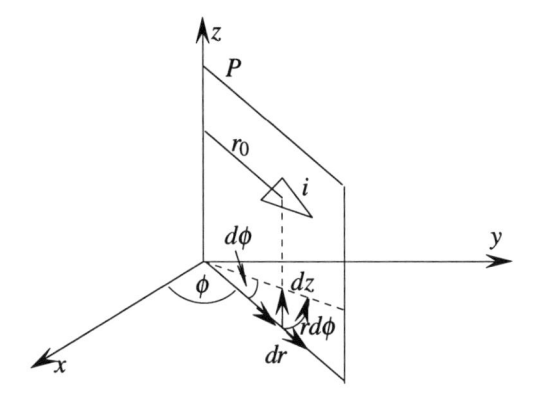

Figure 9.22. The solution domain of Figure 9.20a superimposed on the Cartesian coordinate system. The volume differential is defined as $rdrd\phi dz$.

For problems with axial symmetry, there is no field component in the ϕ direction and any derivative with respect to this direction is therefore zero. There are only fields or flux components in the r and z directions. As an example, B_r and B_z can be nonzero, but B_ϕ is always zero. Using rectangular coordinates, the functional can be written in the following general form

$$F_i = \int_{S_i} f\,ds = \int_{S_i} f\,dxdy \tag{9.77}$$

which is integrated on the surface of element i, because in the perpendicular direction (Oz) there is no field variation.

In axi-symmetric cases, the element volume must be used explicitly, because of its curvature. As an example, for an element near the symmetry axis, this curvature is higher than for an element far from it. For an element relatively far from the axis Oz, the problem is very similar to a problem in rectangular coordinates. With this consideration, the functional can be written as

$$F_i = \int_{v_i} f\,dv = \int_{v_i} f r\,d\phi dr dz$$

where v_i is the element volume. Its value is obtained by making a 360° rotation around the Oz axis. Thus, an axi-symmetric element is a ring with triangular cross-section. A two dimensional element is a triangular prism of unit depth. With the condition that there is no variation in the ϕ direction, F_i becomes

$$F_i = 2\pi \int_{S_i} f r\,dr dz$$

where S_i is the surface of element i.

In evaluating the different expressions of $\partial F_i/\partial V_k$, a difficulty is introduced when using axi-symmetric elements. With first order finite elements, the rectangular coordinate integrands do not contain the x or y variables, and the integration on S_i is done analytically. However, in the axi-symmetric expressions, the product $r\,dr$ appears in the equation and it is necessary to perform a somewhat complex numerical integration on S_i. In order to simplify this operation, we introduce a simple approximation; assuming the integrand f does not vary abruptly within the triangle, we can use the approximation

$$F_i = 2\pi r_0 \int_{S_i} f\,dr\,dz \tag{9.78}$$

where r_0 is the distance of the centroid of the triangle to the Oz axis, as shown in Figure 9.22. This distance is given as: $r_0=(r_1+r_2+r_3)/3$, where r_1, r_2, and r_3 are the distances of the nodes of element i to the Oz axis. This procedure is a good approximation especially if the mesh is made of small elements near the Oz axis (where the relative difference between r_1, r_2, and r_3 is larger). If this precaution is taken, the precision of results will not be affected.

Eqs. (9.78) and (9.77) are of similar form, if z and r are substituted for x and y respectively. The only difference is in the term $2\pi r_0$ multiplying the integrand in Eq. (9.78).

Thus, for the scalar potential, the terms of the elemental matrix are multiplied by $2\pi r_0$. For the problems presented in the previous sections, we have the following generic terms:

a. electrostatic fields

$$S(n,k) = \frac{\pi r_0 \varepsilon}{D}(q_n q_k + r_n r_k) \tag{9.79}$$

and, for the electric charge term,

$$\rho r_0 \pi D/3 \tag{9.80}$$

b. stationary currents

$$S(n,k) = \frac{\pi r_0 \sigma}{D}(q_n q_k + r_n r_k) \tag{9.81}$$

c. magnetic fields

$$S(n,k) = \frac{\pi r_0 \mu}{D}(q_n q_k + r_n r_k) \tag{9.82}$$

Recall that, using the magnetic scalar potential, the magnetic field, defined by $\mathbf{H}=-\nabla\psi$, has the following components

$$H_r = -\frac{\partial \psi}{\partial r}; \quad H_z = -\frac{\partial \psi}{\partial z}; \quad H_\phi = \frac{1}{r}\frac{\partial \psi}{\partial \phi}$$

Since we are working with axial symmetry, the last component is always zero. Thus, there is a complete similarity between the rectangular and cylindrical coordinates formulations.

The vector potential formulation requires more extensive modifications. Since $\mathbf{B}=\nabla\times\mathbf{A}$, the components of $\mathbf{B}$ are

$$B_r = \frac{1}{r}\frac{\partial A_z}{\partial \phi} - \frac{\partial A_\phi}{\partial z} \quad \text{and} \quad B_z = \frac{1}{r}\frac{\partial (rA_\phi)}{\partial r} - \frac{1}{r}\frac{\partial A_r}{\partial \phi}$$

In the axi-symmetric case, $\mathbf{A}$ is perpendicular to the plane of study. Therefore, $A=A_\phi$, $A_z=0$ and $A_r=0$. The expressions above become

$$B_r = -\frac{\partial A}{\partial z} \tag{9.83}$$

$$B_z = \frac{1}{r}\frac{\partial (rA)}{\partial r} = \frac{A}{r} + \frac{\partial A}{\partial r} \tag{9.84}$$

The first term of the functional in Eq. (9.66), which will be called F_{i1}, is

$$F_{i1} = 2\pi r_0 \int_{S_i} \left[\int_0^B H dB\right] dr dz$$

and, calculating $\partial F_{i1}/\partial A_k$ gives

$$\frac{\partial F_{i1}}{\partial A_k} = 2\pi r_0 \int_{S_i} \frac{\partial}{\partial B}\left[\int_0^B H dB\right] \frac{\partial B}{\partial A_k} dr dz$$

$$\frac{\partial F_{i1}}{\partial A_k} = 2\pi r_0 \frac{D}{2}vB\frac{\partial B}{\partial A_k} = \frac{\pi r_0 Dv}{2}\frac{\partial B^2}{\partial A_k} \tag{9.85}$$

Because

$$B^2 = B_r^2 + B_z^2$$

380   9. The Variational Finite Element Method: Some Static Applications

we get

$$\frac{\partial B^2}{\partial A_k} = 2B_r \frac{\partial B_r}{\partial A_k} + 2B_z \frac{\partial B_z}{\partial A_k} \tag{9.86}$$

The second term of this expression is now calculated:

$$B_z \frac{\partial B_z}{\partial A_k} = \left(\frac{A}{r} + \frac{\partial A}{\partial r}\right) \frac{\partial}{\partial A_k} \left(\frac{A}{r} + \frac{\partial A}{\partial r}\right) \tag{9.87}$$

We make the approximation $A/r = A'/r_0$, where $A' = (A_1 + A_2 + A_3)/3$ is the average potential value of the three nodes and r_0 is the distance of the centroid from the Oz axis. This gives

$$\frac{A'}{r_0} = \sum_{l=1}^{3} \frac{A_l}{3r_0} \tag{9.88}$$

Note that, similar to Eq. (9.49), the potential $A(r,z)$ is

$$A(r,z) = \frac{1}{D} \sum_{l=1}^{3} (p_l + s_l r + q_l z) A_l \tag{9.89}$$

where s_l is used here to avoid the ambiguity between r_1 in Eq. (9.49) and the radius r. We have

$$B_z = \frac{A}{r} + \frac{\partial A}{\partial r} = \sum_{l=1}^{3} \frac{A_l}{3r_0} + \frac{1}{D} \sum_{l=1}^{3} s_l A_l = \sum_{l=1}^{3} A_l \left(\frac{1}{3r_0} + \frac{s_l}{D}\right)$$

and

$$\frac{\partial B_z}{\partial A_k} = \frac{1}{3r_0} + \frac{s_k}{D}$$

Equation (9.87) now becomes

$$B_z \frac{\partial B_z}{\partial A_k} = \left[\sum_{l=1}^{3} A_l \left(\frac{1}{3r_0} + \frac{s_l}{D}\right)\right] \left(\frac{1}{3r_0} + \frac{s_k}{D}\right) \tag{9.90}$$

or, similarly,

$$B_z \frac{\partial B_z}{\partial A_k} = \frac{1}{D^2} \sum_{l=1}^{3} \left(\frac{D}{3r_0} + s_l\right) \left(\frac{D}{3r_0} + s_k\right) A_l$$

Defining $f_1 = D/3r_0 + s_1$, we can write

$$B_z \frac{\partial B_z}{\partial A_k} = \frac{1}{D^2} \sum_{l=1}^{3} f_l f_k A_l \qquad (9.91)$$

We now calculate the first term of Eq. (9.86)

$$B_r \frac{\partial B_r}{\partial A_k} = \left(-\frac{\partial A}{\partial z}\right) \frac{\partial}{\partial A_k} \left(\frac{\partial A}{\partial z}\right)$$

which, with Eq. 9.89 gives

$$B_r \frac{\partial B_r}{\partial A_k} = \frac{1}{D^2} \sum_{l=1}^{3} q_l q_k A_l \qquad (9.92)$$

Applying Eqs. (9.91) and (9.92) in Eq. (9.86) results in

$$\frac{\partial B^2}{\partial A_k} = \frac{2}{D^2} \sum_{l=1}^{3} (f_l f_k + q_l q_k) A_l$$

Using this expression in Eq. (9.85), we get

$$\frac{\partial F_{i1}}{\partial A_k} = \frac{\pi r_0 v}{D} \sum_{l=1}^{3} (f_l f_k + q_l q_k) A_l$$

The calculation of the second term in the functional in Eq. (9.66) involves the current density J, (multiplied by $2\pi r_0$)

$$\frac{\partial F_i}{\partial A_k} = \frac{\pi r_0 v}{D} \sum_{l=1}^{3} (f_l f_k + q_l q_k) A_l - \frac{\pi J r_0 D}{3} \qquad (9.93)$$

The generic term of the elemental matrix is

$$S(n,k) = \frac{\pi r_0 v}{D} (f_n f_k + q_n q_k) \qquad (9.94)$$

Finally, we evaluate the permanent magnet case. The functional corresponding to Eq. (9.69) is

$$F_i = 2\pi r_0 \int_{S_i} \frac{v}{2} B_i^2 ds = \pi r_0 v \frac{D}{2} B_i^2$$

382   9. The Variational Finite Element Method: Some Static Applications

and

$$\frac{\partial F_i}{\partial A_k} = \pi r_0 v \frac{D}{2} \frac{\partial B_i^2}{\partial A_k} \qquad (9.95)$$

where

$$B_i^2 = (B_r - B_{rr})^2 + (B_z - B_{rz})^2$$

B_{rr} and B_{rz} are the remnant flux density components of the magnet in the directions Or and Oz. Therefore,

$$\frac{\partial B_i}{\partial A_k} = 2(B_r - B_{rr})\frac{\partial}{\partial A_k}(B_r - B_{rr}) + 2(B_z - B_{rz})\frac{\partial}{\partial A_k}(B_z - B_{rz}) \qquad (9.96)$$

Using Eq. (9.90), after the remnant flux density is introduced, we get

$$(B_z - B_{rz})\frac{\partial}{\partial A_k}(B_z - B_{rz}) = \left[\sum_{l=1}^{3} A_l\left(\frac{1}{3r_0} + \frac{s_l}{D}\right) - B_{zr}\right]\left(\frac{1}{3r_0} + \frac{s_k}{D}\right)$$

with $f_l = s_l + D/3r_0$, this becomes

$$(B_z - B_{rz})\frac{\partial}{\partial A_k}(B_z - B_{rz}) = \frac{1}{D^2}\sum_{l=1}^{3} f_k f_l A_l - B_{rz}\frac{f_k}{D} \qquad (9.97)$$

The remaining term in Eq. (9.96) is

$$(B_r - B_{rr})\frac{\partial}{\partial A_k}(B_r - B_{rr}) = \left(-\frac{\partial A}{\partial z} - B_{rr}\right)\frac{\partial}{\partial A_k}\left(-\frac{\partial A}{\partial z}\right) =$$

$$\left(-\frac{1}{D}\sum_{l=1}^{3} q_l A_l - B_{rr}\right)\left(-\frac{q_k}{D}\right)$$

$$(B_r - B_{rr})\frac{\partial}{\partial A_k}(B_r - B_{rr}) = \frac{1}{D^2}\sum_{l=1}^{3} q_l q_k A_l - \frac{B_{rr} q_k}{D} \qquad (9.98)$$

Using Eqs. (9.97) and (9.98) in Eq. (9.96) and substituting this result in Eq. (9.95), we get

$$\frac{\partial F_i}{\partial A_k} = \frac{\pi r_0 v}{D}\sum_{l=1}^{3}(f_l f_k + q_l q_k)A_l + \pi r_0 v(B_{rr} q_k - B_{rz} f_k) \qquad (9.99)$$

The general term $S(n,k)$ of the elemental matrix is similar to Eq. (9.94), and the permanent magnet source is

$$\pi r_0 v (B_{rr} q_k - B_{rz} f_k) \qquad (9.100)$$

9.8. Nonlinear Applications

Generally, permittivity and conductivity can be considered as constants. However, permeability (or reluctivity) is dependent on the magnetic field intensity. Ferromagnetic materials are characterized by a **B(H)** curve, with permeability varying depending on the location on the **B(H)** curve.

The assembly of the matrix $SS(K,K)$ requires known values of μ (or v) for each element i. However, how can we know the value of μ before the solution is obtained? The system is a nonlinear system and in order to find a solution, it is necessary to establish an iterative procedure. This process is discussed next.

9.8.1. *Method of Successive Approximation*

This method is very simple in that we can use the normal computational procedure in an iterative form. The process is as follows:

a. Make an initial approximation, $\psi=0$ for all nodes.
b. Using the approximation for ψ calculate the field intensity H in the element.
c. Obtain μ from the **B(H)** curve based on the calculated H.
d. Using μ calculate the contribution $S(3,3)$ for the element as well as the source term (current or permanent magnet).
e. Assemble $S(3,3)$ in the global matrix $SS(K,K)$, and the source terms in the matrix Q.
f. Impose the Dirichlet boundary conditions.
g. Solve the system $[SS][\psi]=[Q]$.
h. Compare ψ with ψ of the previous iteration. If the error criterion indicates that the convergence has not yet been obtained go to step b. Repeat steps b. through h. until the solution converges.

Steps b. through g. constitute the normal, or linear solution. It is obvious that if the material property (permeability in this case) is constant, the solution converges in one iteration. With nonlinear material properties, the solution for ψ is convergent after a number of iterations. Generally, this method is convergent, but, sometimes, the iterative procedure is slow. However, it is very simple and easy to implement, since it is based on linearization of the nonlinear dependency.

9.8.2. *The Newton-Raphson Method*

This method is an extension of the Newton method, adapted to matrix systems of equations. Compared to the successive approximation method, it converges faster because the solution is found by tangent functions rather than simple linearization of the **B(H)** curve.

The basis of this method, is a truncated Taylor series, truncated after its first term. For a column matrix $[P]$, which is a function of a vector $[x]$ the truncated Taylor series is, for $[x]$ close to $[x]_m$

$$[P]_m + \left[\frac{\Delta P}{\Delta x}\right]([x] - [x]_m)$$

where $\Delta P/\Delta x$ is the Jacobian of $[P]$ at $[x]_m$. If, for example, $[P]$ and $[x]$ have two components

$$\left[\frac{\Delta P}{\Delta x}\right] = \begin{bmatrix} \frac{\partial P_1}{\partial x_1} & \frac{\partial P_1}{\partial x_2} \\ \frac{\partial P_2}{\partial x_1} & \frac{\partial P_2}{\partial x_2} \end{bmatrix}$$

If $[x]_m$ is not too far from the solution of $[P]$, we find that $[x]_{m+1}$ in the relation

$$[P]_m + \left[\frac{\Delta P}{\Delta x}\right]_m ([x]_{m+1} - [x]_m) = 0$$

is a better approximation to the solution than $[x]_m$. The Newton-Raphson method consists of solving the matrix system in the following form

$$\left[\frac{\Delta P}{\Delta x}\right]_m [\Delta x]_{m+1} = -[P]_m \qquad (9.101)$$

The unknown vector is now $[\Delta x]_{m+1}$, and $[x]_{m+1}$ is easily obtained from

$$[x]_{m+1} = [x]_m + [\Delta x]_{m+1} \qquad (9.102)$$

after the matrix system in Eq. (9.101) is solved.

In our case, using the scalar potential, the column matrix $[P]$ is the matrix product $[SS][V]$ where $[x]$ has been replaced by $[V]$. This matrix product is obtained by summing the terms $\partial F_i/\partial V_k$ as described above. Recall that these terms are, according the Eq. (9.65)

$$\frac{\partial F_i}{\partial V_k} = \sum_{l=1}^{3} \frac{\mu}{2D}(q_l q_k + r_l r_k) V_l \qquad (9.103)$$

The general term of the Jacobian $[\Delta P/\Delta x]$ corresponds to the derivation of Eq. (9.103) related to the mesh node potentials. However, F_i is nonzero only in element i and the derivatives $\partial F_i/\partial V_k$ are nonzero when the derivatives are taken with respect to one of the nodes of element i, as for example, V_n. Performing this general derivation in Eq. (9.103), we get

$$\frac{\partial}{\partial V_n}\frac{\partial F_i}{\partial V_k} = \frac{\mu}{2D}(q_n q_k + r_n r_k) + \frac{1}{2D}\frac{\partial \mu}{\partial H^2}\frac{\partial H^2}{\partial V_n}\sum_{l=1}^{3}(q_k q_l + r_k r_l)V_l \qquad (9.104)$$

Note that μ depends on H^2 and H^2 depends on V_n. We need to calculate $\partial H^2/\partial V_n$. This has already been obtained in Eq. (9.57), which can be written in the form

$$\frac{\partial H^2}{\partial V_n} = \frac{2}{D^2}\left[\sum_{l=1}^{3}(q_l q_n + r_l r_n)V_l\right] \qquad (9.105)$$

With the notation $S(n,k) = (\mu/2D)(q_n q_k + r_n r_k)$, Eq. (9.105) can be written as

$$\frac{\partial H^2}{\partial V_n} = \frac{2}{D^2}\frac{2D}{\mu}\sum_{l=1}^{3}S(l,n)V_l$$

Using the same notation for the last term of Eq. (9.104), we have

$$\frac{\partial}{\partial V_n}\frac{\partial F_i}{\partial V_k} = S(n,k) + \frac{4}{D\mu^2}\left(\frac{\partial \mu}{\partial H^2}\right)\left[\sum_{l=1}^{3}S(k,l)V_l\right]\left[\sum_{l=1}^{3}S(n,l)V_l\right] \qquad (9.106)$$

For use in this equation, μ and $\partial\mu/\partial H^2$ are obtained without any particular difficulties from the material characteristic (**B(H)**) curves. From Eq. (9.106), we note that to obtain the complete Jacobian matrix, only the terms $S(3,3)$ are needed, and these are calculated using the values of V_l in triangle i. Assembly of these terms [Eq. (9.106)] provides the global Jacobian, called here $[SJ]$. The matrix system to be solved is:

$$[SJ][\Delta V] = [R] \qquad (9.107)$$

The right-hand side vector $[R]$ is called a residual vector, because it originates from the matrix product $[SS][V]$ that should be zero. It is not zero because the permeabilities used to construct $[SS]$ are approximations. The more the iterative procedure evolves, the closer the permeabilities are to the final solution, and $[R]$ and $[\Delta V]$ tend to zero. The Newton-Raphson algorithm can be summarized as

386   9. The Variational Finite Element Method: Some Static Applications

a. Make an approximation to the vector V as near the solution as possible.
b. Using the approximation for V, calculate H and, from the **B(H)** curve obtain μ and $\partial\mu/\partial H^2$.
c. Calculate the elemental matrix terms $S(3,3)$ using μ.
d. With $S(3,3)$, $\partial\mu/\partial H^2$ and the potentials V of the previous iteration, calculate the Jacobian terms and the residual.
e. Impose the Dirichlet boundary conditions and solve the system in Eq. (9.107).
f. With the solution and the values of $[\Delta V]$, obtain the new values of V.
g. Repeat steps b. through f. until the convergence criterion is satisfied.

This method is very efficient, especially if the first approximation is close to the solution. A practical rule is to begin the Newton-Raphson loop after five or six iterations are performed with the Successive Approximation method. It is not necessary to dimension a new matrix, since the Jacobian matrix is topologically similar to the global matrix $[SS]$. The same space in computer memory can be used for evaluation of the Jacobian.

The Jacobian terms for axi-symmetric cases with scalar potentials are similar to those for rectangular coordinates. Denoting

$$S(k,n) = \frac{\pi r_0 \mu}{D}(q_k q_n + r_k r_n)$$

we get

$$\frac{\partial^2}{\partial V_n \partial V_k} F_i = S(k,n) + \frac{2}{D\pi r_0 \mu^2}\left(\frac{\partial \mu}{\partial H^2}\right)\left[\sum_{l=1}^{3} S(k,l)V_l\right]\left[\sum_{l=1}^{3} S(n,l)V_l\right] \quad (9.108)$$

For vector potential applications, $\partial F_i/\partial A_k$ in rectangular coordinates is

$$\frac{\partial F_i}{\partial A_k} = \sum_{l=1}^{3} S(k,l)A_l + Q(k)$$

where the general term $S(k,l)$ is

$$S(k,l) = \frac{v}{2D}(q_k q_l + r_k r_l)$$

and $Q(k)$ is the source term. This term takes the form:

$$Q(k) = -D\frac{J}{6} \quad \text{for current sources}$$

$$Q(k) = \frac{v(B_{ry}q_k - B_{rx}r_k)}{2} \quad \text{for permanent magnets}$$

Both of these are independent of A. Therefore, $\partial Q/\partial A_n = 0$ and the general Jacobian term is

$$\frac{\partial^2}{\partial A_n \, \partial A_k} F_i = S(k,n) + \frac{4}{Dv^2}\left(\frac{\partial v}{\partial B^2}\right)\left[\sum_{l=1}^{3} S(k,l)A_l\right]\left[\sum_{l=1}^{3} S(n,l)A_l\right] \quad (9.109)$$

For axi-symmetric applications, the analogous expression is

$$\frac{\partial^2}{\partial A_n \, \partial A_k} F_i = S(k,n) + \frac{2}{D\pi r_0 v^2}\left(\frac{\partial v}{\partial B^2}\right)\left[\sum_{l=1}^{3} S(k,l)A_l\right]\left[\sum_{l=1}^{3} S(n,l)A_l\right] \quad (9.110)$$

where

$$S(k,n) = \frac{\pi v r_0}{D}(f_k f_n + q_n q_k)$$

The residual is

$$[R] = [SS][A] + [Q]$$

and it includes the source $[Q]$, since the source is generated by the sum $\partial F_i/\partial A_k$, and for this, the sources are included.

9.9. The Three-Dimensional Scalar Potential

The use of the vector potential for three-dimensional (3D) problems presents difficulties since each node has three unknowns corresponding to the three components of **A**.

This increases memory requirements and computational time. Several alternative formulations for the solution of 3D problems have been developed (see, for example, Chapter 11 which deals with 3D edge elements). A complete discussion of these methods is beyond the goal of this book, but they will be briefly discussed latter. For more information in this matter, we recommend referring to the bibliography section.

However, the use of the scalar potential V in three-dimensional problems is relatively simple and does not suffer from these limitations. This is described now. The general equation to be solved is

$$\frac{\partial}{\partial x}\alpha\frac{\partial V}{\partial x} + \frac{\partial}{\partial y}\alpha\frac{\partial V}{\partial y} + \frac{\partial}{\partial z}\alpha\frac{\partial V}{\partial z} = 0 \quad (9.111)$$

which can be any of the following:

- Eq. (9.4) with $\alpha=\varepsilon$ and $\rho=0$.
- Eq. (9.8) with $\alpha=\sigma$.
- Eq. (9.13) with $\alpha=\mu$.

Before proceeding with the finite element implementation of this equation we must define a three-dimensional finite element.

9.9.1. The First-Order Tetrahedral Element

As a natural extension of the first-order triangular element, we will use the first order tetrahedral element, shown in Figure 9.23. The potential varies inside this element as

$$V(x,y,z) = a_1 + a_2 x + a_3 y + a_4 z \qquad (9.112)$$

Since this equation has to be satisfied at the four nodes of the element, the coefficients a_1, a_2, a_3, and a_4 are determined from the four equations for the potentials at the four nodes as was done for the triangular element. After evaluation of these constants, Eq. (9.112) can be expressed as

$$V(x,y,z) = \frac{1}{D}\sum_{l=1}^{4}(p_l + q_l x + r_l y + s_l z)V_l \qquad (9.113)$$

where D is six times the volume of the tetrahedron and

$$p_1 = \det\begin{bmatrix} x_2 & y_2 & z_2 \\ x_3 & y_3 & z_3 \\ x_4 & y_4 & z_4 \end{bmatrix} \qquad q_1 = -\det\begin{bmatrix} 1 & y_2 & z_2 \\ 1 & y_3 & z_3 \\ 1 & y_4 & z_4 \end{bmatrix}$$

$$r_1 = -\det\begin{bmatrix} x_2 & 1 & z_2 \\ x_3 & 1 & z_3 \\ x_4 & 1 & z_4 \end{bmatrix} \qquad s_1 = -\det\begin{bmatrix} x_2 & y_2 & 1 \\ x_3 & y_3 & 1 \\ x_4 & y_4 & 1 \end{bmatrix} \qquad (9.114)$$

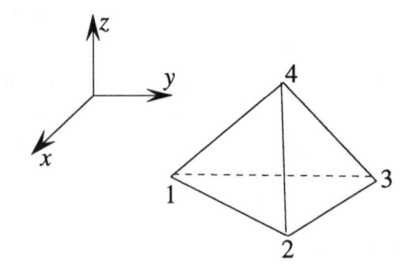

Figure 9.23. The first-order tetrahedral element.

The remaining terms (for $l=2, 3, 4$) are obtained by cyclic permutation of the index in the expressions in Eq. (9.114).

9.9.2. Application of the Variational Method

The functional of this 3D problem is identical to the functional given in Eq. (9.41). This functional takes account of nonlinearity:

$$F = \int_V \left[\int_0^H BdH \right] dxdydz \qquad (9.115)$$

To obtain the elemental matrix $S(4,4)$ (or $\partial F_i/\partial V_k$), the calculation sequence is analogous to that performed for the 2D element, and we obtain

$$\begin{bmatrix} \dfrac{\partial F_i}{\partial V_1} \\ \vdots \\ \dfrac{\partial F_i}{\partial V_4} \end{bmatrix} = \dfrac{\alpha}{6D} \begin{bmatrix} (q_1 q_1 + r_1 r_1 + s_1 s_1) & \cdots & (q_1 q_4 + r_1 r_4 + s_1 s_4) \\ & & \\ \text{Symmetric} & \cdots & (q_4 q_4 + r_4 r_4 + s_4 s_4) \end{bmatrix} \begin{bmatrix} V_1 \\ V_2 \\ V_3 \\ V_4 \end{bmatrix} \qquad (9.116)$$

where α represents ε, σ, or μ, depending on the problem to be solved.

The assembly of this elemental matrix $S(4,4)$ in the global system, the Newton-Raphson application and all others steps are analogous to the 2D formulation and will not be repeated here.

9.9.3. Modeling of 3D Permanent Magnets

In section 9.5.4, the vector potential formulation for permanent magnets was presented. However, for 3D applications, it is possible to use a formulation with the magnetic scalar potential. Assuming the magnet is described as shown in Figure 9.9, where H_c is its coercive field intensity, we have

$$\mathbf{H} = -\nabla \psi$$

Inside the magnet the flux density $\mathbf{B}$ is given by

$$\mathbf{B} = \mu(\mathbf{H} + \mathbf{H}_c) \qquad (9.117)$$

where μ is the magnet permeability. The equation to be solved is

$$\nabla \cdot \left[\mu(-\nabla \psi + \mathbf{H}_c) \right] = 0 \qquad (9.118)$$

and the associated functional is

$$F = \int_v \frac{1}{2} \mu(-\nabla \psi + \mathbf{H}_c) \cdot (-\nabla \psi + \mathbf{H}_c) dxdydz \qquad (9.119)$$

With ψ as given by Eq. (9.113), we get

$$-\nabla \psi = -\frac{1}{D} \sum_{l=1}^{4} (\hat{\mathbf{i}} q_l + \hat{\mathbf{j}} r_l + \hat{\mathbf{k}} s_l) \psi_l \qquad (9.120)$$

We also have

$$\mathbf{H}_c = -\frac{\mathbf{B}_r}{\mu} = \hat{\mathbf{i}} H_{cx} + \hat{\mathbf{j}} H_{cy} + \hat{\mathbf{k}} H_{cz} \qquad (9.121)$$

Substituting Eqs. (9.120) and (9.121) into Eq. (9.119) and calculating $\partial F_i/\partial \psi_k$, we obtain an equation similar to Eq. (9.71)

$$\begin{bmatrix} \frac{\partial F_i}{\partial \psi_1} \\ \vdots \\ \frac{\partial F_i}{\partial \psi_4} \end{bmatrix} = \begin{bmatrix} S_{1,1} & \cdots & S_{1,4} \\ & \vdots & \\ Symm. & \cdots & S_{4,4} \end{bmatrix} \begin{bmatrix} \psi_1 \\ \psi_2 \\ \psi_3 \\ \psi_4 \end{bmatrix} - \frac{\mu}{6} \begin{bmatrix} H_{cx}q_1 + H_{cy}r_1 + H_{cz}s_1 \\ \vdots \\ H_{cx}q_4 + H_{cy}r_4 + H_{cz}s_4 \end{bmatrix}$$

where the terms $S(4,4)$ of the first matrix are the same as those in Eq. (9.116).

9.10. Examples

Some examples of finite element applications are presented in this section. To perform these calculations, the software package EFCAD (developed at the Universidade Federal de Santa Catarina, Brazil) is used.

This software was developed for personal computers and extensive use of the interactive and graphic characteristics of this kind of equipment is made. It is a general purpose software package, for 2D applications in static and dynamic electric and magnetic fields with specialized routines for analysis of electric machines and coupling of circuit equations with magnetic field solutions. Nonlinear, as well as static and transient thermal problems can be solved.

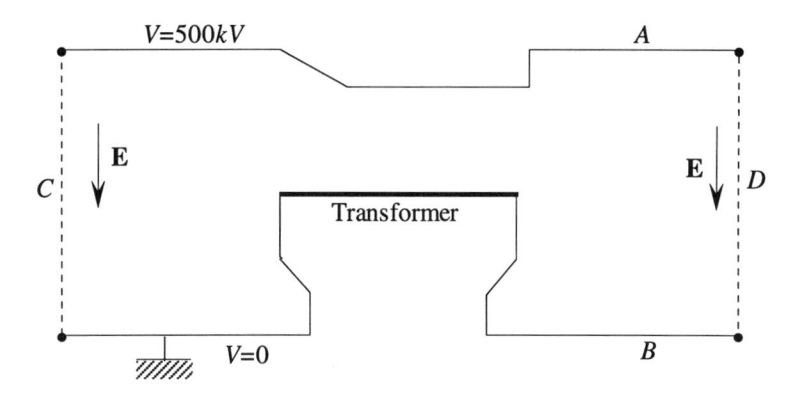

[Figure: electrostatic problem diagram showing V=500kV on top line A, V=0 at bottom, Transformer in middle, boundaries C and D on sides with E field arrows]

Figure 9.24. An electrostatic problem: Determination of the electric field in a domain with specified potentials on boundaries.

The software package has three important parts:

- Pre-processor, in which the general data (geometry, field sources boundary conditions) are furnished by the user. Additionally, the mesh is automatically generated from this input.
- Processor, in which the finite element method is applied to the discretized domain.
- Post-processor, in which graphic (equipotential lines) and numerical (flux, fields, forces, inductances) results are calculated and provided to the user.

The following examples demonstrate the potential and versatility of the finite element method.

9.10.1. Calculation of Electrostatic Fields

Suppose that we wish to determine the electric fields in the region of a high-voltage substation with an electrically grounded piece of equipment such as the transformer shown in Figure 9.24.

Assume that on lines A and B there are imposed boundary conditions ($V=500$ kV on line A and $V=0$ kV on line B). Lines C and D are chosen far from the transformer region, so that we can assume that the fields are approximately vertical and therefore almost tangent to lines C and D. This assumption allows the use of Neumann boundary conditions for the scalar potential on these two boundaries.

The finite element mesh generated is shown in Figure 9.25a. Note that in the region where we expect the field intensity to be higher, a mesh with smaller, denser elements is used. After the finite element calculation, the post-processor displays the equipotential lines, indicated in Figure 9.25b. By visual inspection the regions with higher potential gradients (higher fields) are easily noted.

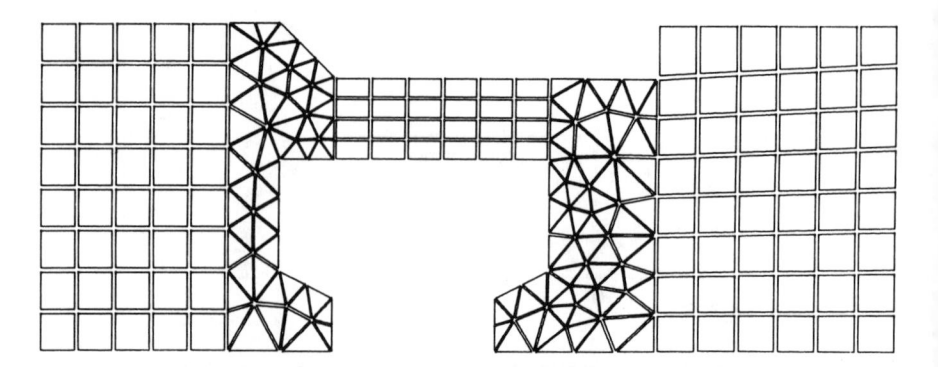

Figure 9.25a. Finite element discretization of the solution domain in Figure 9.24.

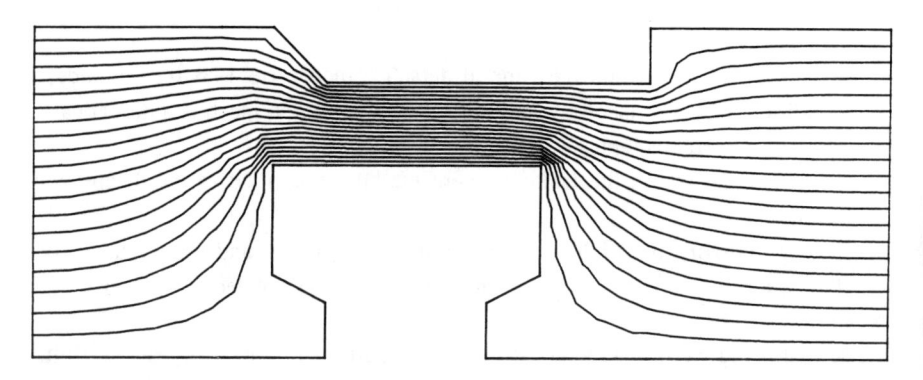

Figure 9.25b. Equipotential lines for the geometry in Figure 9.24.

As numerical results, the module provides the field values, which, if necessary, can be compared with the measured dielectric field strength of the material. These data can then be used for design and safety purposes.

9.10.2. *Calculation of Static Currents*

We wish to obtain the current distribution in a conductor made of a layer of copper and a layer of aluminum, as shown in Figure 9.26. Two different approaches are shown. The first uses the scalar potential formulation, presented in section 9.2.2. Laplace's equation in this case is

$$\frac{\partial}{\partial x}\sigma\frac{\partial V}{\partial x} + \frac{\partial}{\partial y}\sigma\frac{\partial V}{\partial y} = 0$$

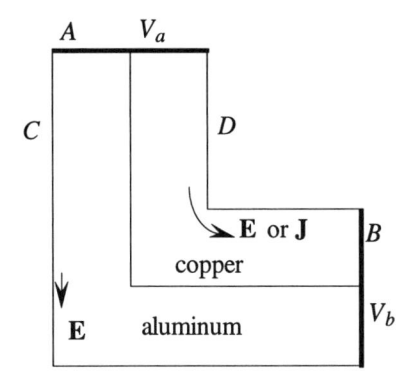

Figure 9.26. A static current problem: Determination of currents using the scalar potential with specified potentials on boundaries.

In this formulation, the potentials V on lines A and B are specified as boundary conditions. Assuming that the field E (or J) is tangential to lines C and D, Neumann boundary conditions are used on these two boundaries.

The mesh generated for this problem is shown in Figure 9.27a and the equipotential line distribution is presented in Figure 9.27b.

This problem can also be treated with the electric vector potential using the formulation in section 9.2.5. The equation to be solved in this case is

$$\frac{\partial}{\partial x}\frac{1}{\sigma}\frac{\partial T}{\partial x} + \frac{\partial}{\partial y}\frac{1}{\sigma}\frac{\partial T}{\partial y} = 0$$

Using this approach, we impose a current difference between lines C and D (for example $I=0$ and $I=I_a$, meaning that the current crossing the conductors is I_a). Recall that these current values are given here in *Amperes/meter*, and they are obtained by dividing the actual current by the width of the device. In this case, on lines A and B, we have Neumann boundary conditions, meaning that J or E are perpendicular to these lines.

The equipotential lines obtained after finite element calculations, are shown in Figure 9.28, showing how the current flux is distributed in the domain. A higher current crosses the copper part of the conductor. In fact, using the numerical data provided by the program, the current in copper is approximately 60% higher than the current in aluminum. This is as it should be, since the conductivity of copper is approximately 60% higher than the conductivity of aluminum.

The choice of formulation depends on the problem to be solved. If the device works with applied voltages, which then can be used as Dirichlet boundary conditions, the electric scalar potential should be used. On the other hand, if the current is known, the electric vector potential should be adopted.

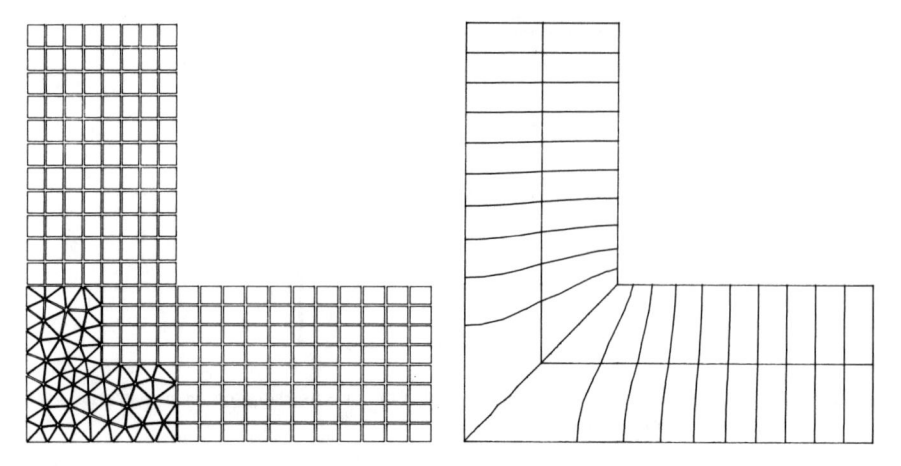

394   9. The Variational Finite Element Method: Some Static Applications

Figure 9.27a. Finite element discretization of the solution domain in Figure 9.26.

Figure 9.27b. Equipotential line distribution for the geometry in Figure 9.26

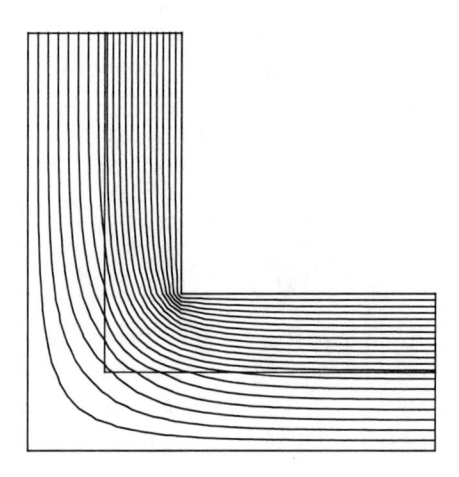

Figure 9.28. Equipotential lines for Figure 9.26 using the electric vector potential **T**. These are lines of current and are perpendicular to the potential lines in Figure 9.27b.

9.10.3. *Calculation of the Magnetic Field: Scalar Potential*

Assume that in the air-gap shown in Figure 9.3a, there are ferromagnetic "poles" (created by slots in the ferromagnetic material in Figure 9.29) and we wish to determine the magnetic flux distribution in this region. Between lines A and B we impose a potential difference related to the magnetomotive force NI of the coil. On lines C and D, a Neumann boundary conditions can be imposed by considering the field as tangential to these lines.

In Figure 9.30a, the equipotential lines obtained after the finite element calculation are shown. In this case, the value of NI is very low and the structure does not reach saturation. The result in Figure 9.30b is for a high value of *NI*, and, since the magnetic flux is large, the material reaches saturation. Thus, we notice that some equipotential lines penetrate into the ferromagnetic material of the poles. To simplify field visualization, we repeat this analysis with the vector potential formulation. Instead of specifying the potential difference between lines *A* and *B*, the magnetic flux difference between lines *C* and *D* is specified.

This approach is similar to that of the previous example. Figures 9.30c and 9.30d show the flux obtained under linear and saturation conditions, respectively. Note that in Figure 9.30d a larger part of the flux crosses through air, since the teeth are saturated. The relative permeability in the highly saturated regions is approximately 10.

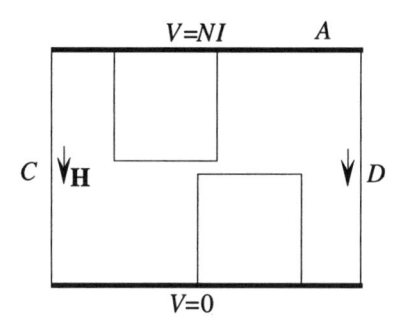

Figure 9.29. A magnetic field problem: calculation of the magnetic field intensity using the magnetic scalar potential. Boundary conditions are also in terms of the magnetic scalar potential.

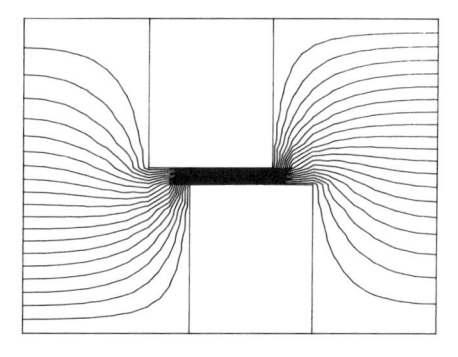

Figures 9.30a. Equipotential lines for Figure 9.29 at low values of *NI* (no saturation). The magnetic scalar potential is used for solution.

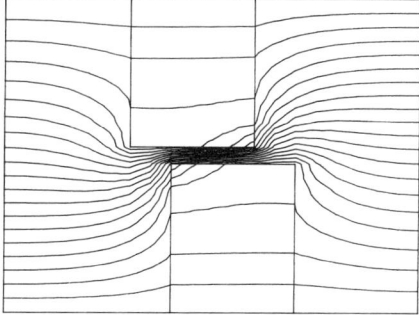

Figures 9.30b. Equipotential lines for Figure 9.29 at high values of *N I* (saturation). The magnetic scalar potential is used for solution.

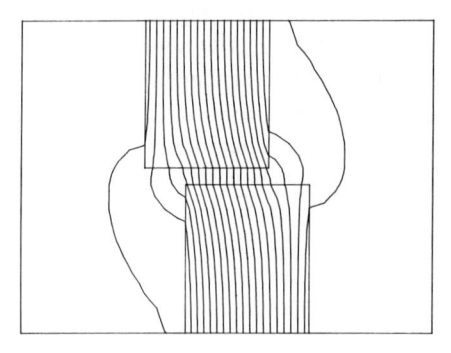

Figures 9.30c. Equipotential lines for Figure 9.29 at low values of NI (no saturation). The magnetic vector potential is used for solution.

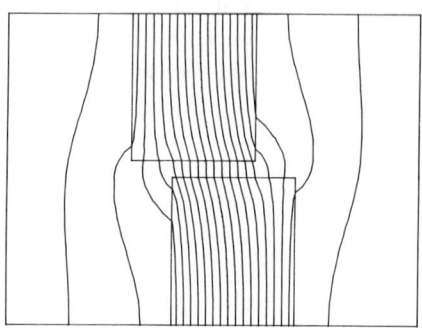

Figures 9.30d. Equipotential lines for Figure 9.29 at high values of NI (saturation). The magnetic vector potential is used for solution.

For the calculation with saturation with the scalar potential formulation, the Newton-Raphson method was used. However, the first five iterations were performed with the Successive Approximation method. The results from the Successive Approximation were used as an initial solution to the Newton-Raphson method. The solution required four Newton-Raphson iterations for convergence. The convergence was obtained with a relative error of 0.001, that is: the value at all nodes of the mesh was within this error.

9.10.4. *Calculation of the Magnetic Field: Vector Potential*

Figure 9.31 shows an axi-symmetric structure, shown in cross-section for visualization purposes. When a current is imposed, the magnetic forces created in the air gap attract the lower, mobile part, to the upper, stationary part of the structure. This attraction force between the two pieces is now calculated.

For the calculation of this structure by EFCAD, the domain data shown in Figure 9.32 is provided to the software package. According to the conventions of this software the symmetry axis must be coincident with the Ox axis. Line A (Figure 9.32) is placed at some distance from the structure because it is known, a priori, that the field has a natural dispersion in the air gap region.

A potential $A=0$ is imposed on lines A, B, C since the magnetic flux does not cross these lines. On line D we also impose $A=0$, since the axi-symmetry of the problem forces these conditions. Figure 9.33a shows the mesh and Figure 9.33b shows the resulting flux distribution.

The important numerical result in this problem is the attraction force on the mobile piece. This force is calculated on line E (Figure 9.32) using the Maxwell tensor method, (discussed in Chapter 6). Normally, the whole mobile piece must be enclosed by the line over which the force is calculated, but considering that the field is significant only on line E, it suffices to calculate the force on this line.

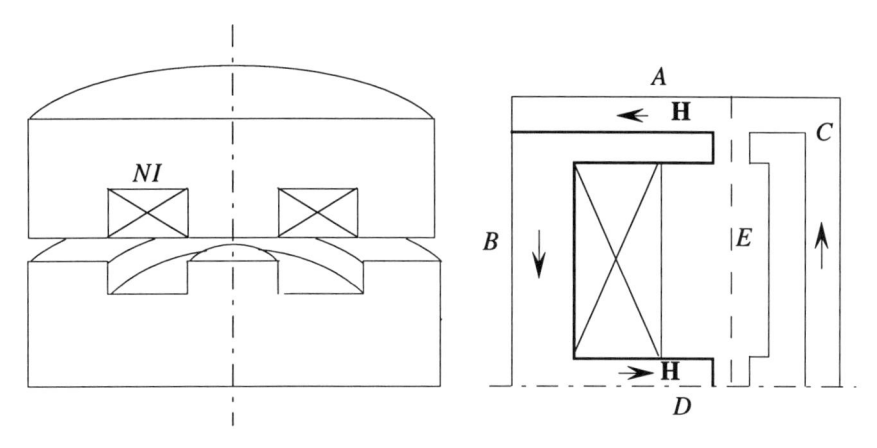

Figure 9.31. An axi-symmetric structure. Two parts of a magnetic circuit are separated by a gap. The upper part contains a coil and is stationary. The lower is free to move.

Figure 9.32. Solution domain for Figure 9.31 as supplied to the finite element program.

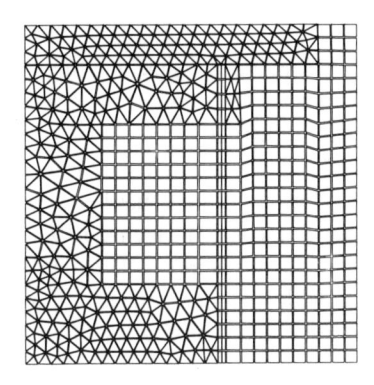

Figure 9.33a. Finite element mesh for the solution domain in Figure 9.32.

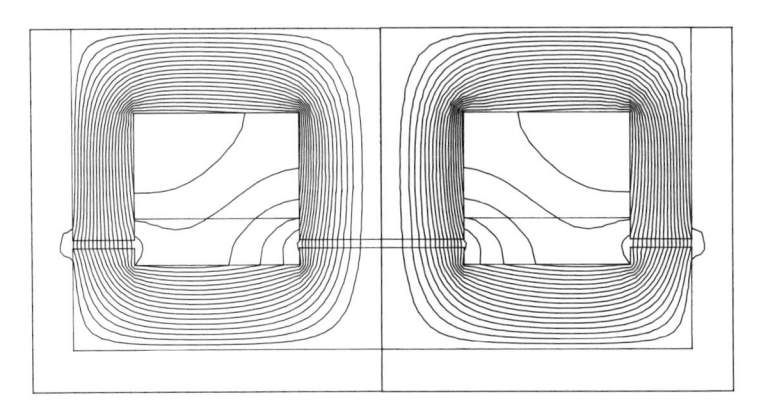

Figure 9.33b. Magnetic flux distribution for the device in Figure 9.31.

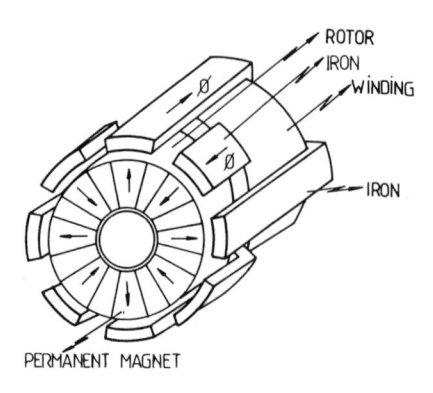

Figures 9.34a. An electric motor with a permanent magnet rotor and a stator made of iron bars.

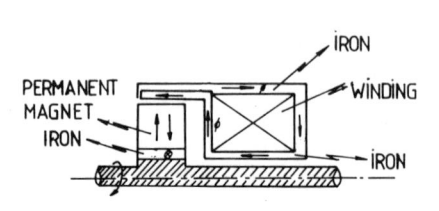

Figures 9.34b. The magnetic path for flux in the motor.

9.10.5. *Three-Dimensional Calculation of Fields of Permanent Magnets*

Figure 9.34a and 9.34b shows an electric machine of particular design, with the rotor made of permanent magnets. The stator of the machine contains a coil, with its turns wound around a ferromagnetic cylindrical ring. Iron bars are placed above and below the coil. These bars are indicated in Figure 9.34a as (1) for the bars over the coil and as (2) for the bars passing below the coil. The bars create, on the left side of the coil, a volume inside which the rotor is placed.

The rotor is made of permanent magnets magnetized in the radial direction. They alternate successively with an equal number of iron bars. The inner part of the rotor is an iron ring. These components create a flux path as shown in Figure 9.34b.

The purpose of this calculation is to determine the magnitude of the flux generated by the magnets through the magnetic circuit and linking the coil. If this flux is large, the torque generated by the machine is also large, resulting in an efficient machine.

We assume the device has four pairs of magnets with its respective four pairs of iron bars. To evaluate the maximum flux we consider the magnets to be below, and aligned with the iron bars. Because of symmetry, only half the geometry is represented. Neumann boundary conditions are used on all surfaces, meaning that, with the formulation developed in section 9.9.3, the flux does not cross these surfaces. This is a good approximation in this case.

Display of graphical results is somewhat difficult because of the relatively complex geometry. These are presented in Figure 9.35a,b.

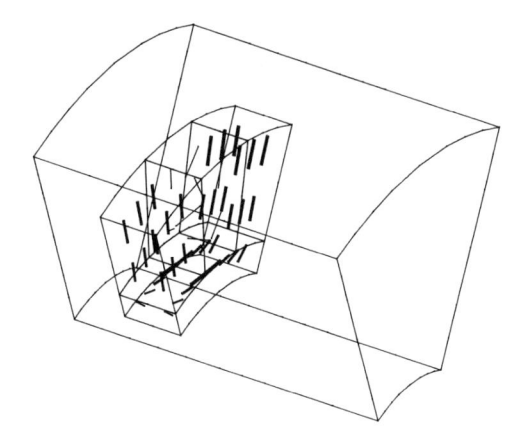

Figures 9.35a. One view of the flux distribution in the motor.

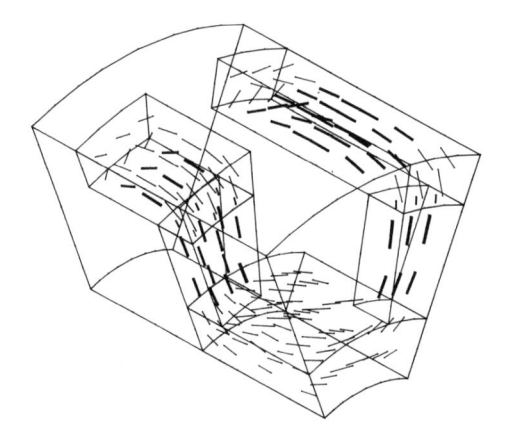

Figures 9.35b. A second view of the flux in the motor.

The plots in Figure (9.35a,b) were obtained directly from the computational package. The plots show the magnetic flux densities in the rotor and stator of the device. The thickness of the lines is proportional to the magnitude |**B**|.

Numerically, 89% of the flux generated by the permanent magnets is useful, indicating that this amount of flux actually links the coil, and that the device is an efficient electric machine.

10
Galerkin's Residual Method: Applications to Dynamic Fields

10.1. Introduction

For some types of problems, the search of functionals is difficult. For these, other methods, using the finite element technique, were developed. The most commonly used of these is the "Galerkin method." It is a particular form of the more general "weighted residual" method, whose principles were outlined in Chapter 8.

There is also a strong tendency to apply this method for any problem, even in cases normally formulated with the variational method, like those presented in the previous chapter. An important advantage of the Galerkin method lies in the fact that the numerical formulation is established from the physical equation defining the phenomenon, instead of the an associated functional which may or may not be easy to formulate.

To understand the principle of the Galerkin method consider the function

$$f_d(v, \nabla v) = 0$$

which depends on the variable v, its partial derivatives (indicated by $\overline{\nabla}v$) and the coordinates of the reference system of the domain D, in which the function f_d is defined. To simplify the notation, this function is denoted as $f(v)$ and represents the exact function. Substituting the solution v with the approximation u, we get an approximate function $f(u)$. The residual R is defined as the difference between the exact and the approximate solutions.

$$R = f(v) - f(u)$$

which is

$$R = -f(u)$$

As this residual tends to zero, the exact solution v is approached. Generally, in weighted residual methods, the residual R must satisfy a certain condition forcing it to tend to be zero. This condition is written as:

$$\int_D W_k R dD = 0, \quad 1 \leq k \leq K \tag{10.1}$$

where W_k is the weighting function defined for the K unknown nodes of the domain D. Galerkin's method is a particular form of the weighted residual method and is characterized by the fact that the weighting function and the approximating function are the same.

10.2. Application to Magnetic Fields in Anisotropic Media

We develop now the numerical formulation for the magnetic field **H** using the magnetic scalar potential ψ, in anisotropic media. For more details on this topic, see Chapter 8. Note that this problem can be treated by the variational method, but the functional for this case is more complex (compared with the isotropic case). Verification by application of Euler's equation is also more complex. On the other hand, the use of Galerkin's method is straightforward. The numerical solution to the isotropic case is identical to that obtained with the variational method for the isotropic case, and is obtained as a particular case of the anisotropic formulation.

Recall that $\mathbf{H} = -\nabla\psi$ and for anisotropic materials $\mathbf{B} = \|\mu\|\mathbf{H}$ where $\|\mu\|$ is the permeability tensor defined in Chapter 2

$$\|\mu\| = \begin{bmatrix} \mu_x & 0 \\ 0 & \mu_y \end{bmatrix}$$

The equation $\nabla \cdot \mathbf{B} = 0$ for the anisotropic case is

$$\nabla \cdot \|\mu\| \nabla\psi = 0$$

The discretized domain contains:

- P nodes in the whole domain.
- M nodes on the boundary with applied potentials (Dirichlet condition).
- $K = P - M$ unknown nodes.

The function $\psi(x,y)$ is represented in the following form:

$$\psi(x,y) = \sum_{k=1}^{K} \psi_k \phi_k \qquad (10.2)$$

where ϕ_k is an approximating function associated with the node k, such that for $x=x_k$ and $y=y_k$, $\psi(x_k,y_k)=\psi_k$. The Galerkin method consists, in this case, of the use of ϕ_k as weighting functions. This yields in domain D:

$$\int_D \phi_k \nabla \cdot \|\mu\| \nabla \psi \, dD = 0 \qquad (10.3)$$

for the K unknown nodes. This representation results in K equations. The approximating function to be used is that given in Eq. (10.2) as we will see shortly.

Applying the formulation discussed in section 8.2, we obtain

$$\oint_{S(D)} \phi_k \|\mu\| \nabla \psi \cdot \hat{\mathbf{n}} \, ds - \int_D (\|\mu\| \nabla \psi) \cdot \nabla(\phi_k) \, dD = 0$$

where $S(D)$ is the surface enclosing the volume D and $\hat{\mathbf{n}}$ is the normal unit vector defined at any point on the surface. Setting the second integral to zero, we can write the following expressions

$$\oint_L \phi_k \|\mu\| \nabla \psi \cdot \hat{\mathbf{n}} \, dl = 0 \qquad (10.4)$$

$$\int_S (\|\mu\| \nabla \psi) \cdot \nabla(\phi_k) \, ds = 0 \qquad (10.5)$$

where, in 2D cases, the domain is the surface S and $S(D)$ becomes the line L enclosing the surface S. The contour L can be divided into two parts:

- L_1, on which the potentials are known; Eq. (10.3) from which Eq. (10.4) originated is not written for these nodes. The functions ϕ_k in Eq. (10.4) are zero at these points verifying this equation.
- $L_2=L-L_1$, which is the remaining boundary. For this part of the boundary we write Eq. (10.3), since the nodal values are unknown. At these points, $\phi_k=1$, and in order to satisfy Eq. (10.4), it is necessary that $\|\mu\|\nabla\psi\cdot\hat{\mathbf{n}}=0$, or, $\mathbf{B}\cdot\hat{\mathbf{n}}=0$, corresponding to the natural Neumann condition for the scalar potential. This is completely equivalent to the representation using the variational method.

Application to Magnetic Fields in Anisotropic Media   403

Now we evaluate Eq. (10.5) and demonstrate that the procedure leads to a matrix system of the form $[SS]\{\psi\}=\{0\}$ as presented in the previous chapter.

The surface S associated with a node k is the sum of the triangles containing the node k, as shown in the example of Figure 10.1. Denoting this surface as S_l, the functions ϕ_{km}, ϕ_{kn}, ϕ_{kq}, ϕ_{kp} are nonzero in this region. The integral in Eq. (10.5) is calculated over S_l.

However, this integration can be done by a different, more convenient method: taking into account that the situation shown in Figure (10.1) is analogous for nodes m and n, the situation for triangle kmn is as shown in Figure 10.2.

Thus, instead of integrating over the surface S_l, containing the node, we can integrate over the triangle surfaces S_i. After the integration in Eq. (10.5) is done for all triangles, we obtain the same result as if the integration was done by nodes.

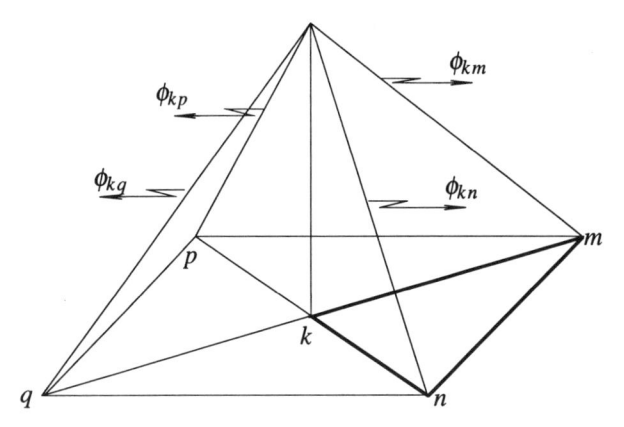

Figure 10.1. The contribution of a node k is the sum of the approximating functions in all elements that share node k.

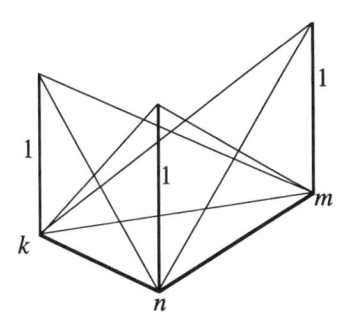

Figure 10.2. Contributions to nodes of an element are calculated in the element itself regardless of contributions from other elements.

This in effect means that the integration is performed as partial sums, and the contribution to a nodal value can be calculated in the elements in which the node is located.

For a first-order triangular element the function $\psi(x,y)$ is

$$\psi(x,y) = \sum_{l=1}^{3} \frac{1}{D}(p_l + q_l x + r_l y)\psi_l \qquad (10.6)$$

As was shown in section 9.5.1, this can be written in the form

$$\psi(x,y) = \sum_{l=1}^{3} \alpha_l \psi_l \qquad (10.7)$$

where, for example, $\alpha_k(x_k, y_k)=1$, $\alpha_k(x_n, y_n)=0$, $\alpha_k(x_m, y_m)=0$, corresponding to the functions shown in Figure 10.2. Assuming that in S_i, α_k corresponds to ϕ_k, Eq. (10.5) becomes

$$\int_{S_i} \nabla \alpha_k \cdot \|\mu\| \nabla \sum_{l=1}^{3} \alpha_l \psi_l \, ds = 0 \qquad (10.8)$$

Since this integration is performed on the surface S_i of element i, $\nabla \alpha_k$ can be extended to the three nodes of the element. Using Eq. (10.6), $\nabla \psi$ is

$$\nabla \psi = \frac{1}{D} \sum_{l=1}^{3} (\hat{i} q_l + \hat{j} r_l)\psi_l \qquad (10.9)$$

and $\nabla \alpha_k$ is

$$\nabla \alpha_k = \frac{1}{D}(\hat{i} q_k + \hat{j} r_k) \qquad (10.10)$$

These integrations do not depend on x or y. The integration over the surface is evaluated directly:

$$\frac{D}{2}\left[\frac{1}{D}\hat{i} q_k + \frac{1}{D}\hat{j} r_k\right] \begin{bmatrix} \mu_x & 0 \\ 0 & \mu_y \end{bmatrix} \frac{1}{D}\left[\hat{i}(q_1\psi_1 + q_2\psi_2 + q_3\psi_3) + \hat{j}(r_1\psi_1 + r_2\psi_2 + r_3\psi_3)\right]$$

which, written in matrix form and extended to the three nodes of the element is

$$\frac{1}{2D}\begin{bmatrix} q_1 & r_1 \\ q_2 & r_2 \\ q_3 & r_3 \end{bmatrix}\begin{bmatrix} \mu_x & 0 \\ 0 & \mu_y \end{bmatrix}\begin{bmatrix} q_1 & q_2 & q_3 \\ r_1 & r_2 & r_3 \end{bmatrix}\begin{bmatrix} \psi_1 \\ \psi_2 \\ \psi_3 \end{bmatrix} \qquad (10.11)$$

where the indices *1*, *2*, and *3* correspond to the nodes k, m, and n of element i. For an isotropic case ($\mu_x=\mu_y=\mu$), Eq. (10.11) is equivalent to Eq. (9.65).

The sum of expressions like Eq. (10.11) for all triangles in the solution domain is the matrix system $[SS]\{\psi\}=\{0\}$, with K equations for the K unknown nodal potentials.

10.3. Application to 2D Eddy Current Problems

The computation of the matrix terms up to this point was relatively simple, primarily because the integration was independent of coordinates. In the following paragraphs we present the formulation of equations involving eddy currents, whose numerical formulation is much more complex. The additional complexity is due to the integration, which depends on the system of coordinates used. More details on this topic may be found in section 8.6.2.

10.3.1. *First-Order Element in Local Coordinates*

Instead of using the global system of coordinates (i.e., Oxy), it is possible to define the element in a "local" or "reference system" of coordinates. This simplifies the algebraic operations involved in integration. The triangle or "local element" is defined in the u,v plane instead of the "real" x,y plane as shown in Figures 10.3 and 10.4. The relationship between these two triangles is defined by a set of "geometric transformation functions" or simply "mapping functions" $N(u,v)$, which, for first order triangular elements are:

$$N_1(u,v) = 1 - u - v \qquad (10.12a)$$
$$N_2(u,v) = u \qquad (10.12b)$$
$$N_3(u,v) = v \qquad (10.12c)$$

Figure 10.3. A finite element defined in a local system of coordinates.

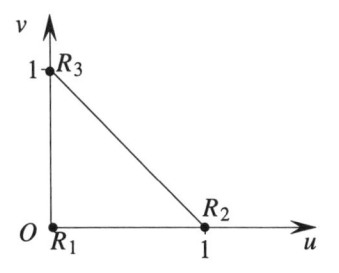

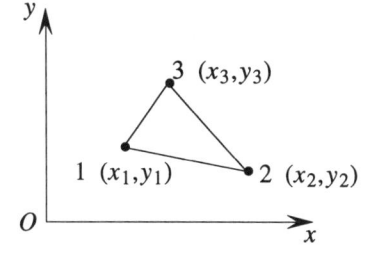

Figure 10.4. The same element after mapping to the global system of coordinates.

406   10. Galerkin's Residual Method: Applications to Dynamic Fields

With these functions, the abscissae x are written as

$$x = \begin{bmatrix} N_1 & N_2 & N_3 \end{bmatrix} \begin{bmatrix} x_1 \\ x_2 \\ x_3 \end{bmatrix} = N_1(u,v)x_1 + N_2(u,v)x_2 + N_3(u,v)x_3 \qquad (10.13)$$

With the functions in Eq. (10.12), this becomes

$$x = \begin{bmatrix} 1-u-v & u & v \end{bmatrix} \begin{bmatrix} x_1 \\ x_2 \\ x_3 \end{bmatrix} \qquad (10.14)$$

For example, at vertex R_1 of the local element, where $u=0$ and $v=0$, we have

$$x = \begin{bmatrix} 1 & 0 & 0 \end{bmatrix} \begin{bmatrix} x_1 \\ x_2 \\ x_3 \end{bmatrix} = x_1$$

If $u=1$, $v=0$ (vertex R_2) we get $x=x_2$ and finally, for $u=0$, $v=1$ (vertex R_3) we get $x=x_3$. Analogously, we can write for y

$$y = \begin{bmatrix} 1-u-v & u & v \end{bmatrix} \begin{bmatrix} y_1 \\ y_2 \\ y_3 \end{bmatrix} \qquad (10.15)$$

The same geometric transformation functions are used as "interpolation functions" for the vector potentials A in the triangular element:

$$A = \begin{bmatrix} 1-u-v & u & v \end{bmatrix} \begin{bmatrix} A_1 \\ A_2 \\ A_3 \end{bmatrix} \qquad (10.16)$$

With intermediate values of u and v (between 0 and 1), the points situated inside the local element, correspond to equivalent points inside the element in the global system of coordinates. As an example, the point $u=0.333$ and $v=0.333$, is mapped into the global system as

$$x = 0.333x_1 + 0.333x_2 + 0.333x_3$$
$$y = 0.333y_1 + 0.333y_2 + 0.333y_3$$
$$A = 0.333A_1 + 0.333A_2 + 0.333A_3$$

The functions N_1, N_2, and N_3, given in Eq. (10.12) describe linear distributions, for both coordinates and potentials. The potentials, in the global element, vary linearly as

$$A(x,y) = a_1 + a_2 x + a_3 y$$

Similarly, in the local element,

Application to 2D Eddy Current Problems   407

$$A(u,v) = a_1 + a_2 u + a_3 v$$

Taking into account the direct transformation between the points in the u,v and x,y planes, the expressions above can be written as

$$A(u,v) = \begin{bmatrix} 1 & u & v \end{bmatrix} \begin{bmatrix} a_1 \\ a_2 \\ a_3 \end{bmatrix}$$

The matrix $[P]$ is called the "polynomial basis" of the element

$$[P] = \begin{bmatrix} 1 & u & v \end{bmatrix}$$

Using the notation $\{a\}$ for the coefficient vector,

$$\{a\} = \begin{bmatrix} a_1 \\ a_2 \\ a_3 \end{bmatrix}$$

we get

$$A(u,v) = [P]\{a\}$$

We use the simplified notation,

$$x(u,v) = Nx \qquad (10.17)$$

$$y(u,v) = Ny \qquad (10.18)$$

$$A(u,v) = NA \qquad (10.19)$$

for the matrix products in Eqs. (10.14), (10.15), and (10.16), where by convention, N is an array corresponding to the geometric transformation functions [Eqs. (10.17) and (10.18)] and the interpolation potential function [Eq. (10.19)]. These are the same for Galerkin's method.

The integration of a function $f(x,y)$ over the global element can be performed using the local element as

$$\int_S f(x,y) dx dy = \int_{S_l} f(x(u,v), y(u,v)) \det [J_l] du dv \qquad (10.20)$$

where S_l is the surface of the local element and $\det[J_l]$ is the determinant of the Jacobian of the coordinates transformation

408   10. Galerkin's Residual Method: Applications to Dynamic Fields

$$J_l = \begin{bmatrix} \dfrac{\partial x}{\partial u} & \dfrac{\partial y}{\partial u} \\ \dfrac{\partial x}{\partial v} & \dfrac{\partial y}{\partial v} \end{bmatrix} \qquad (10.21)$$

The partial derivatives are calculated using Eqs. (10.14) and (10.15)

$$\dfrac{\partial x}{\partial u} = \begin{bmatrix} -1 & 1 & 0 \end{bmatrix} \begin{bmatrix} x_1 \\ x_2 \\ x_3 \end{bmatrix} = x_2 - x_1$$

and, analogously,

$$\dfrac{\partial x}{\partial v} = x_3 - x_1, \qquad \dfrac{\partial y}{\partial u} = y_2 - y_1, \qquad \dfrac{\partial y}{\partial v} = y_3 - y_1$$

The Jacobian is

$$J_l = \begin{bmatrix} x_2 - x_1 & y_2 - y_1 \\ x_3 - x_1 & y_3 - y_1 \end{bmatrix} \qquad (10.22)$$

The determinant of J_l is equal to twice the surface of the triangle, or equal to D, with the notation defined in Chapter 8.

At this point it is useful to calculate the gradient of N, as it is frequently used in subsequent calculations

$$\nabla N = \begin{bmatrix} \dfrac{\partial N}{\partial x} \\ \dfrac{\partial N}{\partial y} \end{bmatrix}$$

Noting that

$$\begin{bmatrix} \dfrac{\partial N}{\partial u} \\ \dfrac{\partial N}{\partial v} \end{bmatrix} = \begin{bmatrix} \dfrac{\partial x}{\partial u} & \dfrac{\partial y}{\partial u} \\ \dfrac{\partial x}{\partial v} & \dfrac{\partial y}{\partial v} \end{bmatrix} \begin{bmatrix} \dfrac{\partial N}{\partial x} \\ \dfrac{\partial N}{\partial y} \end{bmatrix}$$

and using Eq. (10.21) for the Jacobian matrix, we can write

$$\begin{bmatrix} \dfrac{\partial N}{\partial x} \\ \dfrac{\partial N}{\partial y} \end{bmatrix} = [J_l]^{-1} \begin{bmatrix} \dfrac{\partial N}{\partial u} \\ \dfrac{\partial N}{\partial v} \end{bmatrix} \qquad (10.23)$$

The inversion of the matrix in Eq. (10.22) gives $[J_l]^{-1}$ as

$$[J_l]^{-1} = \frac{1}{\det[J_l]} \begin{bmatrix} y_3 - y_1 & y_1 - y_2 \\ x_1 - x_3 & x_2 - x_1 \end{bmatrix} \qquad (10.24)$$

and

$$\begin{bmatrix} \dfrac{\partial N}{\partial u} \\ \dfrac{\partial N}{\partial v} \end{bmatrix} = \begin{bmatrix} -1 & 1 & 0 \\ -1 & 0 & 1 \end{bmatrix} \qquad (10.25)$$

Performing the product of Eq. (10.24) by Eq. (10.25), we obtain for Eq. (10.23)

$$\begin{bmatrix} \dfrac{\partial N}{\partial x} \\ \dfrac{\partial N}{\partial y} \end{bmatrix} = \frac{1}{D} \begin{bmatrix} y_2 - y_3 & y_3 - y_1 & y_1 - y_2 \\ x_3 - x_2 & x_1 - x_3 & x_2 - x_1 \end{bmatrix}$$

which, using our notation becomes

$$\begin{bmatrix} \dfrac{\partial N}{\partial x} \\ \dfrac{\partial N}{\partial y} \end{bmatrix} = \frac{1}{D} \begin{bmatrix} q_1 & q_2 & q_3 \\ r_1 & r_2 & r_3 \end{bmatrix} \qquad (10.26)$$

10.3.2. *The Vector Potential Equation Using Time Discretization*

We analyze here a two-dimensional (2D) case, for which the excitation current is time dependent and where there are conducting materials, as shown in the example of Figure 10.5. A nonconducting magnetic circuit (i.e. a laminated core) and a piece P with nonzero conductivity, allow the generation of eddy currents in the direction perpendicular to the plane of the figure.

The applied current density $\mathbf{J}_s$, is also perpendicular to the plane of the figure, and is externally applied to the coil. $\mathbf{J}_e$ is the induced current in block P. To formulate this problem we use the magnetic vector potential defined as $\mathbf{B} = \nabla \times \mathbf{A}$, where $\mathbf{A} = \hat{\mathbf{k}} A$, and $\hat{\mathbf{k}}$ is the unit vector in the Oz direction, perpendicular to plane of the figure. We also have $\mathbf{J}_s = \hat{\mathbf{k}} J_s$ and $\mathbf{J}_e = \hat{\mathbf{k}} J_e$.

With the equation $\nabla \times \mathbf{H} = \mathbf{J}_t$, where $\mathbf{J}_t = \mathbf{J}_s + \mathbf{J}_e$ is the total current density and $v = 1/\mu$, we have

$$\nabla \times v \nabla \times \mathbf{A} = \mathbf{J}_s + \mathbf{J}_e \qquad (10.27)$$

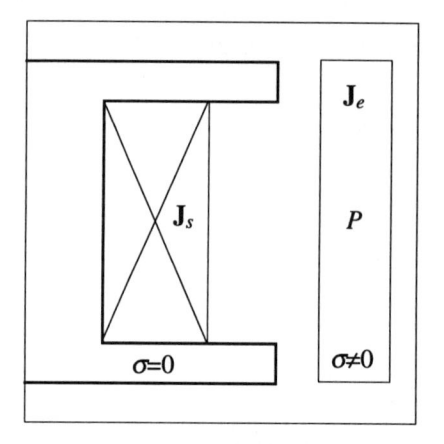

410   10. Galerkin's Residual Method: Applications to Dynamic Fields

Figure 10.5. A magnetic circuit made of a nonconducting part and a conducting part. Eddy currents are generated in the conducting part of the circuit.

Note that $\mathbf{J}_e = \sigma \mathbf{E}$, where $\mathbf{E}$ is the induced electric field intensity in the piece P and σ is its electric conductivity. With these we have

$$\nabla \times \mathbf{E} = -\frac{\partial}{\partial t}(\nabla \times \mathbf{A})$$

or

$$\nabla \times (\mathbf{E} + \frac{\partial \mathbf{A}}{\partial t}) = 0$$

This is,

$$\mathbf{E} + \frac{\partial \mathbf{A}}{\partial t} = \nabla \psi$$

Considering that $\mathbf{E}$ is generated only by the time variation of $\mathbf{B}$, we get $\nabla \psi = 0$, and

$$\mathbf{E} = -\frac{\partial \mathbf{A}}{\partial t} \quad \text{and} \quad \mathbf{J}_e = -\sigma \frac{\partial \mathbf{A}}{\partial t}$$

where all variables are in the Oz direction.

With these, Eq. (10.27) can be written as

$$\nabla \times \nu (\nabla \times \mathbf{A}) + \sigma \frac{\partial \mathbf{A}}{\partial t} - \mathbf{J}_s = 0 \qquad (10.28)$$

Application to 2D Eddy Current Problems   411

According to section 9.2.4, in a 2D case, the first term of Eq. (10.28) can be written as

$$-\frac{\partial}{\partial x} v \frac{\partial A}{\partial x} - \frac{\partial}{\partial y} v \frac{\partial A}{\partial y}$$

where A is the component of **A** in the Oz direction and is, therefore, a scalar. This expression is equal to $\nabla \cdot (v \nabla A)$

$$-\frac{\partial}{\partial x} v \frac{\partial A}{\partial x} - \frac{\partial}{\partial y} v \frac{\partial A}{\partial y} = \nabla \cdot (v \nabla A)$$

Thus, Eq. (10.28), with all vectors in the Oz direction, can be written as

$$\nabla \cdot (v \nabla A) - \sigma \frac{\partial A}{\partial t} + J_s = 0 \qquad (10.29)$$

The time derivative can be expressed as

$$\frac{\partial A}{\partial t} = \frac{A^i - A^{i-1}}{\Delta t}$$

where Δt is a time step and $A^i - A^{i-1}$ is the change in A between time steps $i-1$ and i. Equation (10.29) can now be written as

$$\nabla \cdot (v \nabla A) - \sigma \frac{A^i}{\Delta t} + \sigma \frac{A^{i-1}}{\Delta t} + J_s = 0 \qquad (10.30)$$

Now we apply the Galerkin procedure to Eq. (10.30), where the weighting functions are the functions N given in the previous section. For simplicity, the formulation is performed term by term:

a. The first term in Eq. (10.30). This is

$$\int_S N^t \nabla \cdot (v \nabla A) ds \qquad (10.31)$$

This term, after application of the divergence theorem, results in two integrals in a manner similar to Eq. (10.3).
- The first integral, accounting for boundary conditions, leads to

$$v \nabla A \cdot \hat{\mathbf{n}} = 0 \qquad (10.31a)$$

10. Galerkin's Residual Method: Applications to Dynamic Fields

or, explicitly,

$$\left(\hat{\mathbf{i}}\frac{\partial A}{\partial x} + \hat{\mathbf{j}}\frac{\partial A}{\partial y}\right) \cdot (\hat{\mathbf{i}} n_x + \hat{\mathbf{j}} n_y) = 0$$

With $\mathbf{B} = \nabla \times \mathbf{A}$, we get

$$\left(-\hat{\mathbf{i}} B_y + \hat{\mathbf{j}} B_x\right) \cdot (\hat{\mathbf{i}} n_x + \hat{\mathbf{j}} n_y) = 0$$

or

$$\mathbf{B} \times \hat{\mathbf{n}} = 0$$

meaning that, for the Neumann condition, $\mathbf{B}$ and $\hat{\mathbf{n}}$ are co-linear, as was the case with the variational method.

- The second integral provides the terms of the elemental matrix:

$$-\int_{S_i} \nabla N^t (v \nabla A) ds$$

where S_i indicates that the integral has been applied to element i of the mesh. Noting that A can be expressed as $A(u,v) = NA$ [from Eq. (10.19)], we get

$$-\int_{S_i} \nabla N^t v \nabla N A \, dx dy \qquad (10.31b)$$

Using the local element, we obtain

$$-\int_0^1 \int_0^{1-v} \nabla N^t v \nabla N A \, det\,[J_l] du dv$$

which, using Eq. (10.26) is

$$-\int_0^1 \int_0^{1-v} \begin{bmatrix} q_1 & r_1 \\ q_2 & r_2 \\ q_3 & r_3 \end{bmatrix} \frac{v}{D^2} \begin{bmatrix} q_1 & q_2 & q_3 \\ r_1 & r_2 & r_3 \end{bmatrix} \begin{bmatrix} A_1 \\ A_2 \\ A_3 \end{bmatrix} det[J_l] du dv$$

As there is no dependence on u and v and $det\,[J_l] = D$, we get

$$-\begin{bmatrix} q_1 \ r_1 \\ q_2 \ r_2 \\ q_3 \ r_3 \end{bmatrix} \frac{v}{D} \begin{bmatrix} q_1 & q_2 & q_3 \\ r_1 & r_2 & r_3 \end{bmatrix} \begin{bmatrix} A_1 \\ A_2 \\ A_3 \end{bmatrix} \int_0^1 \int_0^{1-v} du\, dv$$

The integration is performed on the surface of the local element and is equal to 1/2. For Eq. (10.31b), the result is

$$-\frac{v}{2D} \begin{bmatrix} q_1 q_1 + r_1 r_1 & q_1 q_2 + r_1 r_2 & q_1 q_3 + r_1 r_3 \\ & q_2 q_2 + r_2 r_2 & q_2 q_3 + r_2 r_3 \\ \text{symmetric} & & q_3 q_3 + r_3 r_3 \end{bmatrix} \begin{bmatrix} A_1 \\ A_2 \\ A_3 \end{bmatrix} \quad (10.32)$$

This is equivalent to the expression obtained with the variational method. For evaluation of this term using quadratic triangular elements see Eq. (8.137).

b. The second term in Eq. (10.30). The term $-\sigma A^i/\Delta t$ in Eq. (10.30), after applying Galerkin's method, becomes

$$\int_{S_i} N^t \frac{\sigma A^i}{\Delta t} dx\, dy = -\frac{\sigma}{\Delta t} \int_{S_i} N^t A^i dx\, dy \quad (10.33)$$

or, in the local element,

$$-\frac{\sigma}{\Delta t} \int_0^1 \int_0^{1-v} N^t N A^i \det[J_l] du\, dv$$

or

$$-\frac{\sigma D}{\Delta t} \int_0^1 \int_0^{1-v} \begin{bmatrix} 1-u-v \\ u \\ v \end{bmatrix} \begin{bmatrix} 1-u-v & u & v \end{bmatrix} du\, dv \begin{bmatrix} A_1 \\ A_2 \\ A_3 \end{bmatrix}^i$$

Performing the matrix product, we get

$$-\frac{\sigma D}{\Delta t} \int_0^1 \int_0^{1-v} \begin{bmatrix} (1-u-v)^2 & (1-u-v)u & (1-u-v)v \\ u(1-u-v) & u^2 & uv \\ v(1-u-v) & uv & v^2 \end{bmatrix} du\, dv \begin{bmatrix} A_1 \\ A_2 \\ A_3 \end{bmatrix}^i$$

The integration, performed term by term, gives

$$-\frac{\sigma D}{12\Delta t}\begin{bmatrix} 1 & 0.5 & 0.5 \\ 0.5 & 1 & 0.5 \\ 0.5 & 0.5 & 1 \end{bmatrix}\begin{bmatrix} A_1 \\ A_2 \\ A_3 \end{bmatrix}^i \qquad (10.34)$$

For evaluation using second order elements, see Eq. (8.139).

c. The third term in Eq. (10.30). This is similar to the second term, but with the known potential A^{i-1}, of the previous step included. Unlike the second term, this term is a vector. The matrix product in Eq.(10.34), is modified as

$$-\frac{\sigma D}{12\Delta t}\begin{bmatrix} A_1 + 0.5A_2 + 0.5A_3 \\ 0.5A_1 + A_2 + 0.5A_3 \\ 0.5A_1 + 0.5A_2 + A_3 \end{bmatrix}^{i-1} \qquad (10.35)$$

d. The fourth term in Eq. (10.30). This term contains the external current density J_s, and, after applying Galerkin's method, is

$$\int_{S_i} N^t J_s dx dy = \int_0^1 \int_0^{1-v} N^t J_s det[J_l] du dv$$

or

$$J_s D \int_0^1 \int_0^{1-v} \begin{bmatrix} 1-u-v \\ u \\ v \end{bmatrix} du dv$$

which, upon integrations, gives

$$\frac{J_s D}{6}\begin{bmatrix} 1 \\ 1 \\ 1 \end{bmatrix} \qquad (10.36)$$

If second order elements are used, the expression in Eq. (8.143) is obtained.

After evaluation of the terms in Eqs. (10.32), (10.34), (10.35), and (10.36), the following matrix system is established:

$$[SS]\{A\} = \{Q\} \qquad (10.37)$$

where $\{A\}$ is the vector of unknown vector potentials. $\{Q\}$ is the right-hand side vector, containing the source terms resulting from applied currents J_s and induced currents of the previous step [Eq. (10.35)]. The terms that depend on the unknown vector A_i, must be assembled in matrix $[SS]$ as indicated in Chapter 9.

In practice, the procedure for calculation evaluates the terms in Eqs. (10.32), (10.34), (10.35), and (10.36) for each element in turn. If for example $J_S=0$ and $\sigma=0$, only the terms in Eq. (10.32) will be nonzero. For a general case, the total contribution of an element in $[SS]$ is

$$-\begin{bmatrix} S_{1,1} & S_{1,2} & S_{1,3} \\ S_{2,1} & S_{2,2} & S_{2,3} \\ S_{3,1} & S_{3,2} & S_{3,3} \end{bmatrix} - \frac{\sigma D}{12\Delta t}\begin{bmatrix} 1 & 0.5 & 0.5 \\ 0.5 & 1 & 0.5 \\ 0.5 & 0.5 & 1 \end{bmatrix} \quad (10.38)$$

This matrix is multiplied by $\{A\}^i$.

The source terms, placed in vector $\{Q\}$, are

$$\frac{\sigma D}{12\Delta t}\begin{bmatrix} A_1 + 0.5A_2 + 0.5A_3 \\ 0.5A_1 + A_2 + 0.5A_3 \\ 0.5A_1 + 0.5A_2 + A_3 \end{bmatrix}^{i-1} + \frac{J_s D}{6}\begin{bmatrix} 1 \\ 1 \\ 1 \end{bmatrix} \quad (10.39)$$

Generally, in this type of problems, the external current source is time dependent. To establish a calculation procedure, we can assume an initial solution $\{A\}^0=\{0\}$ for the first time step ($i=0$). The matrix system is assembled and the solution $\{A\}^1$ is obtained. With this result, the next step starts by calculating the new matrix system, where in the source expression in Eq. (10.39), both vectors are calculated, modifying J_s to its value at the current time step. Continuing this process establishes the calculation procedure.

Note that, for any step, it is possible to consider nonlinearity as well, by creating an iterative process for each time step. It is possible to apply the Newton-Raphson method as presented in the previous chapter. The Jacobian matrix is calculated as indicated above, noting that the derivation related to A of the first term [Eq. (10.32), with dependence on μ] is given in Eq. (9.106). The derivation of the second term [Eq. (10.34), which contains only constants] gives

$\sigma D/(12\Delta t)$ for diagonal terms

$0.5\sigma D/(12\Delta t)$ for off-diagonal terms

Two important aspects should be noted when using this method:

a. The first aspect concerns the feeding of the device, that is: the method by which the current J_s is applied to the device. Generally, electrical devices are fed by applying a voltage across their inputs. The current J_s depends on the impedance of the structure. A more correct method consists of considering the coupled calculation of the magnetic circuit and the external electrical circuit. This calculation provides J_s and A as results. The formulation above is well adapted to situations in which the impedance of the external source is very high compared to the impedance of the device. Thus we can assume that the applied current, at steady state, has a well-defined shape, corresponding to the shape of the voltage source. In effect, we assume feeding by a current source.

b. The second aspect concerns the partial derivative of A with respect to time, which is approximated as $(A^i - A^{i-1})/\Delta t$. The accuracy of A is acceptable for small values of Δt. A better approach is to use the "θ-algorithm." In this method the calculation is done at time "$t + \theta \Delta t$," and the variable A is given by

$$A(t + \theta \Delta t) = \theta A(t + \Delta t) + (1 - \theta)A(t)$$

and its derivative is approximated by

$$\frac{\partial A}{\partial t}(t + \theta \Delta t) = \frac{A(t + \Delta t) - A(t)}{\Delta t}$$

Using this technique, it is necessary to search for a value for θ (which is unknown a priori and depends on the type of problem being solved) which provides best accuracy. For $\theta = 0.5$, this is the Crank-Nicholson method, normally a good first approximation for θ.

10.3.3. *The Complex Vector Potential Equation*

The purpose of this formulation, as was that of the previous section, is to solve Eq. (10.29). However, if the excitation is sinusoidal and the materials are linear, we can use the complex vector potential, which we denote as $\widetilde{\mathbf{A}}$. Denoting $J_s(t)$ the cosine current source with frequency ω, gives

$$J_s(t) = J_s \cos(\omega t)$$

or, using the complex notation $j = \sqrt{-1}$

$$J_s(t) = Re(J_s e^{j\omega t})$$

The system's response to this excitation is also at steady state, sinusoidal and out of phase, therefore,

$$a(t) = Re(\widetilde{A} e^{j\omega t})$$

where: $\tilde{A} = Ae^{j\alpha}$ is the solution to Eq. (10.29), with α the angle between $a(t)$ and $J_s(t)$. Equation (10.29) can be written, in this case, as

$$\nabla \cdot (\nu \nabla \tilde{A}\, e^{j\omega t}) - \sigma \frac{\partial}{\partial t}(\tilde{A}\, e^{j\omega t}) + J_s e^{j\omega t} = 0 \qquad (10.40)$$

or

$$\nabla \cdot (\nu \nabla \tilde{A}) - \sigma j\omega \tilde{A} + J_s = 0 \qquad (10.41)$$

We have assumed here that the phase angle of the current source is zero. This is often a practical consideration and is commonly used.

When applying Galerkin's method, the first and third terms in Eq. (10.41) we get the same results as in Eqs. (10.32) and (10.36), respectively. However, the second term must be re-evaluated. The elemental matrix for the second term is given by

$$-\sigma j\omega \int_{S_i} N^t \tilde{A}\, dxdy \qquad (10.42)$$

The integrand of this expression is identical to that of Eq. (10.33), and using calculations similar to those used for Eq. (10.33), we get

$$-\frac{\sigma j\omega D}{12} \begin{bmatrix} 1 & 0.5 & 0.5 \\ 0.5 & 1 & 0.5 \\ 0.5 & 0.5 & 1 \end{bmatrix} \begin{bmatrix} \tilde{A}_1 \\ \tilde{A}_2 \\ \tilde{A}_3 \end{bmatrix} \qquad (10.43)$$

This elemental matrix has imaginary terms and must be added to the matrix in Eq. (10.32) to form the global matrix $[SS]$. The solution of the system

$$[SS]\{\tilde{A}\} = \{\tilde{Q}\}$$

results in the complex vector $\tilde{A}$. This type of formulation has an important advantage: solving a single matrix system we obtain the real and imaginary parts of $\tilde{A}$ and, therefore, its magnitude and phase angle in relation to J_s. However, we can not include nonlinearity. If ferromagnetic materials are present, it is necessary to know, a priori, if the excitation current is low enough to avoid nonlinear effects such as saturation in the structure under study. If these conditions are not satisfied, the time discretization formulation of the previous section must be used. While the time required for computation is longer (because an iterative calculation is performed for each step), the time discretization method is the only way to obtain correct results for nonlinear problems of this kind. The main results obtainable with the complex variables formulation are:

418   10. Galerkin's Residual Method: Applications to Dynamic Fields

- Penetration effects can be seen graphically, as shown in Figure 10.6 (this figure is identical to Figure 5.23, reproduced here for convenience).
- Impedance calculations, as seen from the exterior source, are possible.

Noting that $(R+j\omega L)I = j\omega\phi$, we obtain L and R from the expression

$$L - \frac{jR}{\omega} = \frac{\tilde{\phi}}{\tilde{I}}$$

where $\tilde{\phi}$ is equal to $\tilde{A}$ multiplied by the depth of the structure and I is equal to J_s multiplied by the cross-sectional area of the coil.

- The real part L is the inductance of the magnetic circuit. The inductance decreases as ω increases, since the magnetic flux decreases due to lower field penetration (the reluctance increases).
- The imaginary part R/ω represents the resistive term. This is a typical characteristic of eddy currents, induced in conducting media. R increases with ω.

The global losses through eddy currents in the domain are given by

$$\frac{RI^2}{2}$$

The eddy current density J_e is given by

$$J_e = Re(-\sigma j\omega \tilde{A}\, e^{j\omega t})$$

This allows the calculation of average losses by local integration

$$P_e = \frac{1}{2\pi}\int_0^{2\pi}\int_S \frac{J_e^2}{\sigma}\, ds\, d(\omega t)$$

Averaging over a cycle gives

$$\overline{P_e} = \frac{\omega^2}{2}\int_S \sigma |\tilde{A}|^2 ds$$

This must be equivalent to the quantity $RI^2/2$ shown above. If the integration is performed only over the surface S_i of an element, we can obtain the local heating in this element.

Application to 2D Eddy Current Problems 419

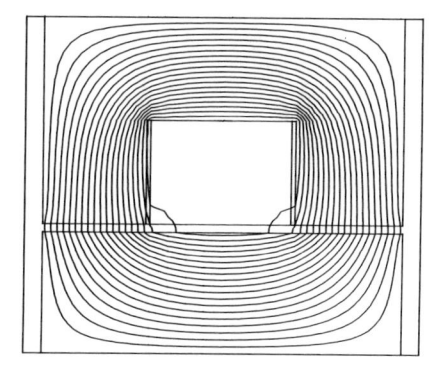

Figure 10.6a Field distribution at 0 *Hz*. The field penetrates freely through the conductor.

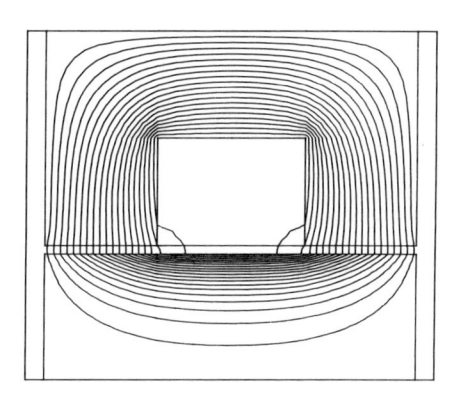

Figure 10.6b. Field distribution at 50 *Hz*.

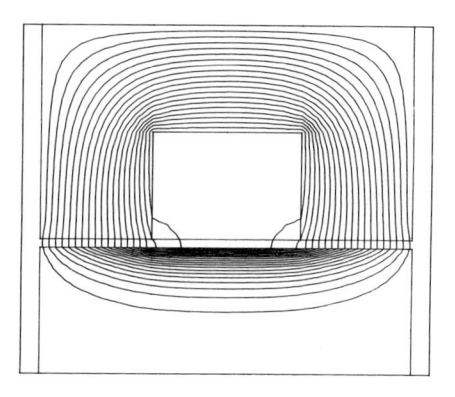

Figure 10.6c. Field distribution at 100 *Hz*.

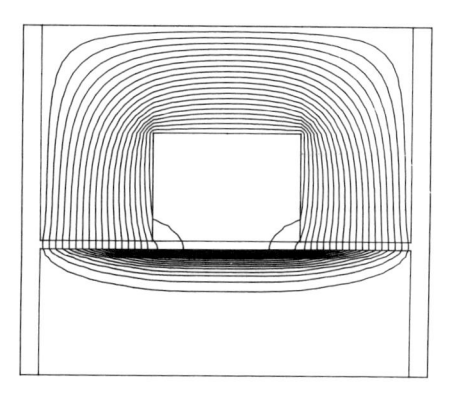

Figure 10.6d. Field distribution at 250 *Hz*.

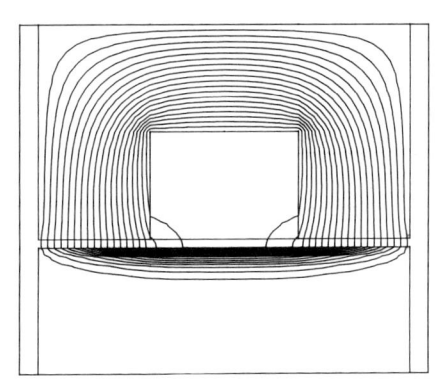

Figure 10.6e. Field distribution at 500 *Hz*.

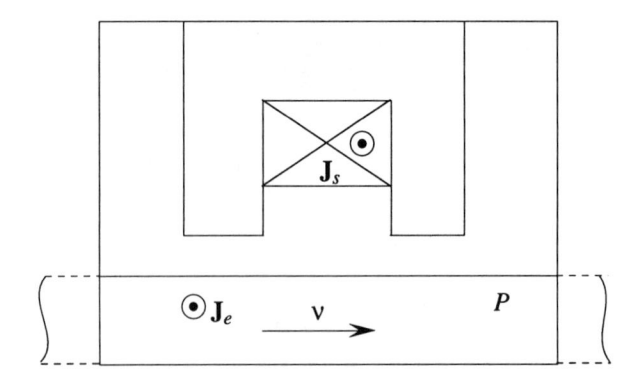

420   10. Galerkin's Residual Method: Applications to Dynamic Fields

Figure 10.7. A magnetic structure with a moving piece. Eddy currents are induced in the moving piece due to velocity.

10.3.4. Structures with Moving Parts

Consider the case of the magnetic brake (given as an example in Chapter 6), where a piece P, placed in a magnetic field moves at a velocity $\mathbf{v}$, as shown in Figure 10.7. This piece is uniform in shape and only the part under the magnetic field influence, is shown. Eddy currents depend on the velocity of the piece P through the expression $\mathbf{E}=\mathbf{v}\times\mathbf{B}$. As $\mathbf{v}$ and $\mathbf{B}$ are in the plane of the figure, $\mathbf{E}$ is perpendicular to this plane:

$$\mathbf{J}_e = \sigma(\mathbf{E} + \mathbf{v}\times\mathbf{B}) = \sigma(-\frac{\partial \mathbf{A}}{\partial t} + \mathbf{v}\times\mathbf{B})$$

where the term $\partial \mathbf{A}/\partial t$ can be described either by the complex formulation (if it is linear and the excitation is sinusoidal) or through time discretization.

If the excitation current is constant, we have a static case (even if velocity $\mathbf{v}$ exists), because, for constant $\mathbf{v}$, the potential $\mathbf{A}$ is also constant in the structure. Using the components of $\mathbf{v}$:

$$\mathbf{v} = \hat{\mathbf{i}}v_x + \hat{\mathbf{j}}v_y$$

and $\mathbf{B}$:

$$\mathbf{B} = \hat{\mathbf{i}}B_x + \hat{\mathbf{j}}B_y = \hat{\mathbf{i}}\frac{\partial A}{\partial y} - \hat{\mathbf{j}}\frac{\partial A}{\partial x}$$

the product $\mathbf{v}\times\mathbf{B}$, for this two-dimensional case (the product is perpendicular to the plane of the figure, that is in the $\hat{\mathbf{k}}$ direction), is

$$|\mathbf{v}\times\mathbf{B}| = -\frac{\partial A}{\partial x}v_x - \frac{\partial A}{\partial y}v_y$$

Eq. (10.29) becomes

$$\nabla \cdot (v \nabla A) - \sigma \left(\frac{\partial A}{\partial x} v_x + \frac{\partial A}{\partial y} v_y \right) + J_s = 0 \qquad (10.44)$$

Applying Galerkin's method to the first and third terms, we obtain the expressions in Eqs. (10.32) and (10.36), respectively. The second term containing the speed v_x is calculated below. We have

$$-\int_{S_i} N^t \sigma \frac{\partial A}{\partial x} v_x dx dy$$

which can be written in the following form

$$- \sigma v_x \int_{S_i} N^t \frac{\partial N}{\partial x} A \, dx dy$$

Using the local element and the expression in Eq. (10.26), which defines $\partial N/\partial x$,

$$- \sigma v_x \int_0^1 \int_0^{1-v} \begin{bmatrix} 1-u-v \\ u \\ v \end{bmatrix} \frac{1}{D} [\, q_1 \quad q_2 \quad q_3 \,] \begin{bmatrix} A_1 \\ A_2 \\ A_3 \end{bmatrix} \det[J_l] du dv$$

$$- \sigma v_x \int_0^1 \int_0^{1-v} \begin{bmatrix} (1-u-v)q_1 & (1-u-v)q_2 & (1-u-v)q_3 \\ uq_1 & uq_2 & uq_3 \\ vq_1 & vq_2 & vq_3 \end{bmatrix} du dv \begin{bmatrix} A_1 \\ A_2 \\ A_3 \end{bmatrix}$$

which, after integration, becomes

$$-\frac{\sigma v_x}{6} \begin{bmatrix} q_1 & q_2 & q_3 \\ q_1 & q_2 & q_3 \\ q_1 & q_2 & q_3 \end{bmatrix} \begin{bmatrix} A_1 \\ A_2 \\ A_3 \end{bmatrix} \qquad (10.45)$$

Analogously, the term with velocity component v_y is

$$-\frac{\sigma v_y}{6} \begin{bmatrix} r_1 & r_2 & r_3 \\ r_1 & r_2 & r_3 \\ r_1 & r_2 & r_3 \end{bmatrix} \begin{bmatrix} A_1 \\ A_2 \\ A_3 \end{bmatrix} \qquad (10.46)$$

The sum of Eqs. (10.45) and (10.46) provides the velocity terms, and these terms must be added to the global matrix $[SS]$. For equivalent expressions for second-order elements, see Eq. (8.141).

One important remark is that, contrary to other formulations we presented previously, the matrices in Eqs. (10.45) and (10.46) are not symmetric. Generally, computation programs take into account the symmetry of the global matrix $[SS]$, storing in memory only half of this matrix. However, in this case the whole matrix must be stored. In practice calculation with velocity terms is normally done in a separate program.

If the excitation current is time dependent, the potential A is time dependent as well. In the term for J_e of the basic equation we now have "$-\sigma \partial A/\partial t$," which, depending on the formulation chosen, is either Eq. (10.34) or Eq. (10.43).

10.3.5. The Axi-Symmetric Formulation

First, we establish the equation $\nabla \times \mathbf{H} = \mathbf{J}_t$ in cylindrical coordinates.

$$\nabla \times \nu \mathbf{B} = \mathbf{J}_t \qquad (10.47)$$

In this system $\mathbf{B} = \nabla \times \mathbf{A}$ has two components

$$B_r = -\frac{\partial A}{\partial z} \qquad \text{and} \qquad B_z = \frac{1}{r}\frac{\partial}{\partial r}(rA)$$

where $\mathbf{A}$ has only a component in the ϕ direction. This is also the direction of $\mathbf{J}_t$. With these conditions, $\nabla \times \nu \mathbf{B} = \mathbf{J}_t$ becomes

$$\frac{\partial}{\partial z}(\nu B_r) - \frac{\partial}{\partial r}(\nu B_z) = J_t$$

$$-\frac{\partial}{\partial z}\nu\left(\frac{\partial A}{\partial z}\right) - \frac{\partial}{\partial r}\left[\nu \frac{1}{r}\frac{\partial}{\partial r}(rA)\right] = J_t \qquad (10.48)$$

and, finally,

$$\frac{\partial}{\partial z}\nu\frac{\partial A}{\partial z} + \frac{\partial}{\partial r}\nu\frac{\partial A}{\partial r} + \frac{\partial}{\partial r}\left(\frac{\nu A}{r}\right) = -J_t \qquad (10.49)$$

The first two terms in Eq. (10.49) are similar to those of Eq. (10.47) for the Cartesian coordinates, if the substitution $r=y$ and $z=x$ is made. However, the term $\partial/\partial r\,(\nu A/r)$ creates an asymmetry in the elemental matrix, when Galerkin's method is applied, because this term depends only on the coordinate r. With the

Application to 2D Eddy Current Problems   423

variational method, this problem does not exist, because its solution is obtained directly from the functionals, as indicated in Eq. (9.77).

To eliminate this inconvenience we introduce a new variable A' related to A:

$$A' = rA \qquad (10.50)$$

Equation (10.48) now becomes

$$\frac{\partial}{\partial z}\frac{v}{r}\frac{\partial A'}{\partial z} + \frac{\partial}{\partial r}\frac{v}{r}\frac{\partial A'}{\partial r} + J_t = 0 \qquad (10.51)$$

From here on we operate as with rectangular coordinates, taking r and z as y and x, respectively. Applying the Galerkin method to Eq. (10.51), we have

$$\int_S N^t \left[\frac{\partial}{\partial r}\frac{v}{r}\frac{\partial A'}{\partial r} + \frac{\partial}{\partial z}\frac{v}{r}\frac{\partial A'}{\partial z} \right] drdz + \int_S N^t J_t drdz = 0 \qquad (10.52)$$

The first term of this equation after integration by parts (similar to the integration performed in section 10.3), is

$$\int_S N^t \left[\frac{\partial}{\partial r}\frac{v}{r}\frac{\partial A'}{\partial r} \right] drdz = \int_S \frac{\partial}{\partial r}\left[N^t \frac{v}{r}\frac{\partial A'}{\partial r} \right] drdz - \int_S \frac{v}{r}\frac{\partial N^t}{\partial r}\frac{\partial A'}{\partial r} drdz$$

Analogously, we obtain for the second term of the first integrand of Eq. (10.52):

$$\int_S \frac{\partial}{\partial z}\left[N^t \frac{v}{r}\frac{\partial A'}{\partial z} \right] drdz - \int_S \frac{v}{r}\frac{\partial N^t}{\partial z}\frac{\partial A'}{\partial z} drdz$$

Adding these two terms, we get

$$\int_S \left[\frac{\partial}{\partial r}\left(N^t \frac{v}{r}\frac{\partial A'}{\partial r} \right) + \frac{\partial}{\partial z}\left(N^t \frac{v}{r}\frac{\partial A'}{\partial z} \right) \right] drdz -$$

$$\int_S \left[\frac{v}{r}\frac{\partial N^t}{\partial r}\frac{\partial A'}{\partial r} + \frac{v}{r}\frac{\partial N^t}{\partial z}\frac{\partial A'}{\partial z} \right] drdz \qquad (10.53)$$

Recalling that

$$H_r = -v\frac{\partial A}{\partial z} = -\frac{v}{r}\frac{\partial A'}{\partial z} \quad \text{and} \quad H_z = \frac{v}{r}\frac{\partial A'}{\partial r}$$

and defining the ∇ operator in the z-r plane, as for the x-y plane,

424   10. Galerkin's Residual Method: Applications to Dynamic Fields

$$\nabla = \hat{\mathbf{i}}\frac{\partial}{\partial z} + \hat{\mathbf{j}}\frac{\partial}{\partial r}$$

the first integral in Eq. (10.53) can be written as

$$\int_S \nabla \cdot (\hat{\mathbf{j}} N^t H_z - \hat{\mathbf{i}} N^t H_r) dr dz$$

Applying the divergence theorem, we get

$$\oint_L N^t (\hat{\mathbf{j}} H_z - \hat{\mathbf{i}} H_r) \cdot \hat{\mathbf{n}} dl$$

where $\hat{\mathbf{n}}$, the unit vector normal to line L is $\hat{\mathbf{n}} = \hat{\mathbf{i}} n_z + \hat{\mathbf{j}} n_r$. This expression is introduced in the equation above:

$$\oint_L N^t (-H_r n_z + H_z n_r) dl$$

Following the procedure used to obtain Eq. (10.31a), we obtain the Neumann condition, since $\mathbf{H} \times \hat{\mathbf{n}} = 0$.

The second integral in Eq. (10.53) provides the elemental matrix. For an element i the relation becomes

$$-\int_{S_i} \left[\frac{v}{r}\frac{\partial N^t}{\partial r}\frac{\partial A'}{\partial r} + \frac{v}{r}\frac{\partial N^t}{\partial z}\frac{\partial A'}{\partial z} \right] dr dz$$

This is equivalent to Eq. (10.31b). In short form notation, we can write

$$-\frac{1}{r_0}\int_{S_i} \nabla N^t v \nabla A' dr dz \qquad (10.54)$$

where we have replaced r by r_0, the centroid of element i. After some algebraic operations, we obtain, as for Eq. (10.31b):

$$-\frac{v}{2Dr_0}\begin{bmatrix} q_1q_1 + r_1r_1 & q_1q_2 + r_1r_2 & q_1q_3 + r_1r_3 \\ & q_2q_2 + r_2r_2 & q_2q_3 + r_2r_3 \\ \text{symmetric} & & q_3q_3 + r_3r_3 \end{bmatrix}\begin{bmatrix} A_1' \\ A_2' \\ A_3' \end{bmatrix} \qquad (10.55)$$

The application of Galerkin's method to the source term J_s yields

$$\int_{S_i} N^I J_s dr dz \qquad (10.56)$$

which takes different forms, depending on the problem being treated; this expression takes any of the forms in Eqs. (10.34), (10.35), (10.36), or (10.43).

Note that, if both Eqs. (10.55) and (10.56) are multiplied by 2π, the matrix system is not changed (since [SS] and {Q} are multiplied by the same factor) and we obtain a formulation very similar to that given by the variational method.

10.3.6. A Modified Complex Vector Potential Formulation for Wave Propagation

In section 10.3.3., we discussed the eddy current equation in the low-frequency limit. We now proceed to show how the eddy current equation can be modified to include high-frequency aspects of field calculation, including wave propagation. We start with Eq. (7.26)

$$\frac{1}{\mu}\nabla^2 \tilde{A} + \tilde{J} + \omega^2 \varepsilon \tilde{A} = 0 \qquad (10.57)$$

The current density $\tilde{J}$ can be viewed as having two parts:

$$\tilde{J} = \tilde{J}_s + \tilde{J}_e = \tilde{J}_s + \sigma \tilde{E}$$

where the first part is the applied current density while the second is the induced or eddy current density. This gives

$$\frac{1}{\mu}\nabla^2 \tilde{A} + \tilde{J}_s + \sigma \tilde{E} + \omega^2 \varepsilon \tilde{A} = 0 \qquad (10.58)$$

Since we are interested in an equation in terms of the magnetic vector, potential, the electric field intensity [from Eq. (7.16)] is

$$\mathbf{E} = -\frac{\partial \mathbf{A}}{\partial t} - \nabla V \qquad (10.59)$$

or, in time-harmonic form

$$\tilde{E} = -j\omega \tilde{A} - \nabla \tilde{V} \qquad (10.60)$$

where $\tilde{V}$ is the electric scalar potential. We note, as was done in Chapter 9 that the gradient of the scalar potential is needed in general so that the electric field does not vanish at zero frequency. Substitution of this in Eq. (10.58) gives

$$\frac{1}{\mu}\nabla^2\widetilde{A} + \widetilde{J}_s - \sigma j\omega\widetilde{A} - \sigma\widetilde{\nabla V} + \omega^2\varepsilon\widetilde{A} = 0$$

Comparing this equation with Eq. (10.41) shows that the first three terms are identical. The fourth term is due to the electric scalar potential, while the last term is due to the displacement current density. In most cases, we can neglect the term due to the scalar potential, and we will do so here. With this assumption

$$\frac{1}{\mu}\nabla^2\widetilde{A} + \widetilde{J}_s - \sigma j\omega\widetilde{A} + \omega^2\varepsilon\widetilde{A} = 0 \qquad (10.61)$$

The Galerkin implementation of this equation is now merely a matter of adding the part of the matrix dealing with the term $\omega^2\varepsilon\widetilde{A}$. For the first three terms we simply use Eqs. (10.32), (10.36), and (10.43) noting that in two dimensions the vector $\widetilde{A}$ has only a z component. Note also that the last term is identical in form to the third term if we interchange $-j\omega\sigma$ with $\omega^2\varepsilon$. Thus, the immediate result of Eq. (10.43) is the following elemental matrix for the displacement currents

$$\frac{\omega^2\varepsilon D}{12}\begin{bmatrix} 1 & 0.5 & 0.5 \\ 0.5 & 1 & 0.5 \\ 0.5 & 0.5 & 1 \end{bmatrix}\begin{bmatrix} \widetilde{A}_1 \\ \widetilde{A}_2 \\ \widetilde{A}_3 \end{bmatrix} \qquad (10.62)$$

Adding this to the global matrix obtained in section 10.3.3. produces the new, modified eddy current system which includes displacement currents. The solution to this system is identical to that for section 10.3.3.

Extension of this type of formulation to 3D geometries is similar to that for any 3D eddy current formulation but, in general, the scalar potential we neglected above must be reintroduced. Similarly, the axi-symmetric formulation of section 10.3.5. can be extended for this equation.

Note that this formulation is deterministic in that the frequency of the fields must be known a priori. This means that, for example, the cutoff frequencies in microwave cavities cannot be calculated but the fields can be evaluated once the frequency is known. The basic idea is to vary the frequency in a range and calculate the propagation constant after each step. If the propagation constant is real, the device (e.g., a waveguide) is below cutoff, while an imaginary propagation constant means it is above cutoff. However, the fields in any domain will be a correct solution regardless of the fact that waves may or may not propagate. Similarly, in cavity resonators, frequency can be scanned until resonance is obtained (as indicated by a dramatic increase in the stored energy in the cavity). This method can then be used to calculate the resonant frequencies of complex cavities or the shift in resonant frequencies of loaded cavities.

10.3.7. *Formulation of Helmholtz's Equation*

In Chapter 7, we presented the theoretical background for fields in waveguides. We now present a Galerkin formulation for *TE* and *TM* modes in a rectangular waveguide. As is normally done, we assume propagation in the *z* direction, so that the waveguide cross section is in the *x-y* plane. Further, we assume source free conditions so that the fields are governed by a Helmholtz equation. We present this formulation here because of the similarity in formulation with the eddy current equation, although the phenomena described by the two equations are very different.

Under the conditions above, the equation to solve is (from section 7.6.1)

$$\frac{\partial^2 E_z}{\partial x^2} + \frac{\partial^2 E_z}{\partial y^2} + (\gamma^2 + k^2) E_z = 0 \qquad (10.63)$$

for *TM* modes and

$$\frac{\partial^2 H_z}{\partial x^2} + \frac{\partial^2 H_z}{\partial y^2} + (\gamma^2 + k^2) H_z = 0 \qquad (10.64)$$

for *TE* modes.

In these equations, E_z and H_z are phasors. The phasor notation used in the previous section was dropped since here we only use H and E in this context.

In the first case, the boundary conditions are specified by $E_z=0$ (Dirichlet type) on the perfectly conducting walls, while in the second, the conditions $\partial H_z/\partial x=0$, $\partial H_z/\partial y=0$ (Neumann type) are used. These boundary conditions are either directly applied to the final coefficient matrix (Dirichlet type boundary conditions) or are implicit in the formulation (Neumann type boundary conditions) and need not be applied. We will assume the waveguide walls to be perfectly conducting but the interior of the waveguide can be lossy, as specified by the complex permittivity in the waveguide.

We start with Eq. (10.63) and rewrite it as

$$\frac{\partial^2 E_z}{\partial x^2} + \frac{\partial^2 E_z}{\partial y^2} + k_c^2 E_z = 0, \qquad k_c^2 = \gamma^2 + k^2 \qquad (10.65)$$

At cutoff, $\gamma^2=0$ and, therefore,

$$k_c^2 = \omega^2 \mu \varepsilon$$

For convenience of representation and interpretation of results we rewrite Eq. (10.65) as

$$\frac{1}{\mu_r}\left(\frac{\partial^2 E_z}{\partial x^2} + \frac{\partial^2 E_z}{\partial y^2}\right) + k_0^2 \varepsilon_r E_z = 0 \qquad (10.66)$$

where ε_r and μ_r are the relative permeability and permittivity, respectively. Now k_0 is the free space wavenumber and once calculated can be directly interpreted as the cutoff frequency of a particular mode of propagation. We further note that ε_r is in general the complex permittivity, depending on the material.

For the TE modes, Eq. (10.64) now becomes

$$\frac{1}{\varepsilon_r}\left(\frac{\partial^2 H_z}{\partial x^2} + \frac{\partial^2 H_z}{\partial y^2}\right) + k_0^2 \mu_r H_z = 0 \qquad (10.67)$$

To apply the Galerkin method, we first define the approximation over the finite element. Since in either of the equations above, the z component of the electric or magnetic field intensity can be viewed as a scalar, the approximation is written as in Eq. (10.2)

$$E_z(x,y) = \sum_{k=1}^{K} E_k \phi_k$$

Following an identical process for the first two terms as in section 10.3.3, we obtain an elemental stiffness matrix as in Eq. (10.32), where the reluctivity v is now replaced by $v_r = 1/\mu_r$. In effect we can write the elemental matrix for the Laplacian as

$$-\frac{v_r}{2D}\begin{bmatrix} q_1 q_1 + r_1 r_1 & q_1 q_2 + r_1 r_2 & q_1 q_3 + r_1 r_3 \\ & q_2 q_2 + r_2 r_2 & q_2 q_3 + r_2 r_3 \\ \text{symmetric} & & q_3 q_3 + r_3 r_3 \end{bmatrix}\begin{bmatrix} E_1 \\ E_2 \\ E_3 \end{bmatrix} \qquad (10.68)$$

where the z component notation has been dropped. Now, E_i are the z components of the electric field intensity at the three nodes of the triangular element.

The third term is treated similarly to eddy currents in Eq. (10.42). After applying Galerkin's method to this term, we have

$$\int_{S_i} k_0^2 \varepsilon_r N^t E\, dx dy$$

and, upon integration, we get the following matrix for the third term of Eq. (10.66)

$$\frac{k_0^2 \varepsilon_r D}{12} \begin{bmatrix} 1 & 0.5 & 0.5 \\ 0.5 & 1 & 0.5 \\ 0.5 & 0.5 & 1 \end{bmatrix} \begin{bmatrix} E_1 \\ E_2 \\ E_3 \end{bmatrix} \qquad (10.69)$$

Now the elemental system to solve is

$$\frac{v_r}{2D} \begin{bmatrix} q_1q_1 + r_1r_1 & q_1q_2 + r_1r_2 & q_1q_3 + r_1r_3 \\ & q_2q_2 + r_2r_2 & q_2q_3 + r_2r_3 \\ \text{symmetric} & & q_3q_3 + r_3r_3 \end{bmatrix} \begin{bmatrix} E_1 \\ E_2 \\ E_3 \end{bmatrix} =$$

$$\frac{k_0^2 \varepsilon_r D}{12} \begin{bmatrix} 1 & 0.5 & 0.5 \\ 0.5 & 1 & 0.5 \\ 0.5 & 0.5 & 1 \end{bmatrix} \begin{bmatrix} E_1 \\ E_2 \\ E_3 \end{bmatrix} \qquad (10.70)$$

Assembly of the elemental equations into a global system of equations gives

$$[SS]\{E\} = k_0^2 [R]\{E\} \qquad (10.71)$$

A similar system can be written for TE modes by replacing v_r by $1/\varepsilon_r$ and ε_r (on the right-hand side) by μ_r

$$\frac{1}{\varepsilon_r 2D} \begin{bmatrix} q_1q_1 + r_1r_1 & q_1q_2 + r_1r_2 & q_1q_3 + r_1r_3 \\ & q_2q_2 + r_2r_2 & q_2q_3 + r_2r_3 \\ \text{symmetric} & & q_3q_3 + r_3r_3 \end{bmatrix} \begin{bmatrix} H_1 \\ H_2 \\ H_3 \end{bmatrix} =$$

$$\frac{k_0^2 \mu_r D}{12} \begin{bmatrix} 1 & 0.5 & 0.5 \\ 0.5 & 1 & 0.5 \\ 0.5 & 0.5 & 1 \end{bmatrix} \begin{bmatrix} H_1 \\ H_2 \\ H_3 \end{bmatrix} \qquad (10.72)$$

This is an eigenvalue problem in complex variables. In the way this particular system was written, the matrix on the left-hand side is real while that on the right-hand side is complex. The solution therefore must be in complex variables. If $\mu_r = 1$, the term v_r can be removed since it is equal to 1. The solution proceeds with calculation of the eigenvalues and eigenvectors of either system using any of the standard methods of solution for this type of eigenvalue problem.

The eigenvalues then represent the cutoff frequencies of the various modes and the eigenvectors, the electric field variation in the cross-section of the waveguide.

From this, the fields anywhere in the waveguide can then be written using the dispersion relation

$$\gamma^2 = -k^2 + \left(\frac{m\pi}{a}\right)^2 + \left(\frac{n\pi}{b}\right)^2 \qquad (10.73)$$

and, using the relation

$$\gamma = \alpha + j\beta$$

the electric field in the waveguide, at any point z, is

$$E_z(x,y,z) = E_0(x,y,z_0)e^{-\alpha z}e^{-j\beta z} \qquad (10.74)$$

where E_0 are the fields calculated from Eq. (10.74) at any point z_0 in the cross section of the waveguide. Normally this is taken at $z=0$. α is the attenuation constant, β the propagation constant.

The same result can be accomplished by using the following functional, rather than using Galerkin's method

$$\int_s \frac{1}{2}\frac{1}{\mu_r}\left[\left(\frac{\partial E}{\partial x}\right)^2 + \left(\frac{\partial E}{\partial y}\right)^2\right]ds - \int_s \frac{1}{2}k_0^2\, E\, ds \qquad (10.75)$$

Similarly, the extension of this functional or Eqs. (10.66) and (10.67) to 3D microwave cavities is relatively straight forward. It essentially involves formulation with a 3D finite element, use of the 3D Laplacian, and 3D (or volume) integration. Aspects of this process will be discussed in the following sections.

10.3.8. Advantages and Limitations of 2D Formulations

Practically all electromagnetic structures are three-dimensional in nature, and some precautions must be taken when 2D approximations to 3D problems are made. Many realistic problems in electrical engineering can be analyzed by 2D methods, if appropriate cut planes are chosen. Note also that, generally, devices are built to avoid the generation of eddy currents, and therefore, we can often work with static methods, even if there are moving parts in the solution domain. Under certain conditions, a sequence of static solutions may provide an answer to the dynamic response of the system.

When eddy currents are present, the 2D formulations presented above require that the currents be perpendicular to the cross-sectional plane over which the elements are generated. This implies that these currents flow from $-\infty$ to $+\infty$. In many cases this is, at best, a poor approximation.

For long structures, this approach may be correct. An example for a good 2D approximation is the case of an induction motor, where the short-circuit bars are connected at the extremities of the rotor, as shown in Figure 10.8a. In this case, we can assume, with little error, that the eddy current loops are closed at infinity, and under this assumptions, the results calculated for the 2D domain in Figure 10.8b are satisfactory.

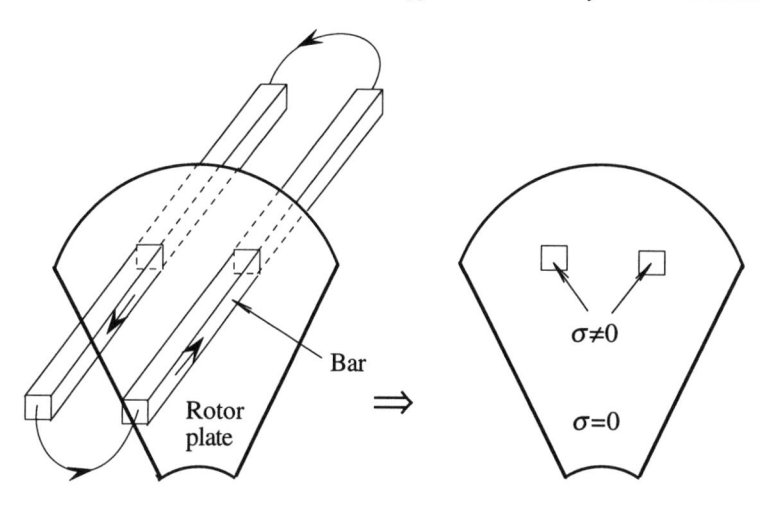

Figure 10.8a. Approximation of a 3D problem geometry. Short-circuited bars in an induction machines are shown together with a section of the rotor plate.

Figure 10.8b. The two-dimensional plane used for analysis assumes the bars are infinitely long.

However, the induction motor, is a "bad" example in other aspects. Its operation involves many complex phenomena such as rotor speed, whose slip generates induced currents, and saturation. A further difficulty is the very small air-gap, affecting accuracy of results. In other words, although the calculation of eddy currents is relatively simple and accurate, a complete, realistic and accurate analysis of an induction motor is one of the more complex of all electromagnetic devices and requires special techniques for numerical calculation of its parameters.

For axi-symmetric problems, the formulation in cylindrical coordinates is very efficient, since the eddy currents (flowing in the ϕ direction) are closed within the structure itself. This is taken into account by the formulation and requires no approximations. In effect, this is a simple solution to what would otherwise be a relatively complex 3D problem.

One important precaution that must be followed concerns the mesh used for analysis of eddy current problems, especially in regions where the currents are significant. The depth of penetration is given by

$$\delta = \sqrt{\frac{2}{\mu\sigma\omega}}$$

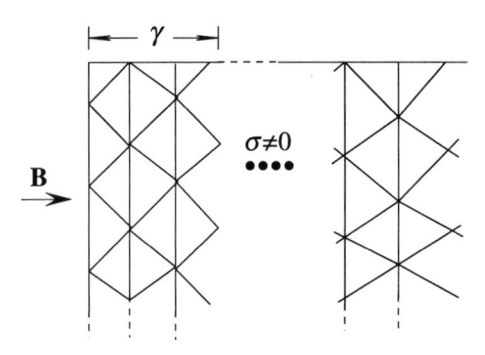

Figure 10.9. Discretization in eddy current domains. Smaller elements are used in eddy current regions.

This is normally very small, especially in ferromagnetic materials. For correct analysis, it is necessary to discretize currents regions with small elements to obtain a good representation of currents in eddy current regions, as indicated in Figure 10.9. For regions far from eddy currents, larger elements can be used. Normally, 1.5 to 2 elements per skin depth are required for correct solution.

Concluding this section, we recall that eddy current problems are normally three-dimensional in nature and the complexity involved in their solution is an important consideration. Many formulations have been proposed for treatment of eddy current problems, each with its own advantages and range of application. A "universal" solution method does not exist. The 2D formulations presented here can, if used judiciously, provide correct results, by relatively simple computation systems. These methods have been frequently used in electrical engineering for a variety of applications.

10.4. Application to the Newton-Raphson Method

The principles of the Newton-Raphson method were described in the previous chapter. Here we present only the calculation of the Jacobian term corresponding to the elemental matrix, where the reluctivity v varies as a function of the magnitude of $\mathbf{B}$. Taking one line of the elemental matrix in Eq. (10.31b), say line m, its derivative with respect to the potential A_n, is

$$-\frac{\partial}{\partial A_n}\int_{S_i} v \nabla N_m^t \nabla N A \, dxdy = -\int_{S_i} v \nabla N_m^t \nabla N_n \, dxdy - \int_{S_i} \nabla N_m^t \nabla N A \frac{\partial v}{\partial B^2} \frac{\partial B^2}{\partial A_n} dxdy$$

The first integral in this equation, was evaluated earlier (see section 10.3.2). To calculate the second integral, note that

$$\frac{\partial B^2}{\partial A_n} = \frac{\partial}{\partial A_n} B^t B = 2B^t \frac{\partial B}{\partial A_n}$$

where

$$B = \nabla N A \quad \text{and} \quad \frac{\partial B}{\partial A_n} = \nabla N_n$$

We have

$$\frac{\partial B^2}{\partial A_n} = 2 \nabla N_n^t \nabla N A$$

and the second integral becomes

$$-2 \int_{S_i} \nabla N_m^t \nabla N A \frac{\partial v}{\partial B^2} \nabla N_n^t \nabla N A \, dxdy$$

This expression is also evaluated by numerical integration. The Jacobian term is

$$J_{m,n} = S_{m,n} + \sum_{i=1}^{r} W_i f(u_i, v_i) \qquad (10.76)$$

where the term $S_{m,n}$ is

$$S_{m,n} = -\sum_{i=1}^{r} W_i v \nabla N_m^t \nabla N_n det[J_{l\,i}]$$

and

$$f(u_i, v_i) = -2 \sum_{i=1}^{r} W_i \nabla N_m^t \nabla N A \frac{\partial v}{\partial B^2} \nabla N_n^t \nabla N A det[J_{l\,i}]$$

The assembly of the matrix system is done analogously to that shown in Chapter 9.

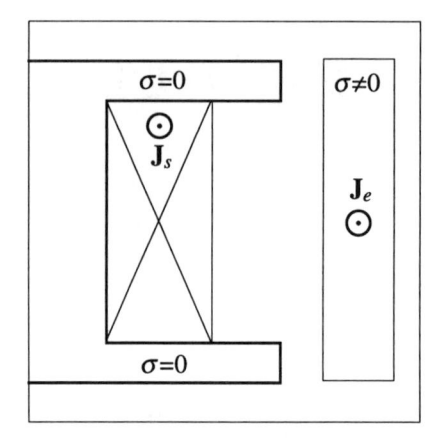

Figure 10.10. A stationary conducting piece in front of an electromagnet. Eddy currents are induced in the piece due to a time-dependent magnetic field.

10.5. Examples

10.5.1. *Eddy Currents: Time Discretization*

Consider the geometry of Figure 10.10, which shows a stationary conducting piece in front of an electromagnet. Eddy currents, indicated by $\mathbf{J}_e$, are induced in the conducting piece due to the time variation of the external current J_s in the excitation coil.

The conducting piece is ferromagnetic with $\mu_r = 1000$ and $\sigma = 10^7$ *(S/m)*. To show a problem that can be easily understood, we apply a current density with the waveform shown in Figure 10.11: it is a current pulse rising from zero to 2 A/mm^2 in 0.01 *sec*. The rising part is sinusoidal in shape; at times beyond 0.01 *sec*, the current density J_s is constant at 2 A/mm^2.

Figure 10.12a, b, c, and d shows the results of the calculation. The number of potential lines is proportional to the flux density, with maximum flux density at steady state.

The behavior of the field and eddy currents are as follows:

a. Figures 10.12a gives the flux distribution at point a ($t=0.002$ *sec*) in Figure 10.11. This is the beginning of the simulation and we note that, by this time, there already are eddy currents in the conducting piece. The potential corresponding to this current is negative, since the induced current is opposite in direction to the applied current. This current describes the reaction to the field penetrating in the conducting piece (Lenz's law).

b. Figure 10.12b was obtained for point e ($t=0.01$ *sec*) in Figure 10.11. The applied current increases and eddy currents increase as well; the induced current is maximum near this point in time.

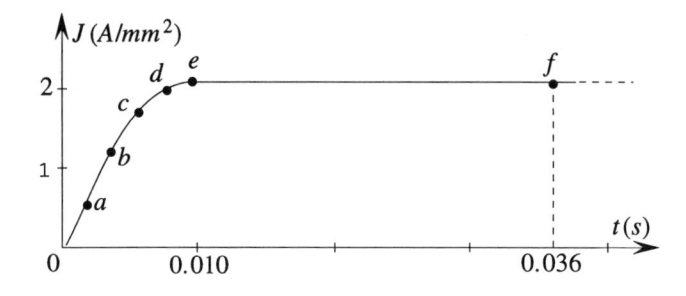

Figure 10.11. The pulse used to drive the geometry in Figure 10.10.

c. Figure 10.12c shows results for point f ($t=0.036$ sec): the applied current is now constant and the induced current is very low, compared with the maximum value; however, it is still present and does not permit complete field penetration into the piece.

d. Figure 10.12d. The field penetrates completely into the conducting piece and the flux distribution is similar to the static case, where the external current density is fixed at 2 A/mm^2

In Figure 10.13a the maximum negative potentials in the conducting piece are shown. They are directly related to the induced currents. These currents decrease after $t=0.010$ sec. By time $t=0.040$ sec, these currents are virtually zero.

Figure 10.13b shows the maximum values of the positive potentials; they correspond to the flux generated by the applied current. This curve tends to the flux value for the static case with a coil current density of $J=2$ A/mm^2.

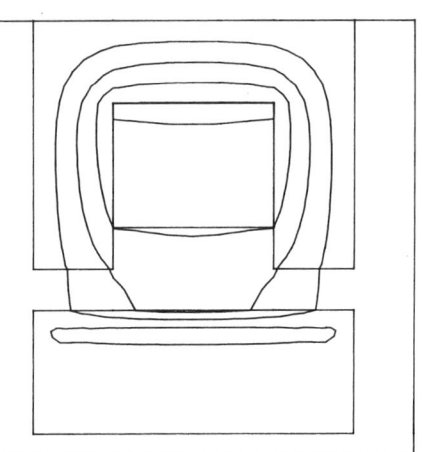

Figure 10.12a. Flux distribution in Figure 10.10 at point a of the pulse in Figure 10.11.

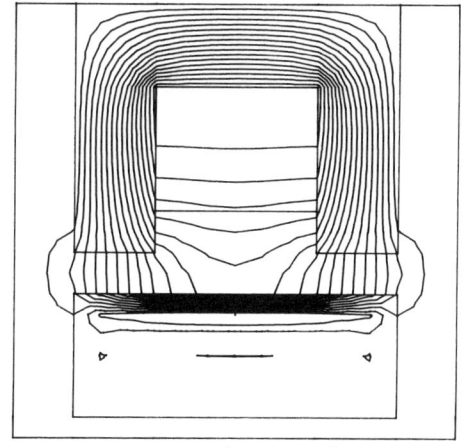

Figure 10.12b. Flux distribution Figure 10.10 at point e of the pulse in Figure 10.11.

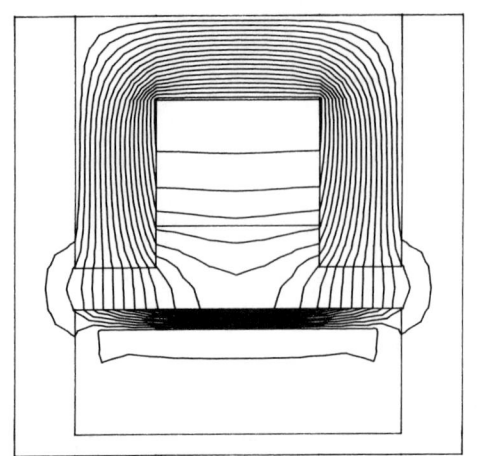

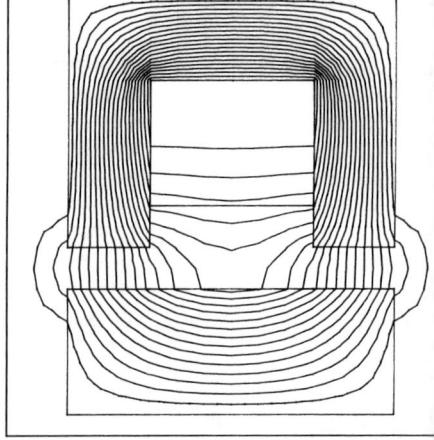

436   10. Galerkin's Residual Method: Applications to Dynamic Fields

Figure 10.12c. Flux distribution in Figure 10.10 at point f of the pulse in Figure 10.11.

Figure 10.12d. Flux distribution in Figure 10.10 at steady state.

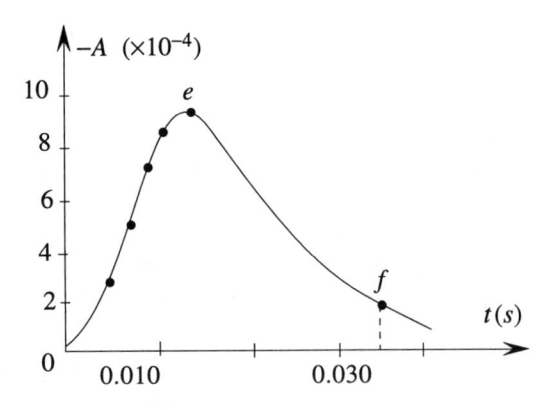

Figure 10.13a. Maximum negative potentials in the conducting piece of Figure 10.10.

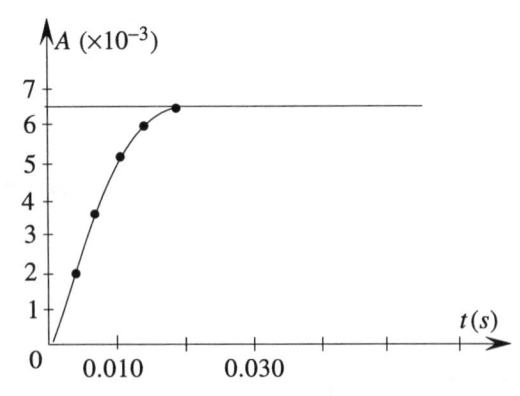

Figure 10.13b. Maximum positive potentials in the conducting piece of Figure 10.10.

10.5.2. *Moving Conducting Piece in Front of an Electromagnet*

Consider a nonferromagnetic conducting piece in front of an electromagnet fed by a constant applied current as shown in Figure 10.7. The behavior of the magnetic field for different velocities of the piece is required. Solutions are shown in Figure 10.14a through 10.14f.

Figure 10.14a shows the solution at zero velocity. The field is essentially that for a static solution. In Figure 10.14b, c, d, e, and f, the field configurations for a moving conducting piece with velocities $v = 1, 5, 10, 20$, and 30 *m/s* respectively are shown. Asymmetries become more pronounced as the speed increases.

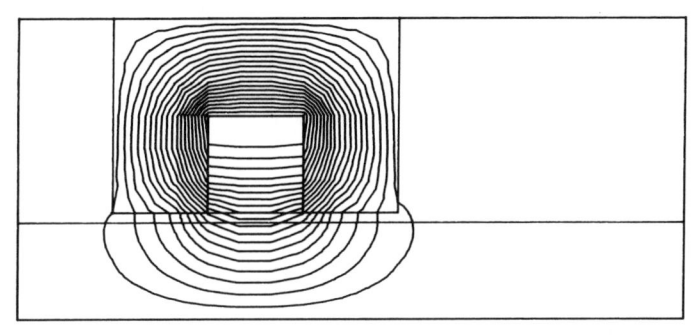

Figure 10.14a. Solution for the geometry in Figure 10.7 at $v=0$ *m/s*.

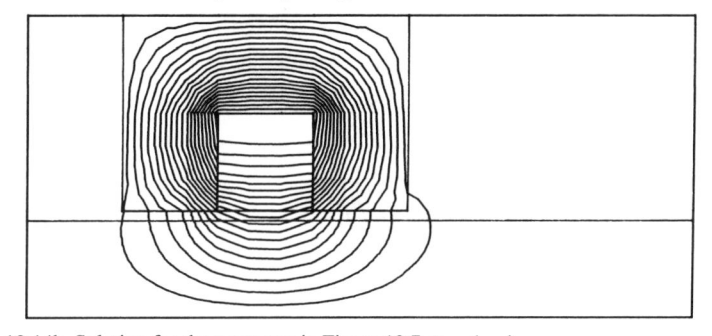

Figure 10.14b. Solution for the geometry in Figure 10.7 at $v=1$ *m/s*.

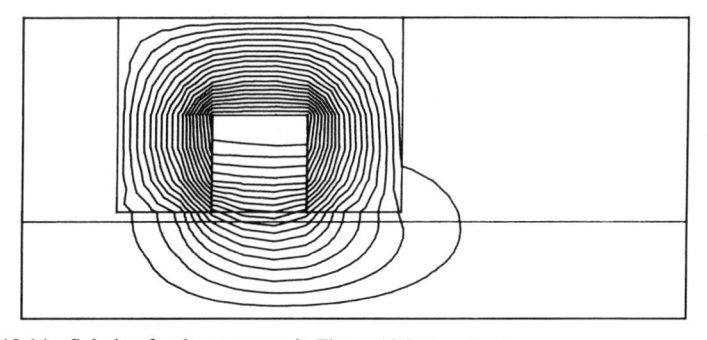

Figure 10.14c. Solution for the geometry in Figure 10.7 at $v=5$ *m/s*.

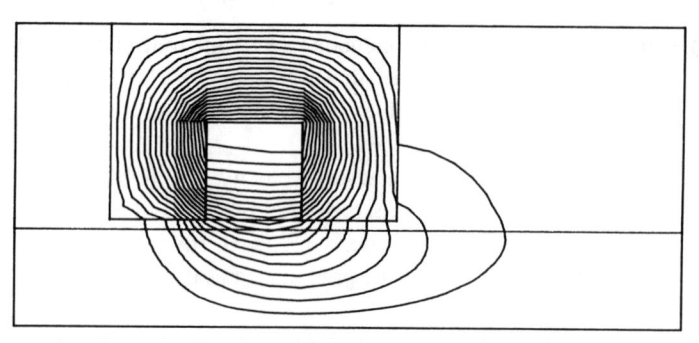

Figure 10.14d. Solution for the geometry in Figure 10.7 at $v=10$ *m/s*.

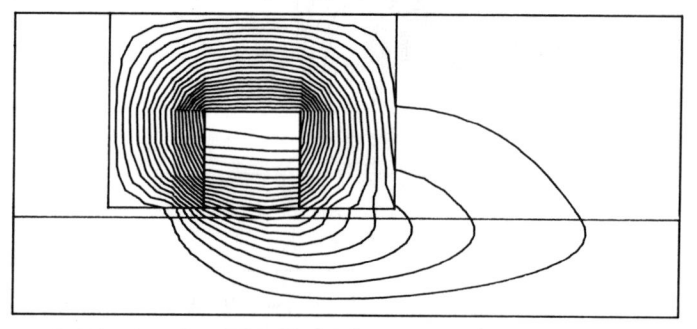

Figure 10.14e. Solution for the geometry in Figure 10.7 at $v=20$ *m/s*.

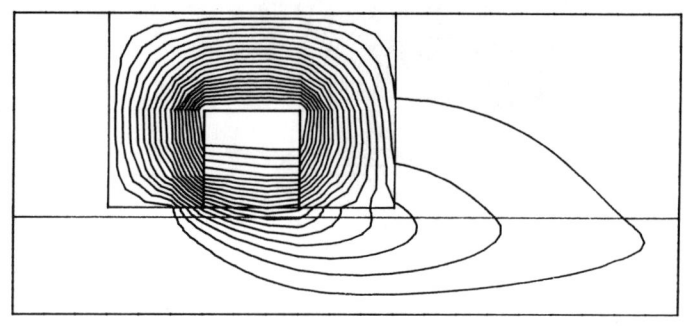

Figure 10.14f. Solution for the geometry in Figure 10.7 at $v=30$ *m/s*.

The following points should be noted in these solutions:

a. **Shape of the field:** The eddy current is given by $\mathbf{J}=\sigma\mathbf{v}\times\mathbf{B}$; in front of the electromagnet (fed by a current perpendicular to the plane of the figure), the situation is that shown in Figure 10.15. The flux densities $\mathbf{B}_1$ and $\mathbf{B}_2$ are in the directions shown in this figure; from the equation for $\mathbf{J}$, the eddy currents $\mathbf{J}_{e1}$ and $\mathbf{J}_{e2}$ are perpendicular to the plane, as indicated. Figure 10.16 shows the three currents: $\mathbf{J}_s$ (the applied current in the coil), $\mathbf{J}_{e1}$ and $\mathbf{J}_{e2}$ (induced currents in the moving piece).

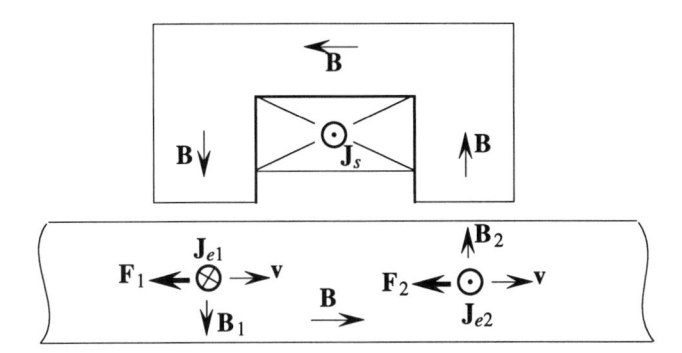

Figure 10.15. Current densities, magnetic field intensities, and forces in the moving piece in relation to velocity.

The general shape of the magnetic fields generated by these currents is shown in Figure 10.16. Based on their summation, we conclude that the total field has the form indicated by the dotted line. This is consistent with the contours in Figures 10.14.

b. Force due to the product J×B: In Chapter 6, the volumetric force density is given by **f=J×B**. Observing the vector directions in Figure 10.15, the forces **F**$_1$ and **F**$_2$ are opposite to the velocity vector **v** of the conducting piece. This is characteristic of the magnetic brake discussed in section 6.9.9.

c. Force due to Maxwell's tensor: According to the contour plots obtained by the finite element method (Figures 10.14) the fields have significant components tangential to the moving piece (Figure 10.17). Using the Maxwell tensor, (presented in Chapter 6), these components generate forces at an angle 2θ with the normal direction. The forces are opposed to the piece movement (as was also concluded in b).

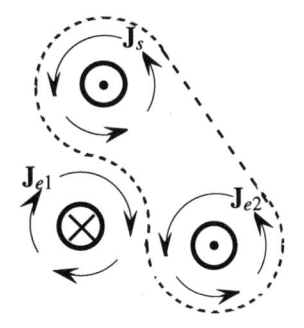

Figure 10.16. Relationship between applied and induced current densities.

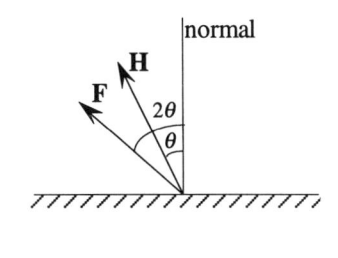

Figure 10.17. Force calculated using Maxwell's force tensor.

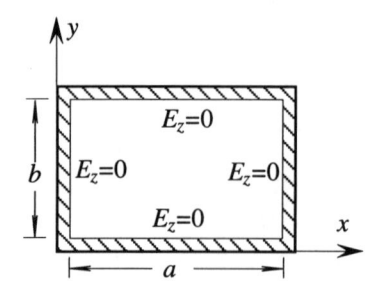

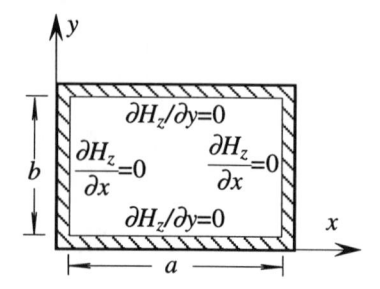

Figure 10.18a. Boundary conditions for *TM* modes in a waveguide.

Figure 10.18b. Boundary conditions for *TE* modes in a waveguide.

10.5.3. *Modes and Fields in a Waveguide*

The cross-section of a waveguide is shown in Figure 10.18. We are interested in the *TE* and *TM* modes of this waveguide and in the field distribution across its cross section. Fig. 10.18a shows the boundary conditions for the *TM* mode (i.e., $E=0$ on the conducting boundaries) while Figure 10.19b shows the boundary conditions for TE modes (i.e., $\partial H/\partial x=0$, $\partial H/\partial y=0$). Although we assume the boundaries to be perfectly conducting, the material within the waveguide is arbitrary and may include lossy or perfect dielectrics, conductors, etc.

Calculation proceeds using the formulation in section 10.3.7. [Eq. (10.67)] and a finite element mesh that is essentially a uniform mesh. The mesh in this case is relatively sparse since the fields are well defined and their gradient is relatively low. Obviously, in the two cases shown above, symmetry lines can be used to simplify the problem. However, to keep the discussion simple, we will not do so here.

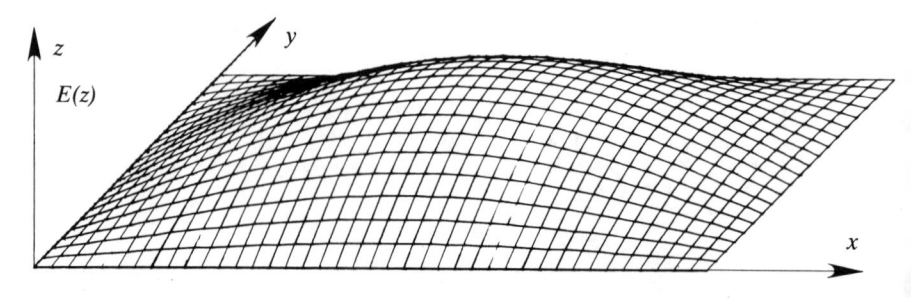

Figure 10.19a. The electric field for TM_{11} mode in a waveguide as obtained from the finite element calculation.

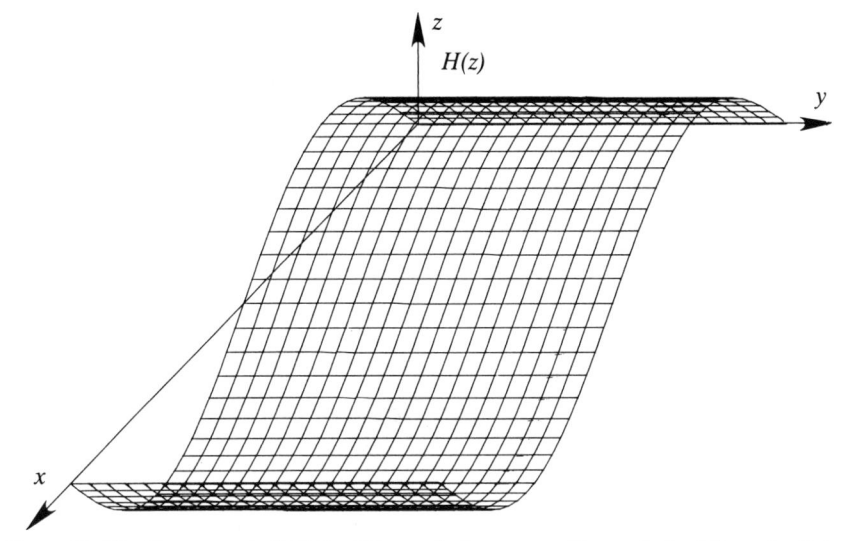

Figure 10.19b. The magnetic field for TE_{20} mode in a waveguide as obtained from the finite element calculation.

Figures 10.19a and 10.19b show the TM_{11} and TE_{20} modes in the waveguide as obtained from the finite element program. Note that the amplitude of the fields is arbitrary. The actual amplitude is determined from the source of the fields as coupled to the waveguide. Since this is not part of the formulation in Eq. (10.67), it cannot be part of the solution.

A more complex problem is shown in Figure 10.20 where a ridged waveguide is shown. Clearly, the shape of the waveguide can be as complex as required as long as the boundary conditions can be applied. The field (z component of the electric field in this case) for the TM_{11} mode is shown in Figure 10.21. While the distinct characteristics of the TM_{11} mode can be seen, the fields around the ridge and its corners are more complex than in a normal rectangular waveguide.

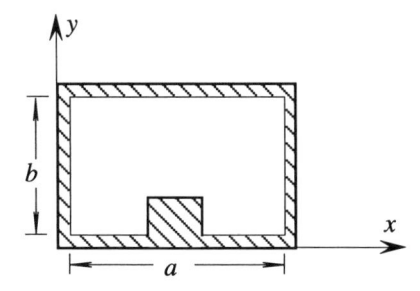

Figure 10.20. Cross-section of a ridged waveguide.

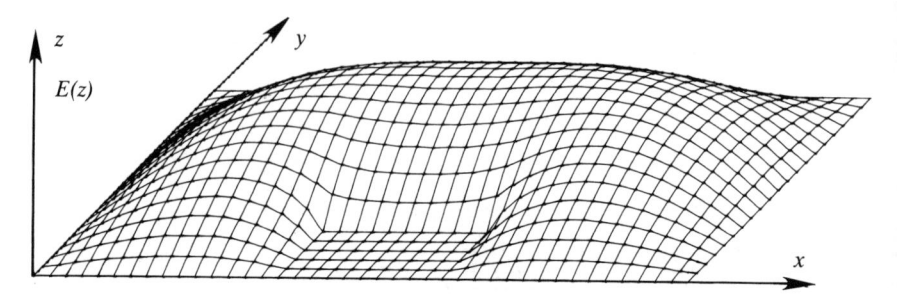

Figure 10.21. The electric field for mode TM_{11} in the ridged waveguide of Figure 10.20.

10.5.4. *Resonant Frequencies of a Microwave Cavity*

A cavity resonator is a three-dimensional structure. The results presented here were obtained using the modified eddy current formulation, in three dimensions, using the linear hexahedral element described in section 10.5.2. The formulation is quite simple and follows the same steps as for the eddy current formulation. In this case, we are not interested in the fields themselves since these are similar to the fields in a waveguide and can be described by Eqs. (7.96) and (7.98). The purpose here is to find the shift in resonant frequency due to changes of material properties in the cavity. This is an important aspect of measurements and testing using cavities. Figure 10.22 shows the cavity modeled. The model uses 125 linear hexahedral elements (5 in each direction) with a total of 216 nodes.

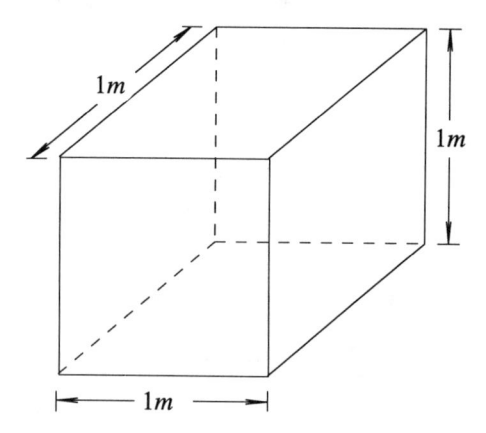

Figure 10.22. Model of a cavity resonator.

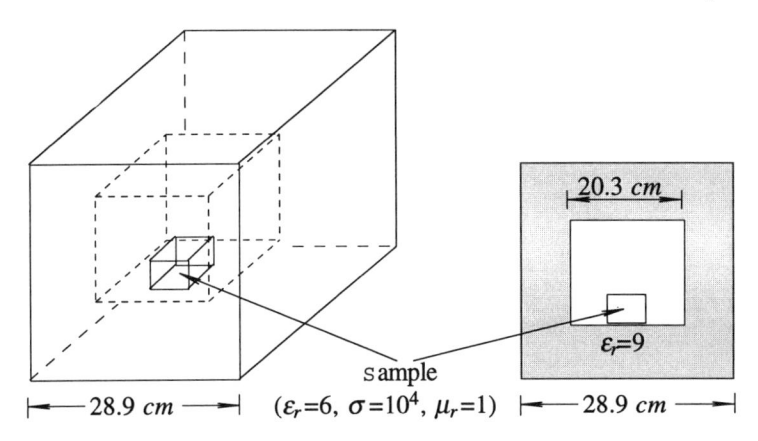

Figure 10.23. Resonant cavity with a material sample shown in two views. The lining inside the cavity is for insulation purposes but must be modeled because it is a dielectric.

The solution is obtained by changing the frequency and solving the problem for each frequency. The source of the fields in the cavity is a current in one element of the cavity. This results in an arbitrary field distribution since the amplitudes are not known, but the shift in resonant frequency is independent of the amplitude. With the given dimensions, the analytic solution for the resonant frequency [given by Eq. (7.100)] is 212 *MHz*. The finite element solution gives 223 *MHz* or about 5% higher than the analytic solution. Filling the cavity with a dielectric (ε_r=1.1) gives a resonant frequency of 213 *MHz*. This can also be calculated analytically from the dispersion relation and the solution is 202 *MHz*.

As an example of a problem that cannot be solved analytically consider Figure 10.23.

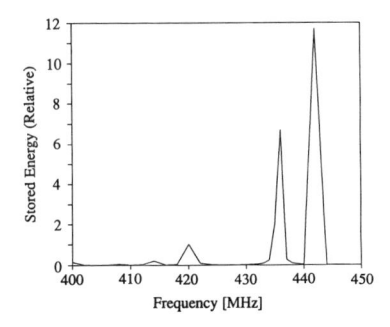

Figure 10.24. Frequency scan obtained by repetitive finite element solution of the geometry in Figure 10.23, showing the resonant frequency.

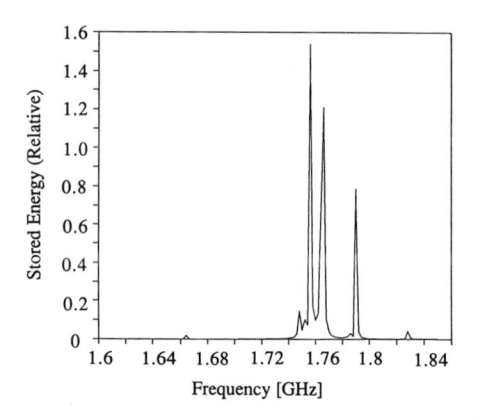

Figure 10.25. Frequency scan obtained by repetitive finite element solution of the geometry in Figure 10.23, showing resonant frequencies at higher modes.

The shift in frequency (2.5 *MHz*) is due to the sample and also depends on the location of the sample. The resonant peaks are clearly distinguishable. The minor peaks are due to discretization and change as the mesh density is changed. Here a sample of a dielectric or lossy dielectric material is located on the bottom of the cavity. The results of the frequency scan are shown in Figures 10.24 and 10.25.

11
Hexahedral Edge Elements – Some 3D Applications

11.1. Introduction

Unlike 2D finite element calculations, in which classical problems are solved using well established formulations, 3D finite element application is still an open area in research. Although some formulations are already widely used, new numerical and physical approaches are still evolving. In fact, there are many difficulties and compromises involved in 3D calculations. Some of these are:

- The number of unknowns can vary from one to four per node of the mesh, depending on the formulation.
- Often, more than one formulation is used in the same problem, each formulation being used in different parts of the physical domain. These formulations must be properly coupled.
- Application of gauges in order to improve the condition of the system matrix (which is normally ill-conditioned). The application of these gauges allows faster convergence in the iterative solution of the system of equations but it also affects accuracy of results.
- The matrix solver itself. The ICCG method of solution (to be discussed later in this chapter) is the most commonly used method for solution of the system of equations. It is an iterative method and divergence can occur, depending on the problem and/or on the condition of the system of equations.

These characteristics and natural difficulties are the motivation for the extensive and active research in this area, resulting in a variety of formulations which, fortunately, provide excellent results in many practical applications.

Two types of finite elements have been employed to perform 3D calculations; nodal and edge elements. The nodal element method is the older and more established method. It dates back to the late 1950s and starting with the early 1980s, software packages have been developed based on this type of elements. One reason for this widespread use of nodal elements is that the work in 2D, which preceded much of the work in 3D was based on nodal elements (2D

446   11. Hexahedral Edge Elements – Some 3D Applications

software still uses primarily nodal elements) and it was therefore natural to extend the use of nodal elements to three dimensions. Although the concept of edge elements is not new, its use evolved much later than that of nodal elements. The first use of edge elements in electromagnetic field computation occurred in the early 1980s. In fact, it required a different approach and way of thinking, with the natural reluctance on the part of those already working in the field using nodal elements. Slowly, edge elements gained acceptance and became widely used. In this regard it is worth mentioning that one factor in their acceptance was the realization that their use in high-frequency applications, allowed solution for cutoff frequencies without "spurious modes," a problem that plagued all solutions based on nodal elements.

Nodal elements, as presented in the previous chapters, are constructed to calculate scalar quantities at the nodes of the finite elements. This is true even when vector quantities such as the magnetic vector potential are used. In 2D formulations, the basic approach has been to calculate the magnitude of the vector potential at the nodes of the finite element, thus preserving the scalar nature of the solution (in the case of the magnetic vector potential, a single component of the potential, usually the z component is sufficient to describe the problem). Therefore, nodal elements are perfectly adapted for calculation of scalar quantities. In 3D calculations, however, the most common equations needed to analyze electromagnetic problems are vector equations. To use nodal elements with vector variables, it is necessary to separate the vector into its components and apply the nodal elements to the components.

Edge elements are defined as "vector elements." As will be seen shortly, the shape functions of these elements represent vector quantities and therefore they can be naturally coupled to vector equations in a very simple and straightforward way. To present the main ideas associated with edge elements, we consider the example below. Suppose that the vector variable $\mathbf{A}$ has to be defined on a rectangular element as shown in Figure 11.1. $\mathbf{A}$ is given as follows:

$$\mathbf{A}(x,y) = \sum_{i=1}^{4} \mathbf{w}_i A_i$$

where $\mathbf{w}_i$ is the shape function related to one of the four edges "i" of the element.

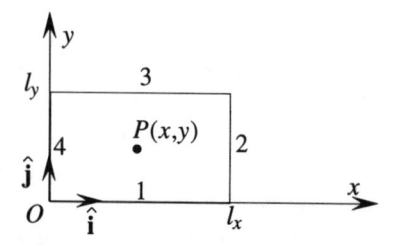

Figure 11.1. A hypothetical rectangular edge element.

The shape functions for the element are

$$\mathbf{w}_1 = \hat{\mathbf{i}}\left[1 - \frac{y}{l_y}\right] \qquad \mathbf{w}_2 = \hat{\mathbf{j}}\left[\frac{x}{l_x}\right] \qquad \mathbf{w}_3 = \hat{\mathbf{i}}\left[\frac{y}{l_y}\right] \qquad \mathbf{w}_4 = \hat{\mathbf{j}}\left[1 - \frac{x}{l_x}\right]$$

We note that these shape functions have a direction ($\hat{\mathbf{i}}$ or $\hat{\mathbf{j}}$, depending on the edge) and also a "placement" or "position" expression (nodal elements only have the latter property). For example, for edge No. 1, the direction is $\hat{\mathbf{i}}$ and the placement is given by $[1-y/l_y]$. From these:

$$\text{For } y=0, \quad \mathbf{w}_1 = \hat{\mathbf{i}}\,1 \quad \text{(at edge 1)}$$
$$\text{For } y=l_y/2, \quad \mathbf{w}_1 = \hat{\mathbf{i}}\,1/2 \quad \text{(half-way between edge 1 and 3)}$$
$$\text{For } y=l_y, \quad \mathbf{w}_1 = \hat{\mathbf{i}}\,0 \quad \text{(at edge 3)}$$

Similarly, for edge No. 3, we get

$$\text{For } y=0, \quad \mathbf{w}_3 = \hat{\mathbf{i}}\,0 \quad \text{(at edge 1)}$$
$$\text{For } y=l_y/2, \quad \mathbf{w}_3 = \hat{\mathbf{i}}\,1/2 \quad \text{(half-way between edge 1 and 3)}$$
$$\text{For } y=l_y, \quad \mathbf{w}_3 = \hat{\mathbf{i}}\,1 \quad \text{(at edge 3)}$$

The analysis for the remaining two edges (parallel to the y axis) is similar.

Now consider the point $P(x,y)$ inside the element ($0 \leq x \leq l_x$ and $0 \leq y \leq l_y$). Suppose that we need to obtain $\mathbf{A}$ at the centroid of the element ($x=l_x/2$, $y=l_y/2$). Applying these conditions to the shape functions, we obtain

$$\mathbf{w}_1 = \hat{\mathbf{i}}\,1/2 \qquad \mathbf{w}_2 = \hat{\mathbf{j}}\,1/2 \qquad \mathbf{w}_3 = \hat{\mathbf{i}}\,1/2 \qquad \mathbf{w}_4 = \hat{\mathbf{j}}\,1/2$$

The value of $\mathbf{A}$ is, therefore,

$$\mathbf{A}(l_x/2, l_x/2) = \hat{\mathbf{i}}\,A_1/2 + \hat{\mathbf{i}}\,A_3/2 + \hat{\mathbf{j}}\,A_2/2 + \hat{\mathbf{j}}\,A_4/2$$

We still have said nothing as to the meaning of A_1, A_2, A_3, and A_4. To see what these represent, take for example, $y=0$. This gives $\mathbf{w}_i = \hat{\mathbf{i}}\,1$ and $\mathbf{w}_3 = 0$. Therefore, we obtain for the x component of $\mathbf{A}$:

$$\mathbf{A}_x(y,0) = \hat{\mathbf{i}}\,A_1$$

which means that A_1 is the projection of the vector field $\mathbf{A}$ on edge No. 1. Similarly, A_3 is the projection of $\mathbf{A}$ on edge No. 3. As y varies from zero to l_y, the "contribution" of A_1 decreases and that of A_3 increases as shown below:

$$\begin{aligned}
y&=0 & A_x &= \hat{\mathbf{i}}(1A_1 + 0A_3)\\
y&=0.3 & A_x &= \hat{\mathbf{i}}(0.7A_1 + 0.3A_3)\\
y&=0.7 & A_x &= \hat{\mathbf{i}}(0.3A_1 + 0.7A_3)\\
y&=1 & A_x &= \hat{\mathbf{i}}(0A_1 + 1A_3)
\end{aligned}$$

Analogously, as x varies from 0 to l_x, we obtain the y component of $\mathbf{A}$. The vector field $\mathbf{A}$ can now be written as:

$$\mathbf{A} = \hat{\mathbf{i}}A_x(y) + \hat{\mathbf{j}}A_y(x)$$

This very simple example was given here in order to describe the idea of edge elements as vector elements. In reality, the shape functions are dimensionally slightly different than what was presented above. The quantities A_1, A_2, A_3, and A_4 are actually the circulation of $\mathbf{A}$ along the edges rather than their projections which requires that the shape functions are also dimensionally modified. Also, the directions of the edges may be in any direction in space rather than parallel to the axes.

In this chapter, we discuss edge element formulations using the hexahedral element. In our experience, these formulation yield reliable results as we will show later. First, the shape functions of these elements are described and then the formulation is applied to physical configurations. We note here that the use of such an element simplifies the geometric construction of the solution domain, but if the geometry is complicated or has very sharp angles, it is difficult to describe the geometry in terms of hexahedral elements. Finally, we point out that we do not claim that the formulation presented here is better or worse than any other available formulation. As already stated, there are many other formulations, each with its own set of advantages, disadvantages and range of applications. More often than not, the performance of a formulation is problem dependent. Nevertheless, our experience with edge elements has been very positive and therefore it is worthy of inclusion here.

11.2. The Hexahedral Edge Element Shape Functions

The hexahedral edge element has six faces, eight nodes and twelve edges as shown in Figure 11.2a and 11.2b.

A vector quantity is given by

$$\mathbf{A} = \sum_{i=1}^{12} \mathbf{w}_i(\mathbf{r})A_i \qquad (11.1)$$

Hexahedral Edge Element Shape Functions   449

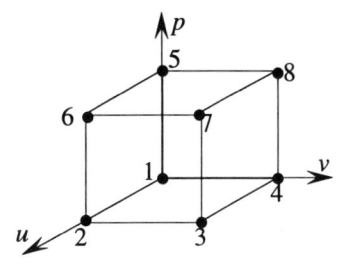

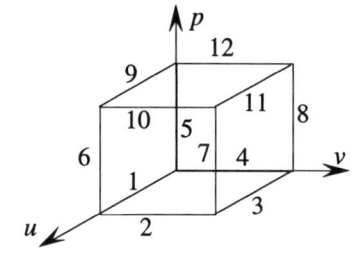

Figure 11.2a. Node numbers.

Figure 11.2b. Edge numbers.

where $\mathbf{w}_i$ is the shape function related to edge "i" and $\mathbf{A}$ is, for instance, the magnetic vector potential or any other vector field. A_i represents the circulation of $\mathbf{A}$ on edge "i." The shape function $\mathbf{w}_i$ is described by the following product

$$\mathbf{w}_i(\mathbf{r}) = \phi_i(u,v,p)\mathbf{q}_i(\mathbf{r}) \qquad (11.2)$$

where

$$\mathbf{r} = \hat{\mathbf{i}}\,x + \hat{\mathbf{j}}\,y + \hat{\mathbf{k}}\,z \qquad (11.3)$$

which represents the position vector of a generic point $M(x,y,z)$. The function $\phi_i(u,v,p)$ depends on the reference coordinates u,v,p and it is the "placement" function of the edge element; it is given in reference coordinates since the numerical integration and other algebraic operations are performed in these coordinate system. The function $\mathbf{q}_i(\mathbf{r})$, which is responsible for the "direction" of the edge, is given in real coordinates, since only these coordinates take into account the actual geometry of the elements. For simplicity, we will use from now on the notation $\mathbf{w}_i$, $\mathbf{q}_i$ and ϕ_i for $\mathbf{w}_i(\mathbf{r})$, $\mathbf{q}_i(\mathbf{r})$ and $\phi_i(u,v,p)$.

As an example, the function ϕ_{11}, which is related to edge 11, is

$$\phi_{11} = vp$$

In fact, on edge No. 11, $v=1$ and $p=1$. Therefore,

$$\phi_{11} = 1.1 = 1 \qquad \text{(on edge 11)}$$

Similarly,

$$\phi_3 = v(1-p) \qquad \text{(edge is located at } v=1, p=0\text{)}$$
$$\phi_1 = (1-v)(1-p) \qquad \text{(edge is located at } v=0, p=0\text{)}$$
$$\phi_9 = (1-v)p \qquad \text{(edge is located at } v=0, p=1\text{)}$$

450   11. Hexahedral Edge Elements – Some 3D Applications

Table 11.1 shows the position function ϕ and the derivatives of the shape functions (which will be employed soon).

The vector quantity $\mathbf{q}_i$ in Eq. (11.2) depends on the direction of the edge. On edge 11, for example, the direction is ∇u since the edge is parallel to the u direction. Therefore,

$$\mathbf{w}_{11} = \phi_{11} \nabla u$$

where ∇u is given as a function of the real coordinates of the element. To obtain ∇u (as well as ∇v and ∇p which are necessary for the edges parallel to v and p respectively), we define the following vectors:

$$\mathbf{v}_u = \frac{\partial \mathbf{r}}{\partial u} = \frac{\partial}{\partial u}\left(\hat{\mathbf{i}} x + \hat{\mathbf{j}} y + \hat{\mathbf{k}} z\right) \tag{11.4a}$$

$$\mathbf{v}_v = \frac{\partial \mathbf{r}}{\partial v} = \frac{\partial}{\partial v}\left(\hat{\mathbf{i}} x + \hat{\mathbf{j}} y + \hat{\mathbf{k}} z\right) \tag{11.4b}$$

$$\mathbf{v}_p = \frac{\partial \mathbf{r}}{\partial p} = \frac{\partial}{\partial p}\left(\hat{\mathbf{i}} x + \hat{\mathbf{j}} y + \hat{\mathbf{k}} z\right) \tag{11.4c}$$

As an example, $\mathbf{v}_u$ can be written as

$$\mathbf{v}_u = \begin{bmatrix} \dfrac{\partial x}{\partial u} \\ \dfrac{\partial y}{\partial u} \\ \dfrac{\partial z}{\partial u} \end{bmatrix}$$

Table 11.1. Shape functions ϕ_i and their derivatives for hexahedral elements.

Edge i	ϕ_i	$\partial \phi_i / \partial u$	$\partial \phi_i / \partial v$	$\partial \phi_i / \partial p$
1	$(1-v)(1-p)$	0	$-(1-p)$	$-(1-v)$
2	$u(1-p)$	$(1-p)$	0	$-u$
3	$v(1-p)$	0	$(1-p)$	$-v$
4	$(1-u)(1-p)$	$-(1-p)$	0	$-(1-u)$
5	$(1-u)(1-v)$	$-(1-v)$	$-(1-u)$	0
6	$u(1-v)$	$(1-v)$	$-u$	0
7	uv	v	u	0
8	$(1-u)v$	$-v$	$(1-u)$	0
9	$(1-v)p$	0	$-p$	$(1-v)$
10	up	p	0	u
11	vp	0	p	v
12	$(1-u)p$	$-p$	0	$(1-u)$

Using this notation, the three vectors in Eq. (11.4) give

$$[\mathbf{v}_u \quad \mathbf{v}_v \quad \mathbf{v}_p] = \begin{bmatrix} \dfrac{\partial x}{\partial u} & \dfrac{\partial x}{\partial v} & \dfrac{\partial x}{\partial p} \\ \dfrac{\partial y}{\partial u} & \dfrac{\partial y}{\partial v} & \dfrac{\partial y}{\partial p} \\ \dfrac{\partial z}{\partial u} & \dfrac{\partial z}{\partial v} & \dfrac{\partial z}{\partial p} \end{bmatrix} \quad (11.5)$$

Transposing the matrix, we obtain a matrix $[J_1]$,

$$[J_1] = \begin{bmatrix} \dfrac{\partial x}{\partial u} & \dfrac{\partial y}{\partial u} & \dfrac{\partial z}{\partial u} \\ \dfrac{\partial x}{\partial v} & \dfrac{\partial y}{\partial v} & \dfrac{\partial z}{\partial v} \\ \dfrac{\partial x}{\partial p} & \dfrac{\partial y}{\partial p} & \dfrac{\partial z}{\partial p} \end{bmatrix} \quad (11.6)$$

This is a Jacobian matrix which expresses a vector in the (x,y,z) system in terms of the (u,v,p) system.

On the other hand, we have

$$\nabla u = \hat{\mathbf{i}} \frac{\partial u}{\partial x} + \hat{\mathbf{j}} \frac{\partial u}{\partial y} + \hat{\mathbf{k}} \frac{\partial u}{\partial z} \quad (11.7)$$

and

$$[\nabla u \quad \nabla v \quad \nabla p] = \begin{bmatrix} \dfrac{\partial u}{\partial x} & \dfrac{\partial v}{\partial x} & \dfrac{\partial p}{\partial x} \\ \dfrac{\partial u}{\partial y} & \dfrac{\partial v}{\partial y} & \dfrac{\partial p}{\partial y} \\ \dfrac{\partial u}{\partial z} & \dfrac{\partial v}{\partial z} & \dfrac{\partial p}{\partial z} \end{bmatrix} = [J_2] \quad (11.8)$$

This is the inverse of $[J_1]$. Therefore, we obtain

$$[J_1][J_2] = [1] \quad (11.9)$$

because

$$vector(x,y,z) = [J_2]\{[J_1]vector(x,y,z)\}$$

Noting that the result in the expression $\{[J_1]vector(x,y,z)\}$ gives the same vector in the (u,v,p) system, we can write

11. Hexahedral Edge Elements – Some 3D Applications

$$\begin{bmatrix} \mathbf{v}_u \\ \mathbf{v}_v \\ \mathbf{v}_p \end{bmatrix} \begin{bmatrix} \nabla u & \nabla v & \nabla p \end{bmatrix} = \begin{bmatrix} 1 & 0 & 0 \\ 0 & 1 & 0 \\ 0 & 0 & 1 \end{bmatrix}$$

Performing the various products we get

$$\begin{bmatrix} \mathbf{v}_u \cdot \nabla u & \mathbf{v}_u \cdot \nabla v & \mathbf{v}_u \cdot \nabla p \\ \mathbf{v}_v \cdot \nabla u & \mathbf{v}_v \cdot \nabla v & \mathbf{v}_v \cdot \nabla p \\ \mathbf{v}_p \cdot \nabla u & \mathbf{v}_p \cdot \nabla v & \mathbf{v}_p \cdot \nabla p \end{bmatrix} = \begin{bmatrix} 1 & 0 & 0 \\ 0 & 1 & 0 \\ 0 & 0 & 1 \end{bmatrix}$$

Inspection of the first column in the equation above indicates that ∇u is perpendicular to $\mathbf{v}_v$ and $\mathbf{v}_p$ because $\nabla u \cdot \mathbf{v}_v = 0$ and $\nabla u \cdot \mathbf{v}_p = 0$. Therefore,

$$\nabla u = k(\mathbf{v}_v \times \mathbf{v}_p) \tag{11.10}$$

where k is a scalar (constant number). Using the first term of the column, we get

$$\nabla u \cdot \mathbf{v}_u = 1$$

and

$$k \mathbf{v}_u \cdot (\mathbf{v}_v \times \mathbf{v}_p) = 1$$

This may also be written as

$$k = \frac{1}{vol} \tag{11.11}$$

since the scalar triple product $\mathbf{v}_u \cdot (\mathbf{v}_v \times \mathbf{v}_p) = 1$ is equal to the volume of the element. Finally, using Eqs. (11.11) and (11.10), we obtain

$$\nabla u = \frac{(\mathbf{v}_v \times \mathbf{v}_p)}{vol} \qquad \nabla v = \frac{(\mathbf{v}_p \times \mathbf{v}_u)}{vol} \qquad \nabla p = \frac{(\mathbf{v}_u \times \mathbf{v}_v)}{vol} \tag{11.12}$$

Now, let us calculate the vectors $\mathbf{v}_u$, $\mathbf{v}_v$ and $\mathbf{v}_p$. To achieve this, we use the shape functions in order to define the coordinates x, y, and z as functions of u, v, and p, as needed for the numerical integration. The vector $\mathbf{r}$ is

$$\mathbf{r} = \hat{\mathbf{i}} x + \hat{\mathbf{j}} y + \hat{\mathbf{k}} z$$

or

$$\mathbf{r} = \sum_{j=1}^{8} N_j(u,v,p) r_j = \hat{\mathbf{i}} \sum_{j=1}^{8} N_j x + \hat{\mathbf{j}} \sum_{j=1}^{8} N_j y + \hat{\mathbf{k}} \sum_{j=1}^{8} N_j z$$

Hexahedral Edge Element Shape Functions 453

Recalling that

$$\mathbf{v}_u = \hat{\mathbf{i}}\frac{\partial x}{\partial u} + \hat{\mathbf{j}}\frac{\partial y}{\partial u} + \hat{\mathbf{k}}\frac{\partial z}{\partial u}$$

and using a simplified notation, we get

$$\mathbf{v}_u = \hat{\mathbf{i}}\frac{\partial N}{\partial u}x + \hat{\mathbf{j}}\frac{\partial N}{\partial u}y + \hat{\mathbf{k}}\frac{\partial N}{\partial u}z \qquad (11.13)$$

where $(\partial N/\partial u)x$ is given as

$$\begin{bmatrix} \dfrac{\partial N_1}{\partial u} & \dfrac{\partial N_2}{\partial u} & \cdots & \dfrac{\partial N_8}{\partial u} \end{bmatrix} \begin{bmatrix} x_1 \\ x_2 \\ \vdots \\ x_8 \end{bmatrix} \qquad (11.14)$$

The terms $(\partial N/\partial u)y$ and $(\partial N/\partial u)z$ are obtained in similar manner. To obtain $\mathbf{v}_v$ and $\mathbf{v}_p$ we use the derivatives $\partial N/\partial v$ and $\partial N/\partial p$, respectively, shown in Table 11.2 (these derivatives were discussed in Chapter 8 in the context of nodal elements).

The following general comments can be made at this point:

- The vectors $\mathbf{v}_u$, $\mathbf{v}_v$, and $\mathbf{v}_p$ are obtained from the coordinates of the eight nodes of the element and the relations in Table 11.12.
- ∇u, ∇v, and ∇p are calculated from Eq. (11.12)
- With the twelve functions ϕ_i (Table 11.1) we obtain the twelve vector shape functions $\mathbf{w}_i$ as follows:

$$\mathbf{w}_i = \phi_i \nabla u \quad \text{(for } i=1,3,9,11) \qquad (11.15a)$$
$$\mathbf{w}_i = \phi_i \nabla v \quad \text{(for } i=2,4,10,12) \qquad (11.15b)$$
$$\mathbf{w}_i = \phi_i \nabla p \quad \text{(for } i=5,6,7,8) \qquad (11.15c)$$

Table 11.2. Nodal shape functions and their derivatives for the first order hexahedral element.

Node	N	$\partial N/\partial u$	$\partial N/\partial v$	$\partial N/\partial p$
1	$a_2b_2c_2/8$	$-b_2c_2/8$	$-a_2c_2/8$	$-a_2b_2/8$
2	$a_1b_2c_2/8$	$b_2c_2/8$	$-a_1c_2/8$	$-a_1b_2/8$
3	$a_1b_1c_2/8$	$b_1c_2/8$	$a_1c_2/8$	$-a_1b_1/8$
4	$a_2b_1c_2/8$	$-b_1c_2/8$	$a_2c_2/8$	$-a_2b_1/8$
5	$a_2b_2c_1/8$	$-b_2c_1/8$	$-a_2c_1/8$	$a_2b_2/8$
6	$a_1b_2c_1/8$	$b_2c_1/8$	$-a_1c_1/8$	$a_1b_2/8$
7	$a_1b_1c_1/8$	$b_1c_1/8$	$a_1c_1/8$	$a_1b_1/8$
8	$a_2b_1c_1/8$	$-b_1c_1/8$	$a_2c_1/8$	$a_2b_1/8$

where: $a_1=(1+u)$ $a_2=1-u$ $b_1=1+v$ $b_2=1-v$ $c_1=1+p$ $c_2=1-p$

454   11. Hexahedral Edge Elements – Some 3D Applications

In working with Maxwell's equations, we will often need to evaluate the expression $(\nabla \times \mathbf{w}_i)$. These expressions are calculated next using edge No. 11 as an example:

$$\nabla \times \mathbf{w}_{11} = \nabla \times (\phi_{11} \nabla u) = \phi_{11}(\nabla \times \nabla u) + \nabla \phi_{11} \times \nabla u$$

The first term on the right-hand side is zero and we get

$$\nabla \times \mathbf{w}_{11} = \nabla \phi_{11} \times \nabla u \qquad (11.16)$$

Similar expressions are obtained for the other edges of the element.

Now we evaluate the expression for ∇, needed to calculate $\nabla \phi$. We have

$$\nabla = \begin{bmatrix} \dfrac{\partial}{\partial x} \\ \dfrac{\partial}{\partial y} \\ \dfrac{\partial}{\partial z} \end{bmatrix} = \begin{bmatrix} \dfrac{\partial u}{\partial x} & \dfrac{\partial v}{\partial x} & \dfrac{\partial p}{\partial x} \\ \dfrac{\partial u}{\partial y} & \dfrac{\partial v}{\partial y} & \dfrac{\partial p}{\partial y} \\ \dfrac{\partial u}{\partial z} & \dfrac{\partial v}{\partial z} & \dfrac{\partial p}{\partial z} \end{bmatrix} \begin{bmatrix} \dfrac{\partial}{\partial u} \\ \dfrac{\partial}{\partial v} \\ \dfrac{\partial}{\partial p} \end{bmatrix}$$

or, noting for example, that $\nabla u = \hat{\mathbf{i}} \partial u/\partial x + \hat{\mathbf{j}} \partial u/\partial y + \hat{\mathbf{k}} \partial u/\partial z$,

$$\nabla = \begin{bmatrix} \nabla u & \nabla v & \nabla p \end{bmatrix} \begin{bmatrix} \dfrac{\partial}{\partial u} \\ \dfrac{\partial}{\partial v} \\ \dfrac{\partial}{\partial p} \end{bmatrix}$$

This gives

$$\nabla = \dfrac{\partial}{\partial u} \nabla u + \dfrac{\partial}{\partial v} \nabla v + \dfrac{\partial}{\partial p} \nabla p \qquad (11.17)$$

Substituting Eq. (11.17) for ∇ in Eq. (11.16), we get

$$\nabla \times \mathbf{w}_{11} = \left(\dfrac{\partial \phi_{11}}{\partial u} \nabla u + \dfrac{\partial \phi_{11}}{\partial v} \nabla v + \dfrac{\partial \phi_{11}}{\partial p} \nabla p \right) \times \nabla u$$

and

$$\nabla \times \mathbf{w}_{11} = -\dfrac{\partial \phi_{11}}{\partial v} (\nabla u \times \nabla v) + \dfrac{\partial \phi_{11}}{\partial p} (\nabla p \times \nabla u) \qquad (11.18)$$

where the first vector product was commutated for convenience. Using the expression in Eq. (11.12), the vector product $\nabla u \times \nabla v$ becomes

$$\nabla u \times \nabla v = (\mathbf{v}_v \times \mathbf{v}_p) \times (\mathbf{v}_p \times \mathbf{v}_u)/vol^2$$

Noting that for general vectors **a**, **b**, and **c**, the following relation holds:

$$\mathbf{a} \times (\mathbf{b} \times \mathbf{c}) = \mathbf{b}(\mathbf{a} \cdot \mathbf{c}) - \mathbf{c}(\mathbf{a} \cdot \mathbf{b})$$

and denoting $\mathbf{a} = \mathbf{v}_v \times \mathbf{v}_p$, $\mathbf{b} = \mathbf{v}_p$, $\mathbf{c} = \mathbf{v}_u$, we obtain

$$\nabla u \times \nabla v = \{\mathbf{v}_p[(\mathbf{v}_v \times \mathbf{v}_p) \cdot \mathbf{v}_u] - \mathbf{v}_u[(\mathbf{v}_v \times \mathbf{v}_p) \cdot \mathbf{v}_p]\}/vol^2$$

This gives

$$\nabla u \times \nabla v = \mathbf{v}_p/vol \qquad (11.19a)$$

Similarly,

$$\nabla v \times \nabla p = \mathbf{v}_u/vol \qquad (11.19b)$$

and

$$\nabla p \times \nabla u = \mathbf{v}_v/vol \qquad (11.19c)$$

Substituting these expressions in Eq. (11.18), we get

$$\nabla \times \mathbf{w}_{11} = \frac{1}{vol}\left(\frac{\partial \phi_{11}}{\partial p}\mathbf{v}_v - \frac{\partial \phi_{11}}{\partial v}\mathbf{v}_p\right)$$

Generalizing this expression, we get:
For the edges parallel to u

$$\nabla \times \mathbf{w}_i = \frac{1}{vol}\left(\frac{\partial \phi_i}{\partial p}\mathbf{v}_v - \frac{\partial \phi_i}{\partial v}\mathbf{v}_p\right) \qquad \text{(edges 1,3,9,11)} \qquad (11.20a)$$

For the edges parallel to v

$$\nabla \times \mathbf{w}_i = \frac{1}{vol}\left(\frac{\partial \phi_i}{\partial u}\mathbf{v}_p - \frac{\partial \phi_i}{\partial p}\mathbf{v}_u\right) \qquad \text{(edges 2,4,10,12)} \qquad (11.20b)$$

For the edges parallel to p

$$\nabla \times \mathbf{w}_i = \frac{1}{vol}\left(\frac{\partial \phi_i}{\partial v}\mathbf{v}_u - \frac{\partial \phi_i}{\partial u}\mathbf{v}_v\right) \qquad \text{(edges 5,6,7,8)} \qquad (11.20c)$$

These expressions are easy to evaluate once the vectors $\mathbf{v}_u$, $\mathbf{v}_v$, $\mathbf{v}_p$, and the expressions in Table 11.1 are calculated.

11.3. Construction of the Shape Functions

In this section, we discuss some program subroutines needed for the calculation of edge element shape functions $\mathbf{w}_i$ and the curl of the shape functions $\nabla \times \mathbf{w}_i$. The primary data for these subroutines are the coordinates (arrays xr, yr, zr) and integration points (u_i, v_i, p_i) necessary for the numerical integration (see Chapter 8). These subroutines are presented and discussed here in some detail because the understanding of the shape functions and the evaluation of their curl is fundamental to the application of edge elements in solving Maxwell's equations. The construction of the edge element shape functions is more complex than that for nodal shape functions and therefore the subroutine listings and the method of construction of the edge element shape functions are presented to clarify the algebraic operations of the previous section.

In the main subroutine, which is responsible for the numerical integration of the elemental matrices (not shown here), there are two subroutines which are called successively: *Dxyzuvp* and *Shapcur*. As can be seen from the listings, most of the variable arrays are transmitted through "common blocks" and some (like the integration point coordinates u, v, p) are transmitted through parameters.

The vectors $\mathbf{v}_u = \partial \mathbf{r}/\partial u$, $\mathbf{v}_v = \partial \mathbf{r}/\partial v$ and $\mathbf{v}_p = \partial \mathbf{r}/\partial p$ are calculated in subroutine *Dxyzuvp* using Eq. (11.4). To perform the required operations, Eqs. (11.13) and (11.14) are evaluated together with subroutine *Funcn*. Because the components of $\mathbf{v}_u$, $\mathbf{v}_v$, and $\mathbf{v}_p$ are the terms of the Jacobian $[J_1]$, the determinant of $[J_1]$, called "den" (and which equals the volume "vol" of the element) is also calculated.

Following the calculations in subroutine *Dxyzuvp*, subroutine *Shapcur* is called to calculate $\mathbf{w}_i$ and $\nabla \times \mathbf{w}_i$. The first task is to call *Funcfi* (also dependent on u,v,p) in which the function ϕ and its derivatives are calculated. Table 11.1 is used to define the variables fi and df [$df(1,12)$ correspond to $\partial \phi/\partial u$, $df(2,12)$ correspond to $\partial \phi/\partial v$ and $df(3,12)$ correspond to $\partial \phi/\partial p$). Upon return from *Funcfi*, function *Cross* is called to calculate the vector products $\mathbf{v}_u \times \mathbf{v}_v$, $\mathbf{v}_v \times \mathbf{v}_p$, and $\mathbf{v}_p \times \mathbf{v}_u$ (resulting in $\mathbf{v}_{12}$, $\mathbf{v}_{23}$, and $\mathbf{v}_{31}$, respectively). With these three vectors, the functions $\mathbf{w}_i$ are calculated in three different groups according to the direction of the edges and Eqs. (11.12) and (11.15). The second task of *Shapcur* is to calculate the terms of $\nabla \times \mathbf{w}_i$ which are also divided in three groups according to the directions of the edges and Eq. (11.20).

Two remarks are appropriate here:

Construction of the Shape Functions   457

- The terms of $\mathbf{w}_i$ and $\nabla \times \mathbf{w}_i$, as calculated in *Shapcur*, are not divided by *vol*; this will be done in another step of the program, to be discussed later.
- In subroutines *Funcfi* and *Funcn*, the integration points are transformed so that the local domain is bounded by the points (0,0,0) and (1,1,1) instead of the local space for nodal elements which was defined between (−1,−1,−1) and (1,1,1).

```fortran
      subroutine Dxyzuvp(u,v,p)
      implicit double precision (a-h), double precision (o-z)
      common /points/xr(8),yr(8),zr(8)
      common /fin/ fi(12),df(3,12),dn(3,8),fn(8),xact,yact,zact
      common /dxuvp/dnxu,dnxv,dnxp,dnyu,dnyv,dnyp,dnzu,dnzv,dnzp,den
c-------------- call Funcn for derivatives of function N
      call Funcn(u,v,p)
c-------------- Calculation of terms dNx/du, dNy/du, dNz/du
c                                    dNx/dv, dNy/dv, dNz/dv
c                                    dNx/dp, dNy/dp, dNz/dp
      dnxu=0.
      do 1 m=1,8
1     dnxu=dnxu+dn(1,m)*xr(m)
      dnyu=0.
      do 2 m=1,8
2     dnyu=dnyu+dn(1,m)*yr(m)
      dnzu=0.
      do 3 m=1,8
3     dnzu=dnzu+dn(1,m)*zr(m)
      dnxv=0.
      do 11 m=1,8
11    dnxv=dnxv+dn(2,m)*xr(m)
      dnyv=0.
      do 12 m=1,8
12    dnyv=dnyv+dn(2,m)*yr(m)
      dnzv=0.
      do 13 m=1,8
13    dnzv=dnzv+dn(2,m)*zr(m)
      dnxp=0.
      do 21 m=1,8
21    dnxp=dnxp+dn(3,m)*xr(m)
      dnyp=0.
      do 22 m=1,8
22    dnyp=dnyp+dn(3,m)*yr(m)
      dnzp=0.
      do 23 m=1,8
23    dnzp=dnzp+dn(3,m)*zr(m)
c-------------- Calculation of the determinant of the Jacobian
      den=dnxu*(dnyv*dnzp-dnzv*dnyp)+
     *    dnyu*(dnzv*dnxp-dnxv*dnzp)+
     *    dnzu*(dnxv*dnyp-dnyv*dnxp)
      return
      end
c-----------------------------------------------------------------
      subroutine Shapcur(u,v,p)
      include 'feec.cmn'
      common /points/xr(8),yr(8),zr(8)
      common /fin/ fi(12),df(3,12),dn(3,8),fn(8),xact,yact,zact
      common /dxuvp/dnxu,dnxv,dnxp,dnyu,dnyv,dnyp,dnzu,dnzv,dnzp,den
      common /shap/vsfx(12),vsfy(12),vsfz(12),csfx(12),csfy(12),csfz(12)
c-------------- call Funcfi for functions Fi and its derivatives
      call Funcfi(u,v,p)
c-------------- Calculation of the 3 vectors v12,v23,v31
      call Cross(dnxu,dnyu,dnzu,dnxv,dnyv,dnzv,vx12,vy12,vz12)
      call Cross(dnxv,dnyv,dnzv,dnxp,dnyp,dnzp,vx23,vy23,vz23)
      call Cross(dnxp,dnyp,dnzp,dnxu,dnyu,dnzu,vx31,vy31,vz31)
c-------------- Calculation of edge shape functions
c-------------- For the edges parallel to u
      vsfx(1)=fi(1)*vx23
      vsfx(3)=fi(3)*vx23
      vsfx(9)=fi(9)*vx23
```

```
      vsfx(11)=fi(11)*vx23
      vsfy(1)=fi(1)*vy23
      vsfy(3)=fi(3)*vy23
      vsfy(9)=fi(9)*vy23
      vsfy(11)=fi(11)*vy23
      vsfz(1)=fi(1)*vz23
      vsfz(3)=fi(3)*vz23
      vsfz(9)=fi(9)*vz23
      vsfz(11)=fi(11)*vz23
c-------------- For the edges parallel to v
      vsfx(2)=fi(2)*vx31
      vsfx(4)=fi(4)*vx31
      vsfx(10)=fi(10)*vx31
      vsfx(12)=fi(12)*vx31
      vsfy(2)=fi(2)*vy31
      vsfy(4)=fi(4)*vy31
      vsfy(10)=fi(10)*vy31
      vsfy(12)=fi(12)*vy31
      vsfz(2)=fi(2)*vz31
      vsfz(4)=fi(4)*vz31
      vsfz(10)=fi(10)*vz31
      vsfz(12)=fi(12)*vz31
c-------------- For the edges parallel to p
      vsfx(5)=fi(5)*vx12
      vsfx(6)=fi(6)*vx12
      vsfx(7)=fi(7)*vx12
      vsfx(8)=fi(8)*vx12
      vsfy(5)=fi(5)*vy12
      vsfy(6)=fi(6)*vy12
      vsfy(7)=fi(7)*vy12
      vsfy(8)=fi(8)*vy12
      vsfz(5)=fi(5)*vz12
      vsfz(6)=fi(6)*vz12
      vsfz(7)=fi(7)*vz12
      vsfz(8)=fi(8)*vz12
c-------------- Construction of the curls of the edges
c-------------- Edges parallel to u
      csfx(1)=dnxv*df(3,1)-dnxp*df(2,1)
      csfx(3)=dnxv*df(3,3)-dnxp*df(2,3)
      csfx(9)=dnxv*df(3,9)-dnxp*df(2,9)
      csfx(11)=dnxv*df(3,11)-dnxp*df(2,11)
      csfy(1)=dnyv*df(3,1)-dnyp*df(2,1)
      csfy(3)=dnyv*df(3,3)-dnyp*df(2,3)
      csfy(9)=dnyv*df(3,9)-dnyp*df(2,9)
      csfy(11)=dnyv*df(3,11)-dnyp*df(2,11)
      csfz(1)=dnzv*df(3,1)-dnzp*df(2,1)
      csfz(3)=dnzv*df(3,3)-dnzp*df(2,3)
      csfz(9)=dnzv*df(3,9)-dnzp*df(2,9)
      csfz(11)=dnzv*df(3,11)-dnzp*df(2,11)
c-------------- Edges parallel to v
      csfx(2)=dnxp*df(1,2)-dnxu*df(3,2)
      csfx(4)=dnxp*df(1,4)-dnxu*df(3,4)
      csfx(10)=dnxp*df(1,10)-dnxu*df(3,10)
      csfx(12)=dnxp*df(1,12)-dnxu*df(3,12)
      csfy(2)=dnyp*df(1,2)-dnyu*df(3,2)
      csfy(4)=dnyp*df(1,4)-dnyu*df(3,4)
      csfy(10)=dnyp*df(1,10)-dnyu*df(3,10)
      csfy(12)=dnyp*df(1,12)-dnyu*df(3,12)
      csfz(2)=dnzp*df(1,2)-dnzu*df(3,2)
      csfz(4)=dnzp*df(1,4)-dnzu*df(3,4)
      csfz(10)=dnzp*df(1,10)-dnzu*df(3,10)
      csfz(12)=dnzp*df(1,12)-dnzu*df(3,12)
c-------------- Edges parallel to p
      csfx(5)=dnxu*df(2,5)-dnxv*df(1,5)
      csfx(6)=dnxu*df(2,6)-dnxv*df(1,6)
      csfx(7)=dnxu*df(2,7)-dnxv*df(1,7)
      csfx(8)=dnxu*df(2,8)-dnxv*df(1,8)
      csfy(5)=dnyu*df(2,5)-dnyv*df(1,5)
      csfy(6)=dnyu*df(2,6)-dnyv*df(1,6)
      csfy(7)=dnyu*df(2,7)-dnyv*df(1,7)
      csfy(8)=dnyu*df(2,8)-dnyv*df(1,8)
      csfz(5)=dnzu*df(2,5)-dnzv*df(1,5)
      csfz(6)=dnzu*df(2,6)-dnzv*df(1,6)
      csfz(7)=dnzu*df(2,7)-dnzv*df(1,7)
```

Construction of the Shape Functions

```fortran
      csfz(8)=dnzu*df(2,8)-dnzv*df(1,8)
      return
      end
c-------------------------------------------------------------
      subroutine Cross(a1,a2,a3,b1,b2,b3,c1,c2,c3)
      implicit double precision (a-h), double precision (o-z)
      c1=a2*b3-a3*b2
      c2=a3*b1-a1*b3
      c3=a1*b2-a2*b1
      return
      end
c-------------------------------------------------------------
      subroutine Funcfi(uu,vv,pp)
      implicit double precision (a-h), double precision (o-z)
      common /fin/ fi(12),df(3,12),dn(3,8),fn(8),xact,yact,zact
c-------------- Change of coords. to work in the element defined by
c-------------- 0,0,0;1,1,1, instead of the element -1,-1,-1;1,1,1
      u=(1.+uu)/2.
      v=(1.+vv)/2.
      p=(1.+pp)/2.
c-------------- fi(12) is the function Fi for the 12 edges
      fi(1)=(1-v)*(1-p)
      fi(2)=u*(1-p)
      fi(3)=v*(1-p)
      fi(4)=(1-u)*(1-p)
      fi(5)=(1-u)*(1-v)
      fi(6)=u*(1-v)
      fi(7)=u*v
      fi(8)=(1-u)*v
      fi(9)=(1-v)*p
      fi(10)=u*p
      fi(11)=v*p
      fi(12)=(1-u)*p
c-------------- df(3,12) is the array for: dFi/du --> df(1,i)
c                                          dFi/dv --> df(2,i)
c                                          dFi/dp --> df(3,i)
      df(1,1)=0.
      df(1,2)=1-p
      df(1,3)=0
      df(1,4)=-(1-p)
      df(1,5)=-(1-v)
      df(1,6)=(1-v)
      df(1,7)=v
      df(1,8)=-v
      df(1,9)=0.
      df(1,10)=p
      df(1,11)=0.
      df(1,12)=-p
      df(2,1)=-(1-p)
      df(2,2)=0.
      df(2,3)=(1-p)
      df(2,4)=0.
      df(2,5)=-(1-u)
      df(2,6)=-u
      df(2,7)=u
      df(2,8)=(1-u)
      df(2,9)=-p
      df(2,10)=0.
      df(2,11)=p
      df(2,12)=0.
      df(3,1)=-(1-v)
      df(3,2)=-u
      df(3,3)=-v
      df(3,4)=-(1-u)
      df(3,5)=0.
      df(3,6)=0.
      df(3,7)=0.
      df(3,8)=0.
      df(3,9)=(1-v)
      df(3,10)=u
      df(3,11)=v
      df(3,12)=(1-u)
      return
      end
```

```
c------------------------------------------------------------------
      subroutine Funcn(uu,vv,pp)
      implicit double precision (a-h), double precision (o-z)
      common /fin/ fi(12),df(3,12),dn(3,8),fn(8),xact,yact,zact
c-------------- Change of coords. to work in the element defined by
c-------------- 0,0,0;1,1,1, instead of the element -1,-1,-1;1,1,1
      u=(1.+uu)/2.
      v=(1.+vv)/2.
      p=(1.+pp)/2.
c-------------- fn(8) is the array for the shape functions [N]
      fn(1)=(1-u)*(1-v)*(1-p)
      fn(2)=u*(1-v)*(1-p)
      fn(3)=u*v*(1-p)
      fn(4)=(1-u)*v*(1-p)
      fn(5)=(1-u)*(1-v)*p
      fn(6)=u*(1-v)*p
      fn(7)=u*v*p
      fn(8)=(1-u)*v*p
c-------------- dn(3,8) is the array for: dN/du --> dn(1,i)
c                                         dN/dv --> dn(2,i)
c                                         dN/dp --> dn(3,i)
      dn(1,1)=-(1-v)*(1-p)
      dn(1,2)=(1-v)*(1-p)
      dn(1,3)=v*(1-p)
      dn(1,4)=-v*(1-p)
      dn(1,5)=-(1-v)*p
      dn(1,6)=(1-v)*p
      dn(1,7)=v*p
      dn(1,8)=-v*p
      dn(2,1)=-(1-u)*(1-p)
      dn(2,2)=-u*(1-p)
      dn(2,3)=u*(1-p)
      dn(2,4)=(1-u)*(1-p)
      dn(2,5)=-(1-u)*p
      dn(2,6)=-u*p
      dn(2,7)=u*p
      dn(2,8)=(1-u)*p
      dn(3,1)=-(1-u)*(1-v)
      dn(3,2)=-u*(1-v)
      dn(3,3)=-u*v
      dn(3,4)=-(1-u)*v
      dn(3,5)=(1-u)*(1-v)
      dn(3,6)=u*(1-v)
      dn(3,7)=u*v
      dn(3,8)=(1-u)*v
      return
      end
```

11.4. Application of Edge Elements to Low-Frequency Maxwell's Equations

Now that the shape functions for the hexahedral edge element have been defined we can turn to the application of edge elements in various cases as described below. Although we will use hexahedral elements throughout, we must point out that the explanations below do not, in any way, depend on this particular element. Other elements may be used just as well although their shape functions must first be defined. To simplify discussion, we use a single formulation in the whole solution domain even for those applications which could benefit from coupling of various formulations. This simplicity comes at a slight increase in memory usage but this increase is justified from the point of view of robustness and simplicity of code implementation.

11.4.1. Static Cases

Maxwell's equation for the static case used here is

$$\nabla \times \mathbf{H} = \mathbf{J} \tag{11.21}$$

Using the magnetic vector potential **A**, defined as

$$\mathbf{B} = \nabla \times \mathbf{A}$$

and $\mathbf{B} = \mu \mathbf{H}$, we obtain

$$\nabla \times \frac{1}{\mu}(\nabla \times \mathbf{A}) = \mathbf{J} \tag{11.22}$$

A is approximated as

$$\mathbf{A} = \sum_{i=1}^{12} A_i \mathbf{w}_i \tag{11.23}$$

The basic concepts of Galerkin's method were discussed in Chapter 8. Therefore, without repeating these steps we show its application to Eq. (11.22). In this case, the weighting function is $\mathbf{w}_m$, which is of the same nature as $\mathbf{w}_i$ (discussed in the previous section). Because both $\mathbf{w}_m$ and Eq. (11.22) are vector expressions, we calculate the scalar product between $\mathbf{w}_m$ and Eq. (11.22) to obtain

$$\int_\Omega \nabla \times \left(\frac{1}{\mu}\nabla \times \mathbf{A}\right) \cdot \mathbf{w}_m d\Omega - \int_\Omega \mathbf{J} \cdot \mathbf{w}_m d\Omega = 0 \tag{11.24}$$

For two general vectors **P** and **Q**, the following identity applies:

$$\int_\Omega (\nabla \times \mathbf{P}) \cdot \mathbf{Q} d\Omega = \int_\Omega \mathbf{P} \cdot (\nabla \times \mathbf{Q}) d\Omega + \int_\Omega \nabla \cdot (\mathbf{P} \times \mathbf{Q}) d\Omega =$$

$$\int_\Omega \mathbf{P} \cdot (\nabla \times \mathbf{Q}) d\Omega + \int_{S(\Omega)} (\mathbf{P} \times \mathbf{Q}) \cdot d\mathbf{s}$$

With this, the first term of Eq. (11.24) becomes

$$\int_\Omega \left(\frac{1}{\mu}\nabla \times \mathbf{A}\right) \cdot (\nabla \times \mathbf{w}_m) d\Omega + \int_{S(\Omega)} \left[\left(\frac{1}{\mu}\nabla \times \mathbf{A}\right) \times \mathbf{w}_m\right] \cdot d\mathbf{s} \tag{11.25}$$

The second term in Eq. (11.25) is related to the boundary conditions and is forced to zero by the residual method.

The boundary $S(\Omega)$ is divided into two parts:
- S_1, to which the magnetic flux density is parallel. Suppose that a local plane xy coincides with S_1. In this case, we impose $A_x=0$ and $A_y=0$ on any edge on S_1. Because these edges are not unknowns, $\mathbf{w}_m$ on this boundary is related to the neighboring edges and is always zero. We can write for $\mathbf{B}$:

$$B_x = \frac{\partial A_z}{\partial y} - \frac{\partial A_y}{\partial z} \qquad B_y = \frac{\partial A_x}{\partial z} - \frac{\partial A_z}{\partial x} \qquad B_z = \frac{\partial A_y}{\partial x} - \frac{\partial A_x}{\partial y}$$

and only the components B_x and B_y exist, matching the desired physical conditions. Strictly speaking, this condition cannot be called a "Dirichlet boundary condition" since the Dirichlet boundary condition was established for scalar partial differential equations, which is not the case here. However, in terms of numerical implementation the procedure is the same.

- $S_2 = S(\Omega) - S_1$. On this complementary boundary, the magnetic flux density $\mathbf{B}$ is perpendicular. By not imposing any value on the edges that belong to S_2, these edges are unknowns and $\mathbf{w}_m$ is nonzero. Therefore, to satisfy the second term of Eq. (11.25), we have

$$\left[\left(\frac{1}{\mu}\nabla\times\mathbf{A}\right)\times\mathbf{w}_m\right]\cdot d\mathbf{s} = 0$$
$$[\mathbf{H}\times\mathbf{w}_m]\cdot d\mathbf{s} = 0$$
$$\mathbf{w}_m\cdot(\mathbf{H}\times d\mathbf{s}) = 0$$

Calling $\hat{\mathbf{n}}$ the normal unit vector and denoting the tangential direction with a "t," we have

$$\mathbf{H}\times\hat{\mathbf{n}} ds = 0$$
$$(\mathbf{H}_n + \mathbf{H}_t)\times\hat{\mathbf{n}} = 0$$
$$\mathbf{H}_n\times\hat{\mathbf{n}} + \mathbf{H}_t\times\hat{\mathbf{n}} = 0$$

This expression shows that $\mathbf{H}_t$ should be zero, since only the first vector product above is zero for any value of $\mathbf{H}_n$. This means that the magnetic field has only normal components on boundary S_2 as desired. In a manner similar to the discussion on the Dirichlet boundary condition above, the term "Neumann boundary condition" is not appropriate for S_2, but it has many similarities with the Neumann boundary condition.

In summary, the boundary conditions above behave exactly the same as for the classical 2D vector potential applications. Therefore, Eq. (11.24) leads directly to the elemental matrix

$$\int_{\Omega}\left(\frac{1}{\mu}\nabla\times\mathbf{A}\right)\cdot(\nabla\times\mathbf{w}_m)d\Omega - \int_{\Omega}\mathbf{J}\cdot\mathbf{w}_m d\Omega = 0 \qquad (11.26)$$

Using the approximation of **A** in Eq. (11.23), the first term of Eq. (11.26) becomes

$$\int_{V_i}\left(\frac{1}{\mu}\nabla\times\sum_{n=1}^{12}\mathbf{w}_n A_n\right)\cdot(\nabla\times\mathbf{w}_m)dv \qquad (11.27)$$

or

$$\int_{V_i}\frac{1}{\mu}\left[(\nabla\times\mathbf{w}_1 A_1)\cdot(\nabla\times\mathbf{w}_m) + (\nabla\times\mathbf{w}_2 A_2)\cdot(\nabla\times\mathbf{w}_m) + ...\right]dv$$

This may be written as

$$\int_{V_i}\frac{1}{\mu}\left[(\nabla\times\mathbf{w}_1)\cdot(\nabla\times\mathbf{w}_m) \quad (\nabla\times\mathbf{w}_2)\cdot(\nabla\times\mathbf{w}_m) \quad \cdots \quad (\nabla\times\mathbf{w}_{12})\cdot(\nabla\times\mathbf{w}_m)\right]\begin{bmatrix}A_1\\A_2\\ \vdots \\A_{12}\end{bmatrix}dv$$
$$(11.28)$$

where V_i is the volume of element i. With $\nabla\times\mathbf{w}_m$ as found in the previous section and allowing m to vary between 1 and 12, we obtain the elemental contribution matrix through numerical integration, as discussed in Chapter 8. A generic term of the elemental matrix is given as

$$a(m,n) = \sum_{i=1}^{8} W_i \frac{1}{\mu}\left[\nabla\times\mathbf{w}_n(u_i,v_i,p_i)\right]\cdot\left[\nabla\times\mathbf{w}_m(u_i,v_i,p_i)\right] det[J(u_i,v_i,p_i)] \qquad (11.29)$$

where $W_i = 1$ and u_i, v_i, p_i are the eight combinations ($\pm 1/\sqrt{3}$, $\pm 1/\sqrt{3}$, $\pm 1/\sqrt{3}$), as was shown in Chapter 8. This matrix is now assembled into the left-hand side of the global matrix system.

The expression for the second term of Eq. (11.26), which relates to the source **J**, can be easily calculated as

$$s(m) = \sum_{i=1}^{8} W_i \mathbf{J}\cdot\mathbf{w}_m(u_i,v_i,p_i) \, det[J(u_i,v_i,p_i)] \qquad (11.30)$$

These contributions are assembled into the right-hand side of the global matrix system.

11.4.2. Listing of the Matrix Construction Code

The listing below is a section of the subroutine responsible for the evaluation of the terms $a(m,n)$ of the elemental matrix and $s(m)$ of the source vector. The actual subroutines employed are those listed in section 11.3 while the listing shown here is a simplified version of the subroutine used in our code.

```
c-------------- Quadrature Gauss integration points and weight w
      w=1.
      va=1/sqrt(3.)
      un(1)=-va
      vn(1)=-va
      pn(1)=-va
      un(2)=va
      vn(2)=-va
      pn(2)=-va
            .
            .
      un(8)=-va
      vn(8)=va
      pn(8)=va
c-------------- Loop on the elements; nel is the number of elements
      do 1 ii=1,nel
      n1=ktri(ii,1)
      n2=ktri(ii,2)
      n3=ktri(ii,3)
      n4=ktri(ii,4)
      n5=ktri(ii,5)
      n6=ktri(ii,6)
      n7=ktri(ii,7)
      n8=ktri(ii,8)
      nmat=ktri(ii,9)
c-------------- xj, yj, zj are the components of the current density J.
      xj=dxj(ii)
      yj=dyj(ii)
      zj=dzj(ii)
c-------------- Defining the arrays xr,yr,zr, coordinates of the 8 nodes
c                of the element
      xr(1)=xcor(n1)
      yr(1)=ycor(n1)
      zr(1)=zcor(n1)
      xr(2)=xcor(n2)
      yr(2)=ycor(n2)
      zr(2)=zcor(n2)
            .
            .
      xr(8)=xcor(n8)
      yr(8)=ycor(n8)
      zr(8)=zcor(n8)
c-------------- Initialization of matrix a(12,12) and source s(12)
      do i=1,12
      s(i)=0.
      do j=1,12
      a(i,j)=0.
      enddo
      enddo
c-------------- Perm gives the permeability of material nmat
      call Perm(nmat,perm)
c-------------- Gauss integration loop
      do 7 l=1,8
c-------------- Calculation of the Jacobian (vectors Vu,Vv and Vp) and det(J)
      call Dxyzuvp(un(l),vn(l),pn(l))
c-------------- Calculation of the shape and curl functions for the edges
      call Shapcur(un(l),vn(l),pn(l))
c-------------- Calculation of terms of matrices a(12,12) and s(12)
      do i=1,12
```

```
        s(i)=s(i)+w*(vsfx(i)*xj+vsfy(i)*yj+vsfz(i)*zj)/8.
        do j=1,12
          a(i,j)=a(i,j)+w*(csfx(i)*csfx(j)+csfy(i)*csfy(j)+
*                        csfz(i)*csfz(j)))/perm/den/8.
        enddo
      enddo
    7 continue
c--------------- Assembling matrices a(12,12) and s(12) in the global system
      call Assemb(ii)
    1 continue
       .
       .
       .
```

Comments:

- Arrays KTRI(II,9) give the node numbers of the eight nodes of the element in its first eight positions. The ninth position gives the material index for the element. Arrays DJX, DJY, and DJZ are the three components of the current density in the element.
- Arrays XCOR, YCOR, ZCOR contain the coordinates of the nodes.
- The calculation of $s(m)$ and $a(m,n)$ takes into account that $\mathbf{w}_i$ and $\nabla \times \mathbf{w}_i$, which were calculated in subroutine *Shapcur* were not divided by DEN [*vol* in Eqs. (11.12) and (11.20] as required. This is done here because the volume of the element equals $det[J]$ which is also used in Eqs. (11.29) and (11.30).
- After matrices $a(12,12)$ and $s(12)$ are evaluated, they are assembled into the global matrix using subroutine ASSEMB.
- The factor 1/8 in the expressions for $a(m,n)$ and $s(m)$ occurs because the integration is performed on an element in the space (0,0,0) to (1,1,1), while the integration points are defined for the space (−1,−1,−1) to (1,1,1). The transformation between the two local systems is

$$u = (1 + uu)/2 \qquad v = (1 + vv)/2 \qquad p = (1 + pp)/2$$

For example, $u(0,1)$ corresponds to $uu(-1,1)$.

The integration requires the definition of another Jacobian J_u to perform the following:

$$\int_0^1 f(u,v,p)\, det[J_u]\, du\, dv\, dp = \int_{-1}^1 f(uu,vv,pp)\, duu\, dvv\, dpp$$

where

$$det[J_u] = det \begin{bmatrix} \dfrac{\partial u}{\partial uu} & \dfrac{\partial v}{\partial uu} & \dfrac{\partial p}{\partial uu} \\ \dfrac{\partial u}{\partial vv} & \dfrac{\partial v}{\partial vv} & \dfrac{\partial p}{\partial vv} \\ \dfrac{\partial u}{\partial pp} & \dfrac{\partial v}{\partial pp} & \dfrac{\partial p}{\partial pp} \end{bmatrix} = det \begin{bmatrix} \dfrac{1}{2} & 0 & 0 \\ 0 & \dfrac{1}{2} & 0 \\ 0 & 0 & \dfrac{1}{2} \end{bmatrix} = \dfrac{1}{8}$$

11.4.3. Modeling of Permanent Magnets

Modeling of permanent magnets can be easily accomplished by assuming there are no currents in the regions occupied by the magnets and taking $\mathbf{B}_r$ as the remnant flux density of the magnet.

The field equation under these assumptions is

$$\mathbf{B} = \mu\mathbf{H} + \mathbf{B}_r \quad \text{or} \quad \mathbf{H} = (\mathbf{B} - \mathbf{B}_r)/\mu \qquad (11.31)$$

Because $\nabla \times \mathbf{H} = 0$, we have

$$\nabla \times \left[\frac{1}{\mu}(\mathbf{B} - \mathbf{B}_r)\right] = 0 \quad \rightarrow \quad \nabla \times \frac{\mathbf{B}}{\mu} - \nabla \times \frac{\mathbf{B}_r}{\mu} = 0$$

Using $\mathbf{B} = \nabla \times \mathbf{A}$ we obtain for the weak form

$$\int_{\Omega_p} \nabla \times \left(\frac{1}{\mu}\nabla \times \mathbf{A}\right) \cdot \mathbf{w}_m d\Omega - \int_{\Omega_p} \nabla \times \mathbf{B}_r \cdot \mathbf{w}_m d\Omega = 0 \qquad (11.32)$$

where Ω_p is the permanent magnet region. The first term in Eq. (11.32) has been discussed before. The second term is the source term related to the permanent magnet. This can be written as

$$\int_{\Omega_p} \nabla \times \mathbf{B}_r \cdot \mathbf{w}_m d\Omega = \int_{\Omega_p} \nabla \times \mathbf{w}_m \cdot \mathbf{B}_r d\Omega$$

This expression is easily evaluated using $\nabla \times \mathbf{w}_m$ as calculated in subroutine *Shapcur*. The calculation of this source term follows the same procedure of implementation as for the current source $s(m)$ discussed in section 11.4.2.

11.4.4. Eddy Currents – the Time-Stepping Procedure

In the equation $\nabla \times \mathbf{H} = \mathbf{J}$, the current density $\mathbf{J}$ may contain one or two terms. In the static case, this is the source current density. In the eddy current case, $\mathbf{J}$ is the sum of the source current density $\mathbf{J}_s$ and the induced or eddy current density $\mathbf{J}_e$. Denoting σ the conductivity of the material in which eddy currents exist, we have

$$\mathbf{J}_e = \sigma \mathbf{E}$$

with

Application of Edge Elements to Low-Frequency Maxwell's Equations

$$\nabla \times \mathbf{E} = -\frac{\partial \mathbf{B}}{\partial t} = -\frac{\partial}{\partial t}\nabla \times \mathbf{A} \qquad (11.33)$$

or

$$\nabla \times \mathbf{E} = -\frac{\partial \mathbf{B}}{\partial t} = -\nabla \times \left(\frac{\partial \mathbf{A}}{\partial t} + \nabla \phi\right)$$

Assuming that $\mathbf{E}$ is an induced electric field and is generated due to the time variation of $\mathbf{A}$, the static term becomes zero and we have

$$\mathbf{E} = -\frac{\partial \mathbf{A}}{\partial t}$$

and

$$\mathbf{J}_e = -\sigma \frac{\partial \mathbf{A}}{\partial t} \qquad (11.34)$$

Therefore, $\nabla \times \mathbf{H} - \mathbf{J} = 0$ becomes

$$\nabla \times \left(\frac{1}{\mu}\nabla \times \mathbf{A}\right) + \sigma \frac{\partial \mathbf{A}}{\partial t} - \mathbf{J}_s = 0 \qquad (11.35)$$

If we wish to treat nonlinear problems with arbitrary excitation sources, the finite difference time discretization is the appropriate procedure. The time derivative becomes

$$\frac{\partial \mathbf{A}}{\partial t} = \frac{\mathbf{A}_i - \mathbf{A}_{i-1}}{\Delta t} \qquad (11.36)$$

where Δt is the time step and "i" and "$i-1$" denote the values of $\mathbf{A}$ at the current calculation step (unknown) and previous calculation step (known and stored), respectively. With this time discretization, Eq. (11.35) becomes

$$\nabla \times \left(\frac{1}{\mu}\nabla \times \mathbf{A}_i\right) + \frac{\sigma \mathbf{A}_i}{\Delta t} - \frac{\sigma \mathbf{A}_{i-1}}{\Delta t} - \mathbf{J}_s = 0 \qquad (11.37)$$

The first and last term of Eq. (11.37) have been calculated above. The second term, after applying Galerkin's method, becomes

$$\int_\Omega \frac{\sigma \mathbf{A}_i}{\Delta t} \cdot \mathbf{w}_m d\Omega$$

Using $\mathbf{A} = \sum_{n=1}^{12} \mathbf{w}_n A_n$, we obtain the following generic term

$$b(m,n) = \int_{v_i} \frac{\sigma \mathbf{w}_n \cdot \mathbf{w}_m}{\Delta t} dv \qquad (11.38)$$

This results in an elemental matrix $b(12,12)$ which must be assembled in the left side of the matrix system. The third term of Eq. (11.37) is obtained by multiplying the matrix $b(12,12)$ by the vector $\mathbf{A}_{i-1}$ (which is known from the previous time step), resulting in a vector of twelve values, to be assembled in the right side of the global matrix system. In this sense, this term is treated as any other source term.

11.4.5. Eddy Currents – The Complex Formulation

If the problem is linear, the excitation is sinusoidal and we are interested in the steady state solution, it is possible to use the following complex formulation. $\mathbf{A}$ is given by the Euler complex form (with $j = \sqrt{-1}$):

$$\mathbf{A} = \mathbf{A}^* e^{j\omega t}$$

with

$$\mathbf{A}^* = A e^{j\alpha}$$

where α is the phase angle of the vector $\mathbf{A}$. We have

$$\frac{d\mathbf{A}}{dt} = j\omega \mathbf{A}^* e^{j\omega t} \qquad (11.39)$$

and Eq. (11.35) becomes

$$\nabla \times \left(\frac{1}{\mu} \nabla \times \mathbf{A}^*\right) e^{j\omega t} + j\omega\sigma \mathbf{A}^* e^{j\omega t} - \mathbf{J}_s^* e^{j\omega t} = 0 \qquad (11.40)$$

Also using $\mathbf{J}_s = \mathbf{J}_s^* e^{j\omega t}$, where $\mathbf{J}_s^* = \mathbf{J}_s e^{j\beta}$, we get

$$\nabla \times \left(\frac{1}{\mu} \nabla \times \mathbf{A}^*\right) + j\omega\sigma \mathbf{A}^* - \mathbf{J}_s^* = 0 \qquad (11.41)$$

When applying Galerkin's method to the second term in Eq. (11.41), we get

$$b^*(m,n) = \int_{v_i} j\omega\sigma \mathbf{w}_n \cdot \mathbf{w}_m dv \qquad (11.42)$$

This is an imaginary term and must be added to the left-hand side of the global matrix. It should be noted here that the source term may, and often has real and imaginary parts given by

$$Re(\mathbf{J}_s^*) = \mathbf{J}_s \cos\beta \qquad (11.43a)$$

$$Im(\mathbf{J}_s^*) = \mathbf{J}_s \sin\beta \qquad (11.43b)$$

The elemental contribution due to the source term is

$$\int_{v_i} \mathbf{J}_s \cos\beta \cdot \mathbf{w}_m dv + j \int_{v_i} \mathbf{J}_s \sin\beta \cdot \mathbf{w}_m dv \qquad (11.44)$$

For additional information on this formulation, see Chapter 10.

11.4.6. The Newton-Raphson Method

The principles of the Newton-Raphson method were discussed in Chapter 9 and its application with edge elements follows the same procedures.

The system to be solved now is

$$[J][\Delta A] = [R] \qquad (11.45)$$

where [J] is the Jacobian due to the Newton-Raphson method (this Jacobian is different than the Jacobian used for numerical integration). [R] is the residual, which is the value of the main equation [Eq. (11.22) or (11.35), for example] when the solution of the previous iteration ($i-1$) is substituted into it.

To obtain the terms of the Jacobian, it is necessary to calculate the derivatives of the terms of the main equation with respect to the unknowns A_l. Obviously, only those terms that depend on $\mathbf{A}$ will contribute to the Jacobian. As an example, the source term does not contribute to the Jacobian. Using $\nu = 1/\mu$, the calculation of the Jacobian term $J(l,m)$ due to the term in Eq. (11.27) is

$$J(l,m) = \frac{\partial}{\partial A_l} \int_{V_i} \nu \nabla \times \left(\sum_{n=1}^{12} \mathbf{w}_n A_n \right) \cdot (\nabla \times \mathbf{w}_m) dv$$

which gives

$$J(l,m) = \int_{V_i} \nu (\nabla \times \mathbf{w}_l) \cdot (\nabla \times \mathbf{w}_m) dv + \int_{V_i} \frac{\partial \nu}{\partial A_l} \left(\sum_{n=1}^{12} \nabla \times \mathbf{w}_n A_n \right) \cdot (\nabla \times \mathbf{w}_m) dv$$

$$(11.46)$$

470   11. Hexahedral Edge Elements – Some 3D Applications

The first integral of Eq. (11.46) is the term $a(l,m)$ defined by Eq. (11.29). The second integral needs the evaluation of $\partial v/\partial A_l$. First, we note that

$$\frac{\partial v}{\partial A_l} = \frac{\partial v}{\partial B^2} \frac{\partial B^2}{\partial A_l}$$

and

$$B^2 = \mathbf{B}\cdot\mathbf{B} = (\nabla\times\mathbf{A})\cdot(\nabla\times\mathbf{A}) = \left(\sum_{n=1}^{12} \nabla\times\mathbf{w}_n A_n\right)\cdot\left(\sum_{n=1}^{12} \nabla\times\mathbf{w}_n A_n\right)$$

Therefore,

$$\frac{\partial B^2}{\partial A_l} = 2\left(\sum_{n=1}^{12} \nabla\times\mathbf{w}_n A_n\right)\cdot(\nabla\times\mathbf{w}_l) \qquad (11.47)$$

Substituting Eq. (11.47) into the second integral in Eq. (11.46), we get

$$\int_{V_i} \frac{\partial v}{\partial B^2} 2\left[\left(\sum_{n=1}^{12} \nabla\times\mathbf{w}_n A_n\right)\cdot(\nabla\times\mathbf{w}_l)\right]\left[\left(\sum_{n=1}^{12} \nabla\times\mathbf{w}_n A_n\right)\cdot(\nabla\times\mathbf{w}_m)\right] dv$$

or, rearranging terms

$$\int_{V_i} \frac{\partial v}{\partial B^2} 2\left[\sum_{n=1}^{12} (\nabla\times\mathbf{w}_n)\cdot(\nabla\times\mathbf{w}_l)A_n\right]\left[\sum_{n=1}^{12} (\nabla\times\mathbf{w}_n)\cdot(\nabla\times\mathbf{w}_m)A_n\right] dv \quad (11.48)$$

The terms inside the brackets are easily expanded with the terms $a(m,n)$ defined in (11.29). $\partial v/\partial B^2$ is obtained directly from the **B(H)** curve of the material. With these considerations, we finally obtain the terms of the Jacobian

$$J(l,m) = a(l,m) + 2\int_{V_i} \frac{\partial v}{\partial B^2} \left(\sum_{n=1}^{12} a(n,l)A_n\right)\left(\sum_{n=1}^{12} a(n,m)A_n\right)\frac{1}{(vvol)^2} dv$$

The factor $(v\,vol)^2$ adjusts the coefficients because $a(n,l)$ and $a(n,m)$ correspond to the terms between brackets in Eq. (11.48), multiplied by v and integrated over the volume vol. The values of A_n in Eq. (11.49) are those obtained in the previous iteration.

The other possible term of the Jacobian, $J_c(l,m)$ is due to the eddy currents in Eq. (11.35) and is calculated as

$$J_c(l,m) = \frac{\partial}{\partial A_l} \int_{V_i} \frac{\sigma}{\Delta t} \left(\sum_{n=1}^{12} \mathbf{w}_n A_n \right) \cdot \mathbf{w}_m dv = \int_{V_i} \frac{\sigma}{\Delta t} \mathbf{w}_l \cdot \mathbf{w}_m dv \quad (11.50)$$

This is added to the expression in Eq. (11.49) when eddy currents exist in the solution domain.

Solution of the matrix system in Eq. (11.45) provides the increment in potential ΔA. This gives the new solution as

$$A_i = A_{i-1} + \Delta A$$

11.4.7. The Divergence of J and Other Particulars

It has been noticed that when the divergence of **J** is not assured to be zero, the iterative method used to solve the system of equations (in our case the ICCG method) does not converge. This aspect of modeling is best illustrated by a coil corner where the flux of **J** (that is, the current in the coil) must be conserved. For example, if a single direction of the current in the corner is imposed (Figure 11.3a), then the current density is divergent and the iterative solution diverges.

In Figure 11.3b, the direction of the current is changed gradually based on the following scheme, shown in Figure 11.4. For any point P in the coil corner, the dimensions a and b are first calculated. The components of **J** at P are then given by

$$J_{0x} = -J_0 \frac{a}{D} \quad \text{and} \quad J_{0y} = J_0 \frac{b}{D}$$

Using this scheme for current density in the corner ensures convergence of the ICCG.

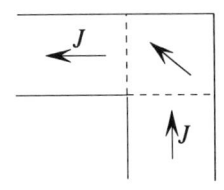

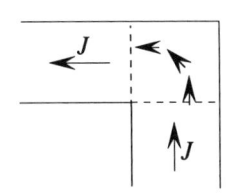

Figure 11.3a. **J** at 45°.

Figure 11.3b. **J** changes direction gradually.

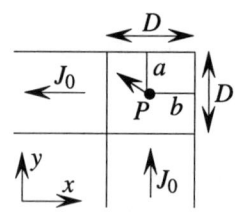

Figure 11.4. A scheme for gradual change in the direction **J** at corners.

The use of single precision variables is another source of difficulties when using iterative solutions in conjunction with edge elements. An improved ICCG algorithm, which uses double precision variables has largely eliminated this difficulty.

Another point of interest concerns the use of edge element formulations without application of gauge conditions. In 3D applications it is often necessary to enforce a gauge condition to ensure uniqueness of solution. While this is a necessity when using primary variables like **B**, it is not crucial when using the magnetic vector potential (as in our formulations). This is because although the magnetic vector potential may not be unique, the magnetic flux density, which is given by derivatives of **A** is unique. The main point here is that in many formulations, if uniqueness is not assured, the iterative solver does not converge. Imposing a gauge condition improves convergence of the solver but it can affect the precision of the results. It has been noticed that the use of hexahedral elements assures convergence without imposition of a gauge. This has two advantages; one is the simplicity of the code and the second is improved accuracy in results. A gauge has been proposed which has the advantage of reducing the number of unknowns but, as has been noticed, it also increases the number of iterative steps in the ICCG solver.

11.5. Modeling of Waveguides and Cavity Resonators

To model fields in waveguides and cavity resonators, we need to write Maxwell's equations in the high-frequency domain using a complex formulation. The equations are (see section 7.2):

$$\nabla \times \mathbf{E} = -j\omega\mu_0\mu_r \mathbf{H} \qquad (11.51)$$

$$\nabla \times \mathbf{H} = j\omega\varepsilon_0\varepsilon_r \mathbf{E} \qquad (11.52)$$

Substituting Eq. (11.52) into Eq. (11.51) and, separately, Eq. (11.51) into Eq. (11.52) we obtain the wave equations:

$$\nabla \times \left(\frac{1}{\mu_r} \nabla \times \mathbf{E} \right) = k^2 \varepsilon_r \mathbf{E} \qquad (11.53)$$

$$\nabla \times \left(\frac{1}{\varepsilon_r} \nabla \times \mathbf{H} \right) = k^2 \mu_r \mathbf{H} \qquad (11.54)$$

where $k^2 = \omega^2 \mu_0 \varepsilon_0$. The analysis here will be made based on Eq. (11.53). Now, the electric field intensity $\mathbf{E}$ is approximated in an edge element as

$$\mathbf{E} = \sum_{n=1}^{12} \mathbf{w}_n E_n \qquad (11.55)$$

Application of Galerkin's method to Eq. (11.53) results in the following

$$\int_\Omega \nabla \times \left(\frac{1}{\mu_r} \nabla \times \mathbf{E} \right) \cdot \mathbf{w}_m d\Omega = \int_\Omega k^2 \varepsilon_r \mathbf{E} \cdot \mathbf{w}_m d\Omega \qquad (11.56)$$

The generic matrix term generated by the left-hand side of Eq. (11.56) is similar to the expression for $a(m,n)$ as described in Eq. (11.30). On the right-hand side of Eq. (11.56), the matrix term follows the procedure used to obtain $b(m,n)$ in Eq. (11.38), adjusting the coefficients as needed.

In the case of cavity resonators, the principal goal is to calculate the resonant frequency of the loaded or empty cavity. In addition, it is often necessary to calculate the quality factor (Q-factor) of the cavity (see section 7.7.4). The electric and magnetic field intensities, electric and magnetic energies as well as wall currents and losses are often needed for analysis. These can all be evaluated from the solution of Eq. (11.56).

11.6. Examples

The formulations for static and time stepping eddy current cases presented in the previous sections have been implemented in program FEECAD. We have used this software regularly for many different applications, and as long as the zero divergence of **J** at corners is satisfied, the code has performed well, providing excellent results. It has proven to be robust in terms of both the nonlinear iteration and ICCG iteration convergence.

Some TEAM problems have been solved successfully using FEECAD. These are:

474   11. Hexahedral Edge Elements – Some 3D Applications

- Problem 10 – eddy current case with strong nonlinearity.
- Problem 13 – static case with strong nonlinearity.
- Problem 20 – calculation of forces in a static case with strong nonlinearity.
- Problem 21 – evaluation of eddy current and losses generated by stray fields, with weak nonlinearity.

In the high-frequency domain, TEAM problem 19 (iris-driven loaded cavity) and problem 18 (waveguide loaded cavity) were also solved using edge elements. The following sections show some of the results for the problems listed above.

11.6.1. Static Calculations (TEAM Problem 13)

The geometry proposed is made of thin steel plates and an exciting coil, arranged as shown in Figure 11.5. The coil current is steady, therefore the problem is static but with a high level of saturation. A vertical steel plate is placed in the middle of the structure and separated from the horizontal plates by a very narrow air-gap. This arrangement results in a small magnetic path reluctance.

Because the structure is symmetric about a horizontal plane crossing the center, the upper part alone is analyzed. The outer boundaries behave as Dirichlet boundary conditions while the symmetry plane behaves as a Neumann boundary condition (see sections 11.4.1). The natural Neumann boundary condition, required by the symmetry condition enforces a normal field across the symmetry plane and is implicit in the formulation (nothing is specified on this boundary).

Because of the field generated by the coil, the problem is highly nonlinear and the Newton-Raphson method was used to obtain a convergent solution.

A new method of relaxation was incorporated in FEECAD to assure convergence. It is based on a local variable relaxation technique. In this method, each edge variable has its own relaxation factor, which varies according to the evolution of convergence.

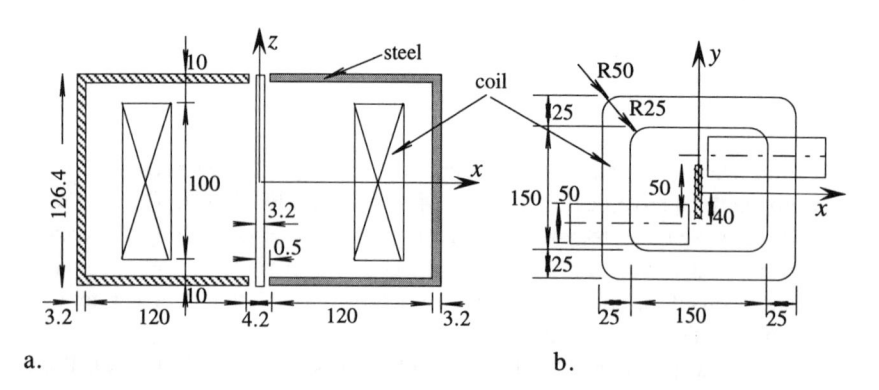

Figure 11.5. Configuration for TEAM problem 13. a. Front view. b. Plan view.

Unknowns with higher stability during the relaxation process tend to work with overrelaxation while those variables that oscillate tend to be underrelaxed. This method is particularly useful when it becomes difficult to define a single, satisfactory relaxation factor (such as in TEAM problem 10, in which convergence based on a single relaxation factor was virtually impossible).

The numerical results obtained are consistent with the experimental data and the graphical results are shown in Figure 11.6.

11.6.2. A Linear Motor with Permanent Magnets

This calculation was performed to obtain the fields generated by permanent magnets using the formulation presented above. This calculation is part of the design of a linear motor in which motion is generated through the interaction between permanent magnets and currents in coils.

Figure 11.7 shows the arrangement of magnets and coils. The mobile part, called a "slider," can move in both directions as indicated. One-half of the structure is shown since there is a symmetry plane at the bottom of the structure. The slider is supported by mechanical linear bearings (not shown in Figure 11.7). The lateral permanent magnets M_1 and M_1' and the central magnet M_2 generate the flux densities which interact with the coil fields. The latter are wound on the fixed iron bars. By changing the direction of the current, the direction of force on the slider changes as well. The fluxes generated by both groups of magnets (lateral and central) add up in the iron bars. The fixed iron bars on both sides are connected at their edges through additional bars without airgaps. A few geometrical simplifications were introduced in the model to simplify the data entry, without affecting the numerical results. In this calculation, only the fields of the permanent magnets are computed.

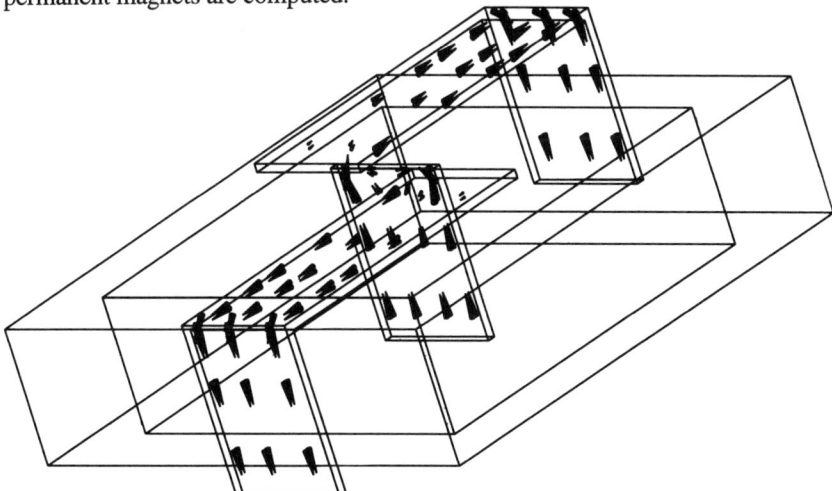

Figure 11.6. Magnetic flux density for TEAM problem 13. The size of the arrows represents the magnitude of **B**.

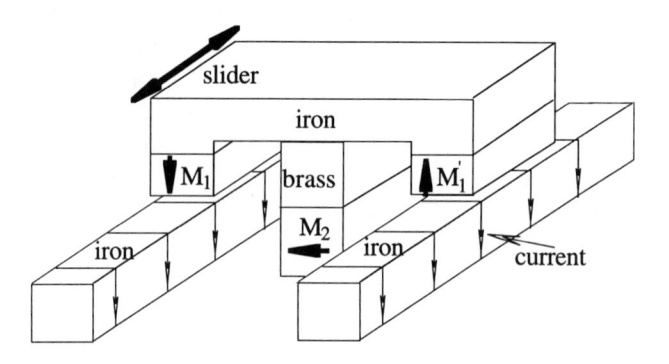

476   11. Hexahedral Edge Elements – Some 3D Applications

Figure 11.7. The slotless linear motor.

Since there are three symmetry planes, it is possible to limit calculation to the domain of Figure 11.8. The fixed iron bar is made larger than the slider because of the flux addition mentioned. A Neumann boundary condition is defined on the plane delimiting the bar since there is no magnetic reluctance beyond this plane. A Neumann boundary condition is also defined on the plane delimiting the central magnet M_2. On all other planes of the domain, Dirichlet boundary conditions are applied.

The finite element mesh for the geometry in Figure 11.8 has 2,295 elements, 2,880 nodes, 8,012 edges and a total of 6,615 unknowns. Computing time required for assembly and solution (with 35 ICCG iterations) is about 61 seconds on a low-end personal computer.

A plot of the magnetic flux density is shown in Figure 11.9. As expected, one can see that the flux of magnets M_1 and M_2 add in the iron bar. Numerically, the magnetic flux density in the gaps (in front of the middle of M_1 and M_2) is 0.55 T. Experimental values are 0.49 T for the lateral magnet and 0.54 T for the central magnet. The respective errors are 12% and 2% which are satisfactory errors.

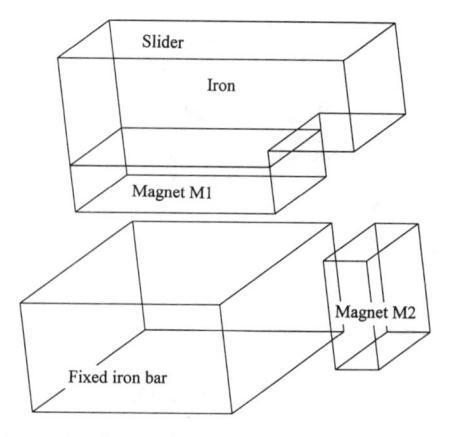

Figure 11.8. Computation region for the slotless linear motor.

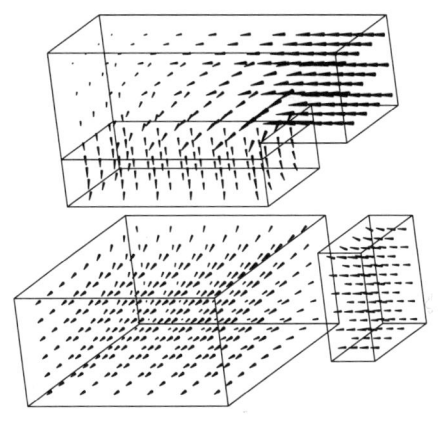

Figure 11.9. Magnetic flux density in the magnets and iron bar.

11.6.3. Eddy Current Calculations (TEAM Problem 21)

This problem was proposed as a TEAM problem to study eddy currents generated by stray fields in transformers. Figure 11.10 shows the structure. There are two coils with currents are flowing in opposite directions. This implies that the flux density is perpendicular to the steel plates the region between the coils.

There are two versions of the problem. In model B there is only a single plate as shown in Figure 11.10a. In model A there are two plates and one of the plates has a hole, as shown in Figure 11.10b. The overall dimensions of the plate in model B are the same as the two plates and gap in model B.

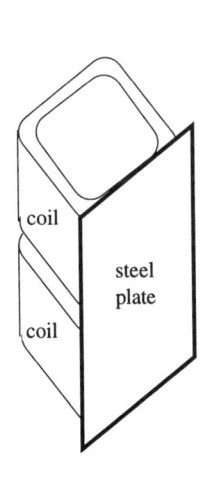

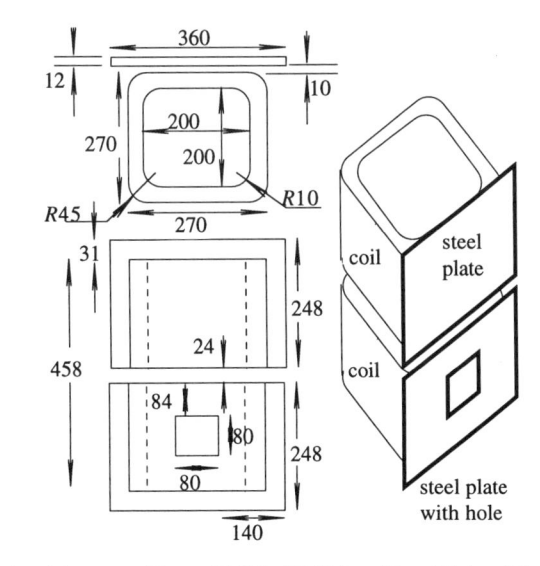

Figure 11.10a. TEAM problem 21 (model B).

Figure 11.10b. TEAM problem 21 (model A).

478   11. Hexahedral Edge Elements – Some 3D Applications

Because there is a double symmetry in model B, only one-fourth of the domain is simulated. In model A, half the geometry is modeled in FEECAD.

The exciting currents are sinusoidal at 50 *Hz*, which obviously generates eddy currents in the plates.

To give some information on the computation, the following data are provided.

- Although the nonlinearity is very weak, the Newton-Raphson method was used. Convergence of the nonlinear iteration was tested by fixing the maximum relative error on all unknowns to 0.001.
- The ICCG method needed an average of 95 iterations for model A and 65 iterations for model B. The relative error in solution was set to 10^{-6}.
- The number of elements in model A is *3,600* and in model B 1,980.
- The number of nodes in model A is 4,400 and in model B 2,496.
- The number of edges in Model A is 12,349 and in model B 6,932.
- The number of unknowns (that is, the number of edges except those with Dirichlet boundary conditions) is 9,349 for model A and 5,024 for model B. In practical terms, the number of unknown edge values is about twice the number of nodes in the mesh. This is rather economical in terms of memory needed for solution.
- Two complete cycles of the source current were calculated and the results from the second cycle are presented to ensure that the procedure has reached numerical steady state. The two cycles were divided into 24 steps.
- Computation time is 144 minutes for model A and 75 minutes for model B, both on a low-end personal computer.

Figure 11.11a and b shows the magnetic flux density distribution for model A. Figure 11.12a and b shows the eddy current density distribution.

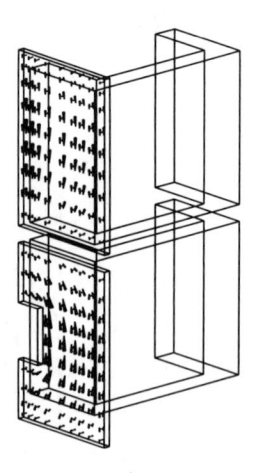

Figure 11.11a. Magnetic flux density distribution for model A.

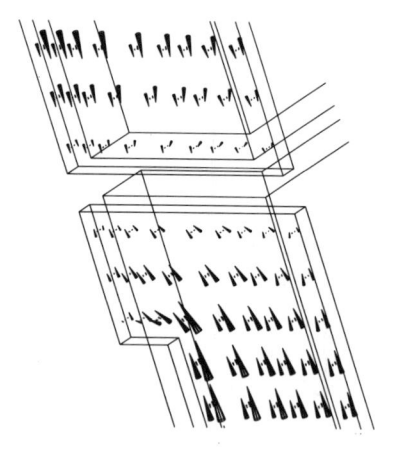

Figure 11.11b. Close view of the magnetic flux density distribution for model A.

Examples 479

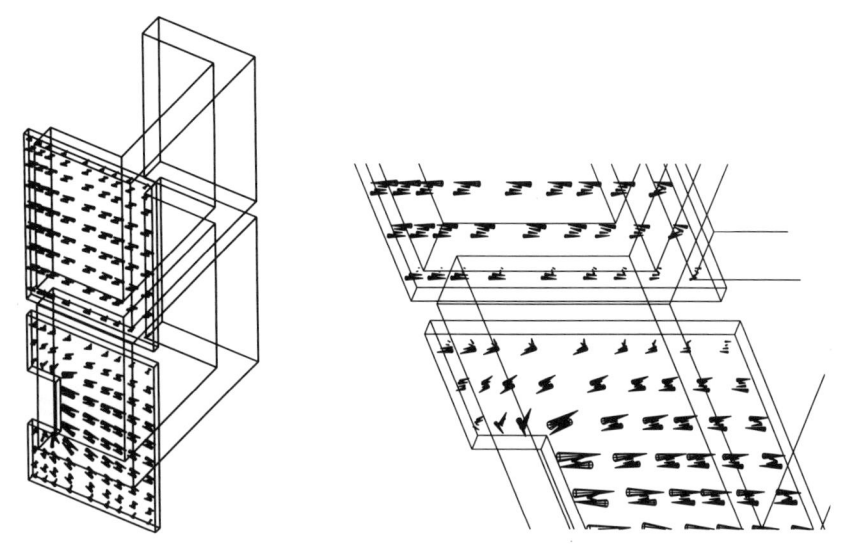

Figure 11.12a. Eddy current density distribution for model A.

Figure 11.12b. Close view of the eddy current density distribution for model A.

Figures 11.13 and 11.14 show the same results for model B.

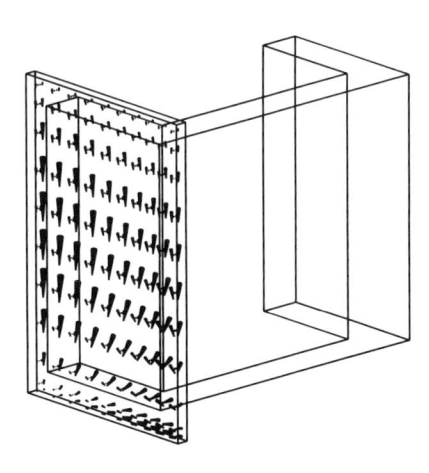

Figure 11.13a. Magnetic flux density distribution for model B.

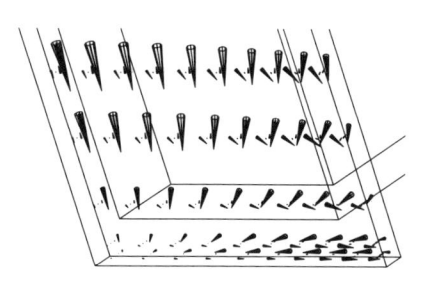

Figure 11.13b. Close view of the magnetic flux density distribution for model B.

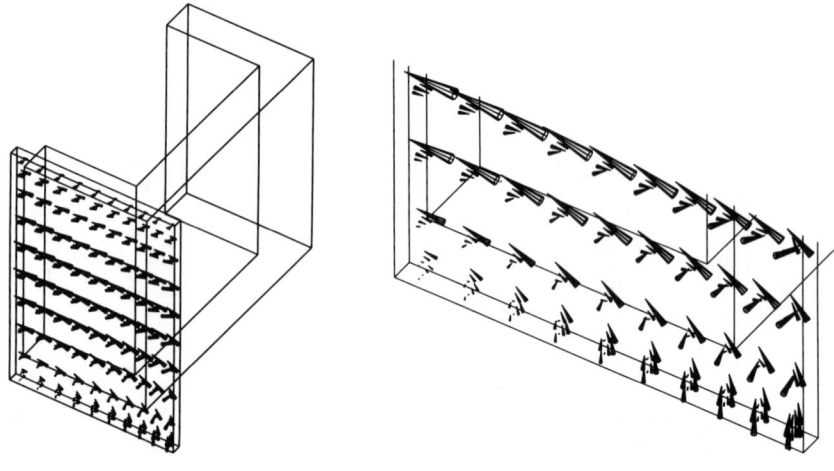

Figure 11.14a. Eddy current density distribution for model B.

Figure 11.14b. Close view of the eddy current density distribution for model B.

The numerical results, such as values of the magnetic flux density at specific locations such as in the steel plates, total losses due to Joule effects, and eddy currents in the plates were in good agreement with the experimental results.

11.6.4. Calculation of Resonant Frequencies (TEAM Problem 19)

TEAM problem 19 was designed to calculate resonant frequencies in an aperture coupled, inhomogeneously loaded cylindrical cavity resonator. The problem is shown schematically in Figure 11.15. Coupling is from a waveguide operating in the TE_{10} mode through a variable width iris. The cavity is loaded with a lossy dielectric rod placed on the axis of the cylinder. The rod is of varying diameter and material properties.

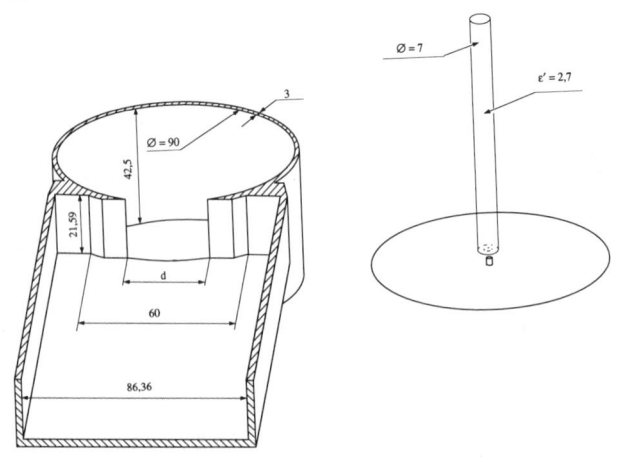

Figure 11.15. Aperture fed loaded cavity resonator. The lossy dielectric rod is shown separately.

The formulation in section 11.5 in conjunction with edge elements is applied here. The fundamental equation is Eq. (11.56), but because the cavity resonator may be seen as a closed structure (provided that the input at any open section is specified), the problem is solved as an eigenvalue problem. To do so, we express Eq. (11.56) in matrix form as

$$[A][E] = k_0^2[B][E] \qquad (11.57)$$

where the coefficients of the matrices A and B are

$$a(m,n) = \int_{v_i} \left[\frac{1}{\mu_r}(\nabla \times \mathbf{w}_n) \cdot (\nabla \times \mathbf{w}_m)\right] dv \qquad (11.58)$$

$$b(m,n) = \int_{v_i} \varepsilon_r (\mathbf{w}_n \cdot \mathbf{w}_m) dv \qquad (11.59)$$

Eq. (11.57) is solved as an eigenvalue problem upon enforcement of boundary conditions. An inhomogeneous Dirichlet boundary condition is applied at the waveguide's entrance while the walls of the cavity and waveguide are homogeneous Dirichlet boundary conditions. Neumann boundary conditions are used on planes of symmetry.

At the input plane in the waveguide, the inhomogeneous boundary condition is set as $E_y = \sin(\pi x/a)$, $E_x = 0$. At resonance, the cavity should operate in the TM_{100} mode because the waveguide is driven at the TE_{10} mode. Three separate cases were studied: (1) empty cavity, (2) cavity loaded with a PVC rod ($\varepsilon_r = 4.0 - j0.05$), and (3) cavity loaded with a Plexiglas rod ($\varepsilon_r = 2.7 - j0.01$). After the resonant frequencies are calculated, a separate, source driven problem is solved by varying the frequency in a small region around resonance to find the quality factor as well as resonance characteristics of the cavity.

Because of symmetry in the electric field, one half of the cavity and waveguide are modeled as shown in Figure 11.16. This figure also shows the mesh (as obtained using PATRAN). Discretization resulted in 576 hexahedral elements with a total of 2,214 edges of which 600 were boundary edges. Solution (on a Cray Y-MP) is approximately 13 seconds.

Figure 11.17 shows the normalized energy as a function of frequency around resonance for an iris width of 15*mm*. The sharp peak is easily distinguishable as the resonant frequency.

Resonant frequencies for different loadings with various iris widths are shown in Figure 11.18. These values show close resemblance with the experimentally determined values. When compared numerically for the 15 *mm* wide iris, the errors were between 0.6 and 1.8%. Contour plots of the E-field distribution are shown in Figures 11.19 through 11.21. In the lossy case in Figure 11.19, there are no losses and the waveguide need not supply any energy into the cavity. With

482   11. Hexahedral Edge Elements – Some 3D Applications

the loaded cavity, energy must be supplied to compensate for the losses. Since the PVC rod is more lossy than the Plexiglas rod, more energy will be needed as shown in Figures 11.20 and 11.21.

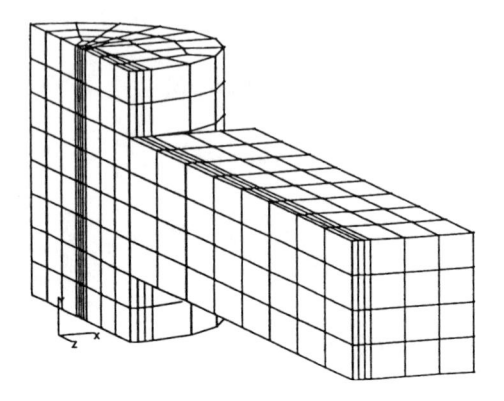

Figure 11.16. Discretization of one-half the geometry into edge elements.

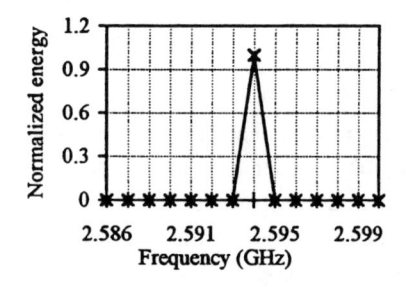

Figure 11.17. Normalized energy in the cavity vs. frequency (*GHz*).

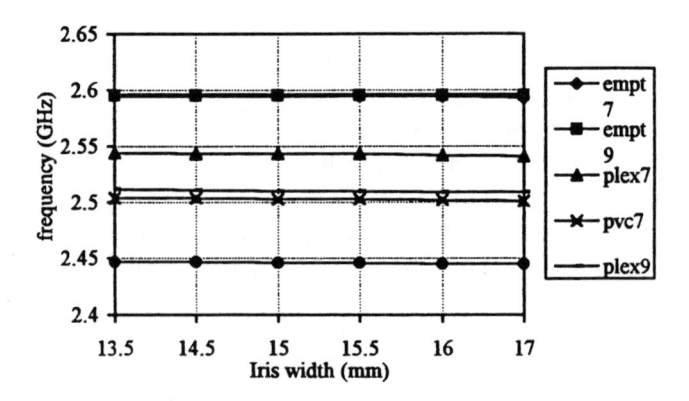

Figure 11.18. Resonant frequency vs. iris width for all cases.

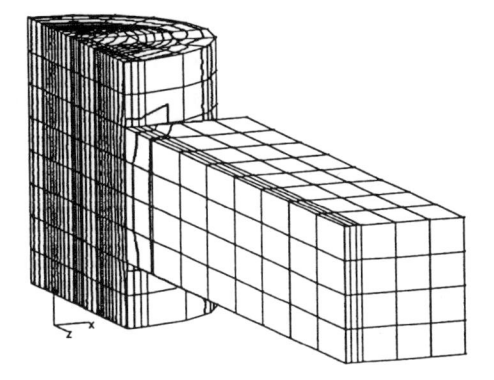

Figure 11.19. |E| field distribution in the unloaded cavity.

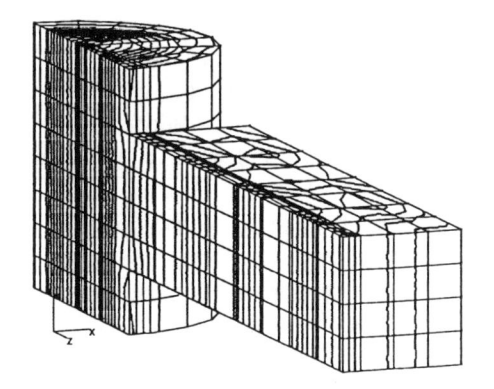

Figure 11.20. |E| field distribution in the PVC loaded cavity ($\varepsilon_r=4.0-j\,0.05$).

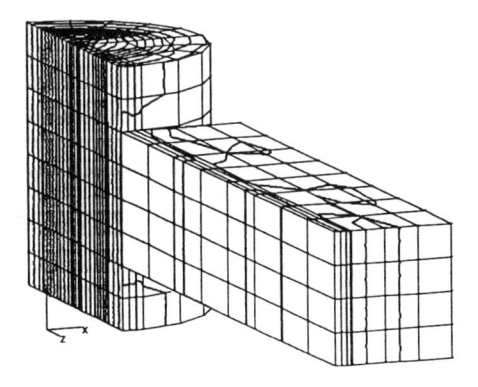

Figure 11.21. |E| field distribution in the Plexiglass loaded cavity ($\varepsilon_r=2.7-j\,0.01$).

12
Computational Aspects in Finite Element Software Implementation

12.1. Introduction

In the previous two chapters, the physical and mathematical basis of the finite element method were described. Next, we discuss some computational aspects relating to the practical implementation of the method.

The implementation of the finite element method in computer programs has two fundamental and complementary aspects: computation speed and memory usage. It has already been shown that this method leads to a system of equations which, generally, has a large number of unknowns.

The storage of the coefficient matrix $[SS]$ of the system

$$[SS]\{V\} = \{Q\}$$

and its solution must take into account the intrinsic characteristics of the system. The judicious use of these characteristics enable the application of the FE method for realistic problems.

In Chapters 9 and 10, it was noted that the matrix $[SS]$ is symmetric (one exception to this is the case of moving bodies, in section 10.3.4). This property, as well as others that are described below, must be considered in the computation program.

12.2. Geometric Repetition of Domains

12.2.1. *Periodicity*

Some problems have geometries that can be composed of a repetitive section of the domain to be analyzed, as for example, the geometry shown in Figure 12.1. In this case, the problem is "periodic," characterized by a geometric repetition of the domain S. If the coil currents (if they exist) or permanent magnets' fields are in the same direction, the potentials on line C are identical to the potentials on line D.

Geometric Repetition of Domains   485

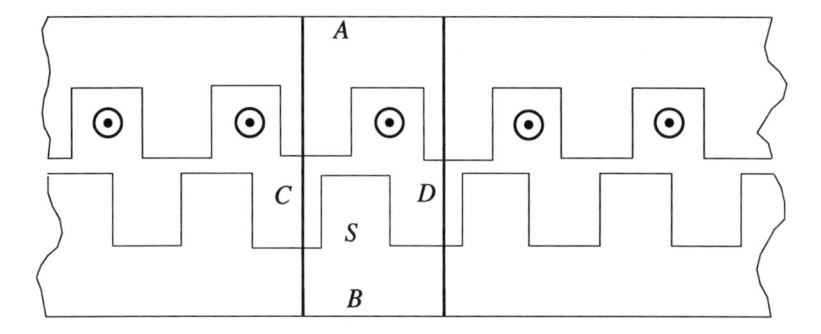

Figure 12.1. A periodic structure. The domain defined by lines A, B, C, and D, is the repetitive domain. Only this part of the structure need be analyzed.

Instead of considering the whole structure, it is sufficient to analyze only the domain S. To do so, we treat the elements on the boundaries between neighboring domains as shown in Figure 12.2. When the elemental matrix $S(3,3)$ for triangle T is generated, the contributions for the nodes i and j must be assembled in nodes i' and j'. This indicates, for line C of the domain, the presence of an identical domain to its left. It is not necessary to consider the nodes i and j in the matrix system; on the other hand, the value at node k does not change. When the system $[SS]\{V\}=\{Q\}$ is solved, we set $V_i=V_i'$ and $V_j=V_j'$. The remaining nodes on line D are treated similarly. The elemental matrix and sources of triangle T are assembled as indicated below

$$\begin{array}{c} \text{line } i' \rightarrow \\ \text{line } j' \rightarrow \\ \text{line } k \rightarrow \end{array} \begin{bmatrix} S_{ii} & S_{ij} & S_{ik} \\ S_{ji} & S_{jj} & S_{jk} \\ S_{ki} & S_{kj} & S_{kk} \end{bmatrix} \text{ and } \begin{bmatrix} Q_i \\ Q_j \\ Q_k \end{bmatrix}$$

This procedure is also valid for the Jacobian and the residual (Newton-Raphson method). Figure 12.3 shows an example of a periodic problem, obtained with program EFCAD.

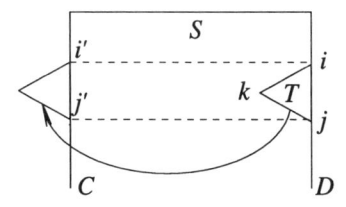

Figure 12.2. Treatment of boundary nodes between neighboring domains in a periodic structure.

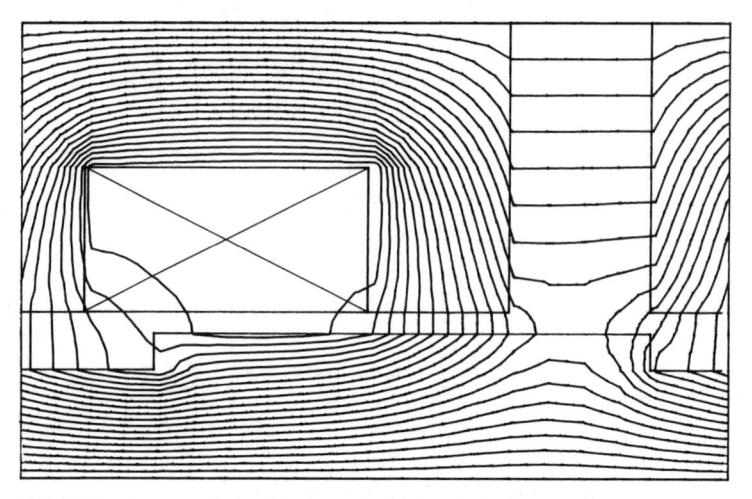

Figure 12.3. Finite element solution for the periodic domain in Figure 12.1.

12.2.2. Anti-Periodicity

Anti-periodicity is similar to the periodic case discussed above: we have geometric repetition of a domain, but the source (current or permanent magnet) has alternately opposing directions, as shown in Figure 12.4.

We again consider only the domain S of Figure 12.4; for the assembly of elemental matrices and sources, the terms must be inserted in the locations for nodes i' and j', instead of nodes i and j, by the rule indicated below:

$$\begin{array}{c} \text{line } i' \to \\ \text{line } j' \to \\ \text{line } k \to \end{array} \begin{bmatrix} S_{ii} & -S_{ij} & -S_{ik} \\ -S_{ji} & S_{jj} & -S_{jk} \\ -S_{ki} & -S_{kj} & S_{kk} \end{bmatrix} \text{ and } \begin{bmatrix} -Q_i \\ -Q_j \\ Q_k \end{bmatrix}$$

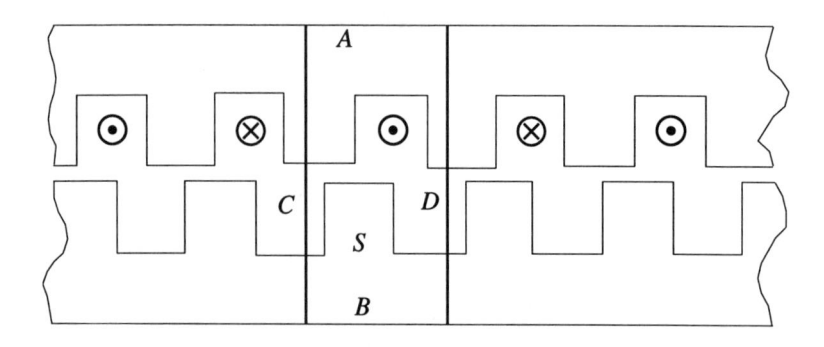

Figure 12.4. An anti-periodic structure. The difference between this and Figure 12.1 is in the alternating directions of currents.

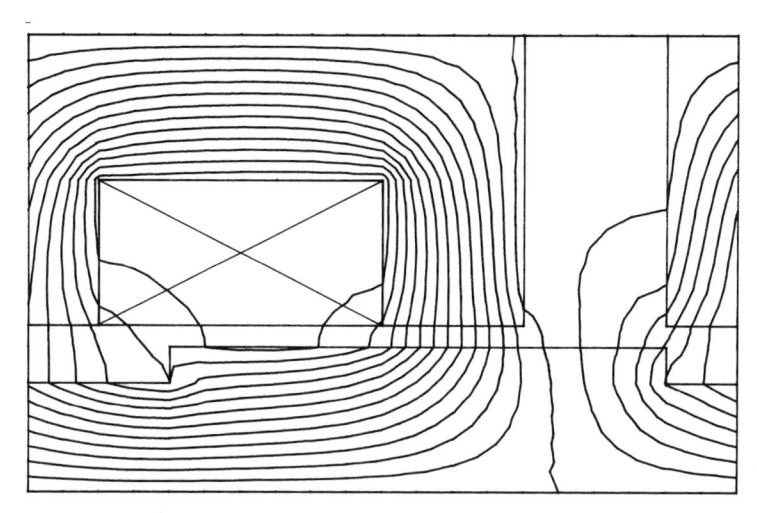

Figure 12.5. Finite element solution for the anti-periodic domain in Figure 12.4.

where the nondiagonal terms are negative. Nodes i and j do not need to be calculated. After solution we set $V_i=-V_i'$ and $V_j=-V_j'$. Figure 12.5 shows an example of an anti-periodic problem.

12.3. Storage of the Coefficient Matrix

12.3.1. *Symmetry of the Coefficient Matrix*

With the exception of moving part problems (section 10.3.4), all elemental matrices treated here are symmetric. The global matrix $[SS]$, formed by the assembly of the matrices $S(3,3)$, is also symmetric. Therefore, it is possible to store in the computer memory only half the matrix $[SS]$, including the diagonal terms. This translates into considerable economy in memory usage.

12.3.2. *The Banded Matrix and Its Storage*

Consider the mesh numbering of Figure 12.6. Note, for example, that node *18* belongs to elements which contain nodes 10, 11, 17, 19, 25, and 26. The most "distant" node numbers are 10 and 26. This means that in line 18 of the matrix $[SS]$ there are nonzero terms only between positions 10 and 26. Similarly, for node *19*, the nonzero terms are located between position 11 and 27. Thus, matrix $[SS]$ has the form shown in Figures 12.7a and 12.7b.

The matrix $[SS]$ in Figure 12.7b is a "banded" matrix. It is only necessary to store the coefficients within the band. Additionally, since it is also symmetric, it is possible to store half the band only.

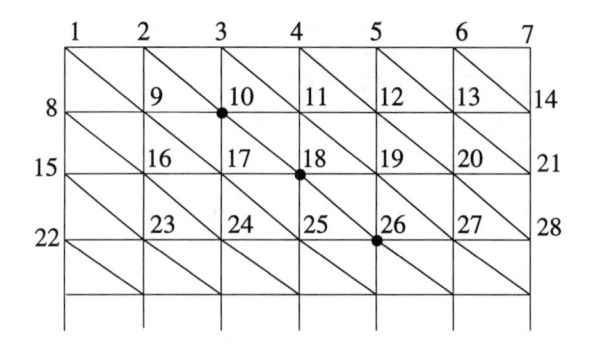

488   12. Computational Aspects in Finite Element Software Implementation

Figure 12.6. Mesh numbering. Only node numbering is shown.

The matrix [SS] can also be stored as a two dimensional array (N,B) as shown in Figure 12.7c. In this array N is the node number and B is the width of the half-band including the diagonal term. A term (i,j) of the complete matrix takes the position (i,l) in the array, with the value of l given by

$$l = j - i - 1$$

Using this method, memory economy is further increased. However, if the finite element mesh is not correctly numbered, it is possible that the matrix cannot be entered in band form, and we lose an important topological advantage. If, as an extreme example, node 1 is in the same element with node N the whole matrix [SS] must be stored.

Another important aspect is the solution of the system [SS]{V}={Q}. The narrower the band of the matrix, the faster its solution, because the solution algorithms operate only within the band.

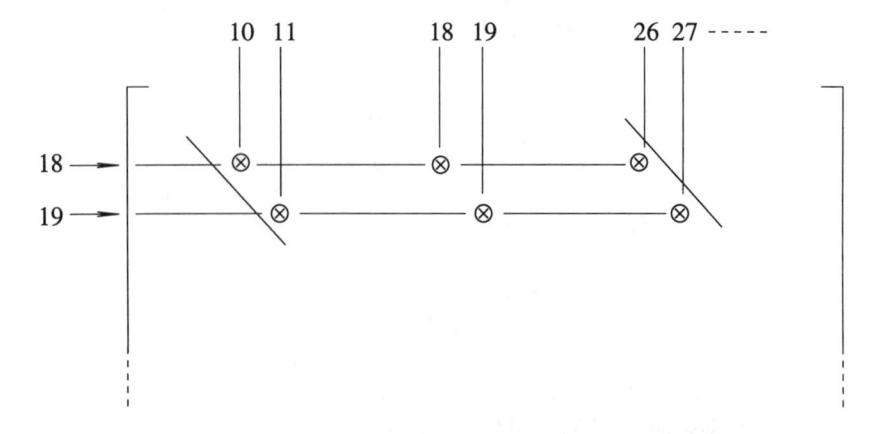

Figure 12.7a. The banded structure of the matrix. The lowest and highest nonzero coefficients for rows 18 and 19, corresponding to nodes 18 and 19 in Figure 12.6 are shown.

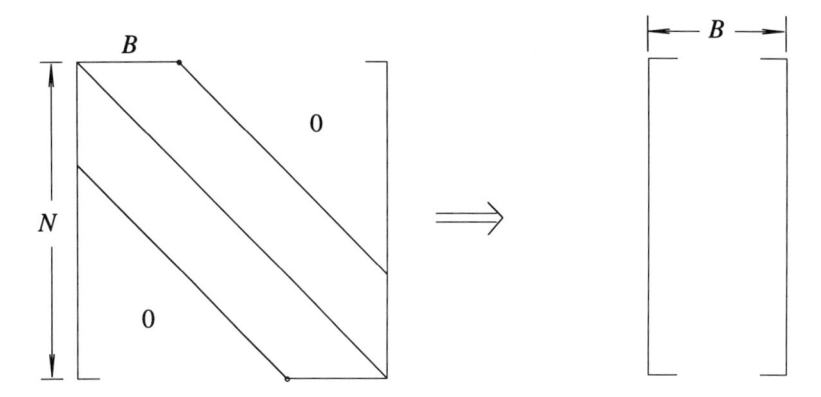

Figure 12.7b. Structure of the global matrix. All coefficients outside the bandwidth are zero.

Figure 12.7c. Only a half bandwidth, including the diagonals is needed for symmetric matrices.

12.3.3. Compact Storage of the Matrix

Even though the storage of the half-band of [SS] is very efficient in memory usage, in some cases, the band may be very sparse. As an example, in the mesh of Figure 12.6, in line 18, column positions 20, 21, 22, 23 and 24 are zero, since node number 18 does not share common elements that contain these nodes. In this case, the matrix [SS] can be stored more efficiently by the use of two arrays: the first is a "pointer" array and the second stores the values of [SS], as shown in Figure 12.8. These arrays are marked as IND and SS.

In array IND (integer pointer variables), for the example above, in the line 18, we have the node numbers of the nodes neighboring node 18. In this example these are nodes 19, 25, and 26. In array SS (real variables) the first position is the diagonal value (this is always present and does not require any indexing information). The values for columns 19, 25, and 26 are placed in the remaining positions in SS. We have:

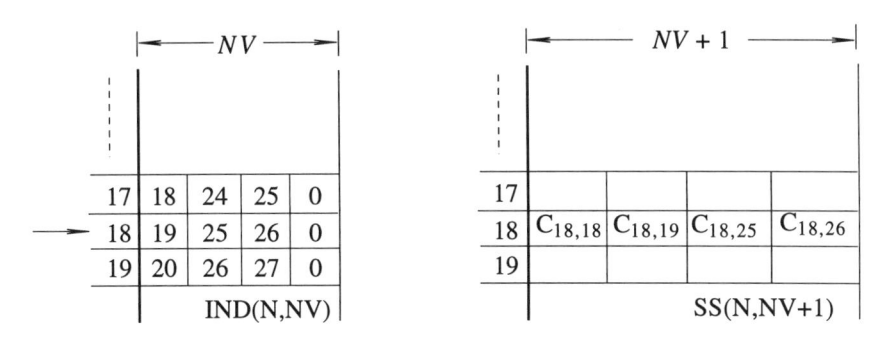

	NV			
17	18	24	25	0
18	19	25	26	0
19	20	26	27	0

IND(N,NV)

	NV + 1			
17				
18	$C_{18,18}$	$C_{18,19}$	$C_{18,25}$	$C_{18,26}$
19				

SS(N,NV+1)

Figure 12.8. Compact storage of the coefficient matrix. SS is the coefficient array, IND the pointer array, pointing to the locations of the coefficients in SS.

$$IND(18,1)=19 \qquad SS(18,2)=C_{18,19}$$
$$IND(18,2)=25 \qquad SS(18,3)=C_{18,25}$$
$$IND(18,3)=26 \qquad SS(18,4)=C_{18,26}$$

and

$$SS(18,1)=C_{18,18}$$

This method of storing [SS] allows random numbering of the nodes in the mesh because the band width has no influence on storage. We need only to estimate the maximum number of neighboring nodes (NV), in order to dimension the tables IND(N,NV) and SS(N,NV+1). NV depends on the type of elements used but, for most 2D applications, it does not exceed 10 and for 3D applications it is normally below 16 (depending on the element used).

In 3D applications, the sparsity of the matrix [SS] is very important because of the large matrices involved and this type of storage is very useful.

12.4. Insertion of Dirichlet Boundary Conditions

After matrix [SS] and vector [Q] are assembled, before the solution step can begin, it is necessary to insert the imposed potentials at the boundary nodes on which Dirichlet conditions are specified.

If, for example, in the system $Ax=b$ with four equations, the variable x_2 has a specified value $x_2=P$, the system is modified as follows:

$$\begin{bmatrix} a_{11} & a_{12} & a_{13} & a_{14} \\ 0 & 1 & 0 & 0 \\ a_{31} & a_{32} & a_{33} & a_{34} \\ a_{41} & a_{42} & a_{43} & a_{44} \end{bmatrix} \begin{bmatrix} x_1 \\ x_2 \\ x_3 \\ x_4 \end{bmatrix} = \begin{bmatrix} b_1 \\ P \\ b_3 \\ b_4 \end{bmatrix} \qquad (12.1)$$

The product of the second line by the unknown vector provides $x_2=P$. This system can be solved under the above form or it can be written as

$$\begin{bmatrix} a_{11} & a_{13} & a_{14} \\ a_{31} & a_{33} & a_{34} \\ a_{41} & a_{43} & a_{44} \end{bmatrix} \begin{bmatrix} x_1 \\ x_3 \\ x_4 \end{bmatrix} = \begin{bmatrix} b_1 - a_{12}P \\ b_3 - a_{32}P \\ b_4 - a_{42}P \end{bmatrix} \qquad (12.2)$$

The form in (12.1) is faster in operation, because there are no algebraic transformations in the global matrix. However, there is a difficulty with this method: the matrix is symmetric and only a half-band is stored (for example, the upper half-band). By zeroing the non diagonal terms, we "lose" their values (a_{23} and a_{24} in the above example). When operating on line 3, we need the term a_{32}, which was not stored, because it is equal to stored term a_{23}, but this last term is lost in this method.

To avoid this inconvenience, we adopt a modified method: the diagonal term is replaced by a large number K. Typical values could be $K=1.0E^{+14}$. In the right hand side vector, the value KP is placed while the non diagonal terms are not modified. This makes the off diagonal coefficients negligible relative to the large diagonal and the effect is as if the off-diagonal terms are zeroed. If K is large compared to the off-diagonal terms, the error due to this method is negligible.

Using this method, we get

$$a_{21}x_1 + Kx_2 + a_{23}x_3 + a_{24}x_4 \approx Kx_2 = KP$$

which provides $x_2=P$. This method is used in the EFCAD software, with no detectable error.

With the Newton-Raphson method, recalling that the vector $\{x\}$ is, in this case, $\{\Delta V\}$ (change in potential), it is obvious that for a node k with a specified potential value, $\{\Delta V_k\}$ is always zero; for the example above, it means that $P=0$ and it is possible to use the form in Eq. (12.1), by zeroing all the nondiagonal terms and imposing a value of 1 in the diagonal position.

12.5. Quadrilateral and Hexahedral Elements

For some two-dimensional applications, the domain geometry is best discretized with quadrilateral elements, as is shown in Figure 12.6. The division of a rectangular region into triangular elements creates an unbalanced division, and errors may be introduced due to asymmetry in discretization. As an example, in Figure 12.9, nodes N_1 and N_3 belong to two elements T_1 and T_2; in the global matrix, lines N_1 and N_3 will contain more contribution terms. Although this decomposition is not theoretically incorrect, the precision in calculation can be affected. At the same time, it is very important to solve for potentials as accurately as possible. Recalling that the magnetic field intensities and flux densities depend on the derivatives of the potential, it is obvious that even if the potential is quite accurate, the flux densities and field intensity may be erroneous. Suppose, for example, that the correct values for nodes N_4 and N_3 (Figure 12.9) should be 95.0 and 94.0, but, after calculations the obtained results are 95.2 and 93.8, respectively. These values are quite good (within about 0.2%), but the field intensity H_y, which is strongly dependent on the potential variation is given by $H_y=(95.2-93.8)/l=1.4/l$, instead of its correct value $H_y=(95.0-94.0)/l=1/l$. This is a 40% error, which is rarely acceptable for design.

A simple and efficient method to avoid inaccuracies due to the mesh asymmetry, is to decompose the element of Figure 12.9 into two triangle sets, as shown in Figure 12.10.

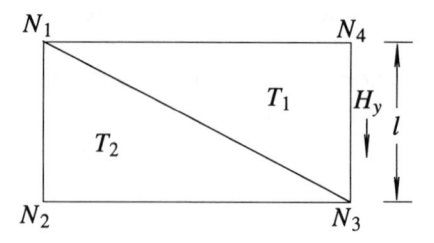

492   12. Computational Aspects in Finite Element Software Implementation

Figure 12.9. The nonsymmetric nature of triangular elements. Nodes N_1 and N_3 have different contributions than nodes N_2 and N_4.

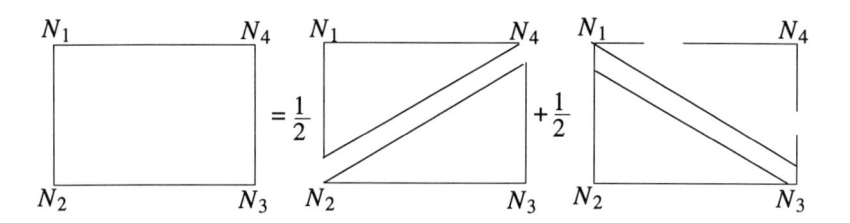

Figure 12.10. A method of symmetrizing the mesh. Two meshes are generated and the final mesh is the sum of the two.

In this division there are no "preferred" triangles and the accuracy is improved. To implement this technique, the triangle's elemental contributions must be multiplied by 1/2, since the surface defined by the nodes N_1, N_2, N_3, and N_4 is integrated twice. This technique is used in EFCAD. If a region requires rectangular elements, it is discretized in this manner.

The same technique can be extended to 3D hexahedral elements. A hexahedral element can be decomposed into six tetrahedra, as shown in Figure 12.11.

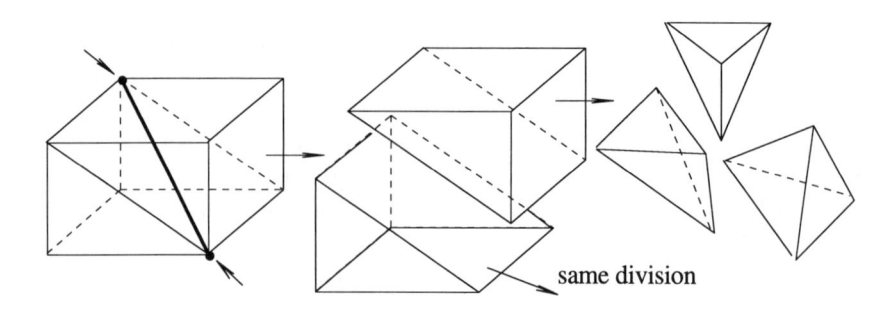

Figure 12.11. Division of a hexahedral element into six tetrahedra. To symmetrize the mesh, four divisions must be made with six tetrahedra each.

The decomposition of Figure 12.11 "prefers" the two arrowed nodes, because all six tetrahedra contribute to these nodes. While in 2D meshes this problem is easily solved, for 3D applications it is necessary to make four decompositions with six tetrahedra each. These must be done as shown in Figure 12.11, taking the four principal diagonals of the hexahedron as the basis for division. Since the volume is integrated four times the contributions must be divided by 4.

12.6. Methods of Solution of the Linear System

After the global matrices

$$[SS]\{V\}=\{Q\} \quad \text{or} \quad [SJ]\{\Delta V\}=\{R\} \qquad (12.3)$$

[given in Eqs. (7.76) and (7.107)] are assembled and the boundary conditions are inserted, either system can be solved.

Before presenting some of the common solution methods, we note that the study of these methods is a subject in itself. A complete, rigorous discussion of solution methods is beyond the scope of this work. We will present here only briefly the most commonly employed methods for finite element programs applied to electromagnetism, with special attention to practical aspects.

We use the classical notation "$Ax=b$" for the matrix systems in Eq. (12.3), where A can be either $[SS]$ or $[SJ]$, x is the unknown vector $\{V\}$ or $\{\Delta V\}$ and b is $\{Q\}$ or $[R]$. The system $Ax=b$ may be solved either by direct or iterative solution methods. The two types of solution methods are discussed next.

12.6.1. Direct Methods

These methods lead directly to the solution x in a single sweep through the matrix. The basis of the methods consists of performing linear transformations of the matrix system such that, when these are completed, the solution is obtained by very simple operations.

12.6.1a. The Gauss Elimination Method

The basic direct method is the "Gauss elimination" method, which is very efficient and widely employed in finite element programs. We outline the steps involved in solution as applied to the banded matrix below

$$\begin{bmatrix} a_{11} & a_{12} & a_{13} & & & \\ a_{21} & a_{22} & a_{23} & a_{24} & & \\ a_{31} & a_{32} & a_{33} & a_{34} & a_{35} & \\ & a_{42} & a_{43} & a_{44} & a_{45} & a_{46} \\ & & & \vdots & & \end{bmatrix} \begin{bmatrix} x_1 \\ x_2 \\ x_3 \\ x_4 \\ \vdots \end{bmatrix} = \begin{bmatrix} b_1 \\ b_2 \\ b_3 \\ b_4 \\ \vdots \end{bmatrix}$$

At the first elimination stage the first line is multiplied by a_{i1}/a_{11} and subtracted from the corresponding terms of the ith row, where i takes the values 2 and 3 for this example. After this step, the system is

$$\begin{bmatrix} a_{11} & a_{12} & a_{13} & & & \\ & a'_{22} & a'_{23} & a'_{24} & & \\ & a'_{32} & a'_{33} & a_{34} & a_{35} & \\ & a_{42} & a_{43} & a_{44} & a_{45} & a_{46} \\ & & & \vdots & & \end{bmatrix} \begin{bmatrix} x_1 \\ x_2 \\ x_3 \\ x_4 \\ \vdots \end{bmatrix} = \begin{bmatrix} b_1 \\ b'_2 \\ b'_3 \\ b_4 \\ \vdots \end{bmatrix}$$

where the terms indicated by ' were modified during the first elimination step as, for example,

$$a'_{32} = a_{32} - (a_{31}/a_{11})a_{12}$$

$$b'_2 = b_2 - (a_{21}/a_{11})b_1$$

Repeating the process, starting with the second row, the matrix, after the second elimination step looks like

$$\begin{bmatrix} a_{11} & a_{12} & a_{13} & & & \\ & a'_{22} & a'_{23} & a_{24} & & \\ & & a''_{33} & a'_{34} & a_{35} & \\ & & a'_{43} & a'_{44} & a_{45} & a_{46} \\ & & & \vdots & & \end{bmatrix} \begin{bmatrix} x_1 \\ x_2 \\ x_3 \\ x_4 \\ \vdots \end{bmatrix} = \begin{bmatrix} b_1 \\ b'_2 \\ b''_3 \\ b'_4 \\ \vdots \end{bmatrix}$$

where, for example,

$$a''_{33} = a'_{33} - (a'_{32}/a'_{22})a'_{23}$$

$$a'_{43} = a_{43} - (a_{42}/a'_{22})a'_{24}$$

$$b''_3 = b'_3 - (a'_{32}/a'_{22})b'_2$$

Continuing this procedure up to the last row, we obtain an upper triangular system. This modified system is denoted $Ux=c$

$$\begin{bmatrix} u_{11} & u_{12} & u_{13} & & & & \\ & u_{22} & u_{23} & u_{24} & & & \\ & & u_{33} & u_{34} & u_{35} & & \\ & & & \cdot & & & \\ & & & & \cdot & & \\ & & & & & \cdot & \\ & & & u_{n-1,n-1} & u_{n-1,n} & \\ & & & & u_{n,n} \end{bmatrix} \begin{Bmatrix} x_1 \\ x_2 \\ x_3 \\ \cdot \\ \cdot \\ \cdot \\ x_{n-1} \\ x_n \end{Bmatrix} = \begin{Bmatrix} c_1 \\ c_2 \\ c_3 \\ \cdot \\ \cdot \\ \cdot \\ c_{n-1} \\ c_n \end{Bmatrix}$$

The system is now solved, starting from the last row. First $x_n = c_n/u_{n,n}$ is calculated. With this the following general recursion formula for x_{n-1} is obtained

$$x_{n-1} = \frac{c_{n-1} - u_{n-1,n} x_n}{u_{n-1,n-1}}$$

This is the back-substitution step in the Gauss elimination method.

It should be pointed out that in order to simplify the presentation we have used the entire band of [A], but, when the matrix is symmetric, the terms of the lower half-band are obtained from the upper half-band. If the matrix is nonsymmetric, the method can be employed as shown, assuming that the entire band is stored.

It is important to note that direct methods modify the original matrix [A]. Terms out of the band are not affected, but inside the band all terms are modified including the zero terms. Therefore, all the half-band (or the complete band for nonsymmetrical system) must be allocated computer memory. At the end of the algebraic transformations of the elimination steps, practically all the band will contain nonzero terms. This fact forces us to use the first mode of storage, in which memory for all the band is dimensioned and the place for the terms generated by the method is predetermined. The form of storage is strongly associated with the solution method to be employed. When the matrix [A] is stored in compact form, iterative methods can be employed, because in these, the structure of [A] is not modified during the iterations, and, since there is no fill-in of terms, the arrays IND and SS are not modified.

12.6.1b. The Choleski Method

Another commonly employed method is the "Choleski decomposition method," for symmetric matrices. It is based on the fact that the matrix [A] can be written as the product of two matrices [L] and $[L]^T$:

$$[A] = [L][L]^T$$

where [L] is a lower triangular matrix (including the diagonals).

The system $Ax = b$ can be written in the form

12. Computational Aspects in Finite Element Software Implementation

$$[L][L]^T\{x\} = \{b\}$$

which can be expressed by the following sequence of equations:

$$[L]\{c\} = \{b\} \tag{12.4a}$$

$$[L]^T\{x\} = \{c\} \tag{12.4b}$$

Eq. (12.4a) has the form

$$\begin{bmatrix} l_{11} & & & 0 \\ l_{21} & l_{22} & & \\ l_{31} & l_{32} & l_{33} & \\ \vdots & \vdots & \vdots & \end{bmatrix} \begin{bmatrix} c_1 \\ c_2 \\ c_3 \\ \vdots \end{bmatrix} = \begin{bmatrix} b_1 \\ b_2 \\ b_3 \\ \vdots \end{bmatrix}$$

from which we can obtain $c_1 = b_1/l_{11}$; and similarly, $c_2 = (b_2 - l_{21}c_1)/l_{22}$. The rest of the terms are calculated similarly.

After calculating the vector $\{c\}$, we can write equation (12.4b) as

$$\begin{bmatrix} l_{11} & l_{21} & l_{31} & & & \\ & l_{22} & l_{32} & l_{42} & & \\ & & \cdot & & & \\ & & & \cdot & & \\ & & & & l_{n-1,n-1} & l_{n-1,n} \\ & & & & & l_{n,n} \end{bmatrix} \begin{bmatrix} x_1 \\ x_2 \\ x_3 \\ \cdot \\ \cdot \\ x_{n-1} \\ x_n \end{bmatrix} = \begin{bmatrix} c_1 \\ c_2 \\ c_3 \\ \cdot \\ \cdot \\ c_{n-1} \\ c_n \end{bmatrix}$$

from which $\{x\}$ is obtained, using a process similar to the back-substitution process in the Gauss elimination method.

The terms l_{ij} of matrix $[L]$, dimensioned (n,n), are given by the expressions

$$l_{11} = (a_{11})^{1/2} \qquad l_{ii} = \left(a_{ii} - \sum_{j=1}^{i-1} l_{ij}^2 \right)^{1/2} \qquad i=2,\ldots,n \tag{12.5a}$$

$$l_{ij} = \frac{1}{l_{jj}} \left(a_{ij} - \sum_{m=1}^{j-1} l_{jm} l_{im} \right) \qquad i=2,\ldots,n; \quad j=i,i+1,\ldots,n \tag{12.5b}$$

$$l_{ij} = 0 \quad i<j \qquad (12.5c)$$

The Choleski method requires fewer operations than the Gauss elimination method and therefore is faster but it is not applicable to nonsymmetric matrices.

There are other direct methods and they are, generally, derived from the two methods described above. These methods have specific advantages for specific problems. However, the two methods presented here are most commonly used in finite element programs because of their good performance.

12.6.2. Iterative Methods

Iterative methods are very useful when the number of nodes is very large and the storage of the matrix in banded form is impractical. In this case the compact storage method is used. As mentioned above, this kind of storage is not amenable to direct solution methods and iterative solution techniques must be employed. We will present here two methods: the first because it is very simple and the second because it is probably the most efficient iterative method for solution of sparse systems arising from the application of the finite element method.

12.6.2a. The Successive Overrelaxation Method

The successive overrelaxation (SOR) method is an extension of the Gauss-Seidel method, which in turn is a modification of the Jacobi method. The matrix system $Ax=b$ can be written in the following form, where we choose to write the vector x explicitly.

$$x_1 = (b_1 - a_{12}x_2 - a_{13}x_3 - \ldots - a_{1n}x_n)/a_{11} \qquad (12.6a)$$

$$x_2 = (b_2 - a_{21}x_1 - a_{23}x_3 - \ldots - a_{2n}x_n)/a_{22} \qquad (12.6b)$$

$$\vdots$$

and generalizing:

$$x_i = (b_i - \sum_{j=1}^{i-1} a_{ij}x_j - \sum_{j=i+1}^{n} a_{ij}x_j)/a_{ii}$$

To establish an iterative procedure, we first assign initial values $x_2^0, x_3^0, \ldots x_n^0$ to calculate x_1^1 using Eq. (12.6a). With $x_1^1, x_3^0, \ldots x_n^0$, x_2^1 is obtained from Eq. (12.6b). The process is repeated successively until x_n is calculated. At this point the first iteration is completed, and the second iteration can commence. The iterations are repeated until convergence to the solution is achieved.

The successive overrelaxation process is a method of accelerating the solutions. At iteration $m+1$, we can write the change in solution as

$$\Delta x_i^{m+1} = x_i^{m+1} - x_i^{m}$$

This change is multiplied by a factor $(\Omega-1)$ and added to the original value of x_i^{m+1} as

$$x_i^{m+1} = x_i^{m+1} + (\Omega-1)\Delta x_i^{m+1}$$

If $\Omega>1$ the procedure is called "over-relaxation"; if $\Omega=0$ there is no relaxation and if $\Omega<1$ we have "under-relaxation." Generally, we use $1.5<\Omega<2$, for this type of matrix systems, although the choice of Ω depends on the specific problem.

We note here that during the iterations, the matrix $[A]$ is not modified; therefore, the use of the arrays IND and SS is well adapted to this method.

The SOR method is very simple and, for some types of problems, it provides excellent results. The performance of this method, as well as any other iterative method, depends strongly on boundary conditions (Dirichlet conditions). If there is a significant number of nodes at which potentials are imposed the convergence is, generally, very fast. Therefore, if the typical problem to be solved falls under this category, successive overrelaxation is a good choice for the solution method.

12.6.2b. The Incomplete Choleski Conjugate Gradient Method

The Incomplete Choleski Conjugate Gradient (ICCG) method, proposed by Meijerink and Van der Vorst in 1977, uses the well known Conjugate Gradient (CG) method with preconditioning of the matrix $[A]$. Preconditioning is based on the fact that if the matrix $[A]$ and the vector $\{b\}$ are multiplied by a matrix $[C]$, $[C]$ is of the same order as $A(n,n)$], the solution of the system is not altered

$$[A]\{x\} = \{b\} \rightarrow [C][A]\{x\} = [C]\{b\}$$

If the matrix $[C]$, when multiplied by $[A]$, generates a product $[C][A]$ close to the unity matrix, the system $[C][A]\{x\} = [C]\{b\}$ is better adapted to the solution by an iterative method. This operation is called "preconditioning" of the matrix $[A]$ If the matrix $[C]$ is equal to $[A]^{-1}$, preconditioning results is the unity matrix and the solution converges in one iteration. However, finding the inverse of $[A]$ is just as complex as finding the solution to the system. Therefore, although it provides a solution in one iteration, this is not a practical preconditioning matrix.

The authors mentioned above proposed the product $[[L][L]^T]^{-1}$ for the matrix $[C]$ where $[L]$ is obtained from the Choleski method, presented above. However, only the terms of $[L][L]^T$ of the existent nonzero coefficients of $[A]$ are calculated. This provides an incomplete approximation of $[L][L]^T$, as the name of the method indicates. The method is valid for any symmetrical matrix $[A]$.

In applying this method, using Eqs. (12.5), the terms of $[L][L]^T$ occupying the nonzero positions in matrix $[A]$, (as indicated by array IND) of section 12.3.3 are calculated. These terms are placed in a new array $[C]$ with the same dimensions as $[SS]$ Since addressing is done by IND, the application of this method needs about 50% more memory allocation than the successive overrelaxation method.

After the matrix $[L][L]^T$ is obtained, the following algorithm is applied. It is the standard Conjugate Gradient algorithm modified to include the preconditioning.

- Initialization step:

$$r_0 = b - Ax_0$$

$$([L][L]^T)p_0 = r_0$$

Obtain p_0 by backward and forward substitution as in the regular Choleski's method.

- Iteration:

$$a_i = \frac{r_i z_i}{p_i A p_i} \quad \text{where:} \quad ([L][L]^T)z_i = r_i$$

$$x_{i+1} = x_i + a_i p_i$$

$$r_{i+1} = r_i - a_i A p_i$$

$$b_i = \frac{r_{i+1} z_{i+1}}{r_i z_i} \quad \text{where:} \quad ([L][L]^T)z_{i+1} = r_{i+1}$$

$$p_{i+1} = z_{i+1} + b_i p_i$$

This method (or one of its many variations) is considered to be the most efficient iterative method for solving matrix systems generated by the finite elements method for electromagnetic problems. The increase in memory needs compared to relaxation methods is largely justified by its efficiency. However, it must be noted that the calculation of $[L][L]^T$, even though it is an incomplete decomposition is lengthy and somewhat difficult to implement.

12.7. Methods of Solution for Eigenvalues and Eigenvectors

In eigenvalue problems, like the solutions for modes in a cavity or waveguide, the system of equations has characteristic values, for which a solution exists. These were called modes in Chapter 7 but, in general, one can view the cutoff frequencies of the modes as the eigenvalues of a system of equations and the eigenvectors, as the corresponding solutions (i.e., fields). The type of equations one needs to solve [e.g., Eqs. [7.93) or (7.94)], is a generalized eigenvalue problem of the form

$$[SS]\{x\} = \lambda [M]\{x\} \qquad (12.7)$$

where $[M]$ and $[SS]$ are normally symmetric, positive definite matrices but, in the case of propagation modes in waveguides and resonant modes in cavities these are complex matrices.

The generalized eigenvalue problem can be reduced to a standard eigenvalue problem by writing

$$[C]\{x\} = \lambda \{x\} \qquad (12.8)$$

where

$$[C] = [M]^{-1}[SS]$$

and $[M]^{-1}$ is the inverse of $[M]$.

For this reason, we will look at methods of finding the eigenvalues and eigenvectors of the standard eigenvalue problem, with complex coefficients.

While there are many methods suitable for the systems normally encountered, the most commonly used for complex systems of equations is the so called QZ algorithm. This, and the equivalent QR algorithm for real matrices will be outlined briefly below. The Jacobi method for eigenvalues and eigenvector is also presented since it is a simple iterative method that illustrates the steps involved in solution. Givens transformation is presented as a more efficient method and also since it is often used in conjunction with the QR and QZ methods.

These and other methods are available in standard libraries like EISPACK and LINPACK and are described in many books, some of which may be found in the bibliography section.

12.7.1. The Jacobi Transformation

The principle of all transformation methods is in a systematic modification of the matrix until its diagonals are the only nonzero coefficients. For a matrix with all

Methods of Solution for Eigenvalues and Eigenvectors 501

zero coefficients except for the diagonal, the diagonals are the eigenvalues. Thus, reduction to a diagonal form provides the eigenvalues of the original matrix, if, in the transformation process, the properties of the matrix have not been changed. This is the case if the transformation is done using so called similarity transformations

$$[C]' = [T]^{-1}[C][T]$$

where $[T]$ is the transformation matrix. If, in addition, the transformation matrix is orthogonal, $[T]^{-1}=[T]^T$ and we may write

$$[C]' = [T]^t[C][T] \qquad (12.9)$$

In the Jacobi method, the transformation matrix is of the form

$$[T] = \begin{bmatrix} 1 & 0 & 0 & 0 & 0 & 0 \\ 0 & \cos\alpha & 0 & -\sin\alpha & 0 & 0 \\ 0 & 0 & 1 & 0 & 0 & 0 \\ 0 & \sin\alpha & 0 & \cos\alpha & 0 & 0 \\ 0 & 0 & 0 & 0 & 1 & 0 \\ 0 & 0 & 0 & 0 & 0 & 1 \end{bmatrix} \qquad (12.10)$$

where $[T][T]^{-1}=1$, indicating an orthogonal transformation matrix. The only nonzero coefficients are in columns I,J and rows I,J. All diagonal values are equal to 1 except for diagonals I,I and J,J as shown.

The principle is to apply this transformation in an iterative fashion. In each iteration, the value of α is calculated such that the off-diagonal terms of the modified matrix $[C]'$ at locations (I,J) and (J,I), become zero. This process is repeated until all off-diagonal terms are zero. If any of the previously zeroed coefficients become nonzero due to the modifications of other coefficients in the transformation in Eq. (12.10), the process is repeated until all off-diagonal terms are within a tolerance parameter from zero. Since the orthogonal transformation retains the symmetry in the matrix only the upper part of the matrix is considered. The algorithm for one Jacobi iteration is:

- To zero the off-diagonal terms we write the condition [from the product in Eq. (12.9)]

$$c'_{I,J} = 0 = (-c_{I,I} + c_{J,J})\cos\alpha\sin\alpha + c_{I,J}(\cos^2\alpha - \sin^2\alpha)$$

From this we write

$$\cos^2\alpha = \frac{1}{2} + \frac{c_{I,I} - c_{J,J}}{2\sqrt{(c_{I,I} - c_{J,J})^2 + 4c_{I,J}^2}}$$

$$\sin^2\alpha = \frac{1}{2} - \frac{c_{I,I} - c_{J,J}}{2\sqrt{(c_{I,I} - c_{J,J})^2 + 4c_{I,J}^2}}$$

$$\sin\alpha\cos\alpha = \frac{c_{I,J}}{\sqrt{(c_{I,I} - c_{J,J})^2 + 4c_{I,J}^2}}$$

After these calculations, the off-diagonal coefficients $c_{I,J}$ and $c_{J,I}$ are set to zero.

- The diagonal coefficients $c_{I,I}$ and $c_{J,J}$ are modified as

$$c'_{I,I} = c_{I,I}\cos^2\alpha + c_{J,J}\sin^2\alpha + 2c_{I,J}\sin\alpha\cos\alpha$$

$$c'_{J,J} = c_{I,I}\sin^2\alpha + c_{J,J}\cos^2\alpha - 2c_{I,J}\sin\alpha\cos\alpha$$

- Similarly, all coefficients in rows I and J, and columns I and J, are modified as

$$c'_{i,I} = c_{i,I}\cos\alpha + c_{i,J}\sin\alpha \qquad i=1,N \quad i\neq I$$

$$c'_{i,J} = -c_{i,I}\sin\alpha + c_{i,J}\cos\alpha \qquad i=1,N \quad i\neq I$$

and, obviously,

$$c'_{i,I} = c'_{I,i} \qquad \text{and} \qquad c'_{i,J} = c'_{J,i}$$

The only remaining difficulty is to decide on the signs of $\sin\alpha$ and $\cos\alpha$. If we arbitrarily choose $\sin\alpha$ to be positive, then $\cos\alpha$ must be chosen to have the same sign as $c_{I,J}$. This procedures leads to a modified coefficient matrix

$$[C]'^{k+1} = [T]^T_k[C]_k[T]_k$$

After n iterations, the matrix $[C]'$ will be diagonal and the eigenvalues are the diagonal values. If the transformation matrices are saved and multiplied together, or if they are multiplied as they are obtained we can write the eigenvectors of the original matrix $[C]$ as the columns of the matrix

$$[X] = [T]_1[T]_2[T]_3....[T]_n$$

where each column is one solution $\{x\}_i$, corresponding to the eigenvector λ_i.

This method, while simple is not very efficient, and many other methods, including modifications of the basic Jacobi method, exist. However, it will serve here as an illustration.

12.7.2. The Givens Transformation

Before continuing with the QR and QZ methods, we will outline the Givens transformation method. While Givens' method is merely a modification of the basic Jacobi method, it is a single sweep method (requires the elimination of off-diagonal terms only once, because terms that have been zeroed to not change after elimination). However, it does not produce a diagonal system but, rather, a tri-diagonal system which must be solved using, for example, the QR or QZ algorithms. The main purpose of discussing it here is that it is often used in conjunction with the QR or QZ methods.

Using the same transformation as for the Jacobi method, each step eliminates the off-diagonal terms $c_{I-1,J}$ and $c_{I,J-1}$. In all other respects the algorithm is the same. The coefficients of the transformation matrix ($sin\alpha$ and $cos\alpha$) are found from the condition

$$c'_{I-1,J} = 0 = -c_{I-1,I} sin\alpha + c_{I-1,J} cos\alpha$$

From this we write

$$cos\alpha = \frac{c_{I-1,I}}{\sqrt{c_{I-1,I}^2 + c_{I-1,J}^2}}$$

$$sin\alpha = \frac{c_{I-1,J}}{\sqrt{c_{I-1,I}^2 + c_{I-1,J}^2}}$$

The remaining steps in the procedure are the multiplication of this matrix with the original matrix as in the Jacobi method. After all possible coefficients have been zeroed, the modified matrix is

$$[C]' = \begin{bmatrix} X & X & 0 & 0 & 0 & 0 \\ X & X & X & 0 & 0 & 0 \\ 0 & X & X & X & 0 & 0 \\ 0 & 0 & X & X & X & 0 \\ 0 & 0 & 0 & X & X & X \\ 0 & 0 & 0 & 0 & X & X \end{bmatrix}$$

where X indicates a nonzero coefficient.

Since the eigenvalues and eigenvectors are yet to be found we only note here that the eigenvalues of the tri-diagonal system above are the same as the eigenvalues of the original matrix and the eigenvectors of the original matrix can be found from those of the tri-diagonal system by their product with the product of the transformations used for the Givens method

$$x_i = [T]_1[T]_2[T]_3...[T]_n x_i''$$

where x_i denotes an eigenvector of the original system and x_i'' denotes the corresponding eigenvector (for the same eigenvalue) of the tri-diagonal system.

12.7.3. *The QR and QZ Methods*

The basic QR algorithm uses the following decomposition for step k of the iteration

$$[C]_k = [Q]_k[R]_k$$

where $[Q]^k$ is an orthogonal matrix and $[R]^k$ is an upper triangular matrix. The modified matrix at step $k+1$ is now written as

$$[C]_{k+1} = [R]_k[Q]_k$$

and $[R]^k$ is related directly to the Givens transformations above. Assuming that n Givens transformations were required to obtain the tri-diagonal system, we have

$$[R]_k = [T]_n^T[T]_{n-1}^T[T]_{n-2}^T...[T]_1^T[C]_k$$

and

$$C_{k+1} = [R]_k[T]_1[T]_2[T]_3...[T]_n$$

where we used the fact that

$$[R]_k = [Q]_k^T[C]_k$$

Obviously, the Givens transformation is not the only one that can be used and, in fact, the Householder transformation is often used for this purpose.

The QZ algorithm is more often used in conjunction with complex matrices of the type we encounter in field computation. The algorithm is based on the following principles:

For a generalized eigenvalue problem

$$[SS]\{x\} = \lambda[M]\{x\}$$

we find a decomposition

$$[Q][SS][Z]\{y\} = \lambda[Q][M][Z]\{y\} \qquad (12.11)$$

such that $[Q][SS][Z]$ and $[Q][M][Z]$ are both upper triangular matrices. Then the eigenvalues of the original matrix are related to the eigenvalues of the transformed matrix by

$$\{x\} = [Z]\{y\} \qquad (12.12)$$

and the eigenvectors are the diagonals of $[Q][A][Z]$.

In the first step, the matrix $[SS]$ is reduced to a complex upper Hessenberg matrix using transformation methods (i.e., by Givens or Householder methods). This can always be done for a general complex matrix. The matrix $[M]$ is triangularized while modifying $[A]$ so that the eigenvalues are retained. These matrices look as follows

$$[SS]' = \begin{bmatrix} X & X & X & X & X & X \\ X & X & X & X & X & X \\ 0 & X & X & X & X & X \\ 0 & 0 & X & X & X & X \\ 0 & 0 & 0 & X & X & X \\ 0 & 0 & 0 & 0 & X & X \end{bmatrix} \qquad [M]' = \begin{bmatrix} X & X & X & X & X & X \\ 0 & X & X & X & X & X \\ 0 & 0 & X & X & X & X \\ 0 & 0 & 0 & X & X & X \\ 0 & 0 & 0 & 0 & X & X \\ 0 & 0 & 0 & 0 & 0 & X \end{bmatrix}$$

The iteration now becomes

$$[SS]_{k+1} = [Q]_k[SS]_k[Z]_k$$

$$[M]_{k+1} = [Q]_k[M]_k[Z]_k$$

When $[SS]$ has been triangularized, the transformation in Eqs. (12.11) and (12.12) are satisfied and the eigenvectors and eigenvalues are obtained.

The discussion above is brief and only includes a few aspects of eigenvalue solution. Many other methods, some much more efficient than those presented exist. The bibliography section points to some of these. There are also special computer libraries like EISPACK dedicated to this aspect of computation. Eigenvalue solvers also exist in all other major computer packages like LINPACK, IMSL, and NAG, to mention only a few.

12.8. Diagram of a Finite Element Program

Figure 12.12 shows a general diagram of a finite element program for solution of static field problems.

In this diagram, if the problem is linear, the variable I is equal to 1 and the assembly/solution of the system is performed only once. For the nonlinear case, the vector V_{ac} is defined as equal to V of the previous iteration, and the convergence is tested by the difference between V and V_{ac}.

It is important to avoid any confusion between the global iterative procedure (for nonlinear problems) with the equations solution method, which can be direct or iterative. It means, for example, that in a linear problem (where $I=1$) the system of equations $[SS]\{V\}=\{Q\}$ can be solved by an iterative method, while in the nonlinear case (i.e., an iterative problem) a direct method can be used to solve the matrix system.

We can use, generally with good results, a relaxation method for potentials if the problem is nonlinear. For this, we define a potential variation $V_{ar}=V-V_{ac}$. We apply the following relaxation expression:

$$V = V + (\Omega-1)V_{ar}$$

For the scalar potential, successive overrelaxation, with $1.5<\Omega<2.0$ may be used to accelerate convergence. With the vector potential, whose convergence is more uncertain, it is possible to use underrelaxation, with $\Omega<1$, but, in some cases the solution may divergence. In any case, the relaxation has to be applied after a few iterations (after the fourth iteration, according to our experience); it is advisable to allow the system to evolve "naturally" in the initial few iterations.

Figure 12.13 shows a flow diagram for the assembly of the matrix system. This diagram corresponds to the block between the points "b" and "c" in Figure 12.12.

For the Newton-Raphson method the sequence of steps is the same as shown in Figure 12.13; however, it is necessary to also calculate $\partial v/\partial B^2$ or $\partial \mu/\partial H^2$ for the Jacobian terms. The global matrix $[SS]$ and $\{Q\}$ are used for $[SJ]$ and $[R]$ respectively, since the memory needs are the same. The 3D systems mentioned in this work use the same type of diagram, for both linear and nonlinear cases.

Eddy current problems with time discretization, presented in the previous chapter, follow the same diagrams; however, an additional loop, for iteration in time, is added between the points "a" and "d" of Figure 12.12.

Diagram of a Finite Element Program   507

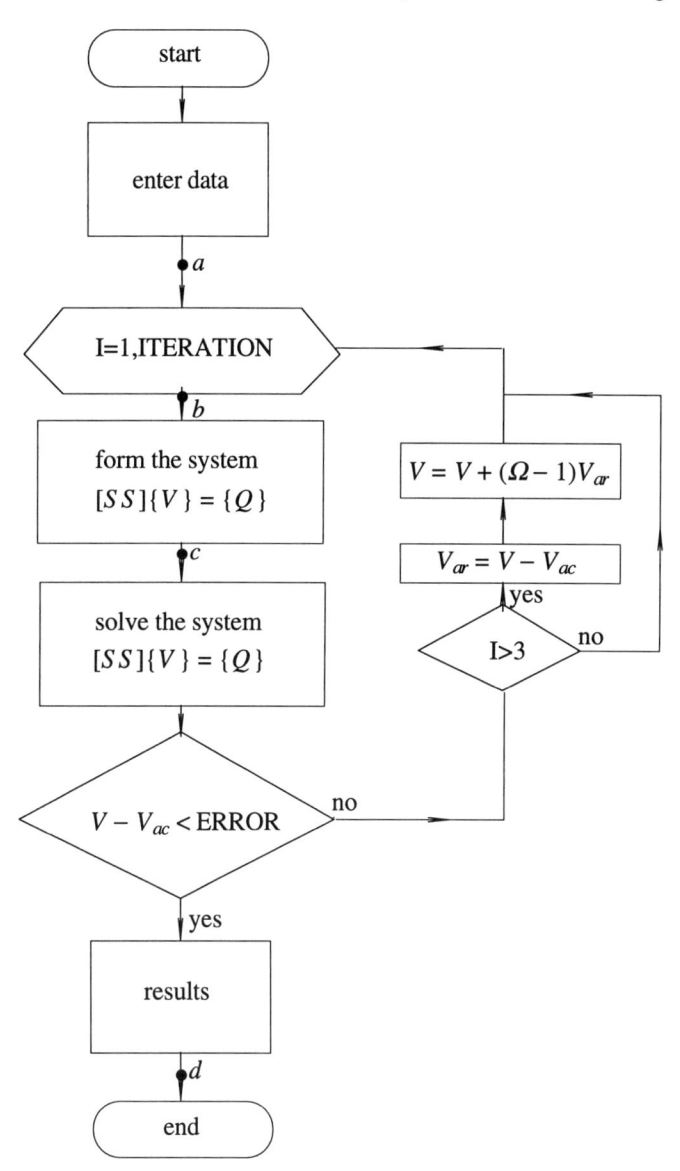

Figure 12.12. General flow diagram of a finite element program for solution of static field problems.

508   12. Computational Aspects in Finite Element Software Implementation

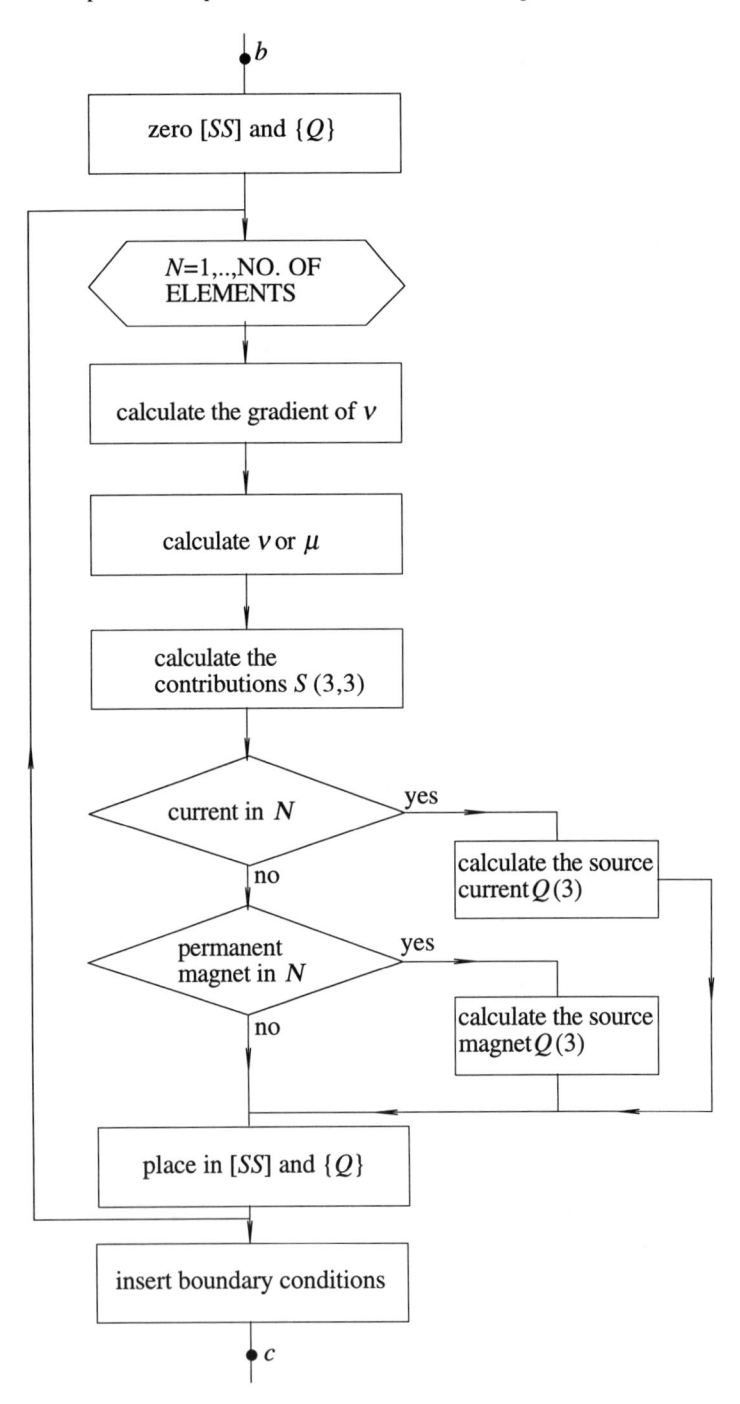

Figure 12.13. Flow diagram for the assembly of the matrix system. This diagram corresponds to the block between points b and c in Figure 12.12.

13
General Organization of Field Computation Software

13.1. Introduction

The main goal of field calculation research in the past was to write specific, special purpose computer programs for the application of the finite element method to computation of electromagnetic fields. At present, there is more interest in building complete computational software packages, providing simple use of the method for the project engineer, regardless of how familiar he may be with the finite element method. This also seems to be the trend for the future, with more emphasis on personal computers and workstations.

These software packages, are part of the computer-aided design (CAD) philosophy. For a given electromagnetic device, they provide a tool for design of the geometry and simulation of working conditions, within realistic approximations. The expensive, time consuming task of designing, prototyping and testing of a device can be first done on a computer and only the final device built and tested. While, in general, computer-aided design of a device cannot replace prototyping, the many steps in design and the need for many prototypes is reduced to at most a few prototypes, very near the final, optimized product.

In addition to the economical value of CAD, the analysis of the electromagnetic device is important in itself. The computation allows analysis of the device behavior, both qualitatively (i.e., plots of equipotential lines which provide a good understanding of the behavior but not exact numerical data) as well as quantitatively (in the form of flux values, inductances, forces, saturation levels, and other useful quantities). This type of analysis is unique to computational software and was not possible in the past.

A CAD package for electromagnetic field analysis, to be useful, must have the following three modules, and, to be convenient, the modules must be integrated in a single, simple, and convenient package:

a. Pre-processor. In this step, the user furnishes the structure to be analyzed (geometric shapes, boundary conditions, sources, material properties). The

task of the computer is to generate an appropriate finite element mesh either automatically, or, as is often the case, semiautomatically.
b. Processor. After discretization, the finite element method can be applied. The primary results of the computation are the potentials at the nodes of the mesh.
c. Post-processor. In this step analysis of the results is performed. Visual representation of data (plotting, equipotential drawing) and numerical results (field values, fluxes, forces) are obtained and presented.

Step b. is the physical-numerical modeling of the phenomenon to be analyzed, and, therefore, is the most important part of the software. However, pre- and post-processing, while essentially incorporated in the software package for the convenience of the user, are becoming more and more important. The reason for this is that the amount of manual work needed for preparation of data for computation is extensive. Similarly, the analysis done by the post-processor is extremely tedious. In fact, for any but the simplest analyses, some forms of automatic data generation and display are necessary because of the size of data sets involved.

The usefulness of a software package is only partly attributed to what it actually calculates and how efficiently it does so. User convenience and confidence is equally, and perhaps, more important. Without these tools, the software cannot be used other than by the designer of the software or a well trained user.

In this chapter, we present briefly the general points associated with the design and implementation of a complete software package for electromagnetic field calculation. In the process we use, for illustration purposes, an existing CAD package. Recent trends in CAD of electromagnetic fields are also presented with discussion of such topics as adaptive meshes generation, and coupled formulations.

13.2. The Pre-Processor Module

13.2.1. *The User/System Dialogue*

The pre-processor step consists of two parts: the first is the user/machine dialogue. The purpose is to provide the user with a tool to define the structure to be analyzed to the computation system. Among all steps in computation, it is here that the user/system interaction is most extensive.

The computational software depends heavily on the available equipment. The availability of graphic tools is essential for design of a good package. Without this, the user interface is incomplete and therefore, the software is lacking its most efficient tools.

During the definition of the structure, it is desirable for the user to be able to graphically visualize it, in order to ensure correctness of the data. It is fundamental that the dialogue be simple, clear, unambiguous, and with adequate options for corrections and modifications. It is also important that the system have partial "back-ups," in order to, for example, save partial data of a big structure, or allow voluntary interruptions of data entry. In cases of external interruptions, like power outages, this procedure minimizes loss of data. These dialogue programs are, generally, very long and complex and their development is time consuming. It is desirable that their graphic section have amplifications or "zooms," windows for drawings and messages, and, in general, any possible tool for convenience and speed. Extensive investment in programming should be done in designing these parts of the software so that the user may use these options with ease.

A particular design problem, with known structure and properties, can be solved by writing a special purpose program which, in many cases is quite simple. However, a general purpose software requires much more than this and its development is a continuous process that can span years.

13.2.2. Domain Discretization

After the structure to be analyzed is defined, the discretization of the solution domain can commence. This part of the pre-processor can be part of the data entry program or it can be a separate module. The choice depends, to a large degree, on the available computer.

Although the data entry and discretization can be done manually, this is not normally an acceptable procedure, because of the reasons mentioned above. The first option for simplifying this task is the semiautomatic mesh generation. Depending on the problem, this option may be the best choice, especially when we wish to force the system to generate a particular type of mesh. A fully automatic mesh generation program may not be capable of generating the required mesh, or the mesh may not be acceptable in some regions of the domain.

Semiautomatic mesh generation can be implemented, for example, as shown in Figure 13.1a and b. In this scheme, the domain is composed of several regions with quadrilateral topology. For every region, the number of divisions on its sides are indicated. In this example, the number of divisions for region R_2 are 3 and 4. The mesh program can generate the mesh shown in Figure 13.1b. Note that this mesh must be compatible with the meshes of regions R_1 and R_3.

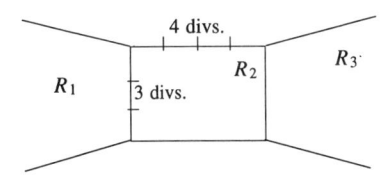

Figure 13.1a. A simple structure to be discretized using a semiautomatic mesh generator. Each subdomain is divided into a preset number of elements.

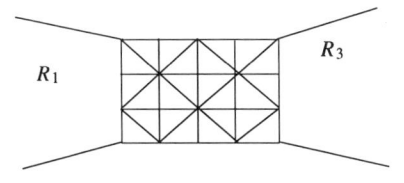

Figure 13.1b. The mesh generated by the mesh generator in one of the subdomains. Neighboring domains will have the same number of elements on shared sides.

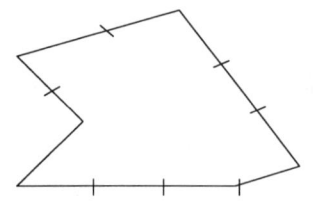

512   13. General Organization of Field Computation Software

Figure 13.2a. Automatic mesh generator: division of geometry sides into segments.

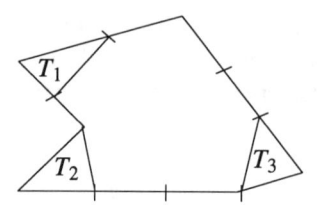

Figure 13.2b. Automatic mesh generation: closing of acute angles.

This procedure is one simple example of semi-automatic mesh generation. It is not difficult to extend this method to other geometrical shapes.

Automatic mesh generation requires more sophisticated programming. There are several methods suitable for automatic mesh generation. We will present here three methods, but we also note that this problem is an extensive subject, and is still an active research topic. The bibliography section provides many sources that treat mesh generation and the reader is encouraged to refer to these.

First, we present the method used in the EFCAD software package (EFCAD stands for **E**lectromagnetic **F**inite element **C**omputer **A**ided **D**esign). The mesh generator in this package is based on three principal operations:

a. Closing of acute angles.
b. Cuts to cancel concavities.
c. Use of element centroids to define a new node or to close a region.

To see how this method works, consider the region shown in Figure 13.2. The region's sides are divided into segments, according to a criterion defined by the user; this criterion indicates to the program if the mesh should have a low, medium, or high element density, depending on the design requirements. Depending on the density required, he program calculates one "standard segment" for the whole domain. The larger the specified density, the smaller the segment. The sides of the region are then divided into "standard segments."

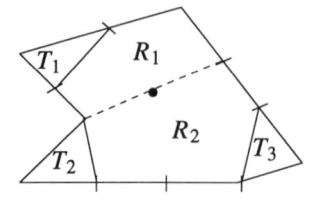

Figure 13.2c. Automatic mesh generation: division of concave regions into all convex regions.

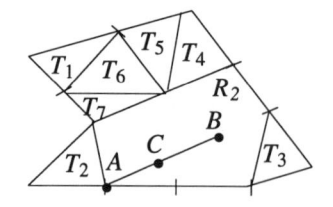

Figure 13.2d. Automatic mesh generation: all remaining acute angles in the new regions are closed.

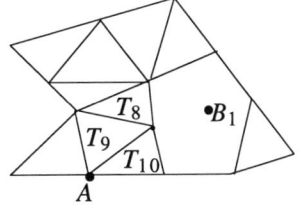

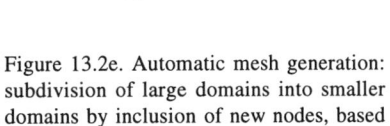

Figure 13.2e. Automatic mesh generation: subdivision of large domains into smaller domains by inclusion of new nodes, based on the size of the standard segment.

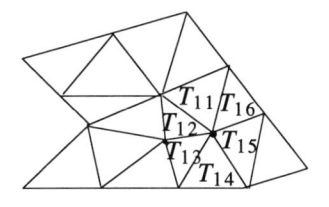

Figure 13.2f. Automatic mesh generation: subdivision of the smaller domains into finite elements.

Discretization of the domain is performed in a sequence of operations. The first operation [operation (1)] consists of closing all acute angles, as shown in Figure 13.2b, where elements T_1, T_2, and T_3 were created by closure of the acute angles. Operation (2) searches for concavities in the remainder of the region. If one exists, the region is cut in two regions, as shown in Figure 13.2c. Region R_1 and R_2 are created. This operation also defines a new node.

Operation (3) verifies that region R_1 has no acute angles and it is fully closed (Figure 13.2d). Operation (4) verifies that region R_2 has no remaining acute angles and concavities.

At this point, the centroid B is calculated and, if the distance BA is large, a new node C is created on BA, such that the distance CA is equal to the "standard segment". In operation (5), the triangles T_8, T_9, and T_{10} are created, to close the existing acute angle (Figure 13.2e).

In operation (6), since there are no more acute angles and concavities in the remainder of the region, a new centroid B_1 is calculated. This centroid is near the remainder of the nodes and the region is closed by connecting the centroid with these nodes, resulting in triangles T_{11} through T_{16}, as shown in Figure 13.2f.

Generally, in automatic mesh generation, some triangles may have a shape far from the ideal form. In this case, the program will, by itself, make "diagonal changes" (shown in Figure 13.3a) and "recentering" of elements (shown in Figure 13.3b).

Figure 13.4a shows a mesh generated by the discretization module in EFCAD, before the application of "smoothing" procedures (diagonal changes and element re-centering). Figure 13.4b shows the same mesh after smoothing. Visually, the mesh in Figure 13.4b is more homogeneous than the mesh in Figure 13.4a. As a consequence, the smoothed mesh will result in better precision when the finite element method is applied.

It is important to note that an ideal mesh would be made of equilateral triangles only. This is virtually impossible for an irregular region. However, a good mesh generator program can generate meshes with good element quality.

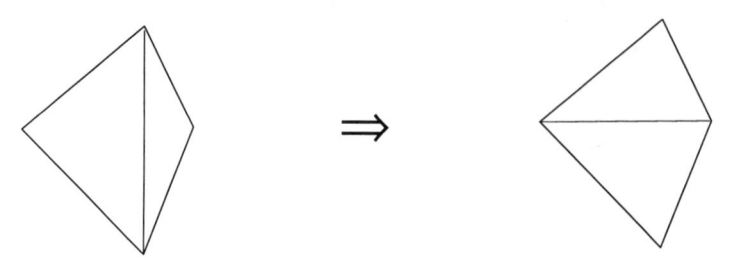

Figure 13.3a. Changes in diagonal connections to improve the aspect ratio of elements.

Figure 13.3b. Recentering of centroids of domains to improve the aspect ratio of elements.

An element "quality factor" q is defined as

$$q = 8(s-a)(s-b)(s-c)/abc$$

where a, b, and c are the lengths of sides of the triangle, and s is given by

$$s = (a+b+c)/2$$

For an equilateral triangle the quality factor is $q=1$. When the complete mesh is generated, it is possible to obtain an average q for the mesh. For a good mesh the average quality factor should be over 0.85. For the meshes shown in Figure 13.4a and 13.4b, the averages q are 0.907 and 0.957, respectively.

Another useful feature of a mesh generator allows the users to insert and delete nodes, and to modify their positions. These procedures, when properly used, result in improvement of the mesh, especially in regions where the analysis of the electromagnetic phenomenon is critical.

Another automatic mesh generation principle works with the scheme shown in Figure 13.5.

Initially the region to be meshed has its sides divided into small segments. Triangles are built following the boundaries from inside the region, until the region is completely filled by triangles. Meshes generated by this method have, in general, good quality factors.

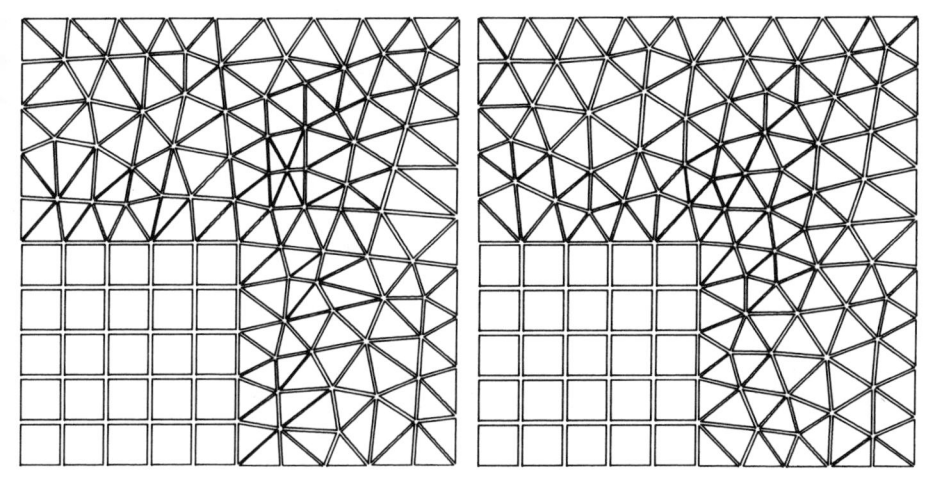

Figures 13.4a. A finite element mesh as generated by EFCAD, before smoothing.

Figures 13.4b. The mesh in Figure 13.4a after smoothing.

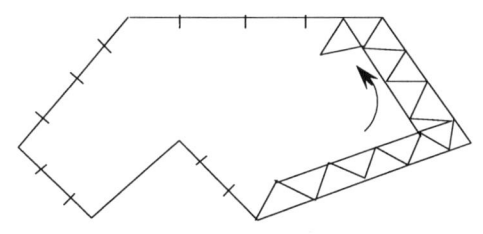

Figure 13.5. A mesh generation scheme based on division of the outer boundary and filling of the inner domain with finite elements by following the boundary division. As each layer is completed, a new inner boundary is created and meshed.

Many mesh generation programs based on the Delaunay method have been developed. The principle of this method is described using the situation shown in Figure 13.6a, where some nodes in a region are shown.

In this method, the triangle defined by nodes 2, 6, and 7 is not accepted as an element in the mesh, because inside the circle defined by these nodes, there is another node (node 1 in Figure 13.6b). The Delaunay algorithm defines the triangles such that this does not occur. Instead, the two triangles shown in Figure 13.6c are generated.

Note that, in these figures, the circles defined by nodes 1,7,2 and 1,6,7 do not contain any other nodes. This principle after extension to all nodes of this region results in the mesh shown in Figure 13.6d.

This method can also be used for three dimensional meshes and the corresponding condition is such that the spheres defined by four nodes (for tetrahedral finite elements) do not include other nodes.

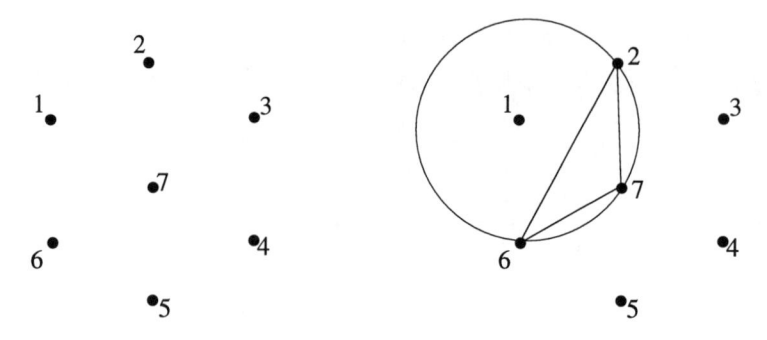

Figure 13.6a. A number of nodes in a region to be subdivided into finite elements.

Figure 13.6b. A possible finite element is rejected because the circle defined by its three nodes contains a fourth node (node 1).

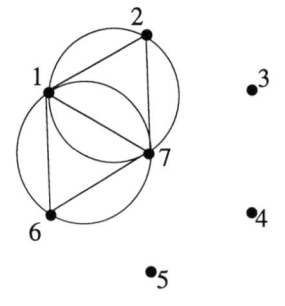

Figure 13.6c. Delaunay mesh generation method. The two triangles shown are accepted as finite elements.

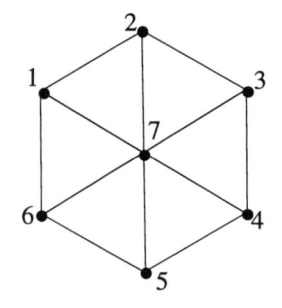

Figure 13.6d. A mesh over the nodes of Figure 13.6a generated with the Delaunay mesh generation method.

Delaunay's method is used in many finite element software systems. However, this method does not create new nodes. This is a procedure to link correctly existing nodes of the mesh. Generally, the Delaunay method must be used in conjunction with some other method whose purpose is to adequately define new nodes.

13.3. The Processor Module

The major computational aspects in this step were presented in Chapter 12. We will comment here on the inclusion of the field calculation program in the software package. This module can be activated only after the domain is discretized and all necessary data have been correctly generated. The necessary data consist of the following:

- The mesh defining the geometry of the domain. Generally, it is specified by a two-dimensional array NT(I,3) defining the numbers of the three nodes of each element in the mesh. This array provides the mesh topology. Its geometrical dimensions are given by arrays X(K) and Y(K). These arrays contain the coordinates of the K nodes of the mesh. An additional array MAT(I) contains a material index for each of the I elements of the mesh.
- Boundary conditions. All nodes for which Dirichlet boundary conditions are specified, are listed. (Neumann boundary conditions are implicit in the formulation and are not specified.)
- Applied potential values at the boundary nodes.
- The number of elements with applied current densities and the value of these current densities.

After the finite element calculation has executed, the program provides only the node potentials. Although the numerical formulations (Chapters 8, 9, and 10) frequently need the field quantities **H** and **B** (or **E** and **D**), the principal results are the potentials at the nodes of the mesh. Fields can be calculated in the post-processor step by very simple subroutines. The use of this procedure keeps the amount of data that needs to be transferred between files to a minimum.

13.4. The Post-Processor Module

This step has, normally, two parts, both very important to the user. The first part draws equipotential lines over the structure under study to visualize the solution in simple, general terms. The second part provides concrete numerical values for required quantities.

13.4.1. *Visualization of Results*

Upon completion of the computation step, the user can request display or plotting of equipotential lines. A fast visual inspection of these plots is useful in verifying that the problem was correctly entered in the pre-processor module. For example, errors in specifying boundary conditions or applied current densities, as well as many other errors can be very easily detected by inspection of equipotential plots. In addition, this visual inspection provides a qualitative analysis of the phenomenon. For example, the flux distribution in the solution domain may indicate regions of saturation, or excessive flux leakages in a magnetic path. Often, it is not necessary to calculate numerical values, because the flux distribution indicates that the device is not working adequately, meaning that the proposed structure does not correspond to an acceptable design.

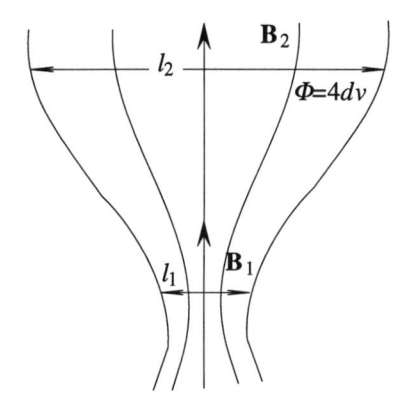

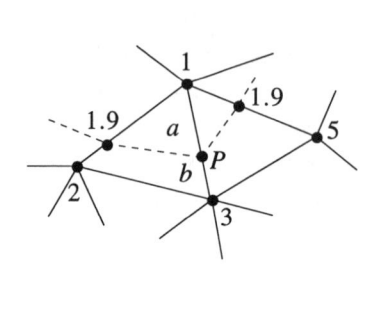

Figures 13.7. Information obtainable from qualitative analysis of equipotential lines. The flux density in section l_2 is lower than in section l_1 because the flux between each two lines is constant.

Figure 13.8. Method of tracing equipotential lines. Linear interpolation between nodes is used.

To obtain a more accurate idea of the magnetic flux distribution (or the electric field intensity), it is necessary to draw several equipotential lines with constant potential differences between the lines. For example, if we wish to draw twenty one lines, supposing that the minimum and maximum values are "V_{min}" and "V_{max}", we calculate $dV=(V_{max}-V_{min})/20$ and trace the twenty one equipotential lines whose values are $V_{min}, V_{min+dV},...,V_{max}$. Enclosed between two neighboring lines, is a flux tube with flux dV. In the regions where these lines are close together, the magnetic field intensity **B** is high. In Figure 13.7, several flux tubes cross the short section l_1 and the magnetic flux density $\mathbf{B}_1$ is high. In section l_2, which is larger than l_1, the total flux is equal to the flux through l_1 ($4dV$ in this case), and the flux density $\mathbf{B}_2$ is lower than $\mathbf{B}_1$ (since flux is conserved $B_1 l_1 = 4dV = B_2 l_2$)

Tracing of the equipotential lines is done by linear interpolation on the sides of the elements of the mesh, as shown in Figure 13.8. In this example we trace an equipotential with potential value 1.9. The potentials shown at the nodes of Figure 13.8 are obtained from the finite element calculation. Point P is placed between the potential values 1.0 and 3.0 such that

$$\frac{1.9-1.0}{3.0-1.0} = \frac{a}{a+b}$$

Noting that the value of $(a+b)$ is known, we can evaluate a and, therefore, the position of the point P can be found.

The calculation of points like P is quite simple, but producing the whole equipotential plot, taking into account all elements in the mesh and the potential values at their nodes is lengthy and complex. The need to evaluate equipotentials on element sides, the comparison tests between real numbers for each point, and other computational difficulties make the development of this kind of program lengthy and tedious. Also, if the equipotential plotting program has been designed for electrical applications (where the differences between potential lines may be large, say between zero and 10,000 V), it can be less than ideal for plotting of equipotential lines in magnetic applications (where typical values of potentials are of the order of 10^{-4} Wb/m), because the comparison test between real variables is quite different. However, these are relatively minor problems in comparison with the design of the mesh generator program.

13.4.2. *Calculation of Numerical Results*

If visualization of results indicates that the geometry and all necessary conditions are correct, we may assume that the solution is correct and continue with a quantitative analysis. While many errors in design can be detected earlier in the process, some errors will show up while evaluating numerical results.

As mentioned in section 13.3, the data required for this step are potentials at the nodes of the mesh and the geometrical data for the mesh. From these, specialized subroutines calculate the potential derivatives, providing the fields **H** and **B** or **E** and **D**. It is also necessary to define the material characteristics in the mesh (μ, ν, ε, or σ) including the $B(H)$ curve (if the problem is nonlinear), in order to obtain saturation levels in ferromagnetic materials. This is normally done in a subroutine (identical to that used in the finite element program).

The numerical results we may wish to calculate are any of the following:

a. Potential value at any point of the solution domain.
b. Fields **H**, **B**, **E**, **D**, and **J**, obtained directly from through derivatives of the potentials.
c. Characteristics of saturable materials; if the problem is linear, this is given as input data, but for nonlinear materials, the characteristics are calculated results.
d. Flux between two points; with the vector potential, we have to calculate the potential difference between these points. With the scalar potential it is possible to obtain the flux by the method shown in Figure 13.9. Line *AB* is divided into several small segments. For each segment we obtain the value of *H* at point *M*, located at the middle of the segment. The normal component of *H* is H_n. The partial flux $d\Phi$, through this segment is given by $\mu H_n l$. The total flux crossing line *AB* is obtained by the sum of the partial fluxes $d\Phi$ of the segments.
e. Calculation of forces. To obtain forces using Maxwell's force tensor, it is necessary first to mark the line enclosing the body subjected to forces. In practice, only those parts of the body that are subjected to considerable field intensities need be included. A procedure similar to that used for flux evaluation can be used.

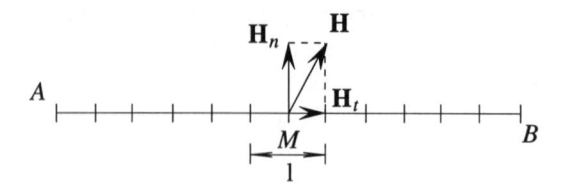

13. General Organization of Field Computation Software

Figure 13.9. Calculation of flux between two points using the scalar potential.

The force $d\mathbf{F}$ on a small segment ds of the enclosing line can be obtained as indicated below. The Maxwell force tensor is given by (see section 6.7)

$$d\mathbf{F} = -\frac{\mu_0}{2} H^2 d\mathbf{s} + \mu_0 (\mathbf{H} \cdot d\mathbf{s})\mathbf{H} \qquad (13.1)$$

with

$$\mathbf{H} = \hat{\mathbf{i}} H_x + \hat{\mathbf{j}} H_y$$

$$d\mathbf{s} = \hat{\mathbf{i}} dx + \hat{\mathbf{j}} dy$$

$$d\mathbf{F} = \hat{\mathbf{i}} dF_x + \hat{\mathbf{j}} dF_y$$

we obtain from Eq. (13.1)

$$dF_x = \frac{\mu_0}{2}\left[(H_x^2 - H_y^2)dx + 2H_x H_y dy\right] \qquad (13.2a)$$

$$dF_y = \frac{\mu_0}{2}\left[(H_y^2 - H_x^2)dy + 2H_x H_y dx\right] \qquad (13.2b)$$

Although the calculations required in Eq. (13.2a,b) are simple, the evaluation of the equivalent complex formulation of Maxwell's force tensor (section 10.3.3) is much more complicated. We calculate now the components of the force in Eq. (13.2a,b) for this case.

The equivalent expression to Eq. (13.2a) in complex variables is

$$dF_x = \frac{\mu_0}{2}\left[\left(R(H_x e^{j\omega t})\right)^2 - \left(R(H_y e^{j\omega t})\right)^2\right]dx + \mu_0 R(H_x e^{j\omega t})R(H_y e^{j\omega t})dy \quad (13.3)$$

where $R(H_x e^{j\omega t})$ is the real part of $H_x e^{j\omega t}$ and H_x and H_y are the complex components of the field. The term in square brackets in Eq. (13.3) is

$$\left(R(H_x e^{j\omega t})\right)^2 - \left(R(H_y e^{j\omega t})\right)^2 = R(H_x e^{j\omega t} + H_y e^{j\omega t})R(H_x e^{j\omega t} - H_y e^{j\omega t}) \quad (13.4)$$

For two general complex numbers A and B and their conjugates A^* and B^*, the following properties apply:

$$R(A) = \frac{1}{2}(A+A^*) \quad \text{and} \quad (AB)^* = A^* B^*$$

the expression in Eq. (13.4) becomes

$$\left(R(H_x e^{j\omega t})\right)^2 - \left(R(H_y e^{j\omega t})\right)^2 = \frac{1}{2} R\Big[\left(H_x e^{j\omega t} + H_y e^{j\omega t}\right)$$
$$\left(H_x e^{j\omega t} - H_y e^{j\omega t} + H_x^* e^{-j\omega t} - H_y^* e^{-j\omega t}\right)\Big]$$

Performing the product and collecting terms, we get

$$\left(R(H_x e^{j\omega t})\right)^2 - \left(R(H_y e^{j\omega t})\right)^2 = \frac{1}{2} R\Big[\left(H_x H_x^* - H_y H_y^*\right) +$$
$$\left(H_x^2 - H_y^2\right) e^{2j\omega t} + \left(H_y H_x^* - H_x H_y^*\right)\Big]$$

Noting that the last term in brackets is imaginary, we get

$$\left(R(H_x e^{j\omega t})\right)^2 - \left(R(H_y e^{j\omega t})\right)^2 = \frac{1}{2} R\Big[\left(H_x H_x^* - H_y H_y^*\right) + \left(H_x^2 - H_y^2\right) e^{2j\omega t}\Big]$$

or

$$\left(R(H_x e^{j\omega t})\right)^2 - \left(R(H_y e^{j\omega t})\right)^2 = \frac{1}{2}\Big[|H_x|^2 - |H_y|^2 + R\left(H_x^2 - H_y^2\right) e^{2j\omega t}\Big] \quad (13.5)$$

The second part of Eq. (13.3) is evaluated in similar steps:

$$R\left(H_x e^{j\omega t}\right) R\left(H_y e^{j\omega t}\right) = \frac{1}{2} R\Big[\left(H_x e^{j\omega t}\right)\left(H_y e^{j\omega t} + H_y^* e^{-j\omega t}\right)\Big] =$$
$$\frac{1}{2} R\Big[H_x H_y^* + H_x H_y e^{2j\omega t}\Big]$$

or

$$R\left(H_x e^{j\omega t}\right) R\left(H_y e^{j\omega t}\right) = \frac{1}{2}\Big[R\left(H_x H_y^*\right) + R(H_x H_y) e^{2j\omega t}\Big] \quad (13.6)$$

Denoting the real and imaginary parts of H_x as h_{xr}, and h_{xi} and the real and imaginary parts of H_y as h_{yr}, and h_{yi}, respectively, we have from Eq. (13.5)

$$|H_x|^2 - |H_y|^2 = h_{xr}^2 + h_{xi}^2 - h_{yr}^2 - h_{yi}^2$$

and

$$R(H_x^2 - H_y^2) = h_{xr}^2 - h_{xi}^2 - h_{yr}^2 + h_{yi}^2$$

Similarly, from equation (13.6)

$$R(H_x H_y^*) = h_{xr} h_{yr} + h_{xi} h_{yi}$$

and

$$R(H_x H_y) = h_{xr} h_{yr} - h_{xi} h_{yi}$$

Using the results in Eqs. (13.5) and (13.6) and applying them in Eq. (13.3), we obtain the following two expressions for the force $dF_x = dF_{x1} + dF_{x2}$:

- Continuous *dc* components

$$dF_{x1} = \frac{\mu_0}{4}\left(h_{xr}^2 + h_{xi}^2 - h_{yr}^2 - h_{yi}^2\right)dx + \frac{\mu_0}{2}\left(h_{xr}h_{yr} + h_{xi}h_{yi}\right)dy \quad (13.7)$$

- Frequency-dependent terms (frequency is 2ω)

$$dF_{x2} = \frac{\mu_0}{4}\left(h_{xr}^2 - h_{xi}^2 - h_{yr}^2 + h_{yi}^2\right)dx + \frac{\mu_0}{2}\left(h_{xr}h_{yr} - h_{xi}h_{yi}\right)dy \quad (13.8)$$

These expressions show that, using this formulation we obtain a continuous component of force which will be superimposed on another variable (frequency 2ω) component.

Equivalent expressions to Eqs. (13.7) and (13.8) can be calculated for dF_y, starting with Eq. (13.2b).

The vector sum of the forces dFx and dFy, for either real or complex formulations, gives the total force over the body.

f. Stored magnetic energy W in the system. This energy is given by one of the following sums:

$$W = \sum_{i=1}^{I} \int_{S_i} \frac{\mu H_i^2}{2} ds \qquad \text{for linear materials}$$

or

$$W = \sum_{i=1}^{I} \int_{S_i} \int_{0}^{B} H \, dB \, ds \qquad \text{for nonlinear materials}$$

The summation is performed over the I elements of the domain, where H_i is the magnetic field intensity on the surface S_i of element i. The field intensity H_i can be constant or variable depending on the type of element (H_i is constant for first-order elements, linear for second order elements and so on). Using an average field intensity value $\overline{H}_i$ in the element, the expression for the linear case becomes

$$W = \sum_{i=1}^{I} \frac{\mu \overline{H}_i^2}{2} ds$$

For nonlinear materials, the integral of HdB must be evaluated directly from the $B(H)$ curve of the material.

If we wish to calculate the force due to a displacement d of a movable piece in a structure, the energies W_1 and W_2 for two positions (around the point of interest) are calculated, from which the force is: $F=(W_2-W_1)/d$, using the idea of energy variation, as presented in Chapter 6.

g. Self- and mutual inductances. To calculate the self-inductance of a coil (see Figure 13.10), for each element k in the coil's cross-section, the current $I_k = JS_k$ is evaluated (J is the current density in the coil). The self-inductance is then, $L_k = \Phi_k/I_k$, where Φ_k is the vector potential A at the centroid of element k (point M). The sum ΣL_k over the elements in the coil's cross-section, provides the self-inductance of the coil. This value, to be valid must be multiplied by the number of turns in the coil.

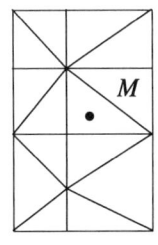

Figure 13.10. Calculation of self-inductance of a coil. Elements in the coil's cross-section are shown.

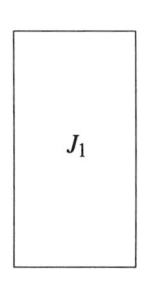

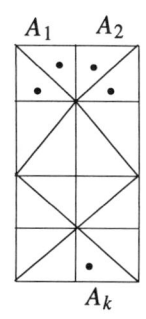

Figure 13.11. Calculation of mutual inductance between two coils. Elements in one coil are shown.

To obtain the correct self-inductance, we impose $J=0$ in all the other coils in the system (if there is more than one), to ensure the vector potential A corresponds to the coil where the inductance is being calculated. For the evaluation of the mutual inductance M_{12}, between coils (1) and (2) (Figure 13.11), the average value of A in coil (2) is calculated as

$$A = \frac{1}{K}\sum_{k=1}^{K} A_k$$

where A_k is the potential at the centroid of element k. This value is divided by the current I_1 in coil (1) to obtain $M_{12}=n_2A/I_1$. In an analogous manner to that used for the calculation of self inductance, we impose $J \neq 0$ only in coil (1).

These are the most commonly required numerical results for classical analysis of an electric or magnetic device. However, for specific applications, a special post-processor can be developed to calculate some other result that might be required. This can be done as long as the result can be derived from potentials and geometric quantities available to the software.

13.5. The Computational Organization of a Software Package

The three steps described above are always present in any field calculation system, in more or less sophisticated forms. However, the manner in which a system is developed depends on three principal factors:

- **Given objectives.** Do we need to develop a general purpose package or a package directed towards special applications? It is clear that for the first case, the software will require complex programming and will take considerable amount of time in programming, testing, and evaluation. In the second case, we can develop a more compact package, including, perhaps, an automated method of use. The limitations of this second type of packages (known a priori) can be used as an operational advantage.
- **Available computational equipment.** It is virtually impossible to build a field calculation system on computational equipment without graphics capabilities. With high-definition monitors, conversational methods using "windows" may be used, adding to the user's comfort and confidence. If the computer has large memory capacity, the meshes can be larger and more accurate (number of unknowns increases). The pre-processor (dialogue, and mesh generation) can be combined in a single program. The same is true for the post-processor module, since this type of program requires considerable

memory. Processing speed is another important factor, especially in nonlinear and/or time varying applications which normally require very long running times. All these equipment aspects must be considered at the development stage of the software.
- **Personal preferences in options.** A software package has its own "personality"; it has strong and weak points. This, of course, is a relative question, since sometimes a weak point for a specific user can be a strong point for another. The important matter is that the software solve correctly the problems presented, with a minimum of user interference. Large computational packages have a large number of options, especially in their pre- and post-processors. Any options chosen for implementation must be well defined, because options characterize the software. A field computation package is constantly changed and extended, but, while it is easy to change a message, a change in the matrix structure of the system is much more involved and any such change must be carefully weighed.

13.5.1. *The EFCAD Software*

As an example of a set of criteria used to develop a software package, we present next, using block diagrams, the EFCAD package, which was developed for DOS compatible personal computers. Figure 13.12 shows the general diagram of EFCAD

a. Generation of Data.

Program EFP creates a file EFMAT.DAT which stores the material properties most often used by users. Once these materials are inserted in the file, it is not necessary to use EFP again unless changes to this file are needed. EFMAT.DAT contains values of μ, ε, σ, the $B(H)$ curves of nonlinear materials, and the remnant flux densities of permanent magnets. This is an archival type of file but the user can always update it. The main purpose is to save the user the time needed to insert this data every time an analysis is performed. The file EFMAT.DAT is used only in finite element calculation programs and in program EFN.

b. Preprocessor Section.

EFCAD is designed to be versatile and user-friendly. Data entry consists of specifying the drawing of the geometry. This is done by entering the segments and points forming the analysis domain. Segments can be furnished in any order. The input file contains two sets of arrays:

```
NP1(i), NP2(i), NC(i)   i=1,NSE segments
X(i), Y(i)              i=1,NPO points
XC(i) YC(i)             i=1,NCC circle centers
```

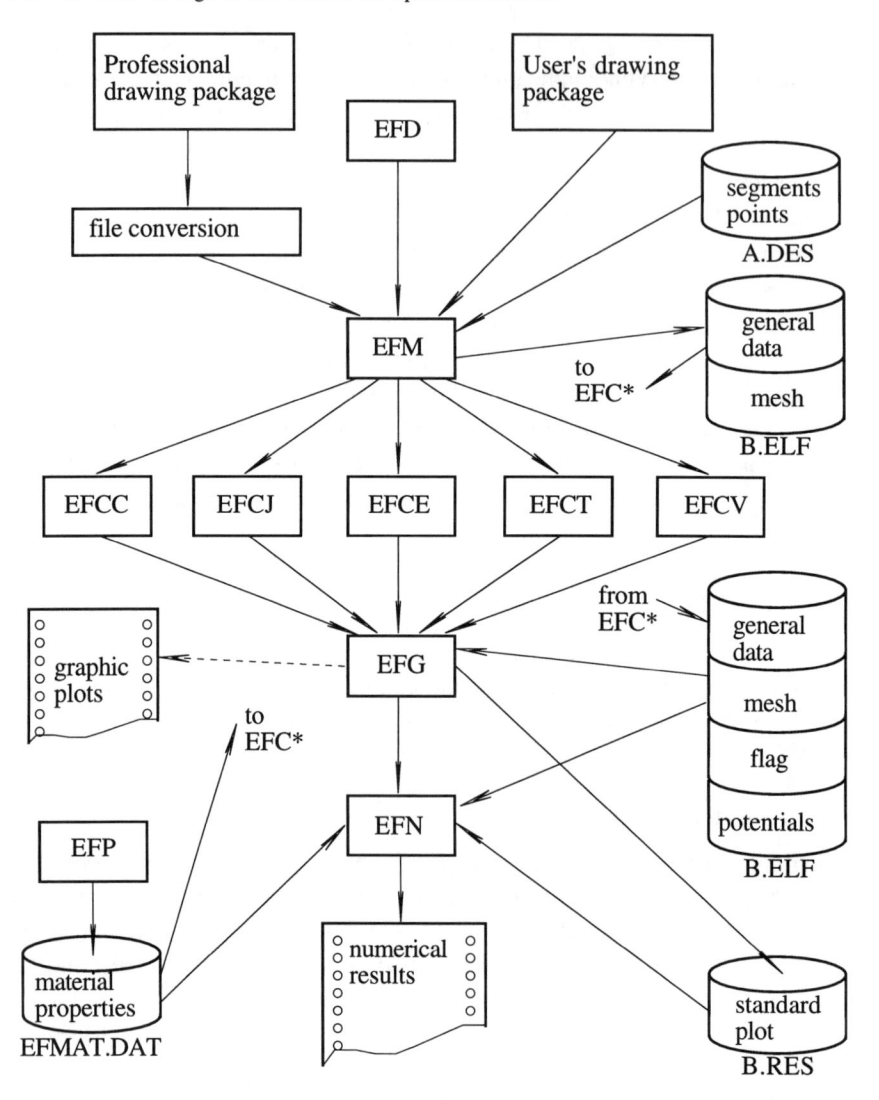

526   13. General Organization of Field Computation Software

Figure 13.12. General diagram of the EFCAD software. * indicates any of the modules EFCC, EFCJ, EFCE, EFCT and EFCV.

Arrays NP1 and NP2 indicate the numbers of the two points defining the segments. Arrays X and Y contain the coordinates of the points. If NC is nonzero, the segment is an arc with the center of the arc numbered NC and coordinates XC and YC. The arc is then defined by the center and the two end points of the arc. The arc starts at point NP1 and ends at point NP2, in the trigonometric direction.

The EFCAD package contains a program called EFD which generates this file (called A.DES in Figure 13.12). Input data entry can also be achieved by means of a professional package such as AUTOCAD™ or VERSACAD™. In this case, a simple file format translation program is needed to convert files to the format required. Since this input is very simple, the user can create his or her own drawing program as long as the format is in the form described above.

The second step in the preprocessor is program EFM. This program reads file A.DES and finds all domain regions and boundaries. Using the cursor or a mouse, the user enters material index numbers of each domain. These indices point to material properties in file EFMAT.DAT. Currents in coils as well as boundary and periodicity conditions are also entered graphically.

The mesh is generated automatically using the method presented in section 13.2.2. A new unformatted file is generated. The file name is user defined. In Figure 13.12, this file is shown as B.ELF. This completes the preprocessor step and the solution can proceed with the finite element calculation step.

c. The Calculation Section.

In this step, the finite element method is applied to the discretized domain. Five different formulations can be used, depending on the type of problem to be treated. Any file created by EFM can be used with any of the formulations. The same is true for the post-processor. The five programs are:

1. EFCE – for static solutions.
2. EFCC – complex formulation, presented in section 10.3.3.
3. EFCJ – variable domain, with induced currents, current-fed coils, using the formulation of section 10.3.2.
4. EFCT – variable domain, with induced currents, voltage-fed coils, using the formulation of section 13.6.4.
5. EFCV – for domains containing moving parts, using the formulation in section 10.3.4.

As a result of the computation, the potentials at the nodes of the mesh are added to the existing file B.ELF. A "flag" (integer number) is stored in the file immediately preceding the potentials to indicate the program used. This is later read and used by the post processor. Depending on the formulation used, the amount of data in this file varies. For time-dependent solutions, as well as for the complex formulation, many sets of potentials are stored, one set for each time step or for each frequency used.

d. The Post-Processor Section.

The first program in the post processor is EFG, whose task is the graphical presentation of results. File B.ELF is read and the plots requested by the user are drawn on the monitor, a plotter, or a printer. Various plots can be drawn at this stage. These include, the outline of the structure, the mesh, or a "standard" plot composed of the outline of the structure superimposed on a plot of twenty one equipotential lines. The last plot is stored in a file named B.RES. This operation is performed automatically if the user requests numerical results. The numerical results requested are calculated in program EFN. File B.ELF and B.RES are read.

The user indicates the required numerical results on the standard plot in graphic-conversational form. The user simply points on the monitor to points or lines where fields, forces, fluxes, and other results are needed.

Note that EFCAD is presented here only as an example of a field computation software. It is currently used by many industrial groups and research teams with good results. The system was specifically developed for personal computers. The advantages and limitations of the equipment were studied in order to take full advantage of the hardware and software.

13.6. Evolving Software

Early field computation packages were limited to finite element programs while the pre- and post-processors were of secondary importance. Part of the reason was the "research" nature of most packages and part of it had to do with the limited availability of simple, good graphics devices. In later stages, it was noticed that the pre- and post-processors should be the goal of further development and complete packages were implemented with efficient and easy to use interfaces.

The natural evolution in this area lead to "dedicated" systems where on the one hand, the operation of the software is facilitated by constant use and, on the other hand, the exploitation of results is extended and optimized to a specific application. This means that it is possible to develop systems, for a particular type of device, giving the user further data entry and result computation option improvements. Solution of coupled problems is also a relatively recent extension of classical electromagnetic finite element software packages.

As examples, we present in this section, in brief terms, a few of the more recent software developments in CAD: the adaptive mesh method; the coupled thermal/electrical problem; a software dedicated exclusively to electrical machines, and, finally, the simultaneous calculation of external electrical and magnetic circuit.

13.6.1. *The Adaptive Mesh Method*

It was noted earlier that it is useful to have high element densities, with small elements, in critical regions of the solution domain. However, it would be even more useful if the solution itself could determinate these regions and automatically discretize them to the required level. This is attempted by the various adaptive mesh generation strategies.

To perform the discretization, the criteria used must be fixed, especially where the mesh has been refined. The study of criteria for adaptive mesh generation is a subject in itself. Much of the current effort in adaptive mesh generation is devoted to this topic. It is not the purpose of this work to deal with this subject in any depth. Instead, we will present two of the criteria commonly used for adaptive mesh generation. These are: the field intensity and the field circulation criteria.

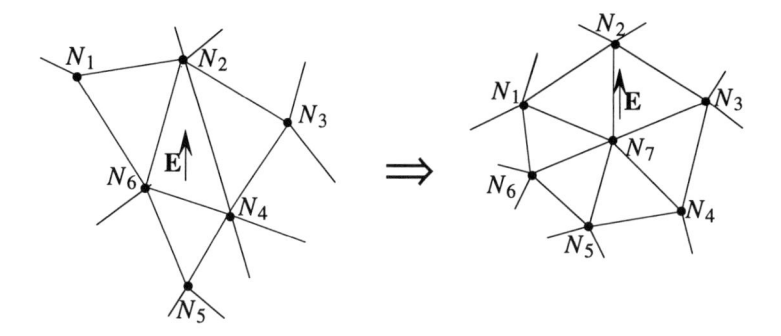

Figure 13.13. Mesh refinement based on the field intensity criterion.

a. The Field Intensity Criterion

Generally, the most critical regions of the domain have the most intense fields such as, for example, air-gaps of electromagnetic devices or, for electrostatic studies, the regions where the electrical field can cause a breakdown of the dielectric material. The present criterion states that, when the field intensity is larger than a specified value, the mesh should be refined in the corresponding region, as shown in the Figure 13.13. This procedure is repeated until the field variation between two successive discretization steps of the domain is lower than a specified error.

In this example, where the magnitude of the field intensity E_1 is large (based on the given criterion), the system generates a new node N_7. The software can now calculate the field E_2 (for example, E_2 can be calculated as the average field in the elements containing node N_7) and test it against the value of E_1. This process is repeated until the difference in field is lower than the error criterion.

b. The Field Circulation Criterion

This criterion is based on the idea of field circulation, which, for magnetic problems is the application of Ampere's law

$$\oint_c \mathbf{H} \cdot d\mathbf{l} = I$$

If for any element in the mesh this law is not satisfactorily verified, the mesh is refined. Consider the example of Figure 13.14, where $\mathbf{H}_1$, $\mathbf{H}_2$, and $\mathbf{H}_3$ are the components of the field intensity $\mathbf{H}$ on the sides of the element. Ampere's law applied to this triangle gives

$$\mathbf{H}_1 \cdot dl_1 + \mathbf{H}_2 \cdot dl_2 + \mathbf{H}_3 \cdot dl_3 - JS_i = \text{error}$$

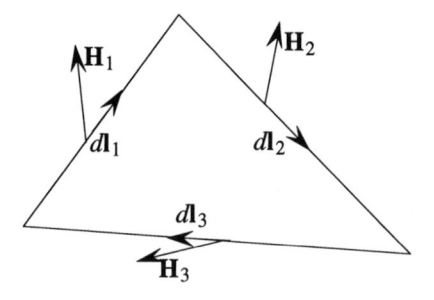

Figure 13.14. Calculation of circulation of **H** around the triangle.

where it is assumed that the element has a current density J and surface S_i. If there is no current in the element, we have

$$\mathbf{H}_1 \cdot d\mathbf{l}_1 + \mathbf{H}_2 \cdot d\mathbf{l}_2 + \mathbf{H}_3 \cdot d\mathbf{l}_3 = \text{error}$$

If the error is lower than the defined error, we consider the field to be correct and accurate enough for our purposes. If not, the process is repeated until the error criterion is satisfied.

For the electrostatic problem, the equivalent error criterion is based on the relation $\nabla \times \mathbf{E} = 0$. This gives

$$\oint_c \mathbf{E} \cdot d\mathbf{l} = 0$$

An example of a mesh generated using the adaptive method is shown in Figure 13.15a through 13.15e. The geometry is shown in Figure 13.15a. An electromagnet made of two pieces of iron and a coil is modeled. Because of symmetry, only the lower half of the geometry (below the symmetry line) is actually modeled. The initial mesh is shown in Figure 13.15b. This mesh contains 88 elements and 58 nodes. Figure 13.15c shows the solution for the initial mesh. Particularly notable is the poor solution in the airgap, characterized by the sharp corners in the flux lines in the gap and its vicinity. Figure 13.15d shows the same mesh after refinement. This mesh has 431 elements and 232 nodes. Note in particular the higher element density in the gap region and at corners of the ferromagnetic material. The flux distribution for the refined mesh is shown in Figure 13.15e. The solution now is much smoother indicating adequate mesh density.

From the computational aspect point of view, the use of this technique requires the linking of the pre-processor module (discretization) and the post-processor module (fields evaluation).

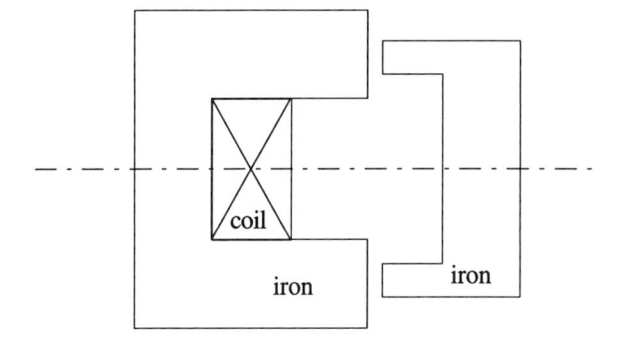

Figure 13.15a. An electromagnet used to demonstrate adaptive mesh solution.

For the implementation of the adaptive mesh method we must modify the finite element matrix system every time the mesh is refined. This can be done using one of two basic methods:

a. Only the matrix entries affected by the elements involved in the refinement are modified. This method has the advantage of being faster since only a limited number of modifications are made at each refinement step. However, the programming necessary for effective modifications is complex because of the strong link between the pre- and post-processors and the calculation module.
b. The whole matrix is redefined and solved again for each refinement step. The use of this method allows incorporation of the method into existing software by relatively simple modifications because the pre- and post-processor as well the finite element program work independently. However, this method requires considerably more operations and unless the computer is fast enough, the solution may be very slow.

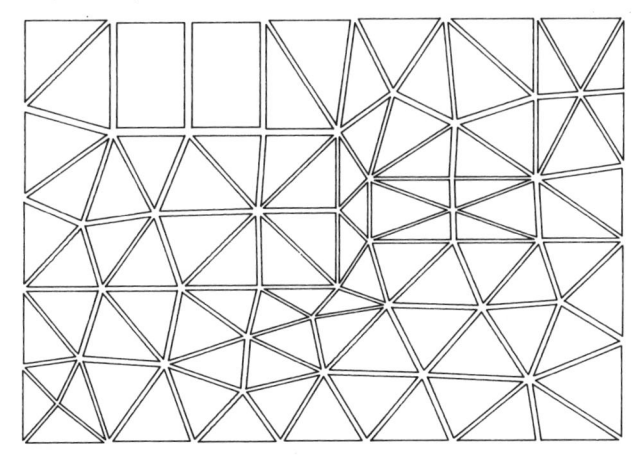

Figure 13.15b. Adaptive mesh generation: initial mesh: 88 elements, 58 nodes.

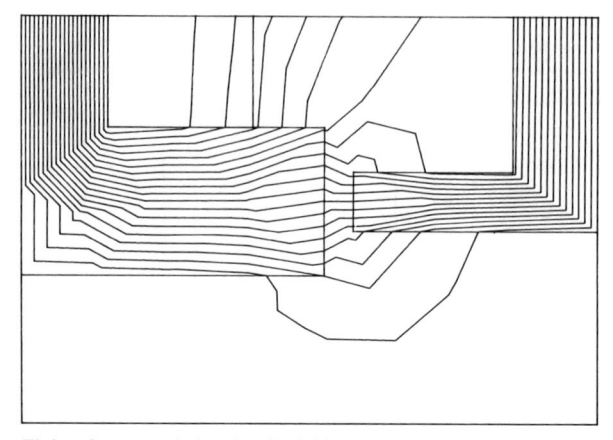

532   13. General Organization of Field Computation Software

Figure 13.15c. Finite element solution for the initial mesh.

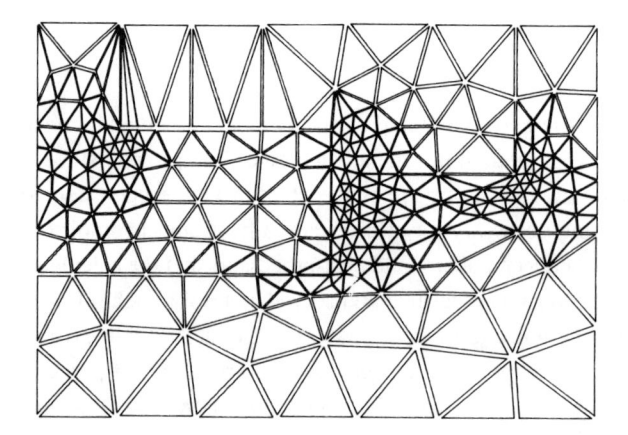

Figure 13.15d. Adaptive mesh generation: refined mesh: 431 elements, 232 nodes.

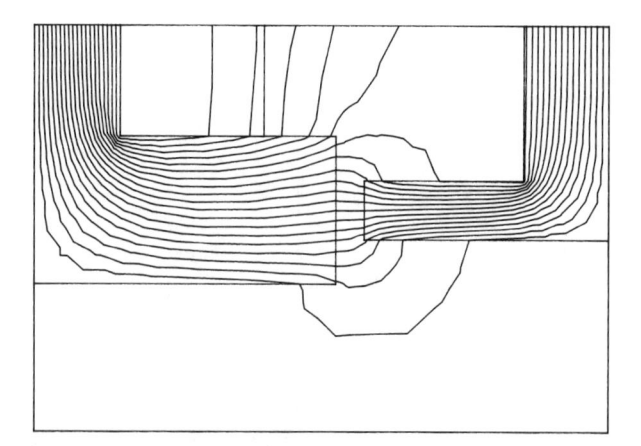

Figure 13.15e. Finite element solution for refined mesh.

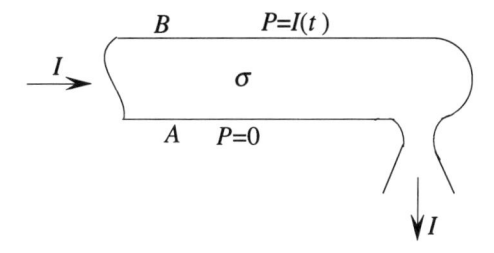

Figure 13.16. Heating in a conducting body due to high currents.

13.6.2. A Coupled Thermal/Electrical System

In electromagnetic devices, heating is a very frequent phenomenon and computation systems to consider its effects on electromagnetic fields were developed. These are called "coupled systems" and are a natural evolution of the classical fields software packages, in which the electromagnetic fields are calculated without regard to other effects.

This section presents an example of coupled modeling, developed to solve a specific problem, as an extension to EFCAD. In this modification, the thermal effects of the currents are evaluated. We wish to study the heating of a conducting body subjected to a strong current. Figure 13.16, shows the conductor, characterized by its conductivity σ. The current is assumed to be time dependent and its conductivity depends on the temperature. Before writing the equations, the following qualitative analysis of the problem is useful:

- At time t_1, the temperature of the conductor is T_1, and the current is I_1. For conductivity σ_1, this current is distributed throughout the conductor.
- Heating is based on Joule's effect; the resistivity $\rho_1 = 1/\sigma_1$ increases to ρ_2.
- At time t_2, with the values of ρ_2 and I_2, a new current distribution exists in the conductor, and because of the heating effect, the temperature increases to T_2. This is then the coupling process between heating and current.

As a result, the temperature in the conductor is a function of time and the applied current. It is, therefore, important to evaluate the correct current, and if the dimensions of the structure can support the heating. For the calculation of the current distribution, we use the electrical vector potential (T) formulation, presented in section 7.2.5, since the problem contains an applied current density. To avoid any ambiguity with temperature, the electric vector potential will be denoted here as **P**. We have $\mathbf{J} = \nabla \times \mathbf{P}$, and the equivalent Laplace equation is

$$\frac{\partial}{\partial x} \rho \frac{\partial P}{\partial x} + \frac{\partial}{\partial y} \rho \frac{\partial P}{\partial y} = 0 \qquad (13.9)$$

Recalling that

$$J_x = \frac{\partial P}{\partial y} \quad \text{and} \quad J_y = -\frac{\partial P}{\partial x} \quad (13.10)$$

and the Dirichlet boundary conditions are such that the imposed potential difference between lines A and B (Figure 13.16) is equal to the applied current (given in A/m in two-dimensional geometries), which may be time dependent.

The thermal equation is

$$\nabla \cdot (\lambda \nabla T) + q = mc\frac{\partial T}{\partial t} \quad (13.11)$$

where λ is the thermal conductivity, T is the temperature, m is the specific mass and c is the specific heat. The term q is the heat source, which, in this case corresponds to the Joule effect. For an element in the mesh, this is equal to $q = \rho J^2 = \rho(J_x^2 + J_y^2)$. The heat source has units of W/m^3. The term $\partial T/\partial t$ can be approximated by

$$\frac{\partial T}{\partial t} = \frac{T^i - T^{i-1}}{\Delta t}$$

where T^i and T^{i-1} are the temperatures at time steps i and $i-1$, respectively. The time step Δt depends strongly on how $I(t)$ varies. The thermal equation becomes

$$\lambda \left(\frac{\partial^2 T^i}{\partial x^2} + \frac{\partial^2 T^i}{\partial y^2} \right) - \frac{mc}{\Delta t} T^i + \frac{mc}{\Delta t} T^{i-1} + q = 0 \quad (13.12)$$

This equation is similar to Eq. (10.30). The calculation of its various terms is done analogously to that presented in Chapter 10 [Eq. (10.30)], with the obvious changes in coefficients. If heating increases very fast, the change can be considered to be adiabatic and, Neumann boundary conditions can be used for temperatures.

The algorithm of the coupled calculation is as follows:

a. At $t=t_0$ (initial time), set $T=T_0$ (initial condition) in the conducting body.
b. Electrical calculation, step i:
- Initialize the global (stiffness) matrix.
- Assemble the global matrix, based on equation (13.9). For each element in the mesh, the resistivity is given by:

$$\rho = \rho_0(1 + \alpha \Delta T^i)$$

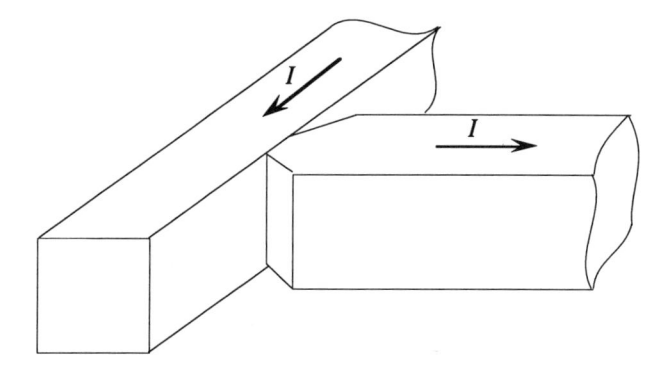

Figure 13.17. Two conducting bodies in contact. The temperature in the conductors, coupled to the current is calculated.

where ρ_0 is the resistivity of the material in the element for $T=0°C$, α is the temperature coefficient and $\Delta T^i = T^i - T^{i-1}$ is the temperature variation.
- Insert of electrical boundary conditions for $I(t^i)$.
- Solve the matrix system.
- Calculate q for all mesh elements, using $q=\rho J^2$.

c. Thermal calculation.
- Initialize the global matrix.
- Assemble the global matrix, based on Eq (13.12).
- Solve the matrix system.
- Store the temperature results T^i, to be used later, in the post-processor module.

d. Increment the time step.

e. If the time is below the given final time, return to b.

An example to the application of this methodology is shown in Figure 13.17. Two bodies made of copper, are in contact at a junction. The bodies are subjected to a constant current density equal to 125 A/mm^2.

The discretization of the solution domain is shown in Figure 13.18, where the dimensions of the geometry are also indicated. The current and thermal equipotentials are shown in Figure 13.19a and b, for a time equal to twenty seconds. Figure 13.20 shows the maximum temperature is a function of time. The location of the maximum temperature is, as expected, near the contact region.

Another example of a coupled system is the computation of thermal and magnetic effects, where the heating is generated by Joule effects due to eddy currents, as in the case presented in the example of section 10.6.1 (Figure 10.13). In parts of the geometry where $\sigma \neq 0$, the heat source term q, can be calculated, and then the magnetic equation [Eq. (10.30)] and the thermal equation [Eq. (13.12)] can be coupled.

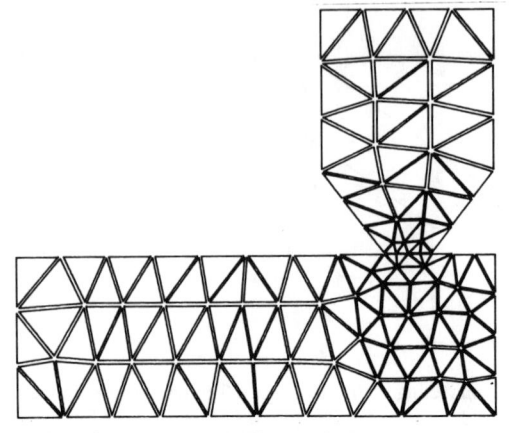

536   13. General Organization of Field Computation Software

Figure 13.18. Discretization of the geometry in Figure 13.17.

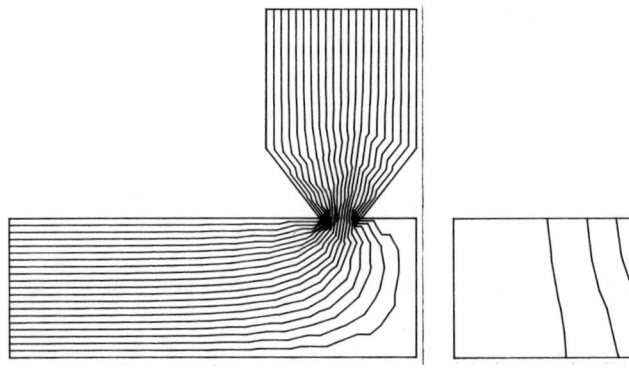

Figure 13.19a. Constant current lines in the conductors of Figure 13.17.

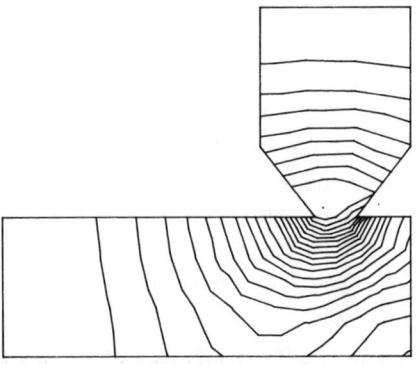

Figure 13.19b. Constant temperature lines in the conductors of Figure 13.17.

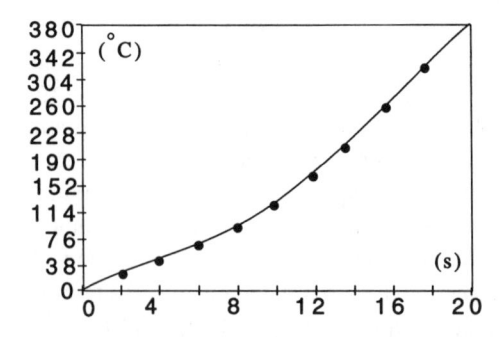

Figures 13.20. Maximum temperature in the conductors as a function of time.

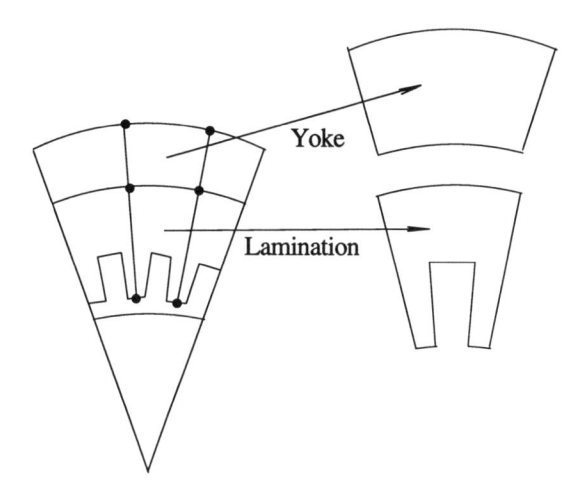

Figure 13.21. Type A regions in an electric machine. This includes yokes and slots.

13.6.3. *A Software Package for Electrical Machines*

As an example of a software exclusively dedicated to a special device, we present the software EFMAQ, developed for analysis of electrical machines.

To provide the geometrical data of an electrical machine to a general purpose software, like EFCAD, is extremely tedious. Unlike other electrical devices, electrical machines have particular geometrical characteristics like slot periodicity, small air-gaps, and saturable magnetic materials to mention a few. In terms of the field calculation, electric machines analysis is characterized by almost exclusive use of the magnetic vector potential, Cartesian coordinates, and well-defined boundary conditions.

These rather specialized requirements for data entry and analysis were the reasons for developing EFMAQ as a special software. This software is characterized by its pre-processor module, in that some of its concepts are different than those found in general purpose pre-processors. The post-processor is also adapted to electrical machines, but, since the problem of data entry is considered to be the major aspect of this software, we concentrate our attention to the pre-processor module. The pre-processor is based on the observation that the geometry of an electrical machine is composed of three types of regions:

a. Periodic regions (type *"A"* regions). These are made primarily of slots and yokes, either in the stator or in the rotor, as shown in Figure 13.21.
b. Miscellaneous regions (type *"B"* regions) made of different, nonperiodic segments. These include holes in the lamination, rotor axis, and any geometric parts that cannot be classified as type *"A"* or *"C"* as indicated in Figure 13.22

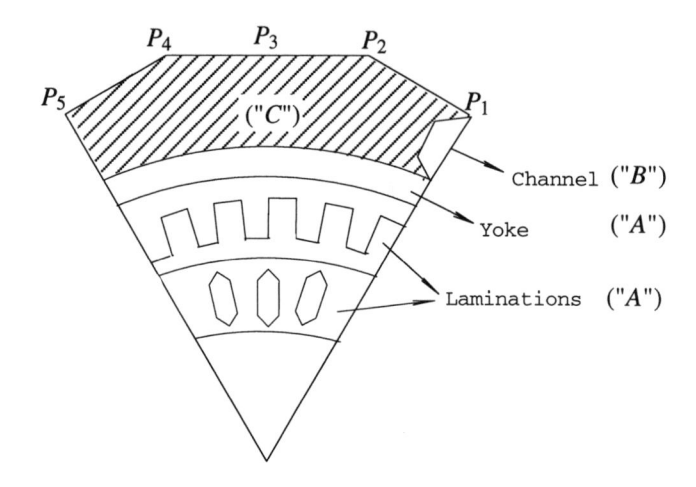

538   13. General Organization of Field Computation Software

Figure 13.22. Miscellaneous regions, classified as type "B" domains include holes and rotor axes. Complementary regions, classified as type "C" domains include the outer shell of the stator.

c. Complementary regions (type "C" regions). These are the region necessary to complete the structure, as shown in Figure 13.22 and may include the outer shell of the stator.

Type "A" regions are prediscretized. When these regions are required in a mesh they are inserted at the correct location. This is a very useful feature since it allows careful preparation of the mesh in structures like slots. These predefined mesh sections are then used repetitively to assemble the whole machine. Not only is time being saved but the resulting mesh is of higher quality. Type "B" and "C" regions are discretized after the machine domain is assembled (i.e., type "A" regions have been assembled).

The first program in the system is MAQP. It generates a file with "A" and "B" elements. The mesh corresponding to element type "A" is included in this file. The principal idea of this software is that the designer stores in this file all slot profiles he or she expects to use. To analyze a prototype, the required slots are taken from this file and appended to the mesh. Several types of elements are supplied by MAQP to the user. The elements are included in MAQP and can be retrieved through its menu. These are:

- Symmetrical profile slots.
- Nonsymmetrical profile slots.
- Slots with double layers of conductors (three-phase machines).
- Arbitrary shaped element, in Cartesian coordinates.

- Arbitrary shaped element, in polar coordinates.
- Rotor slots, open towards the airgap.

The second part of the pre-processor is program MAQF. Its purpose is to assemble the machine domain. First, elements *"A"* and *"B,"* from the file created by MAQP are called and inserted in the domain. Then, complementary regions *"C"* are defined in this program, using points provided by the user. As an example, consider Figure 13.22. Points P_1 through P_5 are given by the user, and with the cursor the user indicates on the monitor how these points are connected to form the complementary region.

After all regions are entered, MAQF (which also contains a mesh generator module), discretizes regions *"B"* and *"C."* The complete mesh is now ready for the finite element calculation.

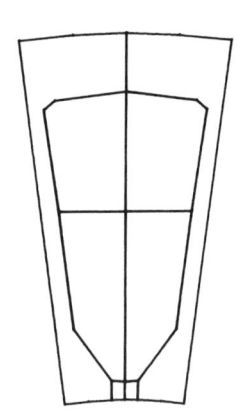

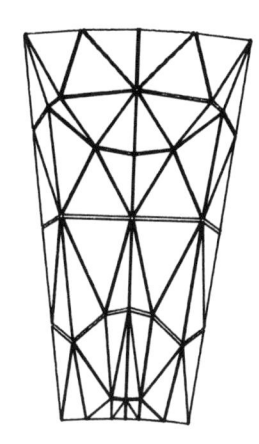

Figure 13.23. Stator slot in a three-phase induction machine and the finite element mesh.

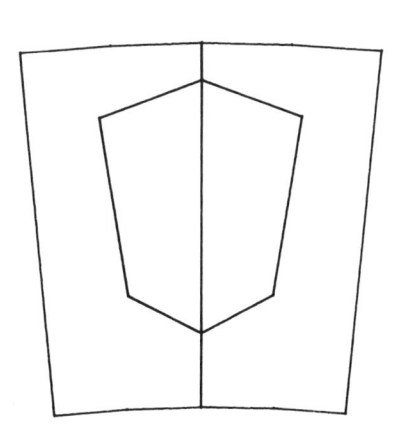

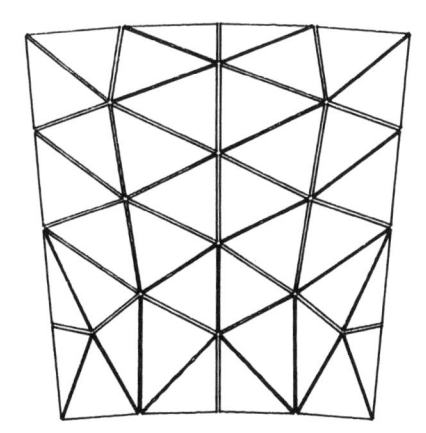

Figure 13.24. Rotor slot in a three-phase induction machine and the finite element mesh.

540   13. General Organization of Field Computation Software

The computation and post-processing are accomplished with programs, MAQC, MAQG, and MAQN. MAQC performs the finite element computation, MAQG displays the results and MAQN performs numerical calculations on the results to obtain numerical values.

As an example to the use of this software, we study a three-phase induction motor whose stator and rotor slots are shown in Figures 13.23 and 13.24.

These type "A" elements, as well the stator and rotor yokes (also type "A" elements but not shown here), are stored in the file generated by MAQP. The hole in the lamination elements (type "B" elements) are also stored in this file. Figure 13.25 shows the machine domain composition and Figure 13.26, the final mesh. The flux distribution is shown in Figure 13.27, as an example of the type of graphical result one can obtain.

This software has additional features, including simplified boundary conditions entry, automatic definition of the airgap region, and meshing and display of results in the airgap.

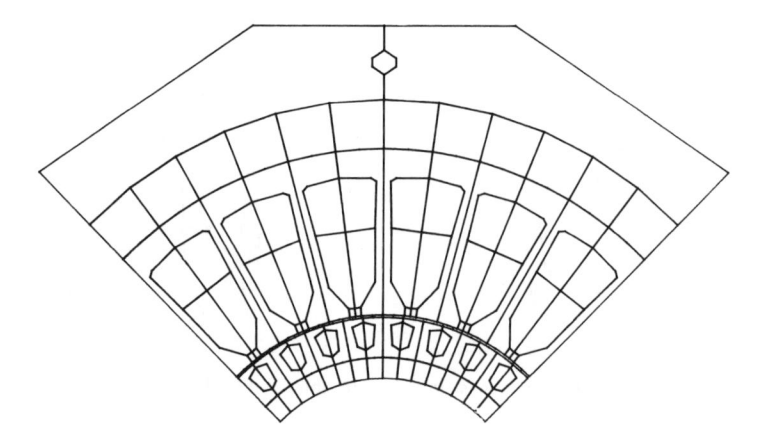

Figure 13.25. The machine domain as composed for analysis.

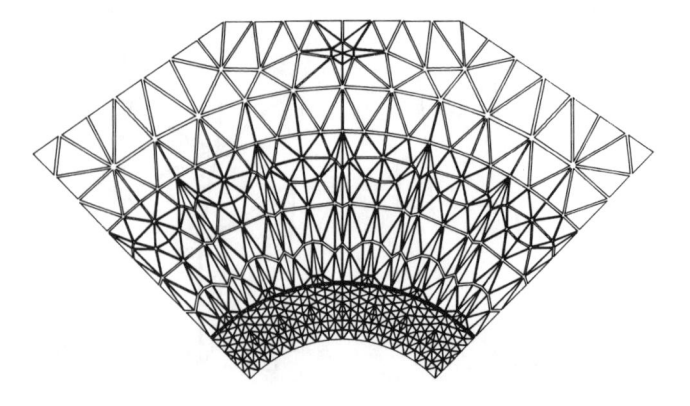

Figure 13.26. Finite element mesh generated over the domain in Figure 13.25.

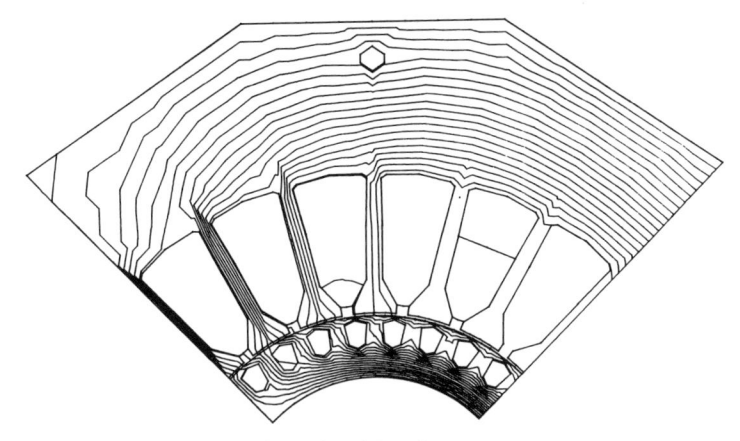

Figure 13.27. Flux distribution in the analyzed domain.

13.6.4. *A System for Simultaneous Solution of Field Equations and External Circuits*

Up to this point, we have assumed that the source of the magnetic field is a current density (i.e., the driving coil is fed by a known, constant current density, as in Chapters 9 and 10). Thus, the coil current density is known before solution begins. However, in many cases, the system is voltage driven and the current in the coil is unknown. It is often useful to be able to solve the problem under these conditions. To achieve this we solve the field equation and the coil circuit equation simultaneously. The circuit equation for the coil is

$$V = RI + L\frac{dI}{dt} + n\frac{d\Phi}{dt}$$

where R and L are the coil resistance and an additional inductance furnished by the user. This inductance is that part of the inductance not included in the field calculation such as end winding or leakage inductance. Φ is the magnetic flux generated in the solution domain and written in terms of the magnetic vector potential A.

The current source contribution of an element of the mesh is given by Eq. (9.68) as

$$\frac{JD}{6}\begin{bmatrix}1\\1\\1\end{bmatrix} \qquad (13.13)$$

where J is the current density and D is equal to twice the surface of the element. A parameter K is now defined as the coil turn density (in *turns/mm²*). If I is the current of one turn, we have

$$J = KI$$

With this, Eq. (13.13) can be written as

$$\frac{KD}{6}\begin{bmatrix}1\\1\\1\end{bmatrix}I$$

which is denoted $[P]I$, and I is considered an unknown, as will be shown below.

For the magnetic field equation, the flux, in terms of the magnetic vector potential is given by (see section 9.2.4)

$$\Phi = l(A_2 - A_1)$$

where l is the depth of the structure (l is supplied by the user). Assuming the potential A_1 on the boundary is zero (as is normally the case), we have

$$\Phi = lA$$

where A is the potential at any point in the solution domain.

Further, assuming that A is the average potential in an element of the mesh that has a specified current, we get

$$A = \frac{A_i + A_j + A_k}{3}$$

where i, j, and k, are the nodes of the element.

If n is the total number of turns within the element, we get

$$n = K \times area\ of\ element = \frac{KD}{2}$$

Therefore,

$$n\Phi = KDlA/2$$

or, denoting $G = KlD/6$, we have in matrix form

$$n\Phi = [G\ G\ G]\begin{bmatrix}A_i\\A_j\\A_k\end{bmatrix} \qquad (13.14)$$

which is denoted as $[G][A]$. Using time discretization, we obtain

$$n\frac{d\Phi}{dt} = \frac{[G]}{\Delta t}\begin{bmatrix} A_i^{m+1} - A_i^m \\ A_j^{m+1} - A_j^m \\ A_k^{m+1} - A_k^m \end{bmatrix}$$

where $m+1$ and m indicate the value of A associated with the time step Δt. Considering the circuit and field equations above, the global system of equations becomes

$$\begin{bmatrix} SS + \dfrac{N}{\Delta t} & -P \\ \dfrac{G}{\Delta t} & R + \dfrac{L}{\Delta t} \end{bmatrix} \begin{bmatrix} A^{m+1} \\ I^{m+1} \end{bmatrix} =$$

$$\begin{bmatrix} \dfrac{N}{\Delta t} & 0 \\ \dfrac{G}{\Delta t} & \dfrac{L}{\Delta t} \end{bmatrix} \begin{bmatrix} A^m \\ I^m \end{bmatrix} + \begin{bmatrix} S^{m+1} \\ V^{m+1} \end{bmatrix} \quad (13.15)$$

where N is the second matrix in equation (10.38), if σ is nonzero.

Performing the products in equation (13.15) we get two equations

a. **Field equation**

$$\left[SS + \frac{N}{\Delta t}\right]A^{m+1} - [P]I^{m+1} = \left[\frac{N}{\Delta t}\right]A^m + [S]^{m+1}$$

or

$$[SS]A^{m+1} + [N]\left[\frac{\Delta A^{m+1}}{\Delta t}\right] - [P]I^{m+1} = [S]^{m+1}$$

Note that the source term $[S]^{m+1}$ can describe the influence of a permanent magnet or another current-fed coil whose current is known at time "$m+1$".

b. **Circuit equation**

$$[G]\frac{A^{m+1}}{\Delta t} + \left(R + \frac{L}{\Delta t}\right)I^{m+1} = [G]\frac{A^m}{\Delta t} + L\frac{I^m}{\Delta t} + V^{m+1}$$

or

$$[G]\frac{\Delta A^{m+1}}{\Delta t} + RI^{m+1} + L\frac{\Delta I^{m+1}}{\Delta t} = V^{m+1}$$

This corresponds to the voltage equation of the coil. Note that the matrices $[P]$ and $[G]$ are very similar. In the process of calculating the elemental matrices for those elements that have prescribed current densities, we also calculate $[P]$ and $[G]$ and assemble these in the appropriate locations in the global matrix. The line of the global system corresponding to the current I (last line) has terms out of the normal band (the band characteristics are preserved in the field part of the system). Therefore, in the solution process, some modifications were performed to take into account the last equation while still using the banded form of the solution. This method can be easily extended to several coils, by adding additional lines in the circuit part of the global system.

As an example to the type of problems that can be solved using this formulation, consider the solenoid shown in Figure 13.28. This device is axisymmetric and consists of two parts, made of iron. We assume the iron to have linear properties and neglect eddy currents in the iron. These restrictions are imposed to allow a comparison with analytical calculations. The numerical data used for this example are:

$$l_1 = 15 \ mm; \quad l_2 = 30 \ mm; \quad e = 2 \ mm;$$
$$r_1 = 20 \ mm, \quad r_2 = 25 \ mm; \quad r_3 = 32 \ mm, \quad r_4 = 38 \ mm$$
$$n \ (\text{number of turns}) = 50 \quad R \ (\text{resistance of coil}) = 2 \ Ohm$$

A step pulse of $10V$ (switched on at $t=0$ as shown in Figure 13.29) is applied on the coil.

Since the airgaps are small, we can calculate the magnetic circuit. This gives

$$h_1 e + h_2 e = nI$$

The flux conservation equations give

$$\Phi_1 = \Phi_2; \quad B_1 S_1 = B_2 S_2 \quad \text{or} \quad \mu_0 h_1 S_1 = \mu_0 h_2 S_2$$

The radii r_1, r_2, and r_3, are dimensioned with the goal of making $B_1 = B_2$. Noting that $\pi r_1^2 \cong \pi (r_3^2 - r_2^2)$, we have: $h_1 = h_2 = h$, and

$$h = \frac{nI}{2e}$$

At steady state, $I = V/R = 5A$. Using the numerical values above, we get

$$h = 62{,}500 \ A/m$$

Evolving Software   545

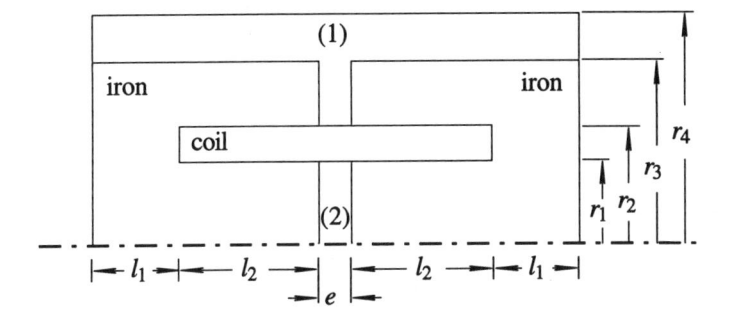

Figure 13.28. A solenoid and its external circuit. The current in the solenoid is dependent on the external circuit and for correct calculation the solenoid and the external circuit must be coupled.

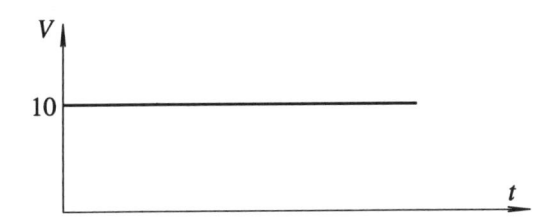

Figure 13.29. The source used to drive the circuit in Figure 13.28.

The flux is

$$\Phi = \mu_0 h S = \mu_0 h \pi r_1^2 = 0.987 \times 10^{-4} \; Wb$$

The inductance is

$$L = n \frac{\Phi}{I} = 0.987 \times 10^{-3} \; H$$

Since the iron properties are linear and there are no eddy currents, the flux is proportional to the current I. Under these conditions, the inductance is constant during the transient state. The theoretical equation for $I(t)$ for this LC circuit is

$$I(t) = \frac{V}{R}\left(1 - e^{-\frac{R}{L}t}\right) \tag{13.16}$$

Or, numerically,

$$I(t) = 5\left(1 - e^{-2026.3\, t}\right) \tag{13.17}$$

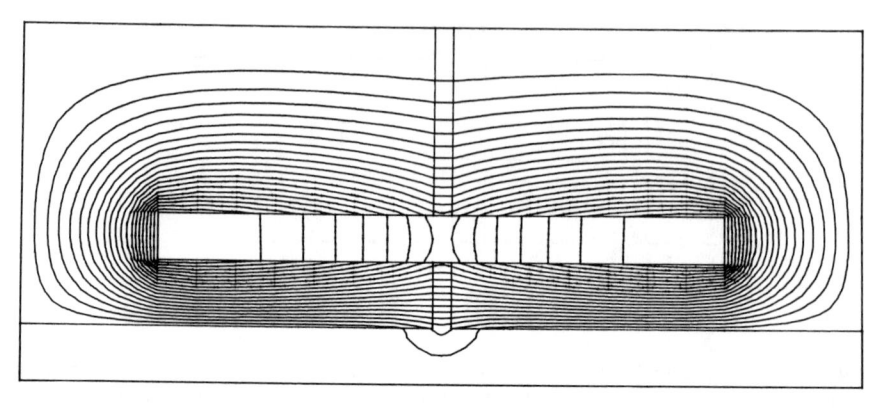

Figure 13.30. Flux distribution in the structure in Figure 13.28.

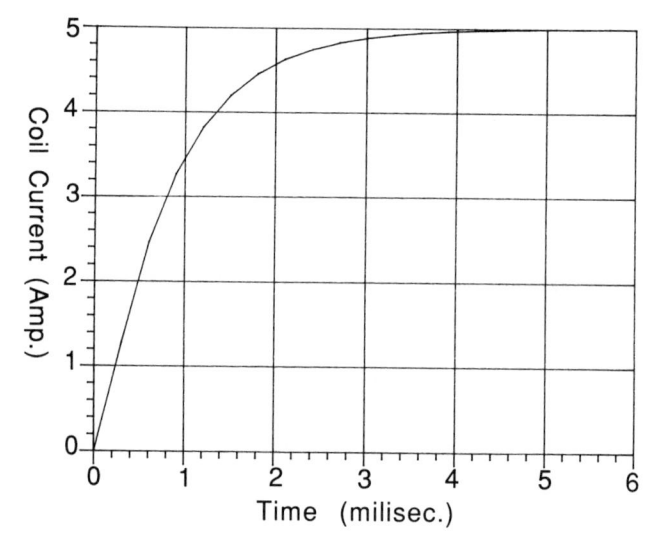

Figure 13.31. Current in the coil of Figure 13.28 as a function of time.

The results are shown in Figures 13.30 and 13.31. The first figure shows the flux distribution in the structure. The second shows the current in the coil as a function of time.

Figure 13.30 is in good agreement with Eq. (13.17) demonstrating the consistency of the formulation. In fact, the theoretical curve increases faster compared with the numerical results. The main reason for this is the fact that the real inductance is higher than the theoretical inductance because the flux crossing the solenoid (Figure 13.31) increases the value of the inductance compared to the theoretical calculation, in which this flux is not considered.

13.6.5. *Computational Difficulties and Extensions to Field Computation Packages*

A field computation software, including all its modules (pre-processor, processor, and post-processor) performs many thousands of computational operations. Many graphical statements are also executed as part of the software. Therefore, it is not surprising that some computational difficulties may arise from time to time. While extensive testing is normally performed before a software package is accepted as "correct," it is obviously not possible to foresee all possible combinations the package may encounter. From time to time, the software will encounter a situation that either cannot be handled or is handled improperly. A piece of software can be characterized by the rather simple, and sometimes funny expression: "a software works well until it doesn't." For these reasons, it is common for large software packages to be superseded by new versions. A new version may contain extensions and enhancements to the original software, or, it may be issued to correct problems that have been detected by users or by the developer.

The interaction between developers and users of software is vital to the development of good software. This can take the form of simply reporting a known "bug" or indicating what the user would like to see in subsequent versions of the software to make it more useful or efficient. A software that does not evolve with time becomes obsolete in short order.

13.7. Recent Trends

First, we note that the primary two dimensional formulations have been fully developed for some time and are widely used by practically all research groups and industrial users involved in field calculation. For 3D scalar potential formulations, the situation is similar. However, it does not mean that these domains are stagnant; dynamic problems (such as transient analysis in induction machines), optimization (automatic design of device parameters), and coupled problems, are only samples of problems that are still under active investigation. It seems improbable, however, that the fundamental 2D formulations will be significantly changed in the future.

Much of the ongoing work with field computation, both theoretical and in software development is in three dimensional modeling and analysis. As mentioned in Chapter 9, the direct use of the vector potential is not the most desirable method, because the number of unknowns is very large (three or four components at each node). The requirements on computer memory and speed are therefore considerable for any realistic discretization. Many alternative formulations were and are being developed to overcome these difficulties. Some of these are briefly presented here in terms of their general principles. The bibliography section contains many references to these and other methods.

a. The first of these uses the concepts of total scalar potential and reduced scalar potential. In this method, the solution domain is divided into three types of regions: ferromagnetic material regions, applied current regions, and regions with eddy currents. In each type of region, a different formulation is applied. The total potential generates the whole magnetic field, and the reduced potential is responsible for the partial field. The different potentials are applied in the corresponding regions and the formulation is globally matched using interface conditions at the interfaces between regions (for example, continuity of normal flux densities and tangential components of field intensities may be imposed). Since only scalar potentials are used, we have fewer unknowns, and, even if the magnetic vector potential is used in eddy current regions, this method minimizes the computational memory needs.

b. Edge or vector finite elements have become popular lately (see Chapter 11 for discussion of hexahedral edge elements and applications). These types of finite elements provide accurate results and, being vector elements, coupling with the differential equations governing many electromagnetic phenomena is very natural and easy to understand. In addition, their implementation is relatively simple. This work also shows that in the majority of practical cases, the use of gages is not necessary, further simplifying the algorithms and avoiding degradation of results often associated with introduction of gages. In low-frequency applications, the magnetic vector potential **A** is preferred while in high frequency applications, the field variables **H** or **E** are used as unknowns. For more detailed information on edge elements we recommend reading Chapter 11 as well as references in the bibliography section.

c. The boundary integral equations method is also extensively used, since it permits the solution of linear problems applying the formulation only on material boundaries. Its fundamental advantage is in reducing a three-dimensional problem to one defined on the surfaces of the three-dimensional geometry (essentially a 2D problem). With this, the complex 3D discretization is avoided and the number of unknowns is reduced considerably.

In terms of applications, a recent trend is in developing advanced systems. Optimization is an important goal in this development. An optimizing software could, based on user provided criteria, automatically decide how the design can be modified to obtain the performance required.

We conclude mentioning that the domain of field computation is so rich in physical, mathematical, numerical, and computational challenges that the evolution of such software packages is essentially nonexhaustible. This, is one of the main reasons why so many continue working in this difficult, but fascinating and highly rewarding area.

Bibliography

General Electromagnetics

1. M. A. Plonus, *Applied Electromagnetics*, McGraw-Hill, New York, 1978.
2. J. D. Kraus, and K. R. Carver, *Electromagnetics*, 2d ed., McGraw-Hill, New York, 1973.
3. J. A. Stratton, *Electromagnetic Theory*, McGraw-Hill, New York, 1952.
4. W. K. H. Panofsky and M. Phillips, *Classical Electricity and Magnetism*, Addison-Wesley, Inc., Reading, Mass., 1956.
5. R. Plosney and R. E. Collin, *Principles and Applications of Electromagnetic Fields*, McGraw-Hill, New York, 1961.
6. P. Lorain and D. R. Corson, *Electromagnetism*, W. H. Freeman, San Francisco, 1978.
7. J. A. Kong, *Theory of Electromagnetic Waves*, John Wiley & Sons, Inc., New York, 1975.
8. R. E. Collin, *Electromagnetic Theory of Guided Waves*, McGraw-Hill, New York, 1991, pp. 564-568.

Books on Computational Electromagnetics

1. C. W. Steele, *Numerical Computation of Electric and Magnetic Fields*, Van Nostrand Reinhold Company, New York, 1987.
2. P. P. Silvester and R. P. Ferrari, *Finite Elements for Electrical Engineers*, Cambridge University Press, Cambridge, 1990.
3. J. C. Sabonnadierre and J. L. Coulomb, *Finite Element Methods in CAD*, Springer Verlag, New York, 1989.
4. S. R. H. Hoole, *Computer-Aided Analysis and Design of Electromagnetic Devices*, Elsevier, New York, 1989.
5. D. A. Lowther and P. P. Sylvester, *Computer Aided Design in Magnetics*, Springer-Verlag, New York, 1986.

Nonlinear Methods

1. P. P. Silvester and M. V. K. Chari, "Finite element solution of saturable magnetic field problems," IEEE Transactions on Power Apparatus and Systems, Vol. 89, 1970, pp. 1642-1651.
2. J. P. A. Bastos and G. Quichaud, "3D modeling of a nonlinear anisotropic lamination," IEEE Transactions on Magnetics, Vol. 21, No. 6, Nov. 1985, pp. 2366-2369.
3. J. Penman and A. M. A. Kamar, "Linearization of saturable magnetic field problems, including eddy currents," IEEE Transactions on Magnetics, Vol. MAG-18, No. 2, March 1982, pp. 563-566.
4. C. S. Holzinger, "Computation of magnetic fields within three-dimensional highly nonlinear media," IEEE Transactions on Magnetics, Vol. MAG-6, No. 1, March 1970, pp. 60-65.
5. J. R. Brauer, R. Y. Bodine and L. A. Larkin, "Nonlinear anisotropic three-dimensional magnetic energy functional," Proceedings of the Compumag Conference, Santa Marguerita, Italy, May 1983.
6. J. R. Brauer, "Saturated magnetic energy function for finite element analysis of electrical machines," IEEE PES Winter Meeting, New York, 1976, paper C75, pp. 151-156.

Solution Methods

1. J. A. Meijerink, and H. A. van der Vorst, "An iterative solution method for linear systems of which the coefficient matrix is a symmetric M-matrix," Mathematics of Computation, Vol. 31, No. 137, January 1977, pp. 148-162.
2. D. M. Young, *Iterative Solution of Large Linear Systems*, Computer Science and Applied Mathematics, Academic Press, New York, 1971.
3. R. V. Southwell, *Relaxation Methods in Theoretical Physics*, Oxford University Press, London, 1946.
4. R. L. Stoll, "Solution of linear steady state eddy current problems by complex successive over-relaxation," IEE Proceedings, Part. A, Vol. 117, 1970, pp. 1317-1323.
5. J. K. Reid, "On the method of conjugate gradients for the solution of large sparse systems of linear equations," in *Large Sparse Sets of Linear Equations*, Reid, J. K., Ed., Academic Press, 1971, pp. 231-252.
6. J. K. Reid, "Solution of linear systems of equations: direct methods (general)," in *Lecture Notes in Mathematics*, Barker, V. A., Ed., Springer-Verlag, Berlin, 1976, No. 572, pp. 102-130.
7. A. Jennings, "A compact storage scheme for the solution of symmetric simultaneous equations," Computer Journal, No. 9, September 1969, pp. 281-285.

8. J. A. George, "Nested dissection of a regular finite element mesh," SIAM Journal of Numerical Analysis, Vol. 10, No. 2, April 1973, pp. 345-363.
9. J. A. George, "On block elimination of finite element systems of equations," in *Sparse Matrices and Their Applications*, Rose, D. J. and Willoughby, R. A., Eds., Plenum Press, New York, 1972, pp. 101-114.
10. J. A. George, "Solution of linear systems of equations: direct methods for finite element problems," in *Lecture Notes in Mathematics*, Barker, V. A., Ed., Springer-Verlag, Berlin, 1976, No. 572, pp. 52-101.
11. J. A. George, "Sparse matrix aspects of the finite element method," Proceedings of the Second International Symposium on Computing Methods in Applied Science and Engineering, Springer-Verlag, 1976.
12. J. F. Gloudeman, "The impact of supercomputers on finite element analysis," Proceedings of the IEEE, Vol. 72, No. 1, January 1984, pp. 80-84.
13. F. G. Gustavson, "Some basic techniques for solving sparse systems of equations," in *Sparse Matrices and Their Applications, Rose*, D. J. and Willoughby, R. A., Eds., Plenum Press, New York, 1972, pp. 41-53.

General Numerical Techniques

1. M. L. James, S. M. Smith and J. C. Wolford, *Applied Numerical Methods for Digital Computation*, Harper & Row, New York, 1985.
2. J. Pachner, *Handbook of Numerical Analysis Applications*, McGraw-Hill, New York, 1984.
3. R. Courant and D. Hilbert, *Methods of Computational Physics*, Vol. 2, Interscience, 1961.

Magnetostatic and Electrostatic Formulations

1. P. Hammond, and T. D. Tsiboukis, "Dual finite-element calculations for static electric and magnetic fields," IEE Proceedings, Vol. 130, Part. A, No. 3, May 1983, pp. 105-111.
2. M. V. K. Chari, M. A. Palmo and Z. J. Czendes, "Axisymmetric and three dimensional electrostatic field solution by the finite element method," Electric Machines and Electromechanics, Vol. 3, 1979, pp. 235-244.
3. M. V. K. Chari, P. P. Silvester and A. Konrad, "Three-dimensional magnetostatic field analysis of electrical machinery by the finite element method," IEEE Transactions on Power Apparatus and Systems, Vol. PAS-100, No. 8, August 1981, pp. 4007-4019.
4. Z. J. Cendes, J. Weiss and S. R. H. Hoole, "Alternative vector potential formulations of 3-D magnetostatic problems," IEEE Transactions on Magnetics, Vol. MAG-18, No. 2, March 1982, pp. 367-372.

5. O. W. Andersen, "Two stage solution of three-dimensional electrostatic fields by finite differences and finite elements," IEEE Transactions on Power apparatus and systems, Vol. PAS-100, No. 8, Aug. 1981, pp. 3714-1721.

Eddy Current Formulations

1. C. J. Carpenter and M. Djurovic, "Three-dimensional numerical solution of eddy currents in thin plates," IEE Proceedings, Vol. 122, No. 6, 1975, pp. 681-688.
2. M. L. Brown, "Calculation of 3-dimensional eddy currents at power frequencies," IEE Proceedings, Vol. 129, Part A, No. 1, January 1982, pp. 46-53.
3. C. S. Biddlecombe, E. A. Heighway, J. Simkin and C. W. Trowbridge, "Methods for eddy current computation in three dimensions," IEEE Transactions on Magnetics, Vol. MAG-18, No. 2, March 1982, pp. 492-497.
4. P. Hammond, "Use of potentials in the calculation of electromagnetic fields," IEE Proceedings, Vol. 129, Part A, No. 2, March 1982, pp. 106-112.
5. T. W. Preston and A. B. J. Reece, "Solution of 3-dimensional eddy current problems: the T - Ω method," IEEE Transactions on Magnetics, Vol. MAG-18, No. 2, March 1982, pp. 486-491.
6. C. R. I. Emson, J. Simkin and C. W. Trowbridge, "Further developments in three-dimensional eddy current analysis," IEEE Transactions on Magnetics, Vol. MAG-21, No. 6, Nov. 1985, pp. 2231-2234.
7. P. I. Koltermann, J. P. A. Bastos and S. S. Arruda, "A model for dynamic analysis of AC contactor," IEEE Transactions on Magnetics, Vol. 28, No. 2, March 1992, pp. 1348-1350.
8. R. C. Mesquita, J. P. A. Bastos, "3D finite element solution of induction heating problems with efficient time-stepping," IEEE Transactions on Magnetics, Vol. 27, No. 5, Sept.. 1991, pp. 4065-4068.
9. N. Ida, "Three Dimensional Eddy Current Modeling," *Review of Progress in Quantitative Nondestructive Evaluation*, D. O. Thompson and D. E. Chimenti, Eds., Plenum Press, New York, Nov. 1987, Vol. 6A, pp. 201-209.
10. H. Song and N. Ida, "An eddy current constraint formulation for 3D electromagnetic field calculation," IEEE Transactions on Magnetics, Vol. 27, No. 5, Sept. 1991, pp. 4012-4015.
11. R. D. Pillsbury, "A three dimensional eddy-current formulation using two potentials: the magnetic vector potential and total magnetic scalar potential," IEEE Transactions on Magnetics, Vol. MAG-19, No. 6, 1983, pp. 2284-2287.

12. R. L. Ferrari, "Complementary variational formulation for eddy-current problems using the field variables E and H directly," IEE Proceedings, Vol. 132, Part A, No. 4, July 1985, pp. 157-164.
13. P. Hammond, "Calculation of eddy currents by dual energy methods," IEE Proceedings, Vol. 125, No. 7, 1978, pp. 701-708.
14. C. R. I. Emson and J. Simkin, "An optimal method for 3-D eddy-currents," IEEE Transactions on Magnetics, Vol. MAG-19, No. 6, 1983, pp. 2450-2452.

Formulations with Velocity Terms

1. N. Ida, "Modeling of velocity effects in eddy current applications," Journal of Applied Physics, Vol. 63, No. 8, April 15, 1988, pp. 3007-3009.
2. N. Ida, "Velocity effects and low level fields in axisymmetric geometries," in COMPEL – International Journal for Computation and Mathematics in Electrical Engineering, Vol. 9, No. 3, September 1990, pp. 169-180.

High-Frequency Formulations

1. A. Konrad, "A direct three-dimensional finite element method for the solution of magnetic fields in cavities," IEEE Transactions on Magnetics, Vol. MAG-21, No. 6, November 1985, pp. 2276-2279.
2. N. Ida, "Microwave NDT," Electrosoft, special issue on NDT Vol. 2, No. 2/3, June/Sept. 1991, pp. 215-237.
3. J. S. Wang and N. Ida, "Eigenvalue analysis in Electromagnetic cavities using divergence free finite elements," in IEEE Transactions on Magnetics, Vol. 27, No. 5, Sept. 1991, pp. 3978-3981.
4. J. S. Wang and N. Ida, "A numerical model for characterization of specimens in a microwave cavity," in *Review of Progress in Quantitative Nondestructive Evaluation*, D. O. Thompson and D. E. Chimenti, Eds., Vol. 10A, Plenum Press, New York 1991, pp. 567-574.

Computation Systems and CAD

5. A. Bossavit and J. C. Verite, "The TRIFOU code: Solving the 3-D eddy currents problem by using H as a sate variable," IEEE Transactions on Magnetics, Vol. MAG- 19, 1983, pp. 2465-2470.
6. N. Ida, "PCNDT - An electromagnetic finite element package for personal computers," IEEE Transactions on Magnetics, Vol. 24, No. 1, January 1988, pp. 366-369.

7. P. Masse, J. L. Coulomb and B. Ancelle, "System design methodology in CAD. programs based on finite element method," IEEE Transactions on Magnetics, Vol. MAG-18, No. 2, March 1982, pp. 609-616.
8. M. L. Barton, V. K. Garg, I. Ince, E. Sternheim and J. Weiss, "WEMAP: a general purpose system for electromagnetic analysis and design," IEEE Transactions on Magnetics, Vol. MAG-19, No. 6, November 1983, pp. 2674-2677.
9. B. Ancelle, E. Gallagher and P. Masse, "ENTREE: a fully parametric preprocessor for computer aided design of magnetic devices," IEEE Transactions on Magnetics, Vol. MAG-18, No. 2, March 1982, pp. 630-632.

Special Aspects

1. A. A. Abdel-Razek, J. L. Coulomb, M. Feliachi and J. C. Sabonnadiere, "Conception of an air-gap element for the analysis of the electromagnetic field in electric machines," IEEE Transactions on Magnetics, Vol. MAG-18, No. 2, March 1982, pp. 655-659.
2. N. Sadowski, Y. Lefevre, M. Lajoie-Mazenc and J. P. A. Bastos, "Calculation of transient electromagnetic forces in axisymmetrical electromagnetic devices with conductive solid parts," ISEF - International Symposium on Electromagnetic Field Calculation, Southampton, England, Sept. 1991.
3. J. P. A. Bastos, N. Sadowski, R. Carlson, "A modeling approach of a coupled problem between electrical current and its thermal effects," IEEE Transactions on Magnetics, Vol. 26, No. 2, March 1990, pp. 536-539.
4. J. S. van Welij, "Calculation of eddy currents in terms of hexahedra," IEEE Transactions on Magnetics, Vol. MAG-21, No. 6, November 1985, pp. 2239-2241.
8. B. Davat, Z. Ren and M. Lajoie-Meznec, "The movement in field modeling," IEEE Transactions on Magnetics, Vol. MAG-21, No. 6, November 1985, pp. 2296-2298.
9. N. Ida, "Efficient treatment of infinite boundaries in electromagnetic field problems," COMPEL - International Journal for Computation and Mathematics in Electrical Engineering, Vol. 6, No. 3, November 1987, pp. 137-149.

Analytic Methods of Solution

1. K. J. Binns and P. J. Lawrenson, *Analysis and Computation of Electric and Magnetic Field Problems*, Pergamon Press, 1973.
2. W. R. Smythe, *Static and Dynamic Electricity*, McGraw-Hill, New York, 1968, pp. 284-287.

General Finite Element Method

1. J. T. Oden, "A general theory of finite elements I: Topological considerations," International Journal for Numerical Methods in Engineering, Vol. 1, No. 2, 1969, pp. 205-221.
2. J. T. Oden, "A general theory of finite elements II: Applications," International Journal for Numerical Methods in Engineering, Vol. 1, No. 3, 1969, pp. 247-259.
3. G. Dhatt and G. Touzot, *The Finite Element Method Displayed*, Wiley Interscience, Chichester, 1984.
4. K. H. Huebner, *The Finite Element Method for Engineers*, Wiley Interscience, John Wiley & Sons, Inc., New York, 1975.
5. C. S. Desai and J. F. Abel, *Introduction to the Finite Element Method*, Van Nostrand Reinhold, 1972.
6. O. C. Zienkiewicz, The *Finite Element Method in Engineering*, Third Edition, McGraw-Hill, London, 1977.
7. M. J. Turner, R. W. Clough, H. C. Martin and L. J. Topp, "Stiffness and deflection analysis of complex structures," Journal of Aeronautical Science, Vol. 23, 1956, pp. 805-823.

Variational Techniques

1. S.G. Mikhlin, *Variational Methods in Mathematical Physics*, Macmillan, New York, 1964.
2. M.E. Gurtin, *Variational principles for linear initial-value problems*, Quarterly of Applied Mathematics, Vol. 22, 1964, pp. 252-256.

Mesh Generation

1. J. R. Adamek, "An automatic mesh generator using two and three-dimensional isoparametric finite elements," M. Sc. Thesis, Naval Postgraduate School, Monterey, Ca. 1973.
2. N. N. Shenton and Z. J. Cendes, "Three-dimensional finite element mesh generation using Delaunay tesselation," IEEE Transactions on Magnetics, Vol. MAG-21, No. 6, November 1985, pp. 2535-2538.
3. K. Sagawa, "Automatic mesh generation for three-dimensional structures based on their three views," in *Theory and Practice in Finite Element Structural Analysis*, Y. Yoshioki and R. H. Gallagher, Eds., 1973, pp. 687-703.

4. M. S. Shephard, "Automatic and adaptive mesh generation," IEEE Transactions on Magnetics, Vol. MAG-21, No. 6, November 1985, pp. 2484-2489.
5. J. P. A. Bastos and R. Carlson, "An efficient automatic 2D mesh generation method," Compumag 1991, Sorrento, Italy, July 7-11, 1991.
6. A. R. Pinchuk and P. P. Silvester, "Error estimation for automatic adaptive finite element mesh generation," IEEE Transactions on Magnetics, Vol. MAG-21, No. 6, November 1985, pp. 2551-2554.
7. S. Pisanetzky, "KUBIK: an automatic three-dimensional finite element mesh generator," International Journal for Numerical Methods in Engineering, Vol. 17, 1981, pp. 255-269.
8. J. Penman and M. D. Grieve, "An approach to self adaptive mesh generator," IEEE Transactions on Magnetics, Vol. MAG-21, No. 6, November 1985, pp. 2567-2570.
9. D. W. Kelly, J. P. De, S. R. Gago. O. C. Zienkiewicz and I. Babuska, "A posteriori error analysis and adaptive processes in the finite element method: I - error analysis," International Journal for Numerical Methods in Engineering, Vol. 19, No. 11, 1983, pp. 1593-1619.
10. D. W. Kelly, J. P. De, S. R. Gago, O. C. Zienkiewicz and I. Babuska, "A posteriori error analysis and adaptive processes in the finite element method: II - adaptive mesh refinement," International Journal for Numerical Methods in Engineering, Vol. 19, No. 12, 1983, pp. 1621-1656.

Edge Elements

1. R. C. Mesquita and J. P. A. Bastos, "An incomplete gauge formulation for 3D nodal finite element magnetostatics," IEEE Transactions on Magnetics, Vol. MAG-28, No. 2, 1992, pp. 1044-1047.
2. J. P. A. Bastos, N. Ida, R. C. Mesquita and J. Gomes, "Use of reduced 3D hexahedral edge elements for 2D TE waveguides and vector potential problems," IEEE Transactions on Magnetics, Vol. MAG-30, No. 5, 1994, pp. 3614-3617.
3. J. P. A. Bastos, N. Ida and R. C. Mesquita, "A variable local relaxation technique in nonlinear problems," IEEE Transactions on Magnetics, Vol. MAG-31, No. 5, 1995, pp. 1733-1736.
4. R. Albanese and G. Rubinacci, "Formulation of eddy current problem," IEE Proceedings, Vol. 137 A, No. 1, January 1990, pp. 16-22.
5. R. Albanese and G. Rubinacci, "Eddy current computation using the $T - \Omega$ method: the Bath cube," COMPEL, Vol. 9, Sup. A, 1990, pp. 206-208.
6. R. Albanese and G. Rubinacci, "Magnetostatic field computations in terms of two component vector potentials," International Journal for Numerical Methods in Engineering, Vol. 29, 1990, pp. 515-532.

7. M. L. Barton and Z. J. Cendes, "New vector finite elements for three dimensional magnetic field computation," Journal of Applied Physics, Vol. 61, No. 2, April, 1987, pp. 3919-3921.
8. O. Biro and K. Preis, "Finite element analysis of 3-D eddy currents," IEEE Transactions on Magnetics, Vol. MAG-26, No. 2, March 1990, pp. 418-423.
9. A. Bossavit, "On the numerical analysis of eddy current problems," Computer Methods in Applied Mechanics and Engineering, Vol. 27, 1981, pp. 303-318.
10. A. Bossavit, "Two dual formulations of the 3-D eddy current problem," COMPEL – The International Journal for Computation and Mathematics in Electrical and Electronic Engineering, Vol. 4, No. 2, 1985, pp. 103-116.
11. A. Bossavit," A rationale for edge elements in 3-D field computations," IEEE Transactions on Magnetics, Vol. MAG-24, No. 1, Jan. 1988, pp. 74-79.
12. A. Bossavit, "Whitney forms: a class of finite elements for three dimensional computations in electromagnetism," IEE Proceedings, Vol. 135-A, No. 3, March 1988, pp. 179-187.
13. A. Bossavit, "Simplicial finite elements for scattering problems in electromagnetism," Computer Methods in Applied Mechanics and Engineering, Vol. 76, 1989, pp. 299-316.
14. A. Bossavit, "Solving Maxwell equations in a closed cavity and the question of 'spurious modes'," IEEE Transactions on Magnetics, Vol. MAG-26, No. 2, March 1990, pp. 702-705.
15. A. Bossavit and J.C. Verite, "A mixed FEM-BIEM method to solve 3-D eddy current problems," IEEE Transactions on Magnetics, VOL. MAG-18, No. 2, March 1982, pp. 431-435.
16. C. R. I. Emson and J. Simkin, "An optimal method for 3-D eddy currents," IEEE Transactions on Magnetics, Vol. MAG-21, No. 6, November 1985, pp. 2231-2234.
17. J. A. Meijerink and H.A. Van der Vorst, "An iterative solution method for linear systems of which the coefficient matrix is a symmetric M-matrix," Mathematics of Computation, Vol. 31, No. 137, January 1977, pp. 148-162.
18. G. Mur, "Optimum choice of finite elements for computing three dimensional electromagnetic fields in strongly inhomogeneous media," IEEE Transactions on Magnetics, Vol. MAG-24, No. 1, January 1988, pp. 330-333.
19. G. Mur, "A mixed finite element method for computing three dimensional time domain electromagnetic fields in strongly inhomogeneous media," IEEE Transactions on Magnetics, Vol. MAG-26, No. 2, March 1990, pp. 674-677.
20. G. Mur and A.T. Hoop, "A finite element method for computing three dimensional electromagnetic fields in inhomogeneous media," IEEE Transactions on Magnetics, Vol. MAG-21, No. 6, Nov. 1985, pp. 2188-2191.
21. T. Nakata, N. Takahashi, K. Fujiwara, T. Imai, and K. Muramatsu, "Comparison of various methods of analysis and finite elements in 3-D

magnetic field analysis," IEEE Transactions on Magnetics, Vol. MAG-27, No. 5, Sept. 1991, pp. 4073-4076.
22. J. C. Nedelec, "Mixed finite elements in R^3," Numerische Mathematik, Vol. 35, 1980, pp. 315-341.
23. D. Rodger, P. J. Leonard and H. C. Lai, "Some recent developments in 3D eddy current computation," COMPEL – The International Journal for Computation and Mathematics in Electrical and Electronic Engineering, Vol. 9, Sup. A, 1990, pp. 25-30.
24. J. Simkin and C. W. Trowbridge, "On the use of the total scalar potential in the numerical solution of field problems in electromagnetics," International Journal for Numerical Methods in Engineering, Vol. 14, 1979, pp. 423-440.
25. C. W. Trowbridge, "Electromagnetic computing: the way ahead?," IEEE Transactions on Magnetics, Vol. MAG-24, No. 1, January 1988, pp. 13-18.
26. J. S. Van Welij, "Calculation of eddy currents in terms of H on hexahedra," IEEE Transactions on Magnetics, Vol. MAG-21, No. 6, November 1985, pp. 2239-2241.
27. K. Fujiwara, "3D magnetic field computation using edge elements," Proceedings of the 5th IGTE Symposium on Numerical Field Computation in Electrical Engineering, Graz, Austria, September 18-22, 1992, pp. 185-212.
28. J. F. Lee, D. K. Sun and Z. J. Cendes, "Tangential vector finite elements for electromagnetic field computation," IEEE Transactions on Magnetics, Vol. 27, No. 5, Sept. 1991, pp. 4032-4035.
29. G. Mur, "The finite element modeling of three dimensional electromagnetic fields using edge and nodal elements," IEEE Transactions on Antennas and Propagation, Vol. 41, No. 7, July 1993, pp. 948-953.
30. G. Mur, "Edge elements, their advantages and their disadvantages," IEEE Transactions on Magnetics, Vol. 30, No. 5, Sept. 1994, pp. 3552-3557.

Software Libraries

1. J. J. Dongarra, C. B. Moller and G. W. Stuart, "LINPACK Users' Guide," SIAM, Philadelphia, 1979.
2. J. J. Dongara, "The LINPACK Benchmark: An explanation – Part I, II," Supercomputing, Spring 1988, pp. 10-18.

Subject Index

adaptive mesh 510, 528
ampere's law 22, 39, 91
anisotropy 35
attenuation 229
attenuation constant 231
axial symmetry 379
axi-Symmetric
   applications 375, 431
   finite element 377
   formulation 422
Biot-Savart
   examples 125, 127
   law 94
boundary conditions
   for electric vector potential 362
   for the magnetic vector potential 360
   for scalar potential 267
   general 271, 278, 360
CAD 509
capacitance 61
   definition 61
   energy in 64
   static electric charge 24, 43, 48
capacitor 61
   spherical 71, 73
   with layered dielectrics 73
cavity resonators 258, 426, 473
   loaded 480
   modeling of fields in 472
characteristic modes 216
charged line 67
charged spherical half-shell 70

Choleski's method 495
circulation
   definition 12
   in electrostatic field 51, 55
   of **A** along edges 448
computation
   aspects of 484, 547
   of electrostatic fields 343, 391
   of magnetic fields 394, 396
   of resonant frequencies 480
   of static currents 392
conductivity
   electric 24, 27, 37
   table of materials 89
continuity
   electric equation 29, 93, 212, 346
   of electric field intensity 56
   of electric flux density 57
   of magnetic field intensity 97
   of magnetic flux density 97
contribution matrix (*see* elemental or stiffness matrix)
constitutive relations 28
Coulomb gage 219
coupling to cavities 263
Crank-Nicholson method 416
current
   conduction 29, 38, 213
   displacement 22, 33, 38, 44, 213
   penetration in conductors 146, 152
cutoff frequency 428
cylindrical coordinates 20

Subject Index

degrees of freedom 267, 298, 313
density
  charge 24, 27
  current 24, 27, 466
  nonuniform 27
  volume charge 27, 30
diamagnetism 99
dielectric Strength
  definition 60
  table of 88
Dirichlet
  boundary condition 80, 275, 279, 286, 354, 356, 360, 375, 393, 401, 427, 462, 476
  homogeneous 481
  inhomogeneous 481
discretization 267, 323, 392, 511
  Delaunay method 515
  of domain 323, 332, 362
divergence 3, 6
  example 11
  theorem 8, 271
  of current density J 471
eddy currents
  2D Problems 405
  brake 209
  calculation of 477
  complex formulation 468, 472
  density 418, 466
  definition of 146
  losses 418
  time-stepping procedure 466
  velocity induced 420, 437
edge elements 446, 460, 548
  applications 445
  hexahedral 448
  shape functions 453
eigenvalues 429, 499
eigenvectors 499
electric field 48
electric field intensity 24, 48
  circulation 51
  computation of 316
  equation of 147
  energy stored in 65
  finite element example 84
electric machines 475, 537
electrostatic 23, 45, 47
  applications 330, 343, 420, 461, 474
  electrostatic problems 391
  field functional 355
  field governing equation 303
elemental matrix 275, 279, 289, 368, 417, 424, 463, 469
energy
  associated with magnets 108
  in cavities 473
  in electric field 64, 355
  in linear systems 119
  stored in magnetic field 122, 180
  with magnetic vector potential 359
energy functional 281, 353
  for permanent magnets 360
  involving scalar potentials 108, 355
  minimization of 374, 354
Euler's equation 286, 354, 358, 361, 468
evolving systems 528
Faraday's law 42, 145
ferromagnetic materials 100
finite element
  1D linear 306
  1D quadratic 314
  3D application 445
  applications 390
  application of edge elements 460
  application of variational method 366
  continuity of 298
    C^0 continuity 298
    C^1 continuity 298
  conforming 363
  cubic element 307
  coupling of different types 323, 324
  definition of 267, 293, 363
  division of domain 75
  edge type (vector) 446, 460, 548
  example of calculation with 84

first order   277, 364
    rectangular   299
    triangular   315
    tetrahedral   314, 388
hexahedral   321, 491
    quadratic   322
high order   292, 296
in local coordinates   405
isoparametric, 2D   296
isoparametric, 3D   296
mesh   268
    node numbers   267, 363
nodal type   445
nonconforming   363
overparametric   298
quadrilateral elements   277, 306, 491
    quadratic   317
subparametric   298
tetrahedral element   307
    quadratic   319
triangular elements   277, 306
    quadratic   315
finite element method   267, 291, 343, 362
    generalization of   290
    introduction to   74
finite element program   81, 330, 333, 506
flux
    conservative   9, 11, 16, 31
    definition   6
    magnetic   25, 40, 93
    per unit depth   350
flux density (induction)
    electric   26
    magnetic   25
    penetration in conductors   146, 152
    remnant   351
force
    density   439
    diamagnetic material   206
    electromechanical   175
    electromotive   52, 162
    ferromagnetic material   205
    Lorentz force   178

    magnetomotive   102, 140
    Maxwell's tensor   188
    on conductor   175, 198
    on moving charge   48, 179
    variation of energy   182
function
    geometric transformation   405
    interpolating   296, 304, 406
    placement (position)   449, 450
gage conditions   472
Galerkin
    basic concepts   266
    extension to 2D   277
    method   266, 272, 304, 327, 400, 411, 421, 426, 428, 461, 467, 473
    applications to dynamic fields   400
Gauss
    elimination   493
    method of integration   307
    theorem   43
Gauss-Legendre integration   316, 320
Givens transformation   503
global
    assembly of matrix   429, 464, 508
    matrix   275, 374, 417
    mesh node numbers   374
    system of coordinates   406
gradient
    definition   2, 3
    example   5
Hall effect   200
Helmholtz equation   427
    formulation of   427
Herz, Heinrich   215
ICCG method   445, 498
incidence plane   241
induced currents   145, 153
    due to change in geometry   166
    due to change in induction   164
inductance
    computation of   523
    definition   119
    energy in   120

mutual 119
   of coaxial cable 128
   of long solenoid 123
   of torus 122
   self- 119
inductive heating 157, 159
infinite element (1D) 323, 325
intrinsic impedance 226
Jacobi transformation 407, 500
Jacobian 300, 316, 328, 451, 465, 469
Joule effect 53, 146
Lagrange elements 298
Laplace
   equation, electric 65, 345
   equation, magnetic 139
   force law 177
Laplacian
   scalar 18
   vector 19
Lenz's law 145
linear generator 203
linear motor 203
local system of coordinates 405
Lorentz force 178
Lorentz gage 219
losses
   conduction 228
   dielectric 228
   eddy currents 153, 418
   hysteresis 157
magnetic circuit
   analog to electric circuit 115
   calculation of the field 130, 133
   with permanent magnet 136
magnetic field 23, 24
   energy 122, 181
   finite element example 125
   in anisotropic media 401
   inside conductor 129
   in waveguides 441
   of circular loop 125
   of rectangular loop 127
   outside conductor 39

penetration in conductors 151, 165
   scalar potential equation 65, 346
   vector potential equation 347, 416
magnetic levitation 207
magnetic vector potential 347, 371, 396
   equation using time discretization 409
   formulation for wave propagation 425
   functionals 359
magnetodynamic fields 23, 143
magnetostatic fields 23, 91
   applications 330, 358
materials
   diamagnetic 99
   ferromagnetic 99, 100
   paramagnetic 99, 100
matrix system 484
   assembly of 373, 374
   banded 487
   compact 487
   direct solution 493
   insertion of boundary conditions 490
   iterative solution 415, 497
   storage 487
   symmetric 487
Maxwell
   approximations to equations 43
   equations 23, 90, 143
   equations in vacuum 29
   force tensor 188, 439
   historical note 22
   in general media 35, 37
   integral form 37
   point form 29
MKS units 45
   table of 46
modes 249, 253, 256, 259, 261, 428
   spurious 446
   TE 249, 251
   TEM 249, 251
   TM 249, 252
      computation of 440, 442
nabla operator 2
Neper 229

Neumann boundary condition   80, 279,
   286, 346, 354, 356, 360, 393, 398, 402,
   412, 427, 462, 476, 481
Newton-Raphson
   algorithm   385, 389
   method   384, 415, 469, 474, 478
nonlinear
   applications   383, 467
   energy functional   359
numerical integration
   accuracy and errors   311
   integration points   307, 316, 320
   integration weights   307, 316, 320
   method   306
Ohm's law   28, 53
operators
   application to functions   19
   nabla   2
   second order   18
organization of software system   509
orthogonal transformation   501
paramagnetism   99
penetration
   depth of   149, 232
   of fields   146
   solution of equations   148
   visual effect   172
periodicity   484
periodicity (anti-)   486
permanent magnets   104
   computation in   372, 381
   dynamic operation of   114
   energy in   108
   examples   136, 169, 475
   principal types   113
   modeling in 3D   389, 398
   modeling of   466
permeability
   magnetic   24, 25, 98
   relative   98, 101, 238
   table of   114, 141
   tensor   35
permittivity

complex   227
   electric   26
   relative   47, 238
   table of   88
phase constant   230
phase velocity   224, 229
phasors   220
plane waves   213
   definition   222
   in conductors   230, 232
   in low-loss dielectrics   229
   in lossy dielectrics   227
   propagation   222
   reflection   235, 236
   refraction   235
   solution for   222
   transmission   235, 236
Poisson's equation   327, 345
   evaluation of terms   329, 330
   for 2D magnetic vector potential   349
   for the electric potential   65
polarization   233
   circular   235
   elliptical   235
   linear   234
   parallel   241
   perpendicular   241
polynomial basis   298, 314, 407
potential
   electric, scalar   49, 201, 387
   electric, vector   352, 393
   energy   186
   magnetic, vector   472
   reduced   548
   total   548
Poynting vector   184
   example of   186
processing   510, 516
   calculation of fields   519
   post-processing   517
   pre-processing   510
   visualization   517
propagation constant   228

QR algorithm   397, 504
quality factor   263, 473
QZ algorithm   500, 504
refraction
  electric field   55
  magnetic field   97
relaxation technique   474
  overrelaxation   475
  underrelaxation   475
residual   270, 400, 469
resonant frequencies   426, 442, 480
rotation (curl)
  definition   2, 12
  example   17
shape functions
  construction of   456
  for six-node, isoparametric triangular elements   316
  for twenty-node hexahedral element   322
  for edge (vector) elements   453
  for first-order hexahedral element   321, 453
  for hexahedral edge elements   448
  for linear 1D elements   314
  for linear tetrahedral element   319
  for quadratic quadrilateral element   318
  for quadratic tetrahedral element   320
  for quadrilateral bi-linear element   317
  for triangular element   315
scalar potential   346, 370, 394, 426
  applications   358
  formulation in terms of   347, 392
shape functions derivatives
  for six-node, isoparametric triangular elements   316
  for twenty-node hexahedral element   322, 323
  for first-order hexahedral element   321
  for linear 1D elements   314
  for linear tetrahedral element   319
  for quadratic quadrilateral element   318
  for quadratic tetrahedral element   320
  for quadrilateral bi-linear element   317

for triangular element   315
similarity transformation   501
skin depth   232
solenoid, long   123
solenoidal   218
solution
  coupled field/circuit   541
  coupled thermal/electric   533
  system of equations   493
spurious modes   446
stiffness matrix   283, 303, 327, 368, 405, 414 (*see also* elemental or contribution matrix)
  nonsymmetric   422
  source contribution   275, 303, 330, 424, 374, 464
Stokes' theorem   14, 350
successive approximation method   383
thermal
  coupling   533
  equation   534
theta algorithm   416
time-harmonic fields   213
time stepping   415
time discretization   434, 467
torque on loop   198
transformer, voltage   172
variational
  application in 3D
  formulation   284, 354
  method   77, 282, 353, 413, 432
    application of   343, 389
    basic concepts   281, 353
    extension to 2D   284
    application to stationary currents   345, 357, 370
    application to electric vector pot.   373
    application to electrostatics   367
    application to magnetic fields   371
    application to finite elements   366
vector
  circulation   12
  derivation   2
  notation   1

wave
   equation   215
   homogeneous equation   216, 219, 221
   nonhomogeneous equation   216, 221
   solution of equation   218, 222
   source free wave equation   221
   time-dependent wave equation   215
   time-harmonic wave equation   220, 252
weak form   271
weighted residual method   270, 271, 401
weighting function   270, 401
waveguides   249, 253
   modeling of fields in   450, 472
wavelength   225
wavenumber   225, 228
Weiss domains   106